Green Chemistry Handbook

Volume I

Green Chemistry Handbook
Volume I

Edited by **Simon Doyle**

NY RESEARCH
PRESS
New York

Published by NY Research Press,
23 West, 55th Street, Suite 816,
New York, NY 10019, USA
www.nyresearchpress.com

Green Chemistry Handbook: Volume I
Edited by Simon Doyle

International Standard Book Number: 978-1-63238-216-0 (Hardback)

Printed in China.

Contents

Preface

Green Sustainable Chemistry is a sub-discipline of Sustainable Development, with a specific focus on, as the name suggests, chemical processes. The origin of this term can be traced to scientists Paul Anastas and John C. Warner, who have been credited for developing the 12 principles of green chemistry. The fundamental premise of all their principles is to minimize the use of chemicals and if at all they're utilized, to reduce their toxicity.

In the present scenario, there's a desperate need to minimize our consumption of anything that degrades our environment. And chemicals, though useful do impact the environment in one way or the other. Thus, green chemistry is an urgent need of the hour, which focuses on satisfying the present needs, while ensuring a healthy environment for the future generations.

Green Sustainable Chemistry covers all branches of Chemistry such as biochemistry, analytical chemistry, organic and inorganic chemistry and also chemical engineering. An important focus of this discipline is on industrial applications. The ultimate aim of Green Sustainable Chemistry is to minimize the hazard and maximize the efficiency of any chemical choice.

I'd like to thank all our contributors who've shared their studies with us in this book, on this emerging field of study.

Editor

Silica-Grafted Ionic Liquids as Recyclable Catalysts for the Synthesis of 3,4-Dihydropyrano[c]chromenes and Pyrano[2,3-c]pyrazoles

Khodabakhsh Niknam*, Abolhassan Piran
Department of Chemistry, Faculty of Sciences, Persian Gulf University, Bushehr, Iran

ABSTRACT

Silica-grafted *N*-propyl-imidazolium hydrogen sulfate ([Sipim]HSO_4) is employed as a recyclable heterogeneous ionic liquid catalyst for the synthesis of 3,4-dihydropyrano[c]-chromenes by the reaction of aromatic aldehydes, malononitrile and 4-hydroxycoumarin at 100°C under solvent-free conditions. Also, heterogeneous ionic liquid catalyst was used for the synthesis of pyrano[2,3-c]-pyrazoles by the reaction of aromatic aldehydes, malononitrile and 3-methyl-l-phenyl-5-pyrazolone at 110°C under solvent-free conditions. The heterogeneous ionic liquid showed much the same efficiency when used in consecutive reaction runs.

Keywords: Silica-Grafted *N*-Propyl-Imidazolium Hydrogen Sulfate; Aldehydes; Pyrano[2,3-c]-Pyrazoles; 3,4-Dihydropyrano[c]-Chromenes; Solvent-Free, Heterogeneous Ionic Liquid Catalysts

1. Introduction

In the recent years, ionic liquids were used as solvents due to their particular properties, such as the ability to dissolve many organic and inorganic substances and undetectable vapor pressure [1]. In addition, Brønsted acidic task-specific ionic liquids (BAILs), such as those possessing HSO_4^- as a counter anion find a broad application in organic synthesis, acting as both solvents and catalysts. Keim and co-workers reported the synthesis of 1-butyl-3-methylimidazolium hydrogen sulfate ([bmim]HSO_4) [2]. In addition, in the year of 2002 ([bmim]HSO_4) was used as a catalyst in the Friedel-Crafts alkylation [3]. The other applications of these acidic ionic liquids such as acetalization and thioacetalization of carbonyl compounds [4], Fischer indole synthesis [5], acetylation of alcohols and phenols [6], preparation of azides from alcohols [7], selective nitration of phenols [8], synthesis of 1,8-dioxo-octahydroxanthenes [9], formylation of alcohols [10], synthesis of polysubstituted quinolines [11], have been proceeded with very good yields and selectivities. Recently, immobilization of acidic ionic liquids on solid supports has been designed and it can offer important advantages in handling, separation and reuse procedures. Based on economic criteria, it is desirable to minimize the amount of ionic liquid utilized in a potential process. Immobilized acidic ionic liquids have been used as novel solid catalysts, e.g., for esterification, nitration reactions [12], acetal formation [13], Baeyer-Villiger reaction [14], synthesis of α-aminonitriles [15] and bis-pyrazolones [16].

Dihydropyrano[c]chromenes and their derivatives are of considerable interest as they possess a wide range of biological properties [17], such as spasmolytic, diuretic, anti-coagulant, anti-cancer, and anti-anaphylactic activity [18,19]. In addition, they can be used as cognitive enhancers, for the treatment of neurodegenerative diseases, including Alzheimer's disease, Parkinson's disease, AIDS associated dementia and Down's syndrome as well as for the treatment of schizophrenia and myoclonus [20]. Also, a number of 2-amino-4H-pyranes are useful as photoactive materials [21]. A number of methods have been reported for the synthesis of 3,4-dihydropyrano[c] chromenes such as piperidine in organic solvent, *i.e.* ethanol and pyridine [22], K_2CO_3 under microwave irradiation [23], diammonium hydrogen phosphate [24], tetrabutylammonium bromide [25], sodium dodecyl sulfate [26], DBU [27], morpholine [28], α-Fe_2O_3 nanopatricles [29], CuO nanopatricles [30], and silica-bonded *N*-propyl-

*Corresponding author.

piperazine sodium propionate [31].

In continuation of our studies on the design and application of acidic ionic liquids or silica-grafted ionic liquids as catalyst in organic transformations **(Scheme 1)** [8-11,15,16], herein, we describe the application of silica-grafted *N*-propyl-imidazolium hydrogen sulfate ([Sipim] HSO_4) in the synthesis of pyrano[c]chromenes and pyrano[3,4c]pyrazoles.

2. Results and Discussion

2.1. Synthesis of Dihydropyrano[c]chromenes

To study the effect of catalyst loading on the synthesis of 2-amino-4-aryl-5-oxo-4*H*,5*H*-pyrano[3,2-c]chromene-3-carbonitrile the condensation reaction of malononitrile, 4-chlorobenzaldehyde, and 4-hydroxycoumarin was chosen as a model reaction (**Table 1**). The results show clearly that ionic liquids and silica-grafted ionic liquids (SGILs) are effective catalysts for this condensation and the optimal amount of SGILs was 0.1 g per 1 mmol of aldehyde under solvent-free conditions at 100°C. The best result was obtained in the presence of [Sipim]HSO_4. This condensation was carried out with the lower amounts of [Sipim]HSO_4 0.05 g and 0.07 g and the corresponding product was obtained in 88% and 91% yield (**Table 1**, entries 4 and 5).

Also, this three-component condensation was accomplished in the presence of [Sipim]HSO_4 in ethanol and water at reflux conditions in longer reaction time and lower yield (**Table 1**, entries 1 and 2). Moreover, the model reaction was examined under solvent-free conditions at 80°C gave **1b** after 90 min in 80% yield (**Table 1**, entry 8). The model reaction was reacted in the presence of *N*-(3-silicapropyl) imidazolium chloride ([Sipim]Cl), *N*-(3-silicapropyl) imidazolium dihydrogen phosphate ([Sipim]H_2PO_4), and *N*-(3-silicapropyl) imidazolium triflate ([Sipim]OTf), under solvent-free conditions at 100°C in 80%, 80%, and 75% yield respectively (**Table 1**, entries 10-12). In addition, the same model reaction was carried out in the presence of ionic liquids such as imidazolium chloride, imidazolium hydrogen sulfate, methylimidazolium hydrogen sulfate, and imidazolium bromide under solvent-free and at 100°C in 85%, 80%, 80%, and 85% yield respectively (**Table 1**, entries 13-17).

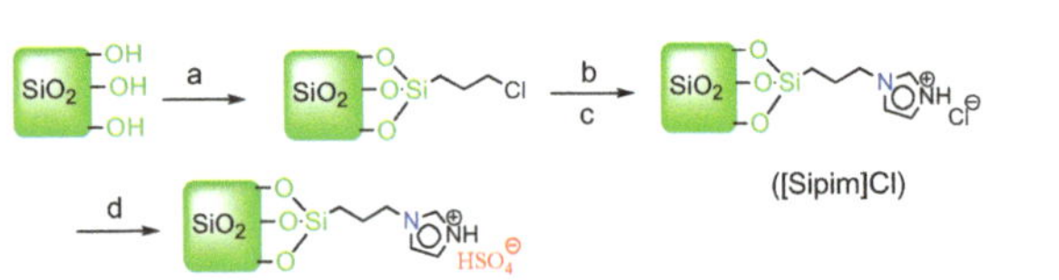

([Sipim]HSO_4)

a) $(MeO)_3Si$-$(CH_2)_3$-Cl, toluene(dry), Et_3N, reflux, 48 h; b) Imidazole, toluene (dry), 24 h, reflux; c) Dry under vaccum (50 °C), 4 h; d) dry CH_2Cl_2 , H_2SO_4 (97%), 48 h, reflux

Scheme 1. Preparation of silica-grafted propyl imidazolium hydrogen sulfate ([Sipim]HSO_4).

Table 1. Investigation the effect of catalyst and solvent on the reaction of 4-chlorobenzaldehyde, malononitrile and 4-hydroxycoumarin.

Entry	Catalyst	Catalyst loading (g)	Solvent/ Temp. (°C)	Time (min)	Yield (%)[a]
1	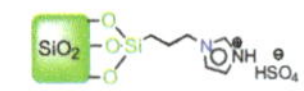	0.1	EtOH/Reflux	75	75
2	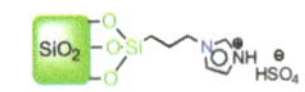	0.1	H_2O/Reflux	75	75
3	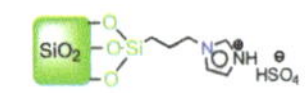	0.1	EtOH/H_2O (1:1)/Reflux	60	86
4	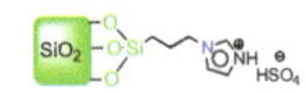	0.05	Solvent-free/100	120	88
5	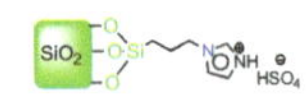	0.07	Solvent-free/100	90	91
6	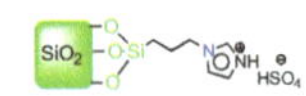	0.1	Solvent-free/100	30	95
7	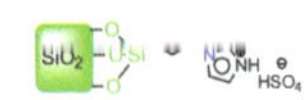	0.15	Solvent-free/100	30	95
8	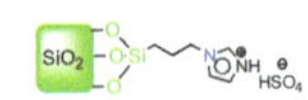	0.1	Solvent-free/80	90	80
9	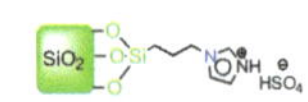	0.1	Solvent-free/120	30	95
10	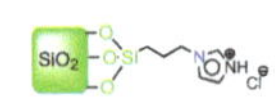	0.1	Solvent-free/100	30	80
11	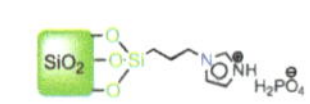	0.1	Solvent-free/100	30	80
12	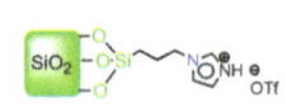	0.1	Solvent-free/100	60	75
13	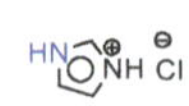	0.2	Solvent-free/100	120	85
14	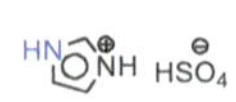	0.2	Solvent-free/100	150	80
15		0.2	Solvent-free/100	210	80
16	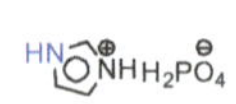	0.2	Solvent-free/100	120	75
17	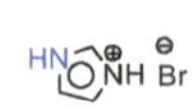	0.2	Solvent-free/100	120	85

[a]Isolated Yield.

The synthesis of 2-amino-4-aryl-5-oxo-4*H*,5*H*-pyrano-[3,2-c]chromene-3-carbonitrile was achieved by the three-component condensation of an aromatic aldehyde, malononitrile, and 4-hydroxycoumarin in the presence of [Sipim]HSO_4 (0.1 g, 0.08 mmol of H^+ [15]) under solvent-free conditions at 100°C (**Table 2**).

Thereafter, a series of different 3,4-dihydropyrano-[c]chromene derivatives were prepared successfully from different aromatic aldehydes bearing electron-withdrawing and electron donating groups, 4-hydroxycoumarin and malononitrile under solvent-free conditions. Electron-withdrawing groups such as 3-nitro, 4-nitro, and 2-nitro-benzaldehyde reacted under optimized conditions into corresponding **1f**, **1g**, and **1h** in 93%, 90%, and 89% yield after 30 min (**Table 2**, entries 6-8). Electrondo-

Table 2. [Sipim]HSO$_4$ catalyzed synthesis of dihydropyrano [c]chromene derivatives.[a]

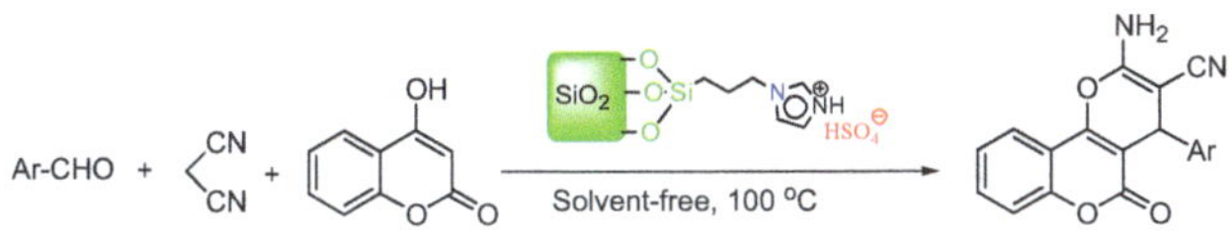

2 3 4 1

Entry	Ar	Product	Yield (%)[b]
1	C_6H_5-	1a	94
2	4-ClC_6H_4-	1b	95, 95, 93, 91, 90[c]
3	3-ClC_6H_4-	1c	93
4	4-BrC_6H_4-	1d	94
5	2,4-$(Cl)_2C_6H_3$-	1e	90
6	3-$O_2NC_6H_4$-	1f	93
7	4-$O_2NC_6H_4$-	1g	90
8	2- $O_2NC_6H_4$-	1h	89
9	4-MeC_6H_4-	1i	94
10	3,4,5-$(CH_3O)_3C_6H_2$-	1j	89
11	4-HO-C_6H_4-	1k	93

[a]Reaction conditions: 4-hydroxycoumarin (1 mmol), malononitrile (1 mmol), aldehyde (1 mmol), solvent-free conditions at 100°C for 30 min. [b]Isolated yield. [c]The recovered [Sipim]HSO$_4$ was used as catalyst.

nating groups such as 4-Me, 3,4,5-(MeO)$_3$-benzaldehyde were treated with malononitrile and 4-hy- droxycoumarin under optimized conditions gave corresponding products **1i** and **1j** in high yields (**Table 2**, entries 9 and 10). The results clearly indicate that reactions can tolerate a wide range of differently substituted aromatic aldehydes.

The possibility of recycling the catalyst was examined using the reaction of malononitrle, 4-chlorobenzaldehyde and 4-hydroxycoumarin under the optimized conditions. Upon completion, the reaction mixture was washed with warm ethanol (3 × 30 mL). The recovered catalyst was dried and reused for subsequent runs. The recycled catalyst could be reused fourth times without any additional treatment. No observation of any appreciable loss in the catalytic activity of [Sipim]HSO$_4$ was made (**Table 2**, entry 2).

2.2. Synthesis of Pyrano[2,3-c]pyrazoles

Condensed pyrazoles are also biologically interesting compounds and their chemistry has recently received considerable attention [32,33]. Several pyrano[2,3-c] pyrazoles are reported to have useful biological effects, such as analgesic and anti-inflammatory activities [34]. Moreover, the biological activity of fused azoles has led to intensive research on their synthesis [35,36]. Recently, three component one-pot condensation of 3-methyl-1-phenyl-1H-pyrazol-5(4H)-one, aldehydes and malononitrile for the construction of 1,4-dihydropy-rano[2,3-c] pyrazole derivatives has been reported under different conditions [37-41]. However, most of the reported methods have their drawbacks. For example, hexadecyltrimethylammonium bromide (HTMAB) is a harmful and irritant catalyst which is dangerous for the environment [39]. In addition, the reusability of the catalysts such as D,L-proline is not reported in a number of cases [40]. Heteropoly acids, silica-supported bases or acids have been reported the synthesis of 1,4-dihydropyrano[2,3-c] pyrazole derivatives *via* multi-component condensation reaction [42-48].

The scope of this methodology was extended to the synthesis of pyrano[2,3-c]pyrazols. Initially three-component condensation reaction of malononitrile, 4-chlorobenzaldehyde, and 3-methyl-1-phenyl-1H-pyrazol-5 (4*H*)-one was chosen as a model reaction and the effect of different catalysts was investigated at 110°C under solvent-free conditions (**Table 3**).

Again, the results show clearly that ionic liquids and silica-grafted ionic liquids (SGILs) are effective catalysts

Table 3. Investigation the effect of catalyst on the reaction of 4-chlorobenzaldehyde, malononitrile and 3-methyl-1-phenyl-1H-pyrazol-5(4H)-one at 110°C under solvent-free conditions.

Entry	Catalyst	Catalyst loading (g)	Time (min)	Yield (%)[a]
1	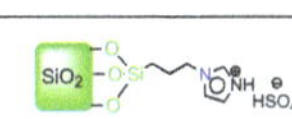	0.05	120	80
2	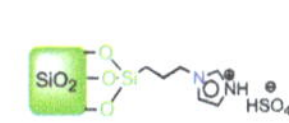	0.07	120	85
3	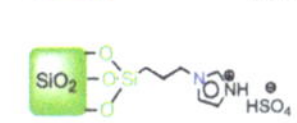	0.1	90	90
4	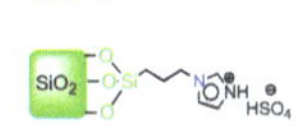	0.15	90	95
5	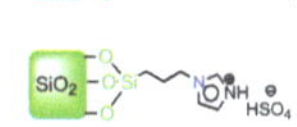	0.2	90	95
6	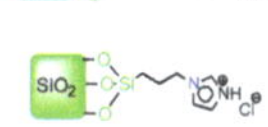	0.15	100	90
7	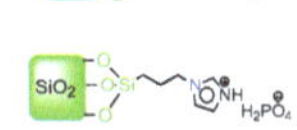	0.15	100	90
8	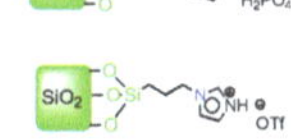	0.15	120	70
13	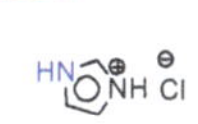	0.2	120	85
14		0.2	150	80
15		0.2	150	80
16	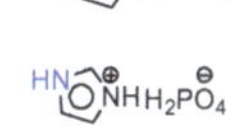	0.2	150	70
17	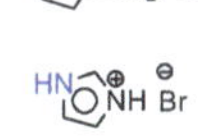	0.2	120	85

[a]Isolated Yield.

for this condensation and the optimal amount of SGILs was 0.15 g per 1 mmol of aldehyde under solvent-free conditions at 110°C.

A range of different substituted groups on aromatic aldehydes involving electron-withdrawing groups such as 3-nitro, 4-nitro, and electron-donating groups such as 3,4-dimethoxy-benzaldehyde reacted under optimized conditions into corresponding **5f**, **5g**, and **5h** in 90%, 88%, and 90% yield after 90 min (**Table 4**, entries 6-8). 4-Hydroxy-benzaldehyde was treated with malononitrile and 3-methyl-1-phenyl-1H-pyrazol-5(4H)-one under optimized conditions gave corresponding products **5i** in 90% yield (**Table 4**, entry 9).

The possibility of recycling the catalyst was examined using the reaction of malononitrle, 4-chlorobenzaldehyde and 3-methyl-1-phenyl-1H-pyrazol-5(4H)-one under the optimized conditions. Upon completion, the reaction mixture was washed with warm ethanol (3 × 30 mL). The recovered catalyst was dried and reused for subsequent runs. The recycled catalyst could be reused fourth times without any additional treatment. No observation of any appreciable loss in the catalytic activity of [Sipim]HSO_4 was made (**Table 4**, entry 2).

3. Experimental Section

3.1. General

Chemicals were purchased from Fluka, Merck and Aldrich Chemical Companies. All the products were charpyl-imidazolium hydrogen sulfate ([Sipim]HSO_4) was acterized by comparison of their IR, ^{1}H NMR and ^{13}C NMR spectroscopic data and their melting points with the reported values [24-31,38-41]. Silica-grafted *N*-proprepared according to our previous reported procedure [15].

Table 4. [Sipim]HSO_4 catalyzed synthesis of dihydropyrano [2,3-c]pyrazole derivatives.[a]

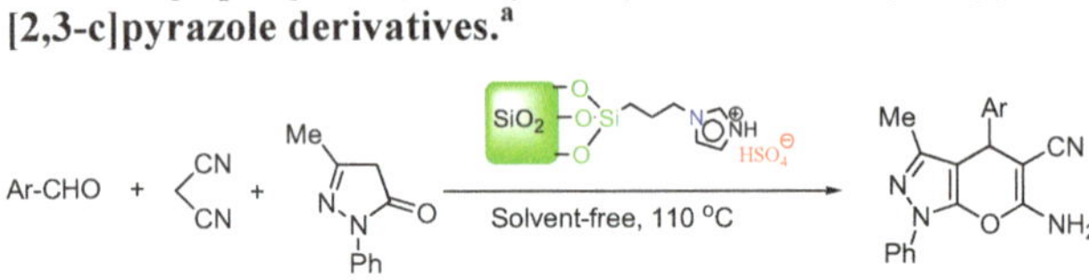

2 3 6 5

Entry	Ar	Product	Yield (%)[b]
1	C_6H_5-	5a	92
2	4-ClC_6H_4-	5b	95, 94, 92, 90, 89[c]
3	2,4-$(Cl)_2C_6H_3$-	5c	89
4	4-BrC_6H_4-	5d	95
5	4-FC_6H_4-	5e	90
6	3-$O_2NC_6H_4$-	5f	90
7	4-$O_2NC_6H_4$-	5g	88
8	3,4-$(CH_3O)_2C_6H_3$-	5h	90
9	4-HO-C_6H_4-	5i	90

[a]Reaction conditions: 3-methyl-1-phenyl-1H-pyrazol-5(4H)-one (1 mmol), malononitrile (1 mmol), aldehyde (1 mmol), solvent-free conditions at 110 °C for 90 min. [b]Isolated yield. [c]The recovered [Sipim]HSO_4 was used as catalyst.

3.2. General Procedure for the Synthesis of 3,4-Dihydropyrano[c]chromenes

To a mixture of aromatic aldehyde (1 mmol), malonitrile (1 mmol), and 4-hydroxycoumarin (1 mmol), catalyst [Sipim]HSO_4 (0.1 g, 0.08 mmol of H^+) was added and the mixture was heated at 100°C under solvent-free conditions. After completion of the reaction, as indicated by TLC, ethanol (10 mL) was added and the reaction mixture was filtered. The remaining was washed with warm ethanol (3 × 5 mL) in order to separate heterogeneous catalyst. After cooling the crude products were precipitated. The crude products were purified by recrystallization from ethanol (95%). The recovered catalyst was dried and reused for subsequent runs.

2-Amino-4-phenyl-4,5-dihydro-5-oxopyrano[3,2-c]chromene-3-carbonitrile 1a: mp 258°C - 260°C, (Lit.: 256°C - 258°C, [24]). ^{1}H NMR (400 MHz, DMSO-d_6): δ (ppm) 4.46 (s, 1 H), 7.23 - 7.27 (m, 3 H), 7.31 - 7.35 (m, 2 H), 7.43 - 7.52 (m, 2H), 7.72 (dt, 1 H, J_1 = 7.8 Hz, J_2 = 1.6 Hz), 7.92 (dd, 1 H, J_1 = 8.0 Hz, J_2 = 0.8 Hz). ^{13}C NMR (100 MHz, DMSO-d_6): δ (ppm) 58.41, 104.47, 113.42, 117.06, 119.68, 122.98, 125.18, 127.62, 128.10, 129.01, 133.45, 143.80, 152.60, 153.89, 158.36, 158.40, 160.05.

2-Amino-4-(4-chlorophenyl)-4,5-dihydro-5-oxopyrano[3,2-c]chromene-3-carbonitrile 1b: mp 263°C - 265°C, (Lit.: 252°C - 255°C, [28]). ^{1}H NMR (400 MHz, DMSO-d_6): δ (ppm) 4.49 (s, 1 H), 7.31 (d, 2 H, J = 8.4 Hz), 7.38 (d, 2H, J = 8.4 Hz), 7.45-7.51 (m, 2 H), 7.72 (t, 1 H, J = 7.8 Hz), 7.91 (d, 1 H, J = 8.0 Hz). ^{13}C NMR (100 MHz, DMSO-d_6): δ (ppm) 57.94, 103.94, 113.39, 117.06, 119.54, 123.01, 125.18, 128.92, 130.12, 132.18, 133.51, 142.80, 152.64, 154.02, 158.34, 158.38, 160.03.

2-Amino-4-(3-chlorophenyl)-4,5-dihydro-5-oxopyrano[3,2-c]chromene-3-carbonitrile 1c: mp 241°C - 243°C, (Lit.: 246°C - 248°C, [26]). ^{1}H NMR (400 MHz, DMSO-d_6): δ (ppm) 4.52 (s, 1 H), 7.26 (d, 1 H, J = 7.2 Hz), 7.31 - 7.38 (m, 3 H), 7.47 - 7.53 (m, 2 H), 7.73 (t, 1H, J = 7.6 Hz), 7.91 (d, 1H, J = 7.6 Hz). ^{13}C NMR (100 MHz, DMSO-d_6): δ (ppm) 57.81, 103.63, 113.43, 117.02, 119.55, 123.06, 125.13, 127.03, 127.66, 128.06, 130.87, 133.48, 133.54, 146.26, 152.66, 154.19, 158.37, 158.42, 160.06.

2-Amino-4-(4-bromophenyl)-4,5-dihydro-5-oxopyrano[3,2-c]chromene-3-carbonitrile (1d): mp 254°C - 256°C, (Lit.: 247°C - 249°C, [26]). ^{1}H NMR (400 MHz,

DMSO-d_6): δ (ppm) 4.48 (s, 1 H), 7.25 (d, 2 H, J = 8.0 Hz), 7.46 - 7.52 (m, 4 H), 7.73 (t, 1 H, J = 7.4 Hz), 7.91 (d, 1 H, J = 7.6 Hz). ^{13}C NMR (100 MHz, DMSO-d_6): δ (ppm) 57.85, 103.88, 113.39, 117.08, 119.54, 120.71, 123.02, 125.20, 130.50, 131.85, 133.54, 143.23, 152.65, 154.03, 158.32, 160.04.

2-Amino-4-(2,4-dichloro-phenyl)-4,5-dihydro-5-oxo pyrano[3,2-c]chromene-3-carbonitrile (1e): mp 258°C - 259°C, (Lit.: 253°C - 255°C, [30]). ^{1}H NMR (400 MHz, DMSO-d_6): δ (ppm) 4.98 (s, 1 H), 7.35 - 7.41 (m, 2 H), 7.48 - 7.54 (m, 4 H), 7.60 (d, 1 H, J = 2.0 Hz), 7.74 (dt, 1H, J_1 = 8.0 Hz, J_2 = 1.6 Hz), 7.92 (dd, 1 H, J_1 = 8.0 Hz, J_2 = 1.6 Hz). ^{13}C NMR (100 MHz, DMSO-d_6): δ (ppm) 34.33, 56.43, 102.96, 113.29, 117.12, 119.18, 123.04, 125.25, 128.36, 129.34, 132.56, 132.88, 133.61, 133.82, 139.91, 152.70, 154.62, 158.57, 159.94.

2-Amino-4-(3-nitrophenyl)-4,5-dihydro-5-oxopyran o[3,2-c]chromene-3-carbonitrile (1f): mp 266°C - 267°C, (Lit.: 262°C - 264°C, [24]). ^{1}H NMR (400 MHz, DMSO-d_6): δ (ppm) 4.74 (s, 1 H), 7.46 - 7.53 (m, 2 H), 7.58 (s, 2H, NH_2), 7.64 (t, 1H, J = 8.0 Hz), 7.74 (dt, 1 H, J_1 = 7.9 Hz, J_2 = 1.4 Hz), 7.82 (d, 1 H, J = 7.6 Hz), 7.93 (dd, 1 H, J_1 = 8.0 Hz, J_2 = 1.2 Hz), 8.12 - 8.15 (m, 2 H). ^{13}C NMR (100 MHz, DMSO-d_6): δ (ppm) 57.36, 103.33, 113.39, 117.07, 119.45, 122.77, 122.93, 123.10, 125.19, 130.56, 133.61, 135.26, 145.96, 148.30, 152.73, 154.35, 158.52, 158.57, 160.10.

2-Amino-4-(4-nitrophenyl)-4,5-dihydro-5-oxopyran o[3,2-c]chromene-3-carbonitrile (1g): mp 259°C - 261°C, (Lit.: 258°C - 260°C, [24]). ^{1}H NMR (400 MHz, DMSO-d_6): δ (ppm) 4.68 (s, 1 H), 7.47 - 7.54 (m, 2 H), 7.58 - 7.61 (m, 2 H), 7.75 (dt, 1 H, J_1 = 8.0 Hz, J_2 = 1.2 Hz), 7.93 (dd, 1H, J_1 = 8.0 Hz, J_2 = 1.2 Hz), 8.21 (d, 2 H, J = 8.4 Hz). ^{13}C NMR (100 MHz, DMSO-d_6): δ (ppm) 57.20, 103.27, 113.36, 117.13, 119.34, 123.09, 124.22, 125.26, 129.67, 133.69, 147.08, 151.23, 152.74, 154.42, 158.43, 158.47, 160.07.

2-Amino-4-(2-nitrophenyl)-4,5-dihydro-5-oxopyran o[3,2-c]chromene-3-carbonitrile (1h): mp 258°C - 260°C, (Lit.: 258°C - 260°C, [31]). ^{1}H NMR (400 MHz, DMSO-d_6): δ (ppm) 5.25 (s, 1 H), 7.46-7.59 (m, 6 H), 7.67 (t, 1 H, J = 7.6 Hz), 7.73 (t, 1 H, J = 7.8 Hz), 7.91 (d, 2 H, J = 7.6 Hz). ^{13}C NMR (100 MHz, DMSO-d_6): δ (ppm) 56.52, 103.75, 113.27, 117.09, 119.22, 123.03, 124.45, 125.21, 128.93, 131.65, 133.57, 134.16, 137.83, 149.66, 152.63, 154.06, 159.08, 160.19.

2-Amino-4-*p*-tolyl-4,5-dihydro-5-oxopyrano[3,2-c]c hromene-3-carbonitrile (1i): mp 253°C - 255°C, (Lit.: 259°C - 260°C, [28]). ^{1}H NMR (400 MHz, DMSO-d_6): δ (ppm) 2.27 (s, 3H), 4.41 (s, 1 H), 7.11 - 7.15 (m, 4 H), 7.40 (s, 2 H, NH_2), 7.46-7.52 (m, 2 H), 7.72 (dt, 1 H, J_1 = 7.8 Hz, J_2 = 1.6 Hz), 7.91 (dd, 1 H, J_1 = 7.8 Hz, J_2 = 1.4 Hz). ^{13}C NMR (100 MHz, DMSO-d_6): δ (ppm) 21.10, 58.53, 104.59, 113.42, 117.03, 119.71, 122.94, 125.15, 128.00, 129.55, 133.38, 136.78, 140.86, 152.56, 153.72, 158.30, 160.01.

2-Amino-4-(3,4,5-trimethoxy-phenyl)-4,5-dihydro-5 -oxopyrano[3,2-c]chromene-3-carbonitrile (1j): mp 236°C - 238°C, (Lit.: 236°C - 238°C, [31]). ^{1}H NMR (400 MHz, DMSO-d_6): δ (ppm) 3.64 (s, 3 H), 3.72 (s, 6 H), 4.44 (s, 1 H), 6.53 (s, 2 H), 7.41 (s, 2 H, NH_2), 7.47-7.52 (m, 2 H), 7.73 (dt, 1H, J_1 = 8.0 Hz, J_2 = 1.6 Hz), 7.91 (dd, 1 H, J_1 = 8.0 Hz, J_2 = 1.6 Hz). ^{13}C NMR (100 MHz, DMSO-d_6): δ (ppm) 56.36, 58.32, 60.39, 104.11, 105.38, 113.54, 117.05, 119.71, 123.04, 125.11, 133.39, 137.03, 139.46, 152.64, 153.30, 153.98, 158.38, 160.14.

2-Amino-4-(4-hydroxy-phenyl)-4,5-dihydro-5-oxop yrano[3,2-c]chromene-3-carbonitrile (1k): mp 266°C - 268°C, (Lit.: 260°C - 263°C, [28]). ^{1}H NMR (400 MHz, DMSO-d_6): δ (ppm) 4.33 (s, 1 H), 6.71 (d, 2 H, J = 8.4 Hz), 7.06 (d, 0.72 H, J = 8.8 Hz), 7.36 (s, 2 H, NH_2), 7.43 - 7.49 (m, 2 H), 7.69 (dt, 1H, J_1 = 7.8 Hz, J_2 = 1.6 Hz), 7.89 (dd, 1 H, J_1 = 7.8 Hz, J_2 = 1.4 Hz), 9.41 (s, 1H, OH). ^{13}C NMR (100 MHz, DMSO-d_6): δ (ppm) 58.84, 104.92, 113.43, 115.59, 116.96, 119.87, 122.89, 125.09, 129.20, 133.25, 134.23, 152.48, 153.42, 156.80, 158.32, 160.02.

3.3. General Procedure for the Synthesis of Dihydropyrano[2,3-c]pyrazoles

To a mixture of aromatic aldehyde (1 mmol), malonitrile (1 mmol), and 3-methyl-1-phenyl-1H-pyrazol-5(4 H)-one (1 mmol), catalyst [Sipim]HSO_4 (0.15 g, 0.12 mmol of H^+) was added and the mixture was heated at 110°C under solvent-free conditions. After completion of the reaction, as indicated by TLC, ethanol (10 mL) was added and the reaction mixture was filtered. The remaining was washed with warm ethanol (3 × 5 mL) in order to separate heterogeneous catalyst. After cooling the crude products were precipitated. The crude products were purified by recrystallization from ethanol (95%). The recovered catalyst was dried and reused for subsequent runs.

6-Amino-3-methyl-1,4-diphenyl-1,4-dihydro-pyran o[2,3-c]pyrazol-5-carbonitrile (5a): mp 170°C - 171°C, (Lit.: 168°C - 170°C, [38]). ^{1}H NMR (400 MHz, $CDCl_3$): δ (ppm) 1.92 (s, 3 H), 4.69 (s, 1H), 4.71 (s, 2 H, NH_2), 7.27 - 7.40 (m, 6 H), 7.49 (t, 2 H, J = 7.8 Hz), 7.68 (d, 2H, J = 8.0 Hz). ^{13}C NMR (100 MHz, $CDCl_3$): δ (ppm) 12.94, 37.44, 63.88, 98.36, 119.14, 121.22, 126.80, 127.62, 127.91, 128.83, 129.31, 137.57, 142.00, 143.84, 146.46, 158.17.

6-Amino-4-(4-chlorophenyl)-3-methyl-1-phenyl-1,4-dihydro-pyrano[2,3-c]pyrazol-5-carbonitrile (5b): mp 174°C - 176°C, (Lit.: 174°C - 177°C, [43]). ^{1}H NMR (400 MHz, $CDCl_3$): δ (ppm) 1.92 (s, 3 H), 4.68 (s, 1 H), 4.74 (s, 2 H, NH_2), 7.22 (d, 2H, J = 8.4 Hz), 7.33-7.37 (m,

3H), 7.49 (t, 2 H, J = 7.8 Hz), 7.67 (d, 2 H, J = 8.0 Hz). ^{13}C NMR (100 MHz, $CDCl_3$): δ (ppm) 12.97, 36.91, 63.36, 97.87, 118.95, 121.25, 126.92, 129.06, 129.28, 129.34, 133.44, 137.45, 140.57, 143.80, 146.28, 158.23.

6-Amino-4-(2,4-dichlorophenyl)-3-methyl-1-phenyl-1,4-dihydro-pyrano[2,3-c]pyrazol-5-carbonitrile (5c): mp 183˚C - 185˚C, (Lit.: 182˚C - 184˚C, [38]). ^{1}H NMR (400 MHz, $CDCl_3$): δ (ppm) 1.92 (s, 3 H), 4.78 (s, 2 H, NH_2), 5.29 (s, 1H), 7.17 (d, 1 H, J = 8.4 Hz), 7.26 - 7.28 (m, 1 H), 7.35 (t, 1 H, J = 7.2 Hz), 7.46-7.51 (m, 3 H), 7.67 (d, 2 H, J = 8.0 Hz). ^{13}C NMR (100 MHz, $CDCl_3$): δ (ppm) 12.77, 33.53, 61.98, 97.58, 118.62, 121.28, 126.97, 128.04, 129.35, 129.68, 131.49, 133.95, 137.41, 137.86, 143.95, 146.04, 158.84.

6-Amino-4-(4-bromophenyl)-3-methyl-1-phenyl-1,4-dihydro-pyrano[2,3-c]pyrazol-5-carbonitrile (5d): mp 182˚C - 184˚C, (Lit.: 176˚C - 177˚C, [40]). ^{1}H NMR (400 MHz, $CDCl_3$): δ (ppm) 1.92 (s, 3H), 4.67 (s, 1 H), 4.73 (s, 2H, NH_2), 7.17 (d, 2H, J = 8.0 Hz), 7.35 (t, 1 H, J = 7.4 Hz), 7.48 - 7.53 (m, 4H), 7.67 (d, 2 H, J = 7.6 Hz). ^{13}C NMR (100 MHz, $CDCl_3$): δ (ppm) 12.98, 36.98, 63.31, 97.78, 118.92, 121.26, 121.60, 126.93, 129.34, 129.63, 132.01, 137.45, 141.08, 146.28, 158.22.

6-Amino-4-(4-fluorophenyl)-3-methyl-1-phenyl-1,4-dihydro-pyrano[2,3-c]pyrazol-5-carbonitrile (5e): mp 170˚C - 171˚C, (Lit.: 167˚C - 168˚C, [38]). ^{1}H NMR (400 MHz, $CDCl_3$): δ (ppm) 1.91 (s, 3H), 4.69 (s, 1 H), 4.73 (s, 2 H, NH_2), 7.07 (t, 2 H, J = 8.6 Hz), 7.24 - 7.28 (m, 2 H), 7.35 (t, 1 H, J = 7.4 Hz), 7.49 (t, 2H, J = 7.8 Hz), 7.67 (d, 2H, J = 8.0 Hz). ^{13}C NMR (100 MHz, $CDCl_3$): δ (ppm) 12.94, 36.78, 63.69, 98.15, 115.86, 119.01, 121.24, 126.88, 129.33, 129.54, 137.49, 143.77, 146.32, 158.14, 162.17 (d, J_{C-F} = 244.0 Hz).

6-Amino-3-methyl-4-(3-nitrophenyl)-1-phenyl-1,4-dihydro-pyrano[2,3-c]pyrazol-5-carbonitrile (5f): mp 190˚C - 191˚C, (Lit.: 190˚C - 192˚C, [43]). ^{1}H NMR (400 MHz, $CDCl_3$): δ (ppm) 1.92 (s, 3 H), 4.84 (s, 3H, CH & NH_2), 7.37 (t, 1 H, J = 7.4 Hz), 7.51 (t, 2 H, J = 7.6 Hz), 7.60 (t, 1 H, J = 8.0 Hz), 7.67 - 7.70 (m, 3 H), 8.14 (s, 1 H), 8.21 (d, 1 H, J = 8.0 Hz). ^{13}C NMR (100 MHz, $CDCl_3$): δ (ppm) 13.01, 37.38, 62.11, 97.21, 118.76, 121.42, 122.87, 127.11, 129.37, 129.89, 134.17, 137.33, 144.49, 145.92, 148.76, 158.75.

6-Amino-3-methyl-4-(4-nitrophenyl)-1-phenyl-1,4-dihydro-pyrano[2,3-c]pyrazol-5-carbonitrile (5g): mp 194˚C - 196˚C, (Lit.: 195˚C - 197˚C, [43]). ^{1}H NMR (400 MHz, $CDCl_3$): δ (ppm) 1.92 (s, 3 H), 4.83 (s, 3 H, CH & NH_2), 7.37 (t, 1 H, J = 7.4 Hz), 7.47 - 7.53 (m, 4 H), 7.78 (d, 2 H, J = 8.0 Hz), 8.27 (d, 2 H, J = 8.4 Hz). ^{13}C NMR (100 MHz, $CDCl_3$): δ (ppm) 12.98, 37.33, 62.26, 97.07, 118.58, 121.35, 124.28, 127.16, 128.88, 129.40, 137.30, 143.83, 146.00, 147.46, 149.19, 158.56.

6-Amino-4-(3,4-dimethoxyphenyl)-3-methyl-1-phenyl-1,4-dihydro-pyrano[2,3-c]pyrazol-5-carbonitrile (5h): mp 167˚C - 169˚C, (Lit.: 193˚C - 195˚C, [38]). ^{1}H NMR (400 MHz, $CDCl_3$): δ (ppm) 1.91 (s, 3 H), 3.85 (s, 3 H), 3.87 (s, 3 H), 4.62 (s, 1 H), 4.73 (s, 2 H, NH_2), 6,73 (d, 1H, J = 1.6 Hz), 6.79 - 6.85 (m, 2H), 7.31 (t, 1 H, J = 8.0 Hz), 7.45 (t, 2H, J = 8.0 Hz), 7.66 (d, 2 H, J = 8.4 Hz). ^{13}C NMR (100 MHz, $CDCl_3$): δ (ppm) 12.98, 37.06, 55.86, 55.98, 63.92, 98.29, 110.82, 111.12, 119.22, 120.18, 121.05, 126.71, 129.28, 134.56, 137.57, 143.80, 146.52, 148.41, 149.20, 158.02.

6-Amino-4-(4-hydroxyphenyl)-3-methyl-1-phenyl-1,4-dihydro-pyrano[2,3-c]pyrazol-5-carbonitrile (5i): mp 206˚C - 208˚C, (Lit.: 206˚C - 207˚C, [40]). ^{1}H NMR (500 MHz, DMSO-d_6): δ (ppm) 1.78 (s, 3 H), 4.55 (s, 1 H), 6.71 (d, 2 H, J = 7.8 Hz), 7.03 (d, 2 H, J = 7.8 Hz), 7.10 (s, 2 H, NH_2), 7.31 (t, 1H, J = 7.0 Hz), 7.48 (t, 2 H, J = 7.3 Hz), 7.77 (d, 2 H, J = 7.7 Hz), 9.31 (s, 1 H). ^{13}C NMR (125 MHz, DMSO-d_6): δ (ppm) 13.45, 36.88, 59.68, 99.91, 116.07, 120.75, 120.96, 126.94, 129.62, 130.18, 134.79, 138.45, 144.65, 146.22, 157.15, 160.305.

4. Acknowledgements

We are thankful to Persian Gulf University Research Council for partial support of this work.

REFERENCES

[1] P. Wasserscheid and T. Welton, "Ionic Liquids in Synthesis," 2nd Edition, Wiley-VCH, Weinheim, 2007.

[2] W. Keim, W. Korth and P. Wasserscheid, "Ionic Liquids," Patent WO/2000/016902, March 30, 2000.

[3] P. Wasserscheid, M. Sesing and W. Korth, "Hydrogensulfate and Tetrakis(Hydrogensulfato)borate Ionic Liquids: Synthesis and Catalytic Application in Highly Brønsted-Acidic Systems for Friedel-Crafts Alkylation," *Green Chemistry*, Vol. 4, No. 2, 2002, pp. 134-138.

[4] N. Gupta, G. L. Sonu, J. Kad and Singh, "Acidic Ionic Liquid [bmim]HSO_4: An Efficient Catalyst for Acetalization and Thioacetalization of Carbonyl Compounds and Their Subsequent Deprotection," *Catalysis Communications*, Vol. 8, No. 9, 2007, pp. 1323-1328.

[5] D. Q. Xu, W. L. Yang, S. P. Luo, B. T. Wang, M. Wu and Z. Y. Xu, "Fischer Indole Synthesis in Brønsted Acidic Ionic Liquids: A Green, Mild, and Regiospecific Reaction System," *European Journal of Organic Chemistry*, No. 6, 2007, pp. 1007-1012.

[6] W. Wang, W. Cheng, L. Shao and J. Yang, "[TMBSA] [HSO_4] Ionic Liquid as Novel Catalyst for the Rapid Acetylation of Alcohols, Hydroxyesters and Phenols under Solvent-Free Conditions," *Catalysis Letters*, Vol. 121, No. 1-2, 2008, pp. 77-80.

[7] A. R. Hajipour, A. Rajaei and A. E. Ruoho, "A Mild and Efficient Method for Preparation of Azides from Alcohols

Using Acidic Ionic Liquid [H-NMP]HSO_4," *Tetrahedron Letters*, Vol. 50, No. 6, 2009, pp. 708-711.

[8] H. Tajik, K. Niknam and F. Parsa, "Using Acidic Ionic Liquid 1-Butyl-3-Methylimidazolium Hydrogen Sulfate in Selective Nitration of Phenols under Mild Conditions," *Journal of the Iranian Chemical Society*, Vol. 6, No. 1, 2009, pp. 159-164.

[9] K. Niknam and M. Damya, "1-Butyl-3-Methylimidazolium Hydrogen Sulfate [bmim]HSO_4: An Efficient Reusable Acidic Ionic Liquid for the Synthesis of 1,8-Dioxo-Octahydroxanthenes," *Journal of the Chinese Chemical Society*, Vol. 56, No. 3, 2009, pp. 659-665.

[10] K. Niknam, M. A. Zolfigol, D. Saberi and M. Khonbazi, "1-Butyl-3-Methylimidazolium Hydrogen Sulfate [bmim] HSO4: An Efficient Reusable Acidic Ionic Liquid for the Formylation of Alcohols," *Chinese Journal of Chemistry*, Vol. 27, No. 8, 2009, pp. 1548-1552.

[11] H. Tajik, K. Niknam and M. Sarrafan, "1-Butyl-3-Me- -thylimidazolium Hydrogen Sulfate ([bmim]-HSO_4) Mediated Synthesis of Polysubstituted Quinolines," *Synthetic Communications*, Vol. 41, No. 14, 2011, pp. 2103-2114.

[12] K. Qiao, H. Hagiwara and C. Yokoyama, "Acidic Ionic Liquid Modified Silica Gel as Novel Solid Catalysts for Esterification and Nitration Reactions," *Journal of Molecular Catalyst A: Chemicals*, Vol. 246, No. 1-2, 2006, pp. 65-69.

[13] R. Sugimura, K. Qiao, D. Tomida and C. Yokoyama, "Immobilization of Acidic Ionic Liquids by Copolymerization with Styrene and Their Catalytic Use for Acetal formation," *Catalysis Communications*, Vol. 8, No. 5, 2007, pp. 770-772.

[14] A. Chrobok, S. Baj, W. Pudlo and A. Jarzebski, "Supported Hydrogensulfate Ionic Liquid Catalysis in Baeyer-Villiger Reaction," *Applied Catalysis A: General*, Vol. 366, No. 1, 2009, pp. 22-28.

[15] M. N. Sefat, D. Saberi and K. Niknam, "Preparation of Silica-Based Ionic Liquid an Efficient and Recyclable Catalyst for One-Pot Synthesis of a-Aminonitriles," *Catalysis Letters*, Vol. 141, No. 11, 2011, pp. 1713-1720.

[16] M. Baghernejad and K. Niknam, "Synthesis of 4,4'-(Arylmethylene)bis(1H-pyrazol-5-ols) Using Silica-Bonded Ionic Liquid as Recyclable Catalyst," *International Journal of Chemistry*, Vol. 4, No. 3, 2012, pp. 52-60.

[17] G. R. Green, J. M. Evans and A. K. Vong, "Comprehensive Heterocyclic Chemistry II," Pergamon Press, Oxford, 1995, p. 469.

[18] W. O. Foye, "Principi di Chemico Farmaceutica," Piccin, Padova, 1991, p. 416.

[19] L. Bonsignore, G. Loy, D. Secci and A. Calignano, "Synthesis and Pharmacological Activity of 2-Oxo-(2H) 1-benzopyran-3-Carboxamide Derivatives," *European Journal of Medicinal Chemistry*, Vol. 28, No. 6, 1993, pp. 517-520.

[20] C. S. Konkoy, D. B. Fick, S. X. Cai, N. C. Lan and J. F. W. Keana, PCT Int Appl, WO 0075123, 2000, [Chem Abstr, Vol. 134, 2001, 29313a].

[21] D. Arnesto, W. M. Horspool, N. Martin, A. Ramos and C. Seaone, "Synthesis of Cyclobutenes by the Novel Photo-Chemical Ring Contraction of 4-Substituted 2-Amino-3,5-Dicyano-6-Phenyl-4H-Pyrans," *The Journal of Organic Chemistry*, Vol. 54, No. 13, 1989, pp. 3069-3072.

[22] R. M. Shaker, "Synthesis and Reactions of Some New 4H-Pyranol[3,2-c]benzopyran-5-One Derivatives and Their Potential Biological Activities," *Pharmazie*, Vol. 51, No. 3, 1996, pp. 148-151.

[23] M. Kidwai and S. Saxena, "Convenient Preparation of Pyrano Benzopyranes in Aqueous Media," *Synthetic Communications*, Vol. 36, No. 18, 2006, pp. 2737-2742.

[24] S. Abdolmohammadi and S. Balalaie, "Novel and Efficient Catalysts for the One-Pot Synthesis of 3,4-Dihydropyrano[c]chromene Derivatives in Aqueous Media," *Tetrahedron Letters*, Vol. 48, No. 18, 2007, pp. 3299-3303.

[25] J. M. Khurana and S. Kumar, "Tetrabutylammonium Bromide (TBAB): A Neutral and Efficient Catalyst for the Synthesis of Biscoumarin and 3,4-Dihydropyrano[c]-Chromene Derivatives in Water and Solvent-Free Conditions," *Tetrahedron Letters*, Vol. 50, No. 28, 2009, pp. 4125-4127.

[26] H. Mehrabi and H. Abusaidi, "Synthesis of Biscoumarin and 3,4-Dihydropyrano[c]chromene Derivatives Catalysed by Sodium Dodecyl Sulfate (SDS) in Neat Water," *Journal of the Iranian Chemical Society*, Vol. 7, No. 4, 2010, pp. 890-894.

[27] J. M. Khurana, B. Nand and P. Saluja, "DBU: A Highly Efficient Catalyst for One-Pot Synthesis of Substituted 3,4-dihydropyrano[3,2-c]chromenes, Dihydropyrano- [4,3-b] pyranes, 2-Amino-4Hbenzo[h]chromenes and 2-Amino-4H benzo[g]chromenes in Aqueous Medium," *Tetrahedron*, Vol. 66, No. 30, 2010, pp. 5637-5641.

[28] M. M. Heravi, M. Zakeri and N. Mohammadi, "Morpholine Catalyzed One-pot Multicomponent Synthesis of Compounds Containing Chromene Core in Water," *Chinese Journal of Chemistry*, Vol. 29, No. 6, 2011, pp. 1163-1166.

[29] H. Nagabhushana, S. S. Saundalkar, L. Muralidhar, B. M. Nagabhushana, C. R. Girija, D. Nagaraja, M. A. Pasha and V. P. Jayashankara, "a-Fe_2O_3 Nanoparticles: An Efficient, Inexpensive Catalyst for the One-Pot Preparation of 3,4-Dihydropyrano[c]chromenes," *Chinese Chemical Letters*, Vol. 22, No. 2, 2011, pp. 143-146.

[30] H. Mehrabi and M. Kazemi-Mireki, "CuO Nanoparticles: An Efficient and Recyclable Nanocatalyst for the Rapid and Green Synthesis of 3,4-Dihydropyrano[c]chromenes," *Chinese Chemical Letters*, Vol. 22, No. 12, 2011, pp. 1419-1422.

[31] K. Niknam and A. Jamali, "Silica-Bonded *N*-Propylpiperazine Sodium n-Propionate as Recyclable Basic Catalyst

for Synthesis of 3,4-Dihydropyrano[c]chromene Derivatives and Biscoumarins," *Chinese Journal of Catalysis*, Vol. 33, No. 11, 2012, pp. 1840-1849.

[32] M. H. Elnagdi, M. R. H. Elmaoghayer and G. E. H. Elgemeie, "Chemistry of Pyrazolopyrimidines," *Advance in Heterocyclic Chemistry*, Vol. 41, 1987, pp. 319-376.

[33] F. Karci and F. Karci, "Synthesis of Some Novel Pyrazolo[5,1-c][1,2,4]triazine Derivatives and Investigation of Their Absorption Spectra," *Dyes and Pigments*, Vol. 76, No. 1, 2008, pp. 97-103.

[34] S. G. Kuo, L. J. Huang and H. Nakamura, "Studies on Heterocyclic Compounds. 6. Synthesis and Analgesic and Antiinflammatory Activities of 3,4-Dimethylpyrano- [2,3-c] pyrazol-6-one Derivatives," *Journal of Medicinal Chemistry*, Vol. 27, No. 4, 1984, pp. 539-544.

[35] A. A. Fadda, A. A. H. Abdel-Rahman, E. A. Hamed and E. H. Khalil, "Utility of Enaminonitriles in Heterocyclic Synthesis: Synthesis and Antimicrobial Activity of Some New Azole and Azine Derivatives," *American Journal of Organic Chemistry*, Vol. 2, No. 2, 2012, pp. 7-13.

[36] N. M. Abunada, H. M. Hassaneen, N. G. Kandile and O. A. Miqdad, "Synthesis and Antimicrobial Activity of Some New Pyrazole, Fused Pyrazolo[3,4-d]-pyrimidine and Pyrazolo[4,3-e][1,2,4]-triazolo[1,5-c]pyrimidine Derivatives," *Molecules*, Vol. 13, No. 7, 2008, pp. 1501-1517.

[37] N. J. Thumar and M. P. Patel, "Synthesis and *in Vitro* Antimicrobial Evaluation of 4H-Pyrazolopyran, Benzopyran and Naphthopyran Derivatives of 1H-Pyrazole," *Arkivoc*, No. 13, 2009, pp. 363-380.

[38] D. Shi, J. Mou, Q. Zhuang, L. Niu, N. Wu and X. Wang, "Three-Component One-Pot Synthesis of 1,4-Dihydropyrano[2,3-c]pyrazole Derivatives in Aqueous Media," *Synthetic Communications*, Vol. 34, No. 24, 2004, pp. 4557-4563.

[39] T. S. Jin, A. Q. Wang, Z. L. Cheng, J. S. Zhang and T. S. Li, "A Clean and Simple Synthesis of 6-Amino-4-Aryl-5-Cyano-3-Methyl-1-Phenyl-1,4-Dihydropyrano[2,3-c]Py razole in Water," *Synthetic Communications*, Vol. 35, No. 1, 2005, pp. 137-143.

[40] S. B. Guo, S. X. Wang and J. T. Li, "D,L-Proline-Catalyzed One-Pot Synthesis of Pyrans and Pyrano[2,3-c]pyrazole Derivatives by a Grinding Method under Solvent-Free Conditions," *Synthetic Communications*, Vol. 37, No. 13, 2007, pp. 2111-2120.

[41] F. Lehmann, M. Holm and S. Laufer, "Three-Component Combinatorial Synthesis of Novel Dihydropyrano[2,3-c] pyrazoles," *Journal of Combinatorial Chemistry*, Vol. 10, No. 3, 2008, pp. 364-367.

[42] H. M. Al-Matar, K. D. Khalil, A. Y. Adam and M. H. Elnagdi, "Green One Pot Solvent-Free Synthesis of Pyrano[2,3-c]-Pyrazoles and Pyrazolo[1,5-a]Pyrimidines," *Molecules*, Vol. 15, No. 9, 2010, pp. 6619-6629.

[43] M. M. Heravi, A. Ghods, F. Derikvand, K. Bakhiari and F. F. Bamoharram, "$H_{14}[NaP_5W_{30}O_{110}]$ Catalyzed One-Pot Three-Component Synthesis of Dihydropyrano[2,3-c]pyrazole and pyrano[2,3-d]pyrimidine Derivatives," *Journal of the Iranian Chemical Society*, Vol. 7, No. 3, 2010, pp. 615-620.

[44] A. Siddekha, A. Nizam and M. A. Pasha, "An Efficient and Simple Approach for the Synthesis of Pyranopyrazoles Using Imidazole (Catalytic) in Aqueous Medium, and the Vibrational Spectroscopic Studies on 6-Amino-4-(4'Methoxyphenyl)-5-Cyano-3-Methyl-1-Phenyl-1,4-Dihydropyrano[2,3-c]pyrazole Using Density Functional Theory," *Spectrochimica Acta Part A: Molecular and Biomolecular Spectroscopy*, Vol. 81, No. 1, 2011, pp. 431-440.

[45] A. Hasaninejad, M. Shekouhy, N. Golzar, A. Zare and M. M. Doroodmand, "Silica Bonded n-Propyl-4-Aza-1-Azoniabicyclo[2.2.2]octane chloride (SB-DABCO): A Highly Efficient, Reusable and New Heterogeneous Catalyst for the Synthesis of 4H-Benzo[b]pyran Derivatives," *Applied Catalysis A: General*, Vol. 402, No. 1-2, 2011, pp. 11-22.

[46] H. Mecadon, M. R. Rohman, M. Rajbangshi and B. Myrboh, "γ-Alumina as a Recyclable Catalyst for the Four-Component Synthesis of 6-Amino-4-Alkyl/Aryl-3-methyl-2,4-Dihydropyrano[2,3-c]pyrazole-5-carbonitriles in aqueous medium," *Tetrahedron Letters*, Vol. 52, No. 19, 2011, pp. 2523-2525.

[47] S. R. Mandha, S. Siliveri, M. Alla, V. R. Bommena, M. R. Bommineni and S. Balasubramanian, "Eco-Friendly Synthesis and Biological Evaluation of Substituted Pyrano [2,3-c]pyrazoles," *Bioorganic and Medicinal Chemistry Letters*, Vol. 22, No. 16, 2012, pp. 5272-5278.

[48] S. H. S. Azzam and M. A. Pasha, "Simple and Efficient Protocol for the Synthesis of Novel Dihydro-1H-pyrano [2,3-c]pyrazol-6-Ones via a One-Pot Four-Component Reaction," *Tetrahedron Letters*, Vol. 53, No. 50, 2012, pp. 6834-6837.

2

Direct Preparation of Hydrogen and Carbon Nanotubes by Microwave Plasma Decomposition of Methane over Fe/Si Activated by Biased Hydrogen Plasma

Katsuya Konno[1,2], Kaoru Onoe[2], Yasuyuki Takiguchi[2], Tatsuaki Yamaguchi[2*]
[1]Hitachi Automotive Systems Ltd., Automotive Systems R&D Laboratory, Kawasaki, Japan
[2]Department of Life and Environmental Sciences, Faculty of Engineering, Chiba Institute of Technology, Chiba, Japan

ABSTRACT

Methane was decomposed to hydrogen and carbon nanotubes (CNTs) by microwave plasma, using Fe/Si catalyst activated by biased (−150 V) hydrogen plasma for various treatment times. Upon exposure to biased hydrogen plasma, the catalyst surface becomes lumpy within 1 min, coheres between 5 and 10 min and forms particles after 20 min. The methane conversion increased up to 93% over the treatment time of 5 min. The hydrogen yield showed as similar tendency as the methane conversion and kept 83% at treatment time of 5 min. The treatment time up to 1 min increased the amount of deposited carbon, and after treatment time of 5 min it dropped; then again after treatment time of 20 min, it increased to reach a maximum value of 22 g_c/g_{cat}. Deposited carbon was found to be consisted of carbon nanotubes. It grew vertically on the catalyst surface and reached a maximum length of 30.7 nm after treatment time of 10 min. Multiple types of CNTs were present, and the CNT diameters decreased with increasing plasma treatment time.

Keywords: Hydrogen; Methane Decomposition; Carbon Nanotubes; Microwave Plasma; Methane; H_2 Plasma Treatment

1. Introduction

Hydrogen production from methane is conventionally preformed with following reactions by steam reforming or partial oxidation.

$$CH_4 + H_2O \rightarrow CO + 3H_2 \quad (1)$$

$$CH_4 + 1/2O_2 \rightarrow CO + 2H_2 \quad (2)$$

Although the reactions produce CO, it is usually removed by the oxidation to CO_2 for the purpose of hydrogen production. However, this is an inconvenient fact to emit large quantities of CO_2, greenhouse effect gas. Therefore, the following reaction proceeded by thermal cracking of methane is expected.

$$CH_4 \rightarrow 2H_2 + C \quad (3)$$

According to the thermodynamics Equation (3) can proceed in the range of temperature from 1000˚C to 2000˚C [1]. This temperature is higher than that of general chemical engineering processes. Therefore Ni [2-7], Fe [8,9] and Co [10] based catalyst has been used to decompose methane at lower temperature. Venugopal *et al.* obtained the methane conversion of 32%, decomposing methane at 600˚C under 30 wt% Ni/SiO_2 [4]. Suelves *et al.* obtained the methane conversion of 67% and hydrogen concentration of 80% at 700˚C under Ni-based catalyst [6]. Shah *et al.* obtained the hydrogen yield of 80 - 90 vol%, decomposing methane at 700˚C - 800˚C under Fe-M (M = Pd, Mo and Ni) catalyst supported on alumina [10]. In these studies, although the hydrogen concentration is high (80% - 90%), the methane conversion is not so high. Onoe *et al.* reported that the high selectivities of acetylene and hydrogen are performed with a high CH_4 conversion by the microwave plasma technique with following reaction [11].

$$CH_4 \rightarrow 1/C_2H_2 2 + 3/2H_2 \quad (4)$$

To obtain higher yield of hydrogen from methane decomposed by microwave plasma, further decomposition of acetylene is necessary. Therefore, the microwave plasma is used together with a catalyst for preceding the following reaction, expecting the resulting carbon become carbon nanotubes

$$C_2H_2 \rightarrow H_2 + 2C \quad (5)$$

In this study, we investigated that the effect of catalyst treatment time of biased hydrogen plasma on microwave

*Corresponding author.

plasma decomposition of methane.

2. Experimental

2.1. Experimental Apparatus

A low-pressure flow type reaction microwave remote plasma apparatus (ULVAC Inc., CN-CVD-100R) was used, which is shown in **Figure 1**. The apparatus consists of a gas supply unit, a microwave generator, a plasma generator, a chamber and a low pressure unit. Microwaves are generated by an air-cooled magnetron and then introduced into a plasma furnace through a horizontally placed waveguide (height 50 mm, width 96 mm) that complied with the appropriate specifications. Reflected waves were reduced using a three-stub tuner and removed by an isolator.

2.2. Experimental Procedure

Market-grade methane and hydrogen were used. A silicon wafer (10 mm × 10 mm) was used as a substrate, onto which Fe was deposited using RF sputtering at 20 W. The thickness of the deposited Fe was 10 nm. The catalyst was set in the chamber at the position of 650 mm from the wave-guide. The hydrogen plasma treatment involved supplying the catalyst at a flow rate of 80 ml/min and an initial pressure of 180 Pa and exposing it to hydrogen plasma generated by a 500 W microwave under a bias of −150 V at 550°C from 1 min to 30 min. The mixed gas of 1:4 molar ratios of methane and hydrogen was introduced into the reactor at a flow rate of 100 ml/min and an initial pressure of 254 Pa. This gas was exposed to a 500 W microwave field for 30 min, and then the catalyst was heated under non-biased conditions to 600°C.

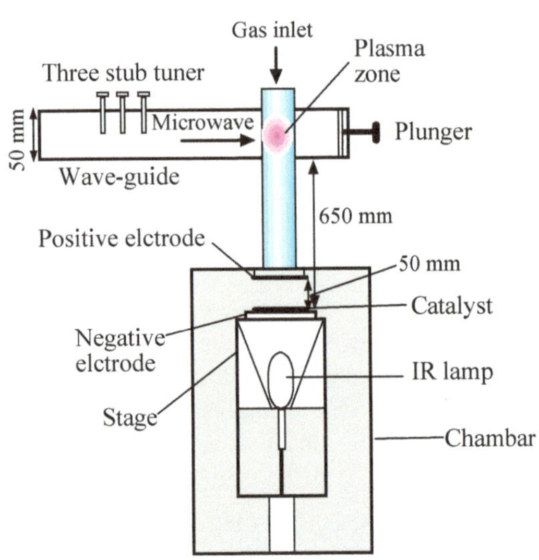

Figure 1. Experimental apparatus for the microwave plasma.

2.3. Calculation of Methane Conversion, Hydrogen Yield, Carbon Yield and Molar Fraction of Output Gas

Gaseous products obtained at 30 min of microwave reaction time was analyzed by gas chromatography. From an analysis of gas chromatography, carbon distribution (CD) of output gases and deposited carbon was calculated. Also hydrogen distribution (HD) of hydrocarbon and H_2 obtained from CH_4 was calculated. Methane conversion (X_{CH_4}) was the total value of the CD of hydrocarbon without CH_4 and deposited carbon. The Carbon yield was calculated from the amount of deposited carbon divided by the weight of Fe/Si catalyst.

2.4. Analysis of Products

Output gases were analyzed using gas chromatograph equipped with a TCD detector (SHIMADZU GC-14B, column: SHINCARBON T). The characteristic of the catalyst was analyzed using scanning electron microscopy (SEM; JSM-6300, JEOL Ltd.). The CNTs were analyzed using SEM, transmission electron microscopy (TEM; H-9000, Hitachi High-Technologies Corporation).

3. Results and Discussion

3.1. SEM Image of the Fe Catalyst Surface Activated by Biased H_2 Plasma for Various Treatment Time

Figure 2 shows SEM images of surface conditions of Fe activated by biased hydrogen plasma for various treatment times. The catalyst surface becomes rough under hydrogen plasma treatment irrespective of the treatment time, indicating that the Fe surface is modified. For up to treatment time of 1 min, the Fe surface is lumpy, and then it coheres over the treatment time span from 5 to 10 min. After treatment time of 20 min, the surface condition of the Fe produces particles whose sizes decrease with increasing treatment time.

In general, when the catalyst was introduced in a hydrogen atmosphere, catalyst particle sizes increased with increasing of microwave power, catalyst thickness, or treatment time. Chen reported that the Cr surface becomes rough and forms particles whose sizes increase with increasing treatment time, when Cr/Si (Cr thickness: 100 nm) was treated by biased hydrogen plasma (−150 V) [12]. Wang *et al.* reported the formation of Ni nanoparticles by hydrogen plasma treatment of Ni/TiN/Si [13], and Choi *et al.* reported that Ni particles were formed by NH_3 plasma treatment of Ni/Si [14].

The formation of catalyst particles is different with previous studies because of the different of catalyst metal, catalyst thickness and microwave power with their condi-

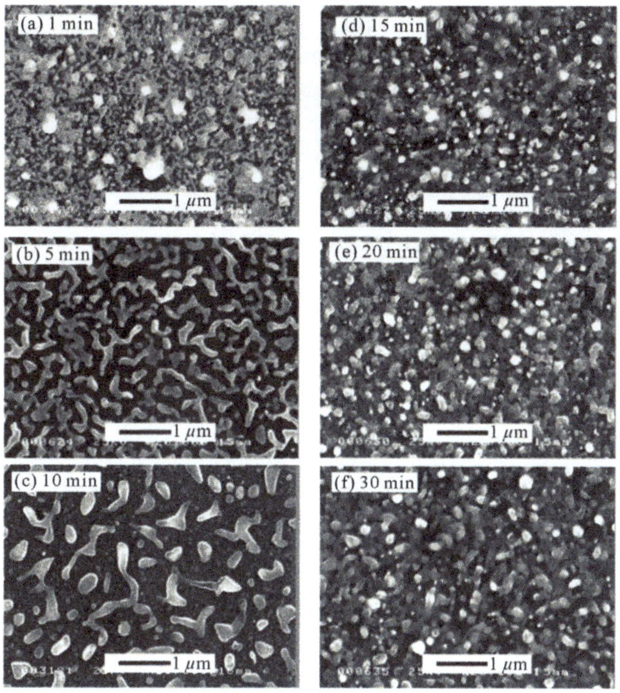

Figure 2. SEM image of Fe/Si surface treated for various times by biased H_2 plasma.

tions. When Fe/Si was used, Fe was cohered over the treatment time span from 5 to 10 min because the melting point of Fe is lower than that of Cr. The decrease in the size of the catalyst particles for plasma treatment times exceeding 10 min might be caused by etching with the hydrogen plasma.

3.2. Plasma Decomposition of Methane over Fe/Si Activated by Biased H_2 Plasma

In order to investigate the effect of catalyst on plasma decomposition of methane, X_{CH_4} and hydrogen distribution were compared when the reaction was preceded under non-catalyst or Fe/Si (without treatment).

Figure 3 shows the comparison of X_{CH_4} and the comparison of the hydrogen distribution of output gases from CH_4, when Fe/Si was not used and Fe/Si was used. In this case, Fe/Si was not treated by biased hydrogen plasma. When Fe/Si was used, X_{CH_4} is slightly decreased from 86% to 83%. The hydrogen distribution of output gases is almost the same value, and carbon was deposited on Fe/Si. Those results found that the both hydrogen and carbon can be obtained using Fe/Si.

In order to active catalyst for the production of hydrogen and deposited carbon, Fe/Si was treated at various treatment times by biased hydrogen plasma.

Figure 4 shows the relationship between treatment time of biased hydrogen plasma and X_{CH_4}. X_{CH_4} is slightly decreased from 83% to 76% with an increase of treatment time.

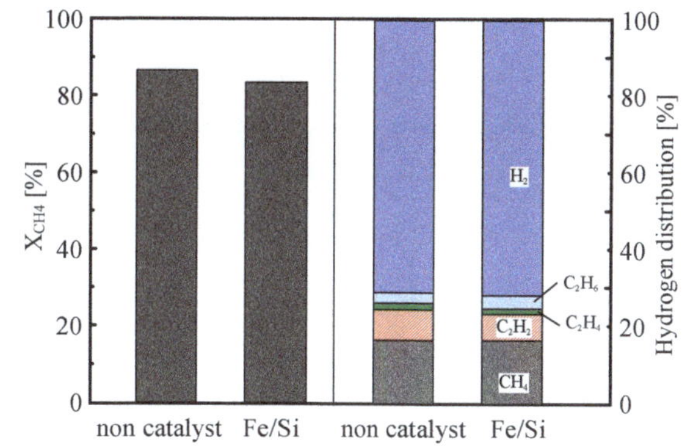

Figure 3. The comparison of methane conversion and hydrogen distribution when the reaction was proceeded under non-catalyst and Fe/Si (without biased hydrogen plasma treatment).

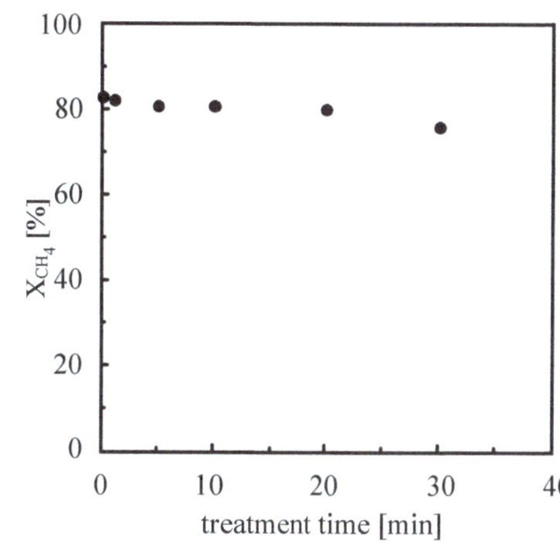

Figure 4. Effect of treatment time of biased hydrogen plasma on methane conversion.

This is probably due to the decrease of catalyst activetion. This factor might be the increase of Fe particle size and deposited carbon.

Figure 5 shows the relationship between treatment time and the hydrogen distribution of output gases from CH_4. Hydrogen element of CH_4 was mainly distributed to H_2 irrespective of treatment time. However, H_2 of hydrogen distribution was not increased with an increase of the treatment time of biased hydrogen plasma. This is due to the decrease of catalyst activation by carbon deposition. Hydrogen distribution of C2 hydrocarbon at treatment time of 1min is larger than that at other treatment time. It is probably because reaction mechanism on catalyst is difference of other catalyst.

Figure 6 shows the relationship between treatment time and the amount of deposited carbon. The amount of deposited carbon decreases until treatment time of 5 min.

Beyond treatment time of 10 min, the amount of carbon deposited increases and attains a maximum value (22.0 g_c/g_{cat}) at treatment time of 20 min.

Venugopal *et al.* also reported that the hydrogen yield

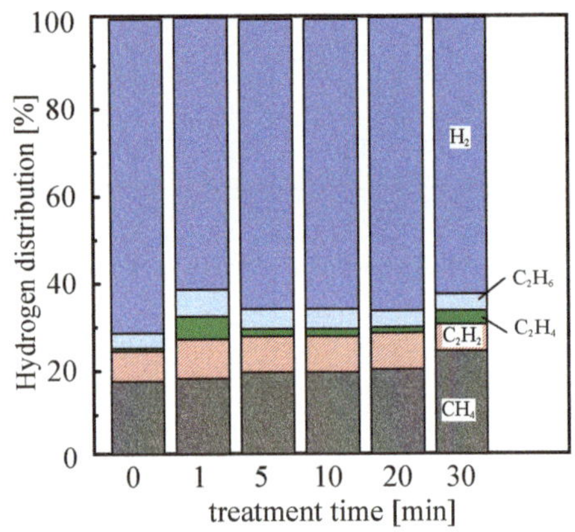

Figure 5. Effect of treatment time of biased hydrogen plasma on hydrogen distribution of output gases.

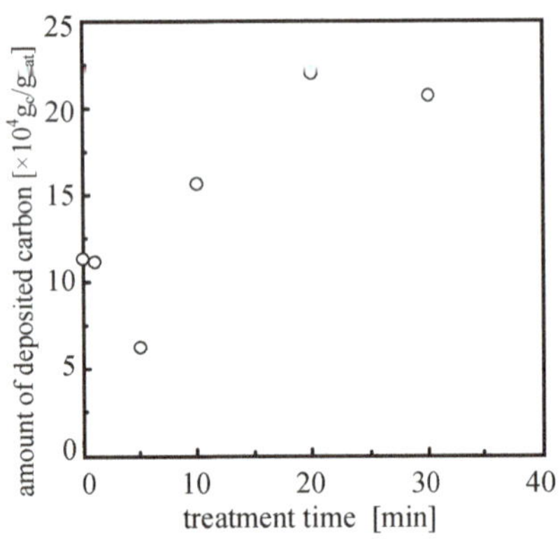

Figure 6. Effect of treatment time of biased H_2 plasma on the amount of deposited carbon.

and carbon yield obtained from methane decomposition over Ni/SiO_2 indicated the maximum value at Ni particle size of 21 nm in the range of Ni particle size from 17 to 40 nm [4]. The decrease of the amount of deposited carbon at treatment time of 5 min is probably due to the decrease of the activated sites on catalyst by cohering Fe. The factor of the maximum value at treatment time of 20 min is due to the highest activated on Fe/Si surface. It is well known that the methane conversion, the hydrogen yield and the amount of deposited carbon depended on catalyst particle size, and catalyst activation generally decreases by the increase of catalyst particle size. As showing **Figure 2**, the Fe/Si surface changed with the different treatment time. The Fe has cohered at the treatment time of 5 min, and the surface condition of the Fe produces particles in the range of treatment time from 10 min to 30 min.

3.3. SEM and TEM Image of Deposited Carbon

Figure 7 shows the SEM images of the multi-walled CNTs (MWCNTs). When Fe/Si without hydrogen plasma treatment was used, the main deposited carbon forms particle. However, MWCNT growth occurs mainly after the treatment time of 1 min.

Koyano *et al.* reported that the catalyst was more activated by the plasma treatment [15] and, as discussed above, the surface condition appears to be due to an increase in catalyst activation. The MWCNTs grow to various vertical heights, which is the same result as obtained in the previous studies that formed CNTs on the catalyst substance by microwave plasma treatment [16-19].

The heights of MWCNTs as a function of plasma treatment times are summarized in **Table 1**. The maximum height of the MWCNTs increases with increasing plasma treatment time, and reaches 30.7 nm at 10 min. This trend is the same as that reported by Sato [19], who observed a maximum height of 11 nm for CNTs prepared on Fe/Si after 10 min of plasma treatment. When using various thicknesses of cobalt as a catalyst, the CNT length decreases with increasing cobalt thickness [19]. Sato used Fe/Si with 70-nm-thick Fe [19]. However here, we used Fe/Si with 10-nm-thick Fe. Therefore, the different MWCNT lengths might be attributed the different Fe thicknesses.

Figure 8 shows the typical TEM images of the MWCNTs, from which we measured the MWCNT diameters. The diameter reaches a maximum value of 22.5 nm at treatment time of 20 min and a minimum value of 9.8 nm at treatment time of 30 min. It is well known that the MWCNT diameter is related to the particle size of the metal catalyst. We have also observed that MWCNT diameters increase with increasing metal catalyst particle

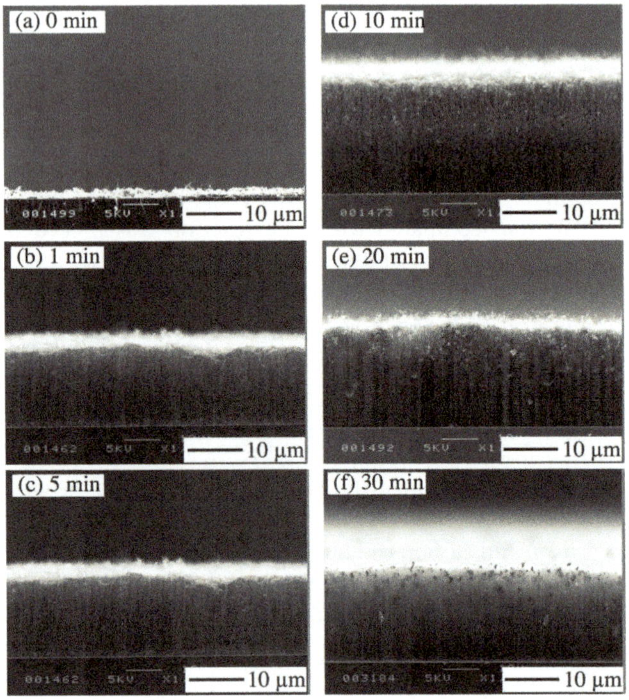

Figure 7. SEM images of carbon nanotubes obtained on the Fe/Si treated for various treatment times by biased H_2 plasma.

Table 1. Summary of yield of deposited carbon and the characteristic of CNTs.

Treatment time	Yield of deposited carbon	Shape	Structure	Length of CNT	Diameter of CNT	I_D/I_G
[min]	[g_c/g_{cat}]			[μm]	[nm]	[-]
0.0	2.6	mainly particle	-	-	-	-
1.0	11.3	fiborous	multi-walled	15.4	15.4	0.71
5.0	6.3	fiborous	multi-walled	21.7	12.6	0.68
10.0	15.8	fiborous	multi-walled	30.7	22.5	0.56
20.0	22.0	fiborous	multi-walled	28.5	15.3	0.52
30.0	20.7	fiborous	multi-walled	24.6	9.8	0.72

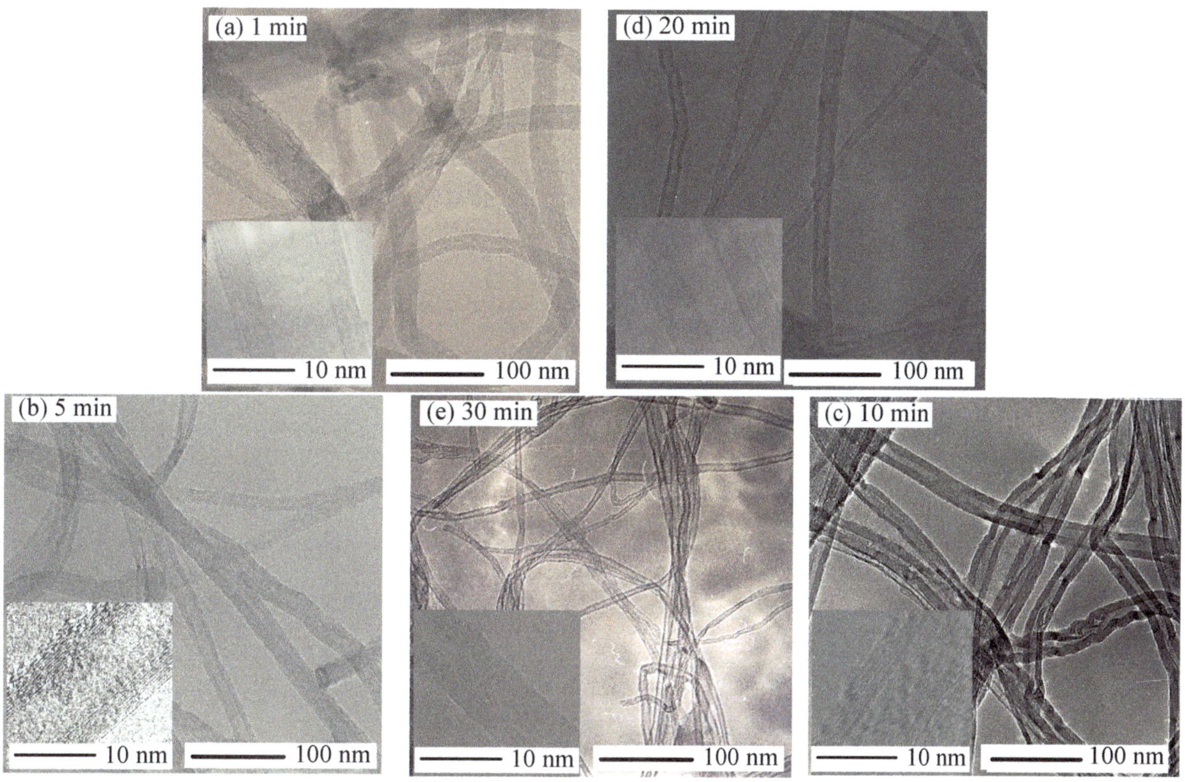

Figure 8. TEM images of carbon nanotubes obtained on the Fe/Si treated for various treatment times by biased H_2 plasma.

size in the microwave plasma technique [20]. In that study, we measured the Fe catalyst particle size using atomic force microscopy and found that the Fe particle size shrinked from 127 nm after treatment time of 20 min to 95 nm after treatment time of 30 min. Consequently, the different MWCNT diameters in the present study may be due to the change in size of the metal catalyst particle because of the varying treatment time.

The high-magnification TEM image shown in **Figure 8** indicates that there are parallel graphite layers between the inner and outer walls of the MWCNTs regardless of plasma treatment time.

However, the number of graphite layers change with plasma treatment time, which suggests that the catalyst particles change as a function of plasma treatment time because it is known that the number of graphite layers depends on the condition of the catalyst, such as whether the catalyst is a particle and the catalyst surface roughness [14,20].

4. Conclusions

Methane was decomposed by microwave plasma with catalyst, which Fe surface condition was activated by biased hydrogen plasma for various treatment times.

X_{CH_4} was over 76% irrespective of the treatment time. The hydrogen distribution of output gas was mainly hydrogen (over 60%) irrespective of the treatment time. It is hopeful to increase the hydrogen yield by the further improtant of reaction conditions, especially the property and volume of catalyst.

The amount of deposited carbon was 22 g_c/g_{cat} at biased hydrogen treatment time of 10 min. The deposited carbon was filamentous, and it grew vertically on Fe surface and was multi-walled carbon nanotubes. The diameter and graphite layer was difference by biased hydrogen treatment time.

REFERENCES

[1] J. R. Fincke, R. P. Snderson, T. A. Hyde and B. A. Detering, "Plasma Pyrolysis of Methane to Hydrogen and Carbon Black," *Industrial and Engineering Chemistry Research*, Vol. 41, No. 6, 2002, pp. 1425-1435.

[2] J. L. Pinilla, I. Suelve, M. J. Lazaro, R. Moliner and J. M. Palacios, "Parametric Study of the Decomposition of Methane Using a $NiCu/Al_2O_3$ Catalyst in a Fluidized Bed Reactor," *International Journal of Hydrogen Energy*, Vol. 35, No. 18, 2010, pp. 9801-9809.

[3] D. Li, J. Chen and Y. Li, "Evidence of Composition Deviation of Metal Particles of a $Ni\text{-}Cu/Al_2O_3$ Catalyst during Methane Decomposition to Cox-Free Hydrogen," *International Journal of Hydrogen Energy*, Vol. 34, No. 1, 2009, pp. 299-307.

[4] A. Venugopal, S. Naveen Kumar, J. Ashok, D. H. Prasad, V. D. Kumari, K. B. S. Prasad and M. Subrahmanyam, "Hydrogen Production by Catalytic Decomposition of Methane over Ni/SiO_2," *International Journal of Hydrogen Energy*, Vol. 32, No. 12, 2007, pp. 1782-1788.

[5] Y. Echegoyen, I. Suelves, M. J. Lazaro, R. Moliner and J. M. Palacios, "Hydrogen Production by Thermocatalytic Decomposition of Methane over Ni-Al and Ni-Cu-Al Catalysts: Effect of Calcination Temperature," *Journal of Power Sources*, Vol. 169, No. 1, 2007, pp. 150-157.

[6] I. Suelves, M. J. Lazaro, R. Moliner, B. M. Corbella and J. M. Palacios, "Hydrogen Production by Thermo Catalytic Decomposition of Methane on Ni-Based Catalysts: Influence of Operation Conditions on Catalyst Deactivation and Carbon Characteristics," *International Journal of Hydrogen Energy*, Vol. 30, No. 15, 2005, pp. 1555-1567.

[7] M. A. Ermakova, D. Y. Ermakov and G. G. Kuvshinov, "Effective Catalysts for Direct Cracking of Methane to Produce Hydrogen and Filamentous Carbon Part I. Nickel catalysts," *Applied Catalysisi A*: *General*, Vol. 201, No. 1, 2000, pp. 61-70.

[8] A. F. Cunha, J. J. M. Orfao and J. L. Figueiredo, "Methane Decomposition on Fe-Cu Raney-Type Catalysts," *Fuel Processing Technology*, Vol. 90, No. 10, 2009, pp. 1234-1240.

[9] A. Konieczny, K. Mondal, T. Wiltowski and P. Dydo, "Catalyst Development for Thermocatalytic Decomposition of Methane to Hydrogen. International Journal of Hydrogen," *Energy*, Vol. 33, No. 1, 2008, pp. 264-272.

[10] N. Shah, D. Panjala and G. P. Huffman, "Hydrogen Production by Catalytic Decomposition of Methane," *Energy & Fuels*, Vol. 15, No. 6, 2001, pp. 1528-1534.

[11] K. Onoe, A. Fujie, T. Yamaguchi and Y. Hatano, "Selective Synthesis of Acetylene from Methane by Microwave Plasma Reaction," *Fuel*, Vol. 76, No. 3, 1997, pp. 281-281.

[12] C. F. Chen, C. L. Lin and C. M. Wang, "Field Emission Properties of Verticall Allgned Carbon Nanotubes Grown on Bias-Enhanced Hydrogen Plasma Pretratment Cr Film," *Thin Solid Films*, Vol. 444, No. 1-2, 2003, pp. 64-69.

[13] W. P. Wang, H. C. Wen, S. R. Jian, J. Y. Juang, Y. S. Lai, C. H. Tsai, W. F. Wu, K. T. Chen and C. P. Chou, "The Effects of Hydrogen Plasma Pretreatment on the Formation of Vertically Aligned Carbon Nanotubes," *Applied Surface Science*, Vol. 253, No. 23, 2007, pp. 9248-9253.

[14] J. H. Choi, T. Y. Lee, S. H. Choi, J. H. Han, J. B. Yoo, C. Y. Park, T. Jung, S. G. Yu, W. Yi, I. T. Han and J. M. Kim, "Chontrol of Carbon Nanotubes Density through Ni Nanoparticle Formation Using the Thermal and NH_3 Plasma Treatment," *Diamond Related Materials*, Vol. 12, No. 3-7, 2003, pp. 794-798.

[15] G. Koyano, H. Watanabe, T. Okuhara, M. Misono, A. Nishijima, N. Matsubayashi and M. Imamura, "Highly Activate Supported Cobalt Oxide Catalysts Prepared by Low Temperature Oxgen Plasma," *Sekiyu Gakkaishi*, Vol. 36, No. 5, 1993, pp. 402-405.

[16] Y. C. Choi, Y. M. Shin, Y. H. Lee, B. S. Lee, G. S. Park, W. B. Choi, N. S. Lee and J. M. Kin, "Controlling the Diameter, Growth Rate, and Density of vertically Aligned Carbon Nanotubes Synthesized by Microwave Plasma-Enhanced Chemical Vapor Deposition," *Applied Physics Letters*, Vol. 76, No. 7, 2000, pp. 2367-2369.

[17] J. Y. Lee and B. S. Lee, "Nitrogen Induced Structure Control of Vertically Aligned Carbon Nanotubes Synthesized by Microwave Plasma Enhanced Chemical Vapor Deposition," *Thin Solid Film*, Vol. 418, No. 2, 2002, pp. 85-88.

[18] M. Taniguchi, H. Nagao, M. Hiramatsu, Y. Ando and M. Hori, "Preparation of Dense Carbon Nanotube Film Using Microwave Plasma-Enhanceed Chemical Vapor Deposition," *Diamond Related Materials*, Vol. 14, No. 3-7, 2005, pp. 855-858.

[19] H. Sato, H. Takegawa and Y. Saito, "Veticall Aligned Carbon Nanotubes Grown by Plasma Enhanced Chemical Vapor Deposition," *Journal of Vacuum Science and Technology B*, Vol. 21, No. 6, 2003, pp. 2564-2568.

[20] C. B. Bower, O. Zhou, W. Zhu, D. J. Werder and J. Sungho, "Nucleation and Growth of Carbon Nanotubes by Microwave Plasma Chemical Vapor Deposition," *Applied Physics Letters*, Vol. 77, No. 17, 2000, pp. 12767-

12769.

[21] K. Konno, K. Onoe and T. Yamaguchi, "Production of Carbon Nanotubes from Methane by Microwave Plasma—Examination of Nickel Catalysis," *Proceeding of 10th Asia Pacific Confederation of Chemical Engineering*, Kitakyusyu, 17-21 October 2004, pp. 705-714.

3

Di(isothiocyanato)bis(4-methyl-4'-vinyl-2,2'-bipyridine) Ruthenium(II) Films Deposited on Titanium Oxide-Coated, Fluorine-Doped Tin Oxide for an Efficient Solar Cell

Yi Liu, Ryuichi Sugimoto, Katsuhiro Sumi*
School of Environmental Science and Engineering, Kochi University of Technology, Kami, Japan

ABSTRACT

Dye-sensitized titanium oxide electrodes were prepared by immobilizing a novel ruthenium complex, di(isothiocyanato)bis(4-methyl-4'-vinyl-2,2'-bipyridine)ruthenium(II) $[(NCS)_2(mvbpy)_2Ru(II)]$ or the ruthenium complex/sodium 4-vinylbenzenesulfonate onto the surface of a titanium oxide-coated, fluorine-doped tin oxide (TiO_2/FTO) electrode through a new electrochemically initiated film formation method, in which the electrolysis step and the film deposition step were individually performed. The incident photon-to-current conversion efficiency (IPCE) of the Ru complex film on a TiO_2/FTO electrode was disappointedly insufficient (1.2% at 440 nm). In sharp contrast, the Ru(II) complex/sodium 4-vinylbenzenesulfonate composite film deposited on the surface of a TiO_2/FTO electrode showed maximum IPCE of 31.7% at 438 nm.

Keywords: Methylvinylbipyridine; Ruthenium Complex; Dye-Sensitized Solar Cell; Photon-to-Current Conversion Efficiency

1. Introduction

Dye-sensitized solar cells (DSSCs) have been attracting abundant attention as technically and economically credible alternatives to traditional silicon-based solar cells [1-5]. As a key component of DSSCs, photosensitizers that can absorb sunlight and inject an electron into the conducting band of nanocrystalline TiO_2 have been mainly focused in DSSC research; photosensitizers having a general structure of $RuL_2(X)_2$, where L stands for functionalized 2,2'-bipyridine such as 4,4'-dicarboxyl-2, 2'-bipyridine and X presents halide, cyanide, thiocyanate, thiolate, thienyl or ancillary water, have been extensively investigated over the past two decades [6-13]. Incident photon-to-current conversion efficiency (IPCE) is one of critical parameters for the evaluation of the performance of solar cells [14,15]. Among solar cells developed so far, that fabricated with N_3 dye (cis-$(dcbpy)_2(NCS)_2$ ruthenium(II): dcbpy = 4,4'-dicarboxyl-2,2'-bipyridine) shows IPCE of about 85% in a visible light region [16].

Dyes are commonly immobilized onto a semiconductor (typically TiO_2) surface with a functional group, such as carboxyl, phosphono or thio group in the dyes. Up to now, Ru(II) complexes without these functional groups cannot be candidates in DSSC research due to impossibility of immobilization; this drawback has limited the development of effective photosensitizers.

Herein, we report a new, electrochemically induced film formation method, which can immobilize di(isothiocyanato)bis(4-methyl-4'-vinyl-2,2'-bipyridine) ruthenium(II), and a novel type of dye-sensitized solar cells with sufficient IPCE.

2. Experimental Section

2.1. Instruments

^{1}H NMR spectra were measured on a Bruker Ascend 400 spectrometer. Chemical shifts were determined with respect to the residual solvent peak (δ in ppm, J in Hz). Matrix-assisted laser desorption/ionization time-of-flight (MALDI-TOF) mass spectra were acquired on a Bruker autoflex speed-KE in a reflection mode with α-cyano-4-hydroxycinnamic acid (CCA) as a matrix. FT-IR spectra were recorded on a JASCO FT/IR-610 spectrometer as KBr pellets. Ultraviolet-visible absorption (UV-vis)

*Corresponding author.

spectra were measured on a Perkin Elmer Lambda 19 UV/vis/NIR spectrometer. Cyclic voltammograms were recorded on a CH instrument 701 electrochemical analyzer at a sweep rate of 100 $mV{\cdot}s^{-1}$ by using 0.1 M tetrabutylammonium perchlorate (TBAP) as an electrolyte solution, Pt wire as a counter electrode, and Ag/AgCl as a reference electrode.

2.2. Materials

All reactions were performed under argon atmosphere. All reagents were purchased from Wako Pure Chemical Industries Ltd except for NH_4SCN and sodium 4-vinylbenzenesulfonate, which were purchased from Nacac Tesque, Inc. The reagents were used as received, except for 4,4'-dimethylbipyridine which was recrystallized from methanol prior to use. Dry tetrahydrofuran (THF) was distilled over Na and benzophenone. 2-(4'-Methyl-(2,2'-bipyridin-4-yl)ethanol [17,18] and 4-methyl-4'-vinyl-2,2'-bipyridine [19] were synthesized according to the procedures in the literatures with some modifications. Di(isothiocyanato)-bis(4-methyl-4'-vinyl-2,2'-bipyridine)-ruthenium(II) was prepared by a method similar to that of bis(2,2'-bipyridine)diisothiocyanatoruthenium(II) [20].

2.2.1. 2-(4'-Methyl-2,2'-bipyridin-4-yl)ethanol

4,4'-Dimethyl-2,2'-bipyridine (6.8 g, 36.91 mmol) was dissolved in dry THF (400 mL), and the solution was cooled down to 0°C by ice/water. Lithium diisopropylamide (20.3 mL, 2 M in THF, 40.60 mmol) was added dropwise to the solution. The purple solution was stirred at 0°C for 2 h, and then a large excess amount of dry paraformaldehyde (5.5 g, 183 mmol as formaldehyde) was added all at once to the solution. The mixture was allowed to warm to room temperature and stir for 24 h. The color of the solution gradually turned from purple to light yellow. The reaction was quenched with distilled water (200 mL), and the solution was concentrated to about 50 mL under reduced pressure. The residue was extracted with dichloromethane (3 × 50 mL), and the extracts were concentrated under reduced pressure. The yellow viscous liquid remained was purified on neutral alumina gel column chromatography; after washing off unreacted 4,4'-dimethyl-2,2'-bipyridine with hexane/ethyl acetate (19:1, v/v), the yellow-colored fraction eluted with hexane/ethyl acetate (2:1, v/v) was collected. The fraction was condensed under reduced pressure to afford light-yellow oil (2.27 g, 29%). ^1H NMR (400 MHz, $CDCl_3$) δ (ppm): 8.56 (d, 1H, $J = 5.0$ Hz, PyH), 8.50 (d, 1H, $J = 5.0$ Hz, PyH), 8.25 (s, 1H, PyH), 8.20 (s, 1H, PyH), 7.18 (dd, 1H, $J_1 = 5.0$ Hz, $J_2 = 1.6$ Hz, PyH), 7.13 (d, 1H, $J = 4.9$ Hz, PyH), 3.96 (t, 2H, $J = 6.5$ Hz, CH_2CH_2OH), 2.95 (t, 2H, $J = 6.4$ Hz, CH_2CH_2OH), 2.44 (s, 3H, CH_3).

2.2.2. 4-Methyl-4'-vinyl-2,2'-bipyridine (mvbpy)

2-(4'-Methyl-2,2'-bipyridin-4-yl)ethanol (2.27 g, 10.7 mmol), 4-t-butylpyrocatechol (0.1 g, 0.6 mmol) and finely powdered sodium hydroxide (4.28 g, 107 mmol) were successively added to benzene (5 mL). The blue mixture was then heated at 115°C for 2 h under reduced pressure (<1 mmHg). The crude product was purified on silica gel column chromatography by using dichloromethane/methanol (100:1, v/v) as an eluent. The corresponding fractions were concentrated under reduced pressure to yield a colorless solid (1.0 g, 48%). ^1H NMR (400 MHz, $CDCl_3$) δ (ppm): 8.62 (d, 1H, $J = 5.0$ Hz, PyH), 8.55 (d, 1H, $J = 5.0$ Hz, PyH), 8.40 (s, 1H, PyH), 8.24 (s, 1H, PyH), 7.30 (dd, 1H, $J_1 = 5.1$ Hz, $J_2 = 1.8$ Hz, PyH), 7.14 (d, 1H, $J = 4.9$ Hz, PyH), 6.77 (dd, 1H, $J_1 = 14.32$ Hz, $J_2 = 6.8$ Hz, PyH), 6.09 (d, 1H, $J = 17.7$ Hz, $CH{=}CH_2$), 5.53 (d, 1H, $J = 10.9$ Hz, $CH{=}CH_2$), 2.44 (s, 3H, CH_3).

2.2.3. Di(isothiocyanato)bis(4-methyl-4'-vinyl-2,2'-bipyridine) Ruthenium(II) [$(NCS)_2(mvbpy)_2Ru(II)$]

4-Methyl-4'-vinyl-2,2'-bipyridine (0.400 g, 2.04 mmol) was added to a solution of $RuCl_3$ (0.211 g, 1.01 mmol) in DMF (20 mL). After the mixture was refluxed for 8 h in the dark, NH_4NCS (6.367 g, 8.36 mmol) was added all at once at room temperature. The solution was refluxed again for 12 h and cooled down to room temperature. Upon adding methanol (200 mL) to the reaction mixture, a purple red powder was deposited, which was collected by filtration over a G4 glass filter. The purple powder was thoroughly washed with cold methanol (20 mL) and diethyl ether (15 mL), and dried for 12 h under reduced pressure to give $(NCS)_2(mvbpy)_2Ru(II)$ (0.24 g, 38%). ν_{max} (KBr, cm^{-1}): 2104 (CN, s), 1400 (CH_3, s). MALDI-TOF MS: calculated for $C_{28}H_{24}N_6S_2Ru$: $[M]^+ = 610.0548$; found 610.0517 with isotope peaks in agreement with those obtained by a calculation.

2.3. Immobilization

The immobilization was performed in three steps under argon atmosphere. All solvents used were purged with argon for 15 min.

Step 1: The electrochemical treatment of an electrode was performed in a one-compartment cell according to the three-electrode method (TiO_2-coated, fluorine-doped tin oxide (TiO_2/FTO) as a working electrode, Pt wire as a counter electrode, and Ag/AgCl as a reference electrode) by using a CH instrument 701 electrochemical analyzer. These electrodes were dipped in an electrolytic solution containing 0.1 M of TBAP, and swept between 0-n-0 V (n = 2, 4, 6, 8, and 10) at a rate of 100 $mV{\cdot}s^{-1}$ with bubbling with argon during the electrolysis.

Step 2: After the electrolysis, the working electrode was soaked in a washing solvent for several seconds to remove adsorbents on the TiO_2/FTO surface.

Step 3: The electrode was dipped into a solution of $(NCS)_2(mvbpy)_2Ru(II)$ (1 mM) or the Ru complex/sodium 4-vinylbenzenesulfonate (1 mM and 1 mM, respectively) in DMF (30 mL), and kept for appropriate hours in the dark under argon atmosphere. Then, the electrode was taken out, washed with acetone (30 mL), and dried at room temperature in air to give the Ru complex film.

2.4. Measurement of Incident Photon-to-Current Conversion Efficiency (IPCE) of the Ru(II) Complex Films

A fluorometer (Horiba Jobin Yvon-Spex fluorolog, Horiba Inc.) equipped with a xenon lamp (Fluorolog instruments S.A., Inc.) was used in order to generate monochromatic lights from 350 nm to 750 nm with a width of 5 nm. The photon number was counted by an optometer (Photometer S370 UDT, Gamma Scientific Inc.). The electrical quantities were measured with an electrometer (Keithley 6514 system electrometer, Keithley Instruments Inc.). The DSSC was fabricated from TiO_2/FTO coated with the Ru(II) complex film (photoanode), FTO (photocathode), and LiI/I_2 (0.5 M/1.1 × 10^{-3} M) in CH_3CN (mediator).

3. Results and Discussion

3.1. Formation of Di(isothiocyanato) bis(4-methyl-4'-vinyl-2,2'-bipyridine) Ruthenium(II) film on the Surface of a Titanium Oxide-Coated, Fluorine-Doped Tin Oxide Electrode

Because the present Ru(II) complex, di(isothiocyanato) bis(4-methyl-4'-vinyl-2,2'-bipyridine)ruthenium(II) [$(NCS)_2(mvbpy)_2Ru(II)$], has no functional group reactive with TiO_2 such as carboxyl, phosphono or thio group, this complex cannot be immobilized onto the surface of TiO_2-coated, fluorine-doped tin oxide (TiO_2/FTO) electrode with conventional film formation methods that have been adopted in the fabrication of DSSCs. Paying our attention to the fact that the Ru complex has vinyl groups, we at first tried to immobilize the Ru complex onto a TiO_2/FTO surface by the electrolysis of a solution of the Ru complex. However, the attempt resulted in failure, most likely due to the decomposition of the Ru complex during the electrolysis. After numerous trials, we finally found a stepwise, electro-chemically induced film formation method to give a film-like deposit of the Ru complex on the surface of TiO_2/FTO. This method is composed of three steps: Step 1, the electrolysis of a TiO_2/FTO electrode only; Step 2, the washing of the electrode with a solvent; Step 3, the immobilization of the Ru complex upon immersing the washed electrode in a solution of the Ru complex. **Figure 1** schematically represents the immo- bilization procedure.

The immobilization was found to be very sensitive to the history of the potential scanned onto a TiO_2/FTO electrode (Step 1). **Figures 2** and **3** show the cyclic voltammograms during the electrolysis of the electrodes in ranges of 0-n-0 V (n = 2, 4, 6, 8, and 10) at a rate of 100 $mV \cdot s^{-1}$ and the UV-vis spectra of the resultant electrodes. As shown in **Figure 2**, an irregular trace in the cyclic voltammograms was observed in a range of 1 - 3.5 V independent on the potential applied, which would correspond to irreversible oxidation.

Moreover, it is clear from **Figure 3** that the electrochemically initiated film deposition of the Ru complex took place only when the electrode, potentially swept between 0-8-0 V at a rate of $100 mV \cdot s^{-1}$, was used. As shown in **Figure 4**, the electrode was colored after Step 3, undoubtedly indicating the film deposition. Moreover, the film deposited was insoluble in any sol- vents including aprotic polar solvents such as DMF and DMSO. These changes strongly support the deposition of the Ru complex through some reaction.

The electrochemically initiated film deposition method was also affected by solvents used in Steps 1-3. As shown in **Figure 5**, the solvent used in Step 1 influenced, to considerable extent, the quantity of the Ru

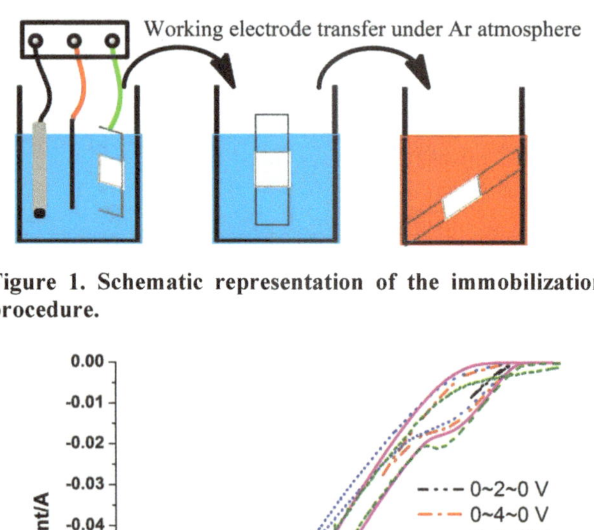

Figure 1. Schematic representation of the immobilization procedure.

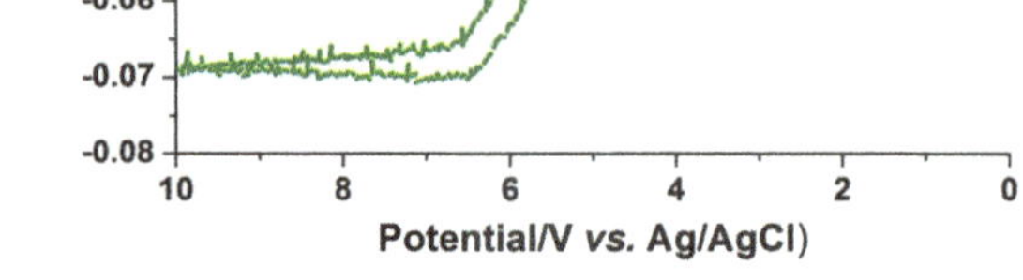

Figure 2. Cyclic voltammograms of TiO_2/FTO electrodes at a scan rate of 100 $mV \cdot s^{-1}$ in Step 1.

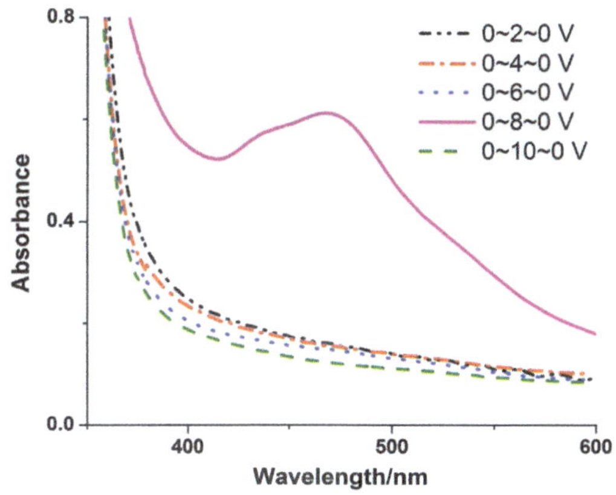

Figure 3. Uv-vis spectra of the resultant TiO_2/FTO electrodes.

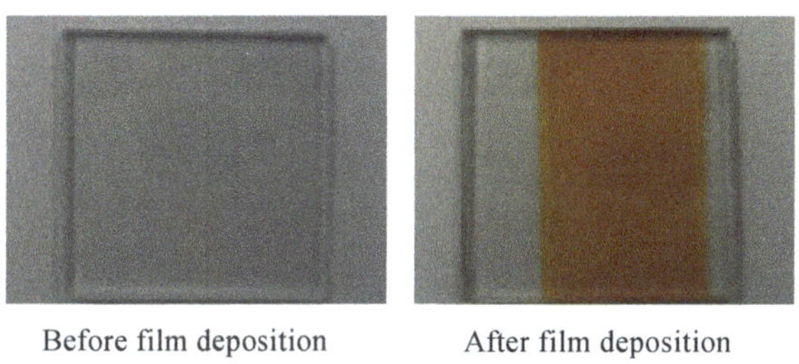

Figure 4. Electrodes before and after film deposition. (The electrodes were potentially swept between 0-8-0 V at a rate of 100 mV·s^{-1} in Step 1).

complex which finally immobilized on the surface of a TiO_2/FTO electrode; although dichloromethane gave a better result than acetonitrile or methanol did, it was very hard to maintain steady conditions for the electrolysis due to the evaporation of dichloromethane upon bubbling argon during the electrolysis. Then, acetonitrile, which was better than methanol, was used in Step 1. In Step 2, no film was formed on the surface of TiO_2/FTO, when the electrode was washed with DMF instead of acetonitrile. Because the Ru complex is soluble only in DMF and DMSO among common organic solvents, the immobilization, Step 3, was carried out by using DMF or DMSO. To our surprise, the film was just formed on a TiO_2/FTO surface, only when DMF was used as the solvent.

Figure 6 shows the effect of the dipping time in Step 3 on the electrochemically initiated film deposition. The absorption arising from the Ru complex on the surface of TiO_2/FTO increased continuously with prolonging the dipping time. In sharp contrast, when a TiO_2/FTO electrode without any electrochemical treatment was directly dipped into a DMF solution of the Ru complex for 24 h, no absorption of the Ru complex was observed, indicating that the film was not formed at all on the TiO_2/FTO surface without electrochemical treatment.

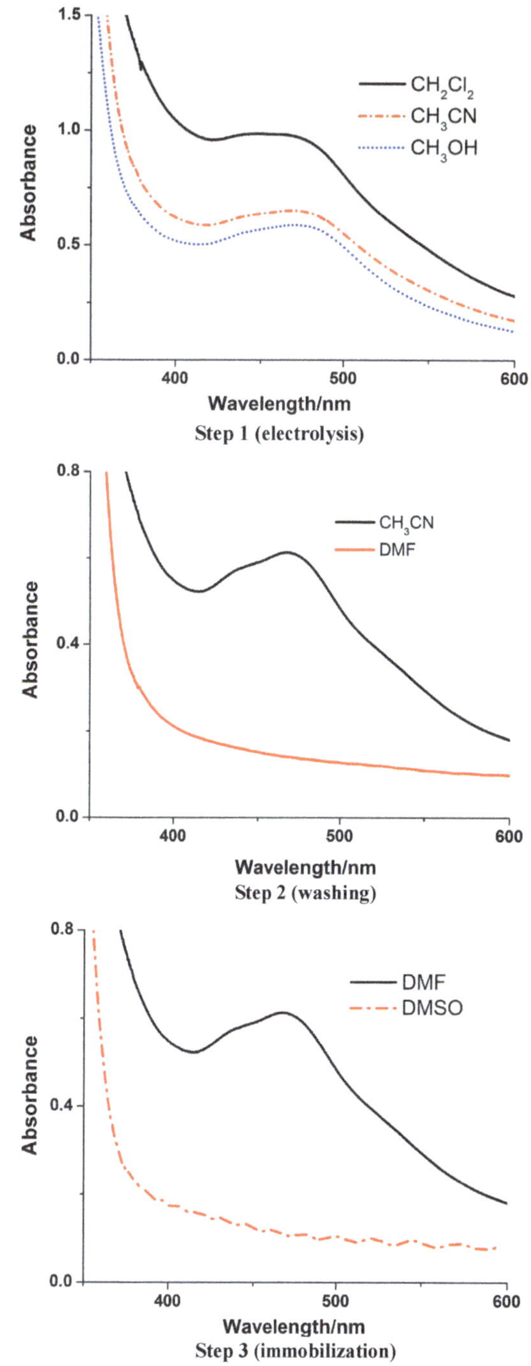

Figure 5. Effect of solvents on the formation of the Ru complex deposit.

3.2. Active Species Generated on the Surface of TiO_2/FTO by the Electrochemical Treatment

In order to know the active species generated on the TiO_2/FTO surface by the electrochemical treatment (Step 1), 2,2,6,6-tetramethylpiperidine-1-oxyl (TEMPO), a typical radical capture, was added to the washing solvent in Step 2 (acetonitrile); after the electrolysis in Step 1, the electrode was transferred into acetonitrile containing

TEMPO (50 mM) for 2 h in Step 2. Then, the electrode was immersed into a DMF solution of Ru complex for 24 h (Step 3). As shown in **Figure 7**, the deposition of the Ru complex was considerably inhibited by TEMPO. On the basis of this result, we supposed the formation of a radical species, which may react with the vinyl group in the Ru complex. The radical formation would be consistent with the fact that irreversible oxidation took place during the electrolysis of a TiO_2/FTO electrode. The real character of the active species on the TiO_2/FTO surface and the reaction mechanism for the deposition of the Ru complex are not clear at present; we will report elsewhere.

3.3. Incident Photon-to-Current Conversion Efficiency of the Ru Complex Film

Incident photon to current conversion efficiency (IPCE), which defines light to electricity conversion efficiency at a certain wavelength (λ), is one of important parameters for the evaluation of a dye-sensitized solar cell (DSSC) performance. The maximum IPCE of the Ru complex film, obtained by the present method, was disappointedly only 1.2% at 440 nm, under the standard AM 1.5 G irradiation conditions. The low efficiency would arise from a fast back electron transfer. In general, electron transport through films takes place in association with a successive electron transfer between neighboring redox centers [21]. After an electron transfer from an excited dye to a neighboring ground-state dye, two kinds of charged moieties, positively and negatively charged moieties, were formed. Strong electrostatic interaction between the two charged dyes can accelerate electron recombination and decrease the IPCE.

3.4. Improvement of Incident Photon-to-Current Conversion Efficiency by Using the Ru Complex/Sodium 4-vinylbenzensesulfonate Composite

Then, we conversely considered that IPCE would be improved when the electron recombination between the layers is blocked. On the basis of the consideration, we coexisted sodium 4-vinylbenzenesulfonate with the Ru complex in the immobilization (Step 3) with an expectation that the sodium cations and the sulfonate groups would partially stabilize electron-accepting dye moieties and electron-donating dye moieties, respectively. Composite films consisting of the Ru complex and sodium 4-vinylbenzenesulfonate were similarly deposited by using a DMF solution containing the Ru complex and sodium 4-vinylbenzenesulfonate (1:1 molar ratio) in the immobilization (Step 3). Although the UV-vis absorption of the composite film decreased with the addition of sodium 4-vinylbenzenesulfonate, the IPCE of the composite film was dramatically increased to show maximum IPCE of 31.7% at 438 nm under the standard AM 1.5 G irradiation conditions. The highly improved IPCE would arise from the electronic neutrality of the composite even after charge separation, which was secured by sodium 4-vinylbennzenesulfonate, that is, sodium cations would neutralize electron-accepting dyes while sulfonate groups would neutralize electron-donating dyes. Thus, the composite film could provide much higher IPCE by the inhibition of a back electron transfer.

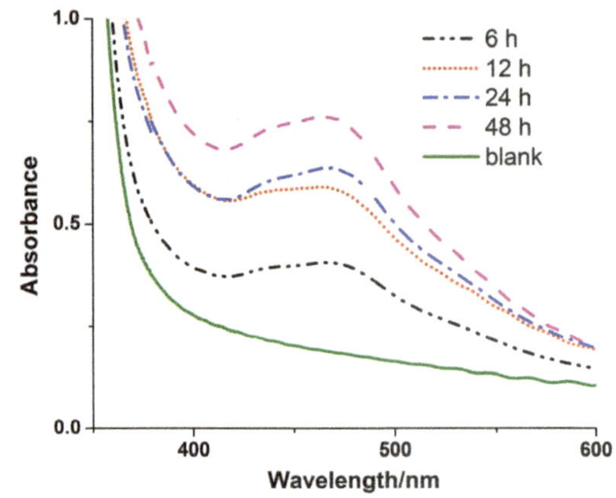

Figure 6. Dependency of the dipping time on the film deposition.

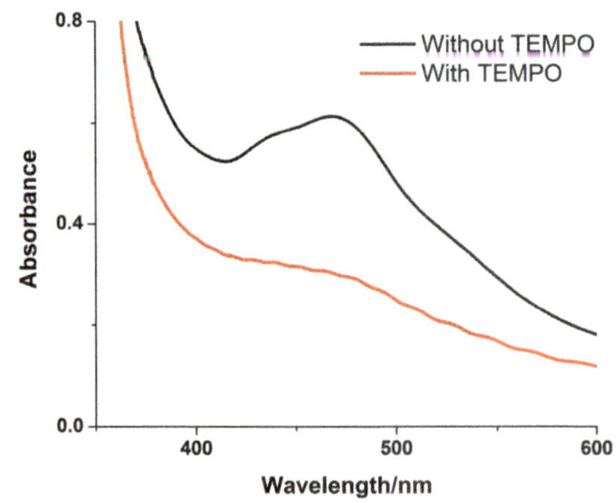

Figure 7. Effect of TEMPO on the deposition of the Ru complex.

4. Conclusion

A novel ruthenium complex di(isothiocyanato)bis(4-methyl-4'-vinyl-2,2'-bipyridine)ruthenium(II) $[(NCS)_2(mvbpy)_2Ru(II)]$ having vinyl groups was synthesized by a method similar to that of $(bpy)_2(NCS)_2Ru(II)$, in which 4-methyl-4'-vinyl-2,2'-bipyridine (mvbpy) was used in place of 2,2'-bipyridine (bpy). The Ru complex could be immobilized on the surface of a TiO_2-coated, fluorine-doped tin oxide (TiO_2/FTO) electrode by a newly developed, electrochemically induced film formation method, which consists of three steps: the electrolysis of an electrode, the washing of the electrode, and the film deposition of the Ru complex upon immersing the electrode in a solution of the Ru complex. This is the first example of the immobilization of a Ru(II) complex without a functional group reactive with TiO_2, such as -COOH, $-PO(OH)_3$, -OH, and -SH. The Ru(II) complex film on TiO_2/FTO thus obtained gave maximum incident photon-to-current efficiency (IPCE) of 1.2%. In contrast, the composite film consisting of the Ru(II) complex and sodium 4-vinylbenzenesulfonate showed much higher maximum IPCE of 31.7% at 438 nm. The new method for the generation of active sites on the surface of a TiO_2/FTO electrode by electrolysis and the new strategy for the stabilization of charge separated dyes would bring about the development of highly efficient dye-sensitized solar cells.

5. Acknowledgements

We acknowledge Dr. Masataka Ohtani and Dr. Yasuhiro Ishida (RIKEN, Advanced Science Institute) for the MALDI-TOF MS measurement.

REFERENCES

[1] B. O'Regan and M. Gratzel, "A Low-Cost, High-Efficiency Solar Cell Based on Dye-Sensitized Colloidal TiO_2 Films," *Nature*, Vol. 353, No. 6346, 1991, pp. 737-740.

[2] K. C. D. Robson, B. D. Koivisto, A. Yella, B. Sporinova, M. K. Nazeeruddin, T. Baumgartner, M. Grätzel and C. P. Berlinguette, "Design and Development of Functionalized Cyclometalated Ruthenium Chromophores for Light-Harvesting Applications," *Inorganic Chemistry*, Vol. 50, No. 12, 2011, pp. 5494-5508.

[3] A. Yella, H.-W. Lee, H. N. Tsao, C. Yi, A. K. Chandiran, M. K. Nazeeruddin, E. W.-G. Diau, C.-Y. Yeh, S. M. Zakeeruddin and M. Grätzel, "Porphyrin-Sensitized Solar Cells with Cobalt (II/III)-Based Redox Electrolyte Exceed 12 Percent Efficiency," *Science*, Vol. 334, No. 6056, 2011, pp. 629-634.

[4] R. Argazzi, N. Y. Murakami Iha, H. Zabri, F. Odobel and C. A. Bignozzi, "Design of Molecular Dyes for Application in Photoelectrochemical and Electrochromic Devices Based on Nanocrystalline Metal Oxide Semiconductors," *Coordination Chemistry Reviews*, Vol. 248, No. 13-14, 2004, pp. 1299-1316.

[5] N. Koumura, Z.-S. Wang, S. Mori, M. Miyashita, E. Suzuki and K. Hara, "Alkyl-Functionalized Organic Dyes for Efficient Molecular Photovoltaics," *Journal of the American Chemical Society*, Vol. 128, No. 44, 2006, pp. 14256-14257.

[6] E. Bae and W. Choi, "Effect of the Anchoring Group (Carboxylate vs Phosphonate) in Ru-Complex-Sensitized TiO_2 on Hydrogen Production under Visible Light," *The Journal of Physical Chemistry B*, Vol. 110, No. 30, 2006, pp. 14792-14799.

[7] P. Liska, N. Vlachopoulos, M. K. Nazeeruddin, P. Comte, M. Graetzel, "cis-Diaquabis(2,2'-bipyridyl-4,4'-dicarboxylate)ruthenium(II) Sensitizes Wide Band Gap Oxide Semiconductors very Efficiently over a Broad Spectral Range in the Visible," *Journal of the American Chemical Society*, Vol. 110, No. 11, 1988, pp. 3686-3687.

[8] K. Hanson, M. K. Brennaman, H. Luo, C. R. Glasson, J. J. Concepcion, W. Song and T. J. Meyer, "Photostability of Phosphonate-Derivatized, Ru(II) Polypyridyl Complexes on Metal Oxide Surfaces," *ACS Applied Materials & Interfaces*, Vol. 4, No. 3, 2012, pp. 1462-1469.

[9] T. Dittrich, B. Neumann and H. Tributsch, "Sensitization via Reversibly Inducible $Ru(dcbpyH_2)_2(NCS)_2$-TiO_2 Charge-Transfer Complex," *The Journal of Physical Chemistry C*, Vol. 111, No. 5, 2007, pp. 2265-2269.

[10] G. Li, P. G. Bomben, K. C. D. Robson, S. I. Gorelsky, C. P. Berlinguette and M. Shatruk, "Ru Complexes of Thienyl-Functionalized Dipyrrins as NCS-Free Sensitizers for the Dye-Sensitized Solar Cell," *Chemical Communications*, Vol. 48, No. 70, 2012, pp. 8790-8792.

[11] P. H. Xie, Y. J. Hou, T. X. Wei, B. W. Zhang, Y. Cao and C. H. Huang, "Synthesis and Photoelectric Studies of Ru(II) Polypyridyl Sensitizers," *Inorganica Chimica Acta*, Vol. 308, No. 1, 2000, pp. 73-79.

[12] M. K. Nazeeruddin, A. Kay, I. Rodicio, R. Humphry-Baker, E. Mueller, P. Liska, N. Vlachopoulos and M. Graetzel, "Conversion of Light to Electricity by cis-X2bis(2,2'-bipyridyl-4,4'-dicarboxylate)ruthenium(II) Charge-Transfer Sensitizers (X = Cl-, Br-, I-, CN-, and SCN-) on Nanocrystalline Titanium Dioxide Electrodes," *Journal of the American Chemical Society*, Vol. 115, No. 14, 1993, pp. 6382-6390.

[13] K. C. D. Robson, B. D. Koivisto, A. Yella, B. Sporinova, M. K. Nazeeruddin, T. Baumgartner, M. Grätzel and C. P. Berlinguette, "Design and Development of Functionalized Cyclometalated Ruthenium Chromophores for Light-Harvesting Applications," *Inorganic Chemistry*, Vol. 50, No. 12, 2011, pp. 5494-5508.

[14] M. Grätzel, "Dye-Sensitized Solar Cells," *Journal of Photochemistry and Photobiology C: Photochemistry Reviews*, Vol. 4, No. 2, 2003, pp. 145-153.

[15] Y.-J. Hou, P.-H. Xie, B.-W. Zhang, Y. Cao, X.-R. Xiao and W.-B. Wang, "Influence of the Attaching Group and Substituted Position in the Photosensitization Behavior of Ruthenium Polypyridyl Complexes," *Inorganic Chemistry*, Vol. 38, No. 26, 1999, pp. 6320-6322.

[16] H.-J. Park, K. H. Kim, S. Y. Choi, H.-M. Kim, W. I. Lee, Y. K. Kang and Y. K. Chung, "Unsymmetric Ru(II) Complexes with N-Heterocyclic Carbene and/or Terpyridine Ligands: Synthesis, Characterization, Ground- and Excited-State Electronic Structures and Their Application for DSSC Sensitizers," *Inorganic Chemistry*, Vol. 49, No. 16, 2010, pp. 7340-7352.

[17] M. Furue, T. Yoshidzumi, S. Kinoshita, T. Kushida, S.-I. Nozakura and M. Kamachi, "Intramolecular Energy Transfer in Covalently Linked Polypyridine Ruthenium(II)/Osmium(II) Binuclear Complexes. Ru(II)$(bpy)_2$Mebpy-$(CH_2)_n$-MebpyOs(II)$(bpy)_2$(n=2, 3, 5, and 7)," *Bulletin of the Chemical Society of Japan*, Vol. 64, No. 5, 1991, pp. 1632-1640.

[18] R. H. Terbrueggen, T. W. Johann and J. K. Barton, Functionalized Rhodium Intercalators for DNA Recognition," *Inorganic Chemistry*, Vol. 37, No. 26, 1998, pp. 6874-6883.

[19] K. Sumi, M. Furue and S.-I. Nozakura, (1985) "Efficiency of Polymer Sensitizer in Photosensitized Reaction of Tris(bipyridine)ruthenium(II) Complex-Containing Polymer," *Journal of Polymer Science*: *Polymer Chemistry Edition*, Vol. 23, No. 12, pp. 3059-3067.

[20] Z. Yu, H. M. Najafabadi, Y. Xu, K. Nonomura, L. Sun and L. Kloo, "Ruthenium Sensitizer with a Thienylvinyl-bipyridyl Ligand for Dye-Sensitized Solar Cells," *Dalton Transactions*, Vol. 40, No. 33, 2011, pp. 8361-8366.

[21] V. Aranyos, J. Hjelm, A. Hagfeldt and H. Grennberg, "Electropolymerisable Bipyridine Ruthenium(II) Complexes. Synthesis and Electrochemical Characterisation of 4-(3-methoxystyryl)- and 4,4[prime or minute]-di(3-methoxystyryl)-2,2[prime or minute]-bipyridine Ruthenium Complexes," *Journal of the Chemical Society, Dalton Transactions*, No. 8, 2001, pp. 1319-1325.

4

Glycerol-Based Carbon-SO_3H Catalyzed Benign Synthetic Protocol for the Acetylation of Alcohols, Phenols and Amines under Solvent-Free Conditions

Katakam N. Gangadhar, Manneganti Vijay, Rachapudi Badari Narayana Prasad, Bhetala Lakshmi Anu Prabhavathi Devi*

Centre for Lipid Research, CSIR-Indian Institute of Chemical Technology, Hyderabad, India

ABSTRACT

A simple and efficient solvent-free method was developed for the acetylation of alcohols, phenols and amines in excellent yields employing glycerol-based sulfonic acid (SO_3H) functionalized carbon catalyst under environmentally benign reaction conditions. The salient features of this protocol are the short reaction time, ease of product isolation and reusability of the carbon catalyst.

Keywords: Acetylation; Glycerol-Based Carbon-SO_3H Catalyst; Solvent-Free Method; Alcohols; Phenols; Amines

1. Introduction

In recent years, the search for environmentally benign chemical processes or methodologies has received much attention [1-4]. Heterogenization of homogeneous catalysts has been an interesting area of research from an industrial point of view; this combines the advantages of homogeneous catalysts (high activity, selectivity, etc.) with the engineering advantages of heterogeneous catalysts (easy catalyst separation, long catalytic life, easy catalyst regeneration, thermal stability, and recyclability) [5-9]. Application of solid acids in organic transformations is important, because solid acids have many advantages such as simplicity in handling, decreased reactor and plant corrosion problems, and more environmentally safe disposal [5-13]. Protection of functional groups plays an important role in the synthesis of complex organic molecules such as natural products. The protection of alcohols, phenols and amines is an important and frequently utilized transformation in synthetic organic chemistry [14]. Among the various protecting groups used for the hydroxyl and amine groups, acetyl is one of the most common groups due to its facile introduction and its stability in the acid reaction conditions and also easy deprotection using mild alkaline hydrolysis [15]. The most commonly used reagent combinations for this reaction are acetic anhydride or acetyl chloride in the presence of Et_3N and pyridine as catalysts [16-18]. 4-(Dimethylamino) pyridine (DMAP) is known to cause remarkable rate acceleration in the reaction [19]. A variety of catalysts such as silica gel supported sulphuric acid [20,21], Ammonium acetate in acetic acid [22], Zinc dust [23], $ZnCl_2$ [24], $CoCl_2$ [25], $Sc(OTf)_3$ [26], $TaCl_5$ [27], Montmorillonit K10 [28], HY Zeolite [29], $In(OTf)_3$ [30], $Cu(OTf)_2$ [31], Yittria-Zirconia based Lewis acid [32], $InCl_3$/Mont. K 10 [33], Manganese (III) bis(2-hydroxyanil) acetylacetonato complex [34], Silica sulfate [35], p-$MeC_6H_4SO_2NBr_2$ [36], DBDMH or TCCA [37], $H_6P_2W_{18}O_{62}\cdot 24H_2O$ [38], $La(NO_3)_3\cdot 6H_2O$ [39], Ionic liquid based on morpholin [40], Borated zirconia [41], 2,4,6-Triacyloxy-l,3,5-triazine (TAT) [42], Sodium dodecyl sulfate (SDS) [43] and DMAP. saccharin [44] are also known to catalyze the acetylation of alcohols, phenols and amines. However, most of these methods suffer from at least one of the following disadvantages like vigorous reaction conditions, high cost and toxicity of the reagent, tedious work-up procedures, unsatisfactory yields, and instability and hygroscopic nature of the reagent.

Carbon-based solid acid catalysts [45-48] have gained importance due to their significant advantages over homogeneous liquid phase mineral acids, such as increased activity and selectivity, longer catalyst life, negligible equipment corrosion, ease of product separation, and reusability. We, recently developed a simple and fast method for the preparation of a similar sulphonic acid

*Corresponding author.

functionalized polycyclic aromatic carbon catalyst from bioglycerol (biodiesel by-product) and also from glycerol-pitch (waste from fat splitting industry) by *in situ* partial carbonization and sulfonation [49,50]. Such catalysts have been shown to be inexpensive, highly stable, robust, recyclable, and easily produced from naturally available bioglycerol, and are demonstrated to be effective for the esterification of fatty acids to its methyl esters [49], THP protection and deprotection of alcohols and phenols [50], and also for the synthesis of highly substituted imidazoles [51], 3,4-dihydropyrimidin-2-(1H)-ones [52], amides from aldehydes [53], and spirooxindole derivatives [54]. In continuation of our efforts towards exploring the applications of the glycerol-based carbon catalyst, here we report a simple and an efficient methodology for the acetylation of alcohols, phenols and amines with acetic anhydride at 65°C under solvent-free conditions in excellent yields (**Scheme 1**).

2. Experimental Details

All chemicals and reagents were procured from suppliers and used without further purification. The isolated products were characterized by chromatographic and spectral studies (GC, GC-MS, FT-IR and ^{1}H NMR). The spectra were compared with those of standard esters. The NMR spectrums of product were obtained using Bruker AC-300 MHz spectrometer with TMS as the internal standard.

2.1. General Experimental Procedure for the Preparation of Glycerol-Based Carbon-SO$_3$H Catalyst

Carbon-SO$_3$H catalyst was prepared as reported [49] by heating a mixture of glycerol (10 g) and concentrated sulfuric acid (30 g) from ambient temperature to 210°C - 220°C for 20 min, to facilitate *in situ* partial carbonization and sulfonation. The reaction mixture was allowed to remain at that temperature for about 5 min (until foaming ceased) to obtain the carbon material. The solid material was cooled to ambient temperature and washed with hot water under agitation until the wash water showed a neutral pH value. The partially crystalline product was filtered and dried in an oven at 120°C for 2 h until it was moisture free to obtain the glycerol-based carbon acid catalyst (4.67 g).

2.2. General Experimental Procedure for the Catalytic Acetylation of Alcohols, Phenols and Amines

To a stirred mixture of the alcohol/amine (1 mmol) and Ac$_2$O (2 mmol), carbon catalyst (15 wt% of substrate) was added and stirring continued at 65°C for 30 to 120

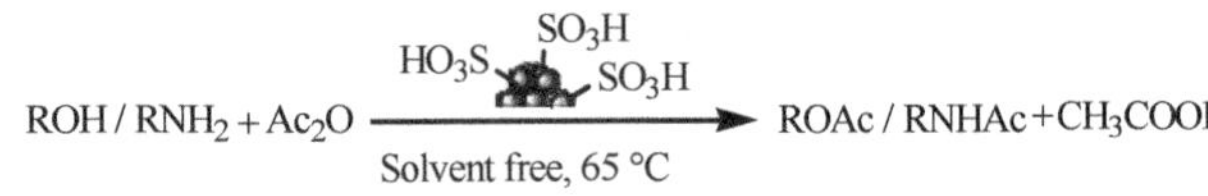

Scheme 1. Acetylation of alcohols, phenols and amines employing carbon-SO$_3$H catalyst.

min. The progress of the reaction was monitored by TLC. After completion of the reaction, ethyl acetate (3 × 5 mL) was added to the reaction mixture and the catalyst was separated by filtration. The organic phase was washed with saturated $NaHCO_3$ solution (15 mL), dried over anhydrous Na_2SO_4 and concentrated under reduced pressure in a rotary evaporator to afford the crude product. The final product was purified by silica gel column chromatography using hexane/ethyl acetate as eluting solvents. The yield was calculated as mmol of purified product with respect to mmol of initial alcohol/amine. The reaction times and yields of the products are presented in **Tables 1** and **2**. All the purified products were characterized by GC, IR and ^{1}H NMR studies and the data is in comparison with authentic samples.

1-Octyl acetate (Entry 1, **Table 1**), ^{1}H NMR (300 MHz, $CDCl_3$) δ 0.88 (t, 3H), 1.20-1.43 (m, 10H), 1-65-1.78 (m, 2H), 2.04 (s, 3H), 4.05 (t, 2H). IR (cm^{-1}): 2928, 2858, 1742, 1238 1039. GC-MS *m/z*: 172 [M]$^+$.

1-Decyl acetate (Entry 2, **Table 1**), ^{1}H NMR (300 MHz, $CDCl_3$) δ 0.88 (t, 3H), 1.20 -1.45 (m, 30H), 2.0 (m, 2H), 2.04 (s, 3H), 4.05 (t, 2H). IR (cm^{-1}): 2925, 2854, 1741, 1239. GC-MS *m/z*: 200[M]$^+$.

1-Octadecyl acetate (Entry 3, **Table 1**), ^{1}H NMR (300 MHz, $CDCl_3$) δ 0.89 (t, 3H), 1.20 -1.45 (m, 30H), 2.0 (m, 2H), 2.04 (s, 3H), 4.05 (t, 2H). IR (cm^{-1}): 2928, 2858, 1742, 1238, 1039. GC-MS *m/z*: 312 [M]$^+$.

2-Ethyl-1-hexyl acetate (Entry 4, **Table 1**), ^{1}H NMR (300 MHz, $CDCl_3$) δ 0.9 (t, 6H), 1.2-1.4 (m, 7H), 2.1 (s, 3H); 3.9 (t, 2H). IR (cm^{-1}): 3020, 2962, 1726, 1215, 763. GC-MS *m/z* 172 [M]$^+$.

Cyclohexyl acetate (Entry 5, **Table 1**), ^{1}H NMR (300 MHz, $CDCl_3$) δ 1.25-1.85 (m, 10H), 2.00 (s, 3H), 4.65 (m, 2H). IR (cm^{-1}): 3020, 2940, 1721, 1256, 1215, 756. GC-MS *m/z:* 142 [M]$^+$.

2-Ethoxy ethyl acetate (Entry 6, **Table 1**), ^{1}H NMR (300 MHz, $CDCl_3$) δ 1.23 (t, 3H), 2.04 (s, 3H), 3.40 (m, 2H), 3.58 (t, 2H), 4.51 (t, 2H). IR (cm^{-1}): 3020, 2962, 1726, 1215, 763. GC-MS *m/z*: 132 [M]$^+$.

Acetylated sun flower Sterols (Entry 7, **Table 1**), ^{1}H NMR (300 MHz, $CDCl_3$): δ 0.70 (s, 3H), 0.88 (d, 6H), 0.94 (d, 3H), 1.04 (s, 3H), 2.05 (s, 3H), 4.64 (m, 1H), 5.39 (d, 1H). IR (cm^{-1}): 2940, 1730, 1262, 1038. GC-MS *m/z*: 396 [M]$^+$.

Benzyl acetate (Entry 8, **Table 1**), ^{1}H NMR (300 MHz, $CDCl_3$) δ 2.07 (s, 3H), 5.05 (s, 2H), 7.26 -7.36 (m, 5H). IR (cm^{-1}): 3020, 1733, 1216, 756. GC-MS *m/z*: 150 [M]$^+$.

Phenyl acetate (Entry 9, **Table 1**), ^{1}H NMR (300 MHz, $CDCl_3$) δ 2.21 (s, 3H), 7.0-7.4 (m, 5H). IR (cm^{-1}): 1763, 1193, 748. GC-MS *m/z*: 136 $[M]^+$.

4-Methoxyphenyl acetate (Entry 10, **Table 1**), ^{1}H NMR (300 MHz, $CDCl_3$) δ 2.1 (s, 3H), 3.8 (s, 3H), 6.70-7.0 (m, 4H). IR (cm^{-1}): 3019, 2960, 1758, 1506, 1217, 1193, 755. GC-MS *m/z*: 166 $[M]^+$.

4-Methylphenyl acetate (Entry 11, **Table 1**), ^{1}H NMR (300 MHz, $CDCl_3$) δ 2.01 (s, 3H), 2.25 (s, 3H), 6.92 (d, 2H), 7.04 (d, 2H). IR (cm^{-1}): 3027, 2953, 1738, 1160, 754. GC-MS *m/z*: 150 $[M]^+$.

1-Naphthyl acetate (Entry 12, **Table 1**), ^{1}H NMR (300 MHz, $CDCl_3$) δ 2.4 (s, 3H), 7.2-7.9 (m, 7H). IR (cm^{-1}): 3061, 2924, 1767, 1368, 1200, 773. GC-MS *m/z*: 186 $[M]^+$.

4-Chlorophenyl acetate (Entry 13, **Table 1**), ^{1}H NMR (300 MHz, $CDCl_3$) δ 1.98 (s, 3H), 7.0 (d, 2H), 7.25 (d, 2H); IR (cm^{-1}): 1763, 1216, 1198, 756; GC-MS *m/z*: 170 $[M]^+$.

2-Chlorophenyl acetate (Entry 14, **Table 1**), ^{1}H NMR (300 MHz, $CDCl_3$) δ 2.2 (s, 3H), 7.0 (d, 2H), 7.5 (d, 2H). IR (cm^{-1}): 1763, 1216, 756. GC-MS *m/z*: 170 $[M]^+$.

N-phenylacetamide (Entry 1, **Table 2**), ^{1}H NMR (300 MHz, $CDCl_3$) δ 2.19 (s, 3H), 7.16-7.11 (m, 1H), 7.28 (broad s, 1H), 7.36-7.30 (m, 2H), 7.55-7.52 (m, 2H). IR (cm^{-1}): 3293, 1662, 1598, 1557, 1500, 1431, 1368, 1325, 1262, 1040, 1012, 962, 906, 750. GC-MS *m/z*: 135 $[M]^+$.

N-(2-hydroxyphenyl)acetamide (Entry 2, **Table 2**), ^{1}H NMR (300 MHz, $CDCl_3$) δ 2.24 (s, 3H), 6.90 (m, 1H), 7.02 (m, 2H), 7.42 (m, 1H), 8.80(s, 1H),9.11 (bs, 1H). IR (cm^{-1}): 3403, 3150, 1658, 1587, 1539, 1446, 1397, 1287, 1103, 1037, 891, 767. GC-MS *m/z*: 151$[M]^+$.

N-(3-hydroxyphenyl)acetamide (Entry 3, **Table 2**), ^{1}H NMR (300 MHz, $CDCl_3$) δ 2.10 (s, 3H), 6.54 (d, 1H), 6.88 (d, 1H), 7.05 (t, 1H), 7.30 (s, 1H), 8.97(s, 1H), 9.30 (bs, 1H). IR (cm^{-1}): 3300, 3100, 1653, 1600, 1571, 1520, 1441, 1380, 1254, 1110, 1050, 820, 760, 720. GC-MS *m/z*: 151$[M]^+$.

N-(4-hydroxyphenyl)acetamide (Entry 4, **Table 2**), ^{1}H NMR (300 MHz, $CDCl_3$) δ 2.03 (s, 3H), 7.00 (d, 1H), 7.50 (d, 1H), 8.97(s, 1H), 9.40 (bs, 1H). IR (cm^{-1}): 3326, 3164, 1652, 1611, 1565, 1507, 1442, 1372, 1327, 1259, 1228, 1173, 1108, 969, 837, 808, 715, 687. GC-MS *m/z*: 151$[M]^+$.

N-benzylacetamide (Entry 5, **Table 2**), ^{1}H NMR (300 MHz, $CDCl_3$) δ 7.36-7.29 (m, 5H), 6.06 (bs, 1H), 4.43 (d, 2H), 2.03 (s, 3H). IR (cm^{-1}): 3294, 1646, 1548, 1500, 1283. GC-MS *m/z*: 149 $[M]^+$.

N-(pyridin-2-yl)acetamide (Entry 6, **Table 2**), ^{1}H NMR (300 MHz, $CDCl_3$) δ 2.19 (s, 3H), 7.03 (m,1H), 7.60 (t, 1H), 8.18 (d, 1H), 8.26 (d, 1H), 8. 50 (bs, 1H). IR (cm^{-1}): 3388, 2923, 1685, 1434, 1303, 744. GC-MS *m/z*: 137 $[M]^+$.

3. Results & Discussion

Acetylation of 1-octanol (1 mmol) with acetic anhydride (2 mmol) in presence of 15 wt% of carbon-SO_3H catalyst was studied at 65°C to afford 1-octyl acetate in 96% yield in 30 min. However, the acetylation of 1-octanol with acetic anhydride at room temperature resulted <50% acetylated product even after 4 h. After the successful acetylation of 1-octanol with excellent yield to octyl acetate, the effect of catalyst loading for the acetylation of 1-octanol with acetic anhydride (1:2 mmol) to octyl acetate was studied by varying the catalyst dosage from 2 to 20wt% of alcohol (**Table 3**). The reaction was found to be slow at room temperature and hence the reaction temperature was increased to 65°C. At this temperature, with the increase of the catalyst loading from 2 to 15 wt% of alcohol decreased the reaction time substantially from 120 min to 30 min with conversions ranging from 65% to 96% (entries 1 - 4, **Table 3**). However, further increase of the catalyst loading to 20 wt% decreased the reaction time marginally from 30 min to 25 min with 98% conversion. The reactions were monitored by GC

Table 1. Acetylation of alcohols, phenols by employing carbon-SO_3H catalyst[a].

Entry	Alcohol	Product	Time (min)	Yield (%)[b]
1	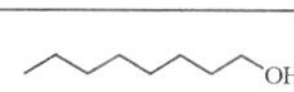	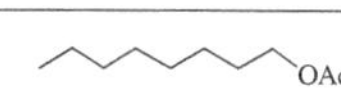	30	96
2	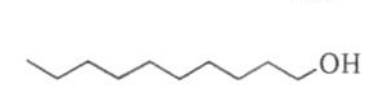	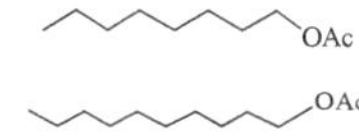	30	96
3	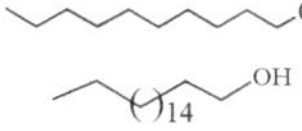	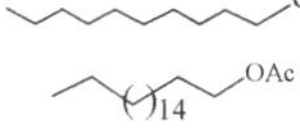	30	95
4	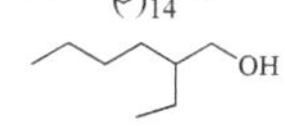	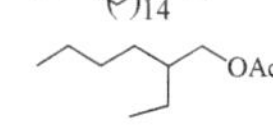	30	95
5	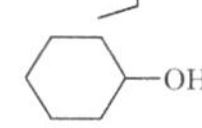	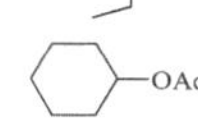	30	96
6	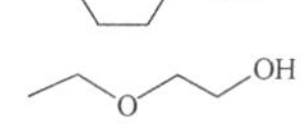	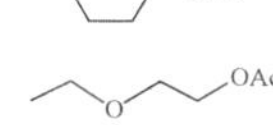	45	95
7	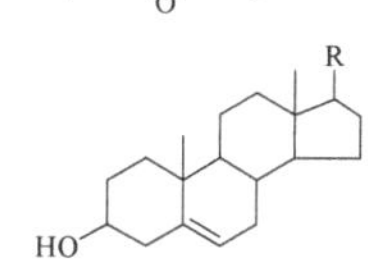	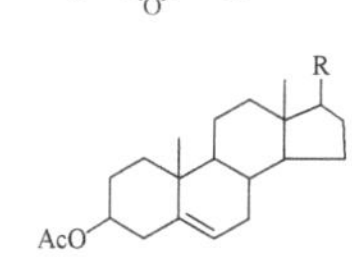	60	95
8	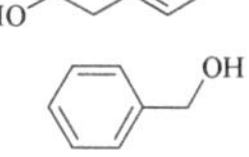	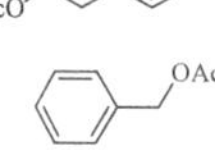	45	92
9	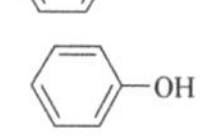	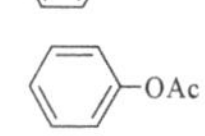	30	91
10	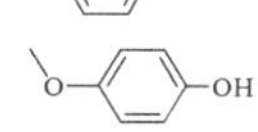	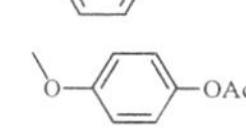	60	93
11	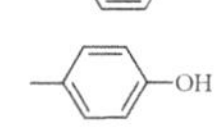	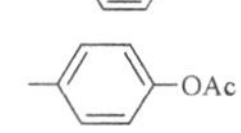	45	92
12	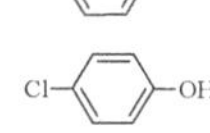	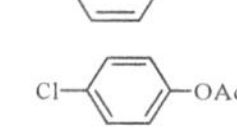	120	92
13	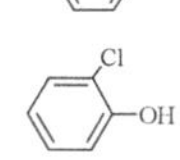	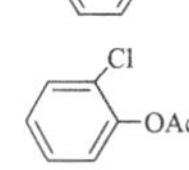	120	75
14	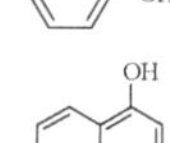	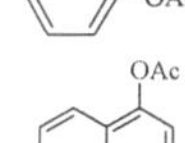	45	91

[a]Reaction conditions: Alcohol (1 mmol), Ac_2O (2 mmol), catalyst (15 wt%), 65°C; [b]GC yield.

Table 2. Acetylation of aromatic amines by employing carbon-SO_3H catalyst[a].

Entry	Amine	Product	Time (min)	Yield (%)[b]
1	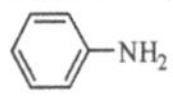	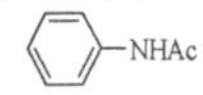	30	97
2	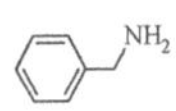	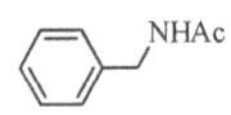	30	95
3	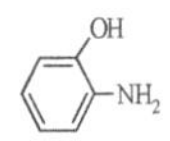	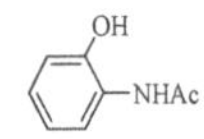	30	95
4	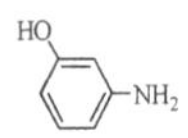	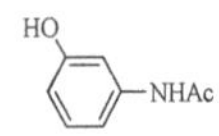	30	96
5	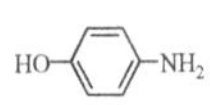	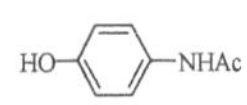	30	95
6	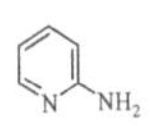	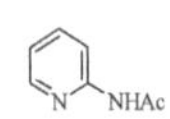	30	92

[a]Reaction conditions: Amine (1 mmol), Ac_2O (2 mmol), catalyst (15 wt%); 65°C; [b]GC yield.

Table 3. Effect of carbon-SO_3H catalyst loading on the acetylation of 1-octanol[a].

Entry	Catalyst loading (wt% of alcohol)	Time (min)	Yield (%)[b]
1	2	120	65
2	5	90	78
3	10	60	90
4	15	30	96
5	20	25	98

[a]Reaction conditions: 1-Octanol (1 mmol), Ac_2O (2 mmol), 65°C; [b]GC yield.

using HP-1 capillary column. Based on this study, 15 wt% of the carbon catalyst was found to be optimum for the acetylation of 1-octanol with 96% conversion at 65°C in 30 min.

In order to establish the effectiveness and the acceptability of the method in a wider context of synthetic organic chemistry, acetylation of various alcohols and phenols with electron donating and electron withdrawing groups were studied in presence of acetic anhydride under optimum conditions and the results are given in **Table 1**. The study revealed that primary and secondary alcohols (entries 1-5, **Table 1**) are acetylated with similar yields (95% - 96%) within 30 min, where as acetylation of ethoxy alcohol (entry 6, **Table 1**) and phytosterols (entry 7, **Table 1**) resulted the corresponding acetylated products in 95% yields in 45 and 60 min respectively. Acetylation of phenols was found to be slow when compared to aliphatic alcohols. Phenols with electron donating and electron withdrawing groups (entries 9 - 12 and 14, **Table 1**) were acetylated in 91% - 93% yields within 45 - 60 min and similar observation was made in case of benzyl alcohol also (entry 8, **Table 1**). However acetylation of 2-chlorophenol resulted 2-chlorophenyl acetate (75%, entry 13, **Table 1**) in less yields when compared to 4-chlorophenyl acetate (92% yield, entry 12, **Table 1**) due to steric effect.

In a similar fashion acetylation of aromatic amines also executed, and resulted the corresponding acylated products in good yields (92% - 97%, entries 1 - 6, **Table 2**) in 30 min. To examine the chemo selectivity of the present method, bi-functional substrates containing $-NH_2$ and -OH groups were studied (entries 3-5, **Table 2**). Selective acetylation of the $-NH_2$ group in the presence of the -OH group was observed at room temperature with 2 equivalent of acetic anhydride to give corresponding N-acetate product, and no O-acetate product was observed under these conditions. This might be due to more nucleophilicity of amines than phenols. The present protocol is excellent for the acetylation of alcohols (primary and secondary), phenols, amines, and bi-functional compounds containing $-NH_2$ and -OH groups.

To check the reusability of the carbon catalyst, benzyl alcohol was employed for the acetylation for five cycles under the optimum conditions (**Figure 1**). After each cycle the catalyst was recovered by filtration, washed with methanol, dried in oven at 120°C for 1h and reused and the yields were found to be reduced marginally from 92 to 89% after 5th cycle.

4. Conclusion

In conclusion, we have demonstrated a simple, efficient and eco-friendly protocol for the solvent-free acetylation of alcohols, phenols and amines with acetic anhydride employing a novel glycerol-based $-SO_3H$ functionalized carbon as a solid acid catalyst. The low cost and simple preparation of the catalyst, and the easy procedure and work-up indicate that this catalyst is very attractive for this type of reactions. The yields are very good, and in addition, the carbon-based solid acid catalyst can be recovered by simple filtration for reuse without any pretreatment. The carbon-based solid acid can be used in place of sulfuric acid in the synthesis of organic compounds.

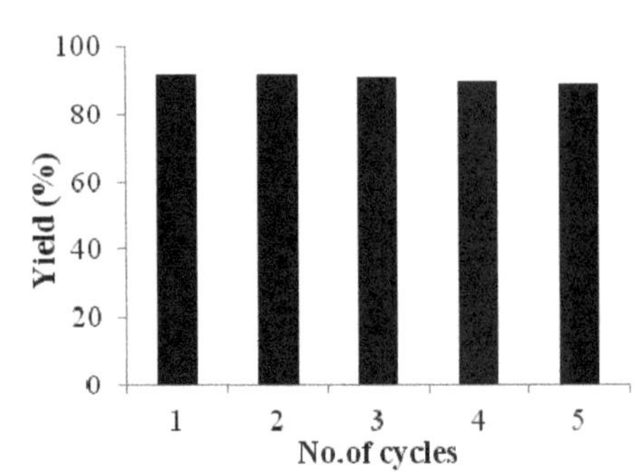

Figure 1. Recycling study of the carbon-SO_3H catalyst for the acetylation of benzyl alcohol with acetic anhydride.

5. Acknowledgments

Katakam N. Gangadhar and Manneganti Vijay thanks to CSIR New Delhi for SRF.

REFERENCES

[1] P. A. Grieco, "Organic Synthesis in Water," Blackie Academic and Professional, London, 1998.

[2] C. J. Li and T. H. Chan, "Organic Reactions in Aqueous Media," John Wiley and Sons, New York, 1997.

[3] U. M. Lindstorm, "Stereoselective Organic Reactions in Water," *Chemical Reviews*, Vol. 102, No. 8, 2002, pp. 2751-2772.

[4] S. Kobayashi and K. Manabe, "Development of Novel Lewis Acid Catalysts for Selective Organic Reactions in Aqueous Media," *Accounts of Chemical Research*, Vol. 35, No. 4, 2002, pp. 209-217.

[5] A. E. C. Collis and I. T. Horvàth, "Heterogenization of Homogeneous Catalytic Systems," *Catalysis Science & Technology*, Vol. 1, No. 6, 2011, pp. 912-919.

[6] D. Choudhary, S. Paul, R. Gupta and J. H. Clark, "Catalytic Properties of Several Palladium Complexes Covalently Anchored onto Silica for the Aerobic Oxidation of Alcohols," *Green Chemistry*, Vol. 8, No. 5, 2006, pp. 479-482.

[7] Z. Li, X. Ma, J. Liu, X. Feng, G. Tian and A. Zhu, "Silica-Supported Aluminum Chloride: A Recyclable and Reusable Catalyst for One-Pot Three-Component Mannich-Type Reactions," *Journal of Molecular Catalysis A: Chemical*, Vol. 272, No. 1-2, 2007, pp. 132-135.

[8] B. Karimi and M. Ghoreishi-Nezhad, "Highly Chemiselective Acetalization of Carbonyl Compounds Catalyzed by a Novel Recyclable Ammonium Triflate Functionalized Silica," *Journal of Molecular Catalysis A: Chemcal*, Vol. 277, No. 1-2, 2007, pp. 262-265.

[9] J. A. Melero, R. G. Van Grieken and G. Morales, "Advances in the Synthesis and Catalytic Applications of Organosulfonic Functionalized Mesostructured Materials," *Chemical Reviews*, Vol. 106, No. 9, 2006, pp. 3790-3812.

[10] K. Niknam, B. Karami and M. A. Zolfigol, "Silica Sulfuric Acid Promoted Aromatization of 1,2-Dihydroquinolines by Using $NaNO_2$ as Oxidizing Agent under Mild and Heterogeneous Conditions," *Catalysis Communication*, Vol. 8, No. 9, 2007, pp. 1427-1430.

[11] K. Niknam, D. Saberi and M. Nouri Sefat, "Silica-Bonded S-Sulfonic Acid as a Recyclable Catalyst for Chemo Selective Synthesis of 1,1-Diacetates," *Tetrahedron Letters*, Vol. 50, No. 28, 2009, pp. 4058-4062.

[12] K. Niknam, D. Saberi and M. Mohagheghnejad, "Silica Bonded *S*-Sulfonic Acid: A Recyclable Catalyst for the Synthesis of Quinoxalines at Room Temperature," *Molecules*, Vol. 14, 2009, pp. 1915-1926.

[13] F. Rashedian, D. Saberi and K. Niknam, "Silica-Bonded *N*-Propyl Sulfamic Acid: A Recyclable Catalyst for the Synthesis of 1,8-Dioxodecahydroacridines, 1,8-Dioxooctahydroxanthenes and Quinoxalines," *Journal of Chinese Chemical Society*, Vol. 57, 2010, pp. 998-1006.

[14] T. W. Green and P. G. M. Wuts, "Protective Groups in Organic Synthesis," 4th Edition, John Wiley and Sons, New York, 2007, p. 223.

[15] A. L. Pearson and W. J. Roush, "Reagents for Organic Synthesis: Activating Agents and Protecting Groups," John Wiley, Chichester, 1999, p. 9.

[16] W. Steglich and G. Hofle, "N,N-Dimethyl-l-pyridine Amine, A Very Effective Acylation Catalyst," *Angewandte Chemie International Edition England*, Vol. 8, No. 12, 1969, p. 981.

[17] D. Horton, "Organic Syntheses Collection," John Wiley and Sons, New York, Vol. 5, 1973, p. 1.

[18] R. I. Zhdanov and S. M. Zhenodarova, "Chemical Methods of Oligonucleotide Synthesis," *Synthesis*, Vol. 4, 1975, pp. 222-245.

[19] G. Hofle, W. Steglich and H. Vorbruggen, "4-Dialkylaminopyridines as Highly Active Acylation Catalysts," *Angewandte Chemie International Edition England*, Vol. 17, No. 8, 1978, pp. 569-583.

[20] M. G. Nascimento, S. P. Zanotto, M. Scremin and M. C. Rezende, "Carboxylic Acid Supported on Silica: A Smooth Acylation Agent for Alcohols," *Synthetic Communications*, Vol. 26, No. 14, 1996, pp. 2715-2721.

[21] F. Shirini, M. A. Zolfigol and K. Mohammadi, "Silica Sulfuric Acid as a Mild and Efficient Reagent for the Acetylation of Alcohols in Solution and under Solvent Free Conditions," *Bulletin of Korean Chemical Society*, Vol. 25, No. 2, 2004, pp. 325-327.

[22] H. S. Prasad, G. R. Srinivasa and D. C. Gowda, "Convenient, Cost-Effective, and Mild Method for the N-Acetylation of Anilines and Secondary Amines," *Synthetic Communications*, Vol. 35, No. 9, 2005, pp. 1189-1195.

[23] M. A. Pasha, M. B. M. Reddy and K. Manjula, "Zinc Dust: An Extremely Active and Reusable Catalyst in Acylation of Phenols, Thiophenol, Amines and Alcohols in a Solvent-Free System," *European Journal of Chemistry*, Vol. 1, No. 4, 2010, pp. 385-387.

[24] R. Gupta, V. Kumar, M. Gupta, S. Paul and R. Gupta, "Silica Supported Zinc Chloride Acetylation of Amines, Alcohols and Phenols," *Indian Journal of Chemistry*, Vol. 47B, 2008, pp. 1739-1743.

[25] J. Iqbal and R. R. Srivastava, "Cobalt (II) Chloride Catalyzed Acylation of Alcohols with Acetic Anhydride: Scope and Mechanism," *Journal of Organic Chemistry*, Vol. 57, No. 7, 1992, pp. 2001-2007.

[26] K. Ishihara, M. Kubota, H. Kurihara and H. Yamamoto,

"Scandium Trifluoromethanesulfonate as an Extremely Active Acylation Catalyst," *Journal of American Chemical Society*, Vol. 117, No. 15, 1995, pp. 4413-4414.

[27] S. Chandrasekhar, T. Ramachander and M. Takhi, "Acylation of Alcohols with Acetic Anhydride Catalyzed by $TaCl_5$: Some Implications in Kinetic Resolution," *Tetrahedron Letters*, Vol. 39, No. 20, 1998, pp. 3263-3266.

[28] T. Li and A. Li, "Montmorillonite Clay Catalysis. Part 10.1 K-10 and KSF-Catalysed Acylation of Alcohols, Phenols, Thiols and Amines: Scope and Limitation," *Journal of Chemical Society, Perkin Transactions* 1, No. 12, 1998, pp. 1913-1917.

[29] R. Ballini, G. Bosica, S. Carloni, L. Ciaralli, R. Maggi and G. Sartori, "Zeolite HSZ-360 as a New Reusable Catalyst for the Direct Acetylation of Alcohols and Phenols under Solvent less Conditions," *Tetrahedron Letters*, Vol. 39, No. 33, 1998, pp. 6049-6052.

[30] K. K. Chauhan, C. G. Frost, I. Love and D. Waite, "Indium Triflate: An Efficient Catalyst for Acylation Reactions," *Synlett*, No. 11, 1999, pp. 1743-1744.

[31] P. Saravanan and V. K. Singh, "An Efficient Method for Acylation Reactions," *Tetrahedron Letters*, Vol. 40, No. 13, 1999, pp. 2611-2614.

[32] P. Kumar, R. K. Pandey, M. S. Bodas, S. P. Dagade, M. K. Dongare and A. V. Ramaswamy, "Acylation of Alcohols, Thiols and Amines with Carboxylic Acids Catalyzed by Yttria-Zirconia-Based Lewis Acid," *Journal of Molecular Catalysis A*: *Chemical*, Vol. 181, No. 1-2, 2002, pp. 207-213.

[33] V. R. Choudhary, K. Y. Patil and S. K. Jana, "Acylation of Aromatic Alcohols and Phenols over $Incl_3$/Montmorillonite K-10 Catalysts," *Journal of Chemical Sciences*, Vol. 116, No. 3, 2004, pp. 175-177.

[34] M. S. Niasari, S. Hydarzadeh, A. Amiri and S. Salavati, "Manganese(III) bis(2-hydroxyanil) Acetyl Acetonato Complex as Effective Catalyst for Acylation of Alcohols, Amines and Phenols with Acetic Anhydride," *Journal of Molecular Catalysis A*: *Chemical*, Vol. 231, No. 1-2, 2005, pp. 191-195.

[35] Y. Zhao, L. B. Liu, Z. Chen, T. S. Li and T. S. Jin, "Rapid and Efficient Method for Acetylation of Alcohols and Phenols with Acetic Anhydride Catalyzed by Silica Sulfate," *Synthetic Communications*, Vol. 36, 2006, pp. 1221-1227.

[36] Z. Duan, Y. Gu and Y. Deng, "Neutral Ionic Liquid [Bmim] BF_4 Promoted Highly Selective Esterification of Tertiary Alcohols by Acetic Anhydride," *Journal of Mo- lecular Catalysis A*: *Chemical*, Vol. 246, No. 1-2, 2006, pp. 70-75.

[37] M. A. Zolfigol, A. Khazaei, A. G. Choghamarani, A. Rostami and M. Hajjami, "Acylation of Alcohols Catalyzed by Using 1,3-Dibromo-5,5-Dimethyl Hydentoin or Trichloroisocyanuric Acid," *Catalysis Communications*, Vol. 7, No. 6, 2006, pp. 399-402.

[38] G. P. Romanelli, D. Bennardi, J. Autino, G. Baronetti and H. Thomas, "A Simple and Mild Acylation of Alcohols, Phenols, Amines, and Thiols with a Reusable Heteropoly Acid Catalyst ($H_6P_2W_{18}O_{62}\cdot 24H_2O$)," *European Journal of Chemistry*, Vol. 5, 2008, pp. 641-647.

[39] B. Akhlaghinia and Z. P. Seyyedlary, "Lanthanum Nitrate hexahydrate ($La(NO_3)_3\cdot 6H_2O$), An Efficient and Highly Selective Catalyst for Solventless Acetylation of Alcohols and Phenols with Acetyl Chloride," *Journal of the Iranian Chemical Research*, Vol. 2, No. 1, 2009, pp. 31-38.

[40] C. Yue, Q. Liu, T. Yi and Y. Chen, "Acetylation of Alcohols and Phenols with Acetic Anhydride under Solvent-Free Conditions Using an Ionic Liquid Based on Morpholine as a Recoverable and Reusable Catalyst," *Monatshefte fur Chemie-Chemical Monthly*, Vol. 141, No. 9, 2010, pp. 975-978.

[41] L. Osiglio, G. Romanelli and M. Blanco, "Alcohol Acetylation with Acetic Acid Using Borated Zirconia as Catalyst," *Journal of Molecular Catalysis A*: *Chemical*, Vol. 316, No. 1-2, 2010, pp. 52-58.

[42] S. S. Gholap, G. B. Tambe and C. H. Gill, "Chemoselective Acylation of Amines, Thiols and Phenols Using 2,4,6-Triacyloxy-l,3,5-Triazine (TAT) as a New and Effective Reagent Under Mild Condition," *Journal of Indian Chemical Society*, Vol. 86, No. 2, 2009, pp. 179-182.

[43] S. Naik, G. Bhattacharjya, B. Talukdar, and B. K. Patel, "Chemoselective Acylation of Amines in Aqueous Media," *European Journal of Organic Chemistry*, Vol. 2004, No. 6, 2004, pp. 1254-1260.

[44] N. Lu, W. H. Chang, W. H. Tu and C. K. Li, "A Salt Made of 4-*N,N*-Dimethylaminopyridine (DMAP) and Saccharin as an Efficient Recyclable Acylation Catalyst: A New Bridge Between Heterogeneous and Homogeneous Catalysis," *Chemical Communications*, Vol. 47, No. 25, 2011, pp. 7227-7229.

[45] M. Hara, T. Yoshida, A. Takagaki, T. Takata, J. N. Kondo, S. Hayashi and K. Domen, "A Carbon Material as a Strong Protonic Acid," *Angewandte Chemie International Edition*, Vol. 43, No. 22, 2004, pp. 2955-2958.

[46] M. Toda, A. Takagaki, M. Okamura, J. N. Kondo, S. Hayashi, K. Dome and M. Hara, "Green Chemistry: Biodiesel Made with Sugar Catalyst," *Nature*, Vol. 438, No. 7065, 2005, p. 178.

[47] A. Takagaki, M. Toda, M. Okamura, J. N. Kondo, S. Hayashi, K. Domen and M. Hara, "Esterification of Higher Fatty Acids by a Novel Strong Solid Acid," *Catalysis Today*, Vol. 116, No. 2, 2006, pp. 157-161.

[48] M. H. Zong, Z. Q. Duan, W. Y. Lou, T. J. Smith and H. Wu, "Preparation of a Sugar Catalyst and its Use for Highly Efficient Production of Biodiesel," *Green Chemistry*, Vol. 9, No. 5, 2007, pp. 434-437.

[49] B. L. A. Prabhavathi Devi, K. N. Gangadhar, P. S. Sai Prasad, B. Jagannadh and R. B. N. Prasad, "A Glycerol-

Based Carbon Catalyst for the Preparation of Biodiesel," *ChemSusChem*, Vol. 2, No. 7, 2009, pp. 617-620.

[50] B. L. A. Prabhavathi Devi, K. N. Gangadhar, K. L. N. Siva Kumar, K. Shiva Sanker, R. B. N. Prasad and P. S. Sai Prasad, "Synthesis of Sulfonic Acid Functionalized Carbon Catalyst from Glycerol Pitch and Its Application for Tetrahydropyranyl Protection/Deprotection of Alcohols and Phenols," *Journal of Molecular Catalysis A: Chemical*, Vol. 345, No. 1-2, 2011, pp. 96-100.

[51] K. Ramesh, S. Narayana Murthy, K. Karnakar, Y. V. D. Nageswar, K. Vijayalakhsmi, B. L. A. Prabhavathi Devi and R. B. N. Prasad, "A Novel Bioglycerol-Based Recyclable Carbon Catalyst for an Efficient One-Pot Synthesis of Highly Substituted Imidazoles," *Tetrahedron Letters*, Vol. 53, No. 9, 2012, pp. 1126-1129.

[52] K. Karnakar, S. Narayana Murthy, K. Ramesh, Y. V. D. Nageswar, T. Vijai Kumar Reddy, B. L. A. Prabhavathi Devi and R. B. N. Prasad, "Revisit to the Biginelli Reaction: A Novel and Recyclable Bioglycerol-Based Sulfonic Acid Functionalized Carbon Catalyst for One-Pot Synthesis of Substituted 3,4-Dihydropyrimidin-2-(1H)-ones," *Tetrahedron Letters*, Vol. 53, No. 15, 2012, pp. 1968-1973.

[53] K. Ramesh, S. Narayana Murthy, K. Karnakar, K. Harsha Vardhan Reddy, Y. V. D. Nageswar, M. Vijay, B. L. A. Prabhavathi Devi and R. B. N. Prasad, "A Mild and Expeditious Synthesis of Amides from Aldehydes Using Bio Glycerol-Based Carbon as a Recyclable Catalyst," *Tetrahedron Letters*, Vol. 53, No. 21, 2012, pp. 2636-2638.

[54] K. Karnakar, S. Narayana Murthy, K. Ramesh, K. Harsha Vardhan Reddy, Y. V. D. Nageswar, U. Chandrakala, B. L. A. Prabhavathi Devi and R. B. N. Prasad, "A Novel One-Pot Synthesis of Spiro[indoline-3,4'-pyrazolo[3,4-*e*][1,4]thiazepine]diones Using Recyclable Bioglycerol-Based Sulfonic Acid Functionalized Carbon Catalyst," *Tetrahedron Letters*, Vol. 53, No. 27, 2012, pp. 3497-3501.

Garlic Clove Catalyzed Biginelli Reaction in Water at Ambient Temperature

Hishame Abd-Elnabi[1], Afaf Mohamed Abdel Hameed[1], Ramadan Ahmed Mekheimer[1,2], Reham R. Awed[1], Kamal Usef Sadek[1*]

[1]Department of Chemistry, Faculty of Science, El-Minia University, El-Minia, Egypt
[2]Department of Chemistry, Faculty of Science for Girls, King Abdulaziz University, Jeddah, KSA

ABSTRACT

A simple, green and highly efficient procedure for the *Biginelli* condensation reaction of aldehydes, β-ketoesters, urea or thiourea catalyzed by crushed garlic clove at ambient temperature is reported.

Keywords: Garlic Glove; Biocatalysis; Water; Biginelli Reaction

1. Introduction

Pyrimidines and their derivatives play an important role in human vital functions. The pyrimidine skeleton is the component of a series of natural compounds (*vitamin* B_1, nucleic acids), chemotherapeutic drugs (*Flurouracil*) and synthetic medicines (*Barbiturates*). The biological importance of pyrimidine derivatives caused a significant interest in their synthesis [1-3]. Dihydropyrimidines (DHMPs) and their derivatives are pharmaceutically important as Calcium channel blockers, α_1-1-a-antagonists, antihypertensive agents, inhibitors of the fatty acid transporter, and mitotic kinesin inhibition [4,5]. These compounds have also been found to posses antiviral, antitumor and antibacterial properties [6]. Moreover, the biological activity of some isolated marine natural products and alkaloids have been attributed to the dihydropyrimidine moiety [7]. For example, the anti-cancer agent monostrol (**Figure 1**) has been shown to specifically effect mitosis *via* a new mechanism consisting of the specific and reversible inhibition of the motiliry of the motor protein mitotic kinesin [8-10].

Since more than one hundred year, *Biginelli* suggested a dihydropyrimidine ring construction based on the use of β-dicabonyl compounds as a source of two carbon fragment with aldehydes and urea or thiourea as a N-C-N fragment [11]. It is worth mentioning that *Biginelli* reaction is one of the most named reactions and his collaboration is still considered one of the important pyrimidine synthesis. In the past 10 years, several one-pot methodologies for the synthesis of DHPM derivatives were developed and several modifications have been introduced. Most of them are based on Lewis acid-catalyzed reactions [12-19] which permits the reaction to proceed under milder conditions and with higher yields, than those outlined by *Biginelli* in the original procedure. Microwave irradiation has also proved beneficial [20]. Natural acidic catalysts have been also utilized [21]. Very recently, *Biginelli* reaction has been conducted under basic conditions. This involves the use of PPh_3, under solvent free conditions [22], *t*-BuOK at 70°C [23], chiral primary amines [24] and ammonium carbonate in water [25].

It is worth mentioning that many of these existing methods displayed drawbacks, such as environmental pollution caused by utilizing catalysts in stoichiometric

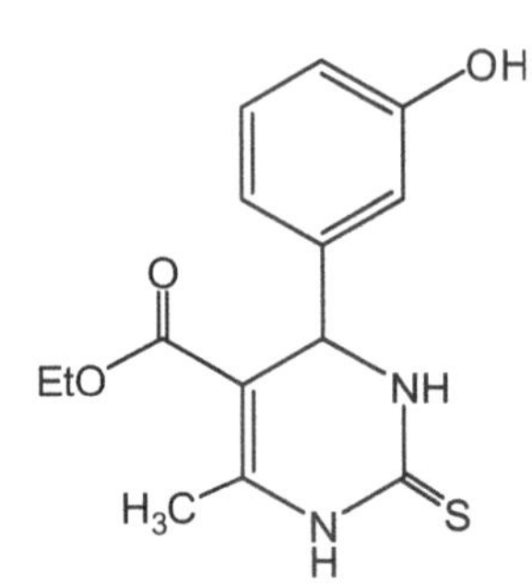

Figure 1. Monostrol.

*Corresponding author.

quantities, exotic reaction conditions, unsatisfactory yields and complicated operations while others posses some advantages overcoming these drawbacks. With the aim for the development of environmentally friendly technique, we investigated herein the utility of bio-catalysis in *Biginelli* reaction at ambient temperature. To the best of our knowledge, such approach has not been reported in the literature. Employing bio-catalysis is advantageous since it is green, proceeded at ambient temperature and is attractive from economical, environmental and handling point of view. In continuation of our work on the synthesis of azines and fused azines [26-28], we report herein the first attempt to synthesis DHMPs **4** using aryl aldehydes **1**, ethyl acetoacetate **2**, urea or thiourea **3** catalyzed by crushed garlic clove at ambient temperature (**Scheme 1**).

2. Results and Discussion

The structure of the products **4** were confirmed by comparison (TLC) with authentic samples [29-33], prepared by refluxing the reaction mixture in ethanol containing 1 mL of hydrochloric acid. In order to examine the effect of substituted aryl aldehyde **1** on the reaction rate and overall yield, various functionalized aryl aldehydes were used under the above reaction conditions. It has been found that the reaction proceeds smoothly to give DHMPs **4** in high yields with a slight decrease in the yield when the aryl substituent involves a strong electron donating group (cf. **Table 1**).

A proposed mechanism to account for the formation of **4** is demonstrated in **Scheme 2** which involves enzymatic transformation of allin[(+)-(S)-allyl-cysteine-sulfoxide] influenced by alliinase producing allyl sulfenic acid, pyruvic acid and ammonia. Both could catalyze *Biginelli* reaction in the well established mechanism either acidic or basic conditions. At room temperature, these enzymatic transformations occur in 10 - 15 minutes (**Scheme 2**).

3. Conclusion

We have successfully developed an easy, high yielding and versatile protocol for the synthesis of DHMPs from the reaction of aryl aldehydes, ethyl acetoacetate, urea or thiourea catalyzed by crushed garlic clove at ambient temperature. The process does not require the use of any hazardous or harmful catalysts and thus it is a simple, environmentally friendly technique of high atom economy and yields.

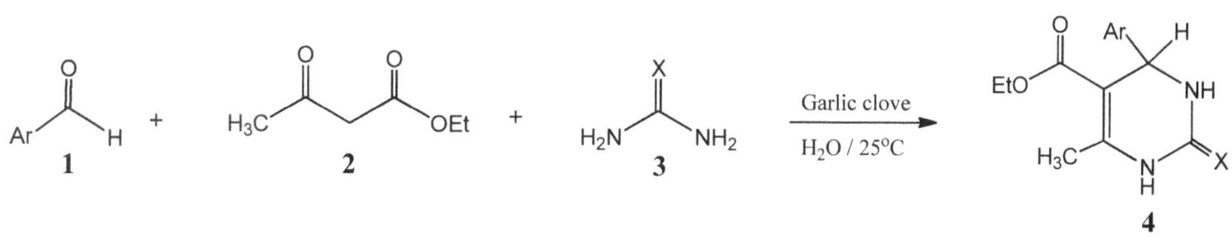

1, 4	Ar	X
a	C_6H_5	O
b	4-MeO-C_6H_4	O
c	4-Me-C_6H_4	O
d	3-NO_2-C_6H_4	O
e	4-Cl-C_6H_4	O
f	C_6H_5	S
g	2-MeO-C_6H_4	S
h	3-NO_2-C_6H_4	S
i	4-Cl-C_6H_4	S
j	4-MeO-C_6H_4	S

Scheme 1. Synthesis of dihydropyrimidines.

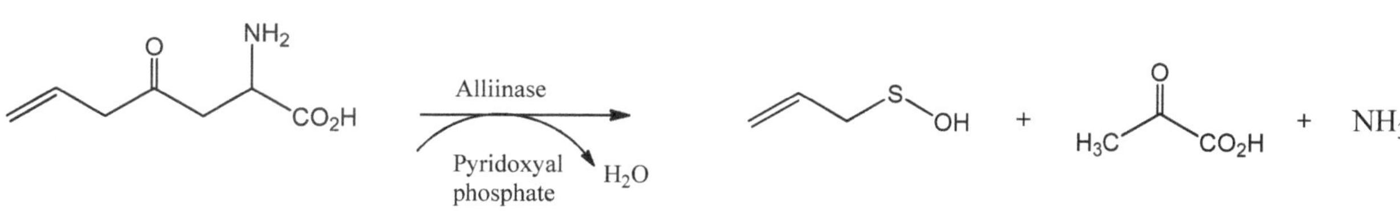

Scheme 2. A proposed mechanism for the formation of dihydropyrimidines 4.

Table 1. Garlic clove synthesis of *Biginelli* DHMPs 4a-j.

Product	Ar	X	Time/h	Yield	Mp (°C)	
					Found	Lit.
4a	C_6H_5	O	3	92	202 - 203	201 - 204 [29]
4b	4-MeO-C_6H_4	O	2.5	85	203 - 205	204 - 206 [29]
4c	4-Me-C_6H_4	O	2.5	80	213 - 215	212 - 213 [29]
4d	3-NO_2-C_6H_4	O	3	75	225 - 227	226 - 228 [29]
4e	4-Cl-C_6H_4	O	3	83	212 - 214	212 - 214 [30]
4f	C_6H_5	S	2.5	90	207 - 209	207 - 209 [30]
4g	2-MeO-C_6H_4	S	1.5	83	190 - 191	188 - 189 [29]
4h	3-NO_2-C_6H_4	S	2.5	74	210 - 212	210 - 212 [33]
4i	4-Cl-C_6H_4	S	2	75	182 - 183	180 - 182 [33]
4j	4-MeO-C_6H_4	S	1.5	82	150 - 152	148 - 150 [31]

4. Experimental

4.1. General

Melting points were measured on a Gallenkamp apparatus and were not corrected. IR spectra were recorded with a Schimadzu 470 spectrophotometer in KBr disks. Peaks are reported in cm^{-1}. ^1H and ^{13}C NMR spectra were recorded on a Bruker AM400 (400 MHz for ^1H NMR and 100 MHz for ^{13}C NMR) spectrometer in DMSO-d_6 using TMS as an internal standard; the chemical shifts are given in δ units (ppm). Mass spectra were measured on a GCMS-QP1000EX (EI, 70 eV) mass spectrometer. Analytical thin-layer chromatography (TLC) was performed on aluminum sheets precoated with silica gel (Merck, Kieselgel 60F-254). Visualization was accomplished by UV light. Micro-analyses were performed at the Microanalytical Data Unit at Cairo University and analytical values obtained were within ±0.4% of the calculated values.

4.2. Synthesis of Dihydropyrimidines 4a-j; General Procedure

To a mixture of each of aldehydes **1a-j** (10 mmol), ethyl acetoacetate **2** (10 mmol) was added crushed garlic glove (100 mg every 15 min.). Then, the reaction mixture was stirred at room temperature (25°C) for the cited time (TLC) control. Then, the reaction mixture was dissolved in EtOH, filtered and the filtrate was concentrated under reduced pressure. The resulting solid product was collected by filtration, dried and recrystallized from EtOH.

4.3. *Ethyl* 6-*methyl*-2-*oxo*-4-*phenyl*-1,2,3,4-*tetrahydro*-5-*pyrimidine carboxylate* (4*a*)

^1H NMR (400 MHz, DMSO-d_6): δ = 1.12 (t, J = 7.2 Hz, 3H), 2.33 (s, 3H), 4.02 (q, J = 7.2 Hz, 2H), 5.08 (d, J = 4 Hz, 1H), 6.35 (t, J = 7.5 Hz, 2H), 6.87 (t, J = 7.3 Hz, 1H), 7.33 (d, J = 7.5 Hz, 2H), 9.66 (s, 1H), 10.44 (s, 1H). ^{13}C NMR (100 MHz, DMSO-d_6): δ = 13.2, 61.5, 111.2, 127.8, 128.2, 130.7, 158.5, 166.0. IR (KBr): ν = 3231, 3207, 1689, 1664 cm^{-1}. MS: m/z = 278 (M^+, 100). Found: C, 64.63; H, 6.22; N, 10.87; anal. calcd for $C_{14}H_{16}N_2O_3$: C, 64.60; H, 6.20; N, 10.76.

4.4. *Ethyl* 4-(4-*methoxyphenyl*)-6-*methyl*-2-*oxo*-1,2,3,4-*tetrahydro*-5-*pyrimidine carboxylate* (4*b*)

^1H NMR (400 MHz, DMSO-d_6): δ = 1.12 (t, J = 7.2 Hz, 3H), 2.22 (s, 3H), 3.73 (s, 3H), 4.20 (q, J = 7.2 Hz, 2H), 5.09 (d, J = 3.2 Hz, 1H), 6.88 (d, J = 8.4 Hz, 2H), 7.14 (d, J = 8.4 Hz, 2H), 9.60 (s, 1H), 10.15 (s, 1H). IR (KBr): ν = 3230, 3204, 1688, 1665 cm^{-1}. MS: m/z = 290 (M^+, 100). Found: C, 61.96; H, 6.27; N, 9.59; anal. calcd for $C_{15}H_{18}N_2O_4$: C, 62.05; H, 6.24; N, 9.64.

4.5. *Ethyl* 4-(4-*methoxyphenyl*)-6-*methyl*-2-*thioxo*-1,2,3,4-*tetrahydro*-5-*pyrimidine carboxylate* (4*j*)

^1H NMR (400 MHz, DMSO-d_6): δ = 1.12 (t, J = 7.1 Hz, 3H), 2.29 (s, 3H), 3.72 (s, 3H), 4.00 (q, J = 7.1 Hz, 2H), 5.06 (d, J = 3.5 Hz, 1H), 6.92 (d, J = 8.5 Hz, 2H),7.13 (d, J = 8.5 Hz, 2H), 9.22 (s, 1H), 10.32 (s, 1H). IR (KBr): ν = 3230, 3205, 1688, 1664 cm^{-1}. MS: m/z = 306 (M^+, 100). Found: C, 58.76; H, 5.82; N, 9.23, S, 10.52; anal. calcd for $C_{15}H_{18}N_2O_3S$: C, 58.80; H, 5.92; N, 9.14, S, 10.47.

REFERENCES

[1] Y. R. Mirzaei, H. Azamat and H. Namazi, "Investigation the Reactivity of Positions N-3, C-5 and C_6-Methyl Group in Biginelli Type Compounds and Synthesis of New Dihydropyrimidine Derivatives," *Journal of Heterocyclic Chemistry*, Vol. 38, No. 5, 2001, pp. 1051-

1054.

[2] M. R. Shaaban, T. S. Saleh, A. S. Mayhoub, A. Mansour and A. M. Farag, "Synthesis and Analgesic/Anti-Inflammatory Evaluation of Fused Heterocyclic Ring Systems Incorporating Phenylsulfonyl Moiety," *Bioorganic & Medicinal Chemistry*, Vol. 16, No. 12, 2008, pp. 6344-6352.

[3] H. Cho, M. Ueda, K. Shima, A. Mizuno, M. Hayashimatsu, Y. Ohnaka, Y. Takeuchi, M. Hamaguchi, K. Aisaka and T. Hidaka, "Dihydropyrimidines: Novel Calcium Antagonists with Potent and Long-Lasting Vasodilative and Anti-Hypertensive Activity," *Journal of Medicinal Chemistry*, Vol. 32, No. 10, 1989, pp. 2399-2406.

[4] G. C. Rovnyak, S. D. Kimball, B. Beyer, G.Cucinotta, J. D. DiMarco, J. Gougoutas, A. Hedberg, M. Malley, J. P. McCarthy, R. Zhang and S. J. Moreland, "Calcium Entry Blockers and Activators: Conformational and Structural Determinants of Dihydro-Pyrimidine Calcium Channel Modulators," *Journal of Medicinal Chemistry*, Vol. 38, No. 1, 1995, pp. 119-129.

[5] M. C. Van Zandt, M. L. Jones, D. E. Gunn, L. S. Geraci, J. H. Jones, D. R. Sawicki, J. Sredy, J. L. Jacot, A. Th. DiCioccio, T. Petrova, A. Mitschler and A. D. Podjarny, "Discovery of 3-[(4,5,7-Triflurobenzothiazol-2-yl)Methyl] Indole-*N*-Acetic Acid (Lidorestat) and Congeners as Highly Potent and Selective Inhibitors of Aldose Reeducates for Treatment of Chronic Diabetic Complications," *Journal of Medicinal Chemistry*, Vol. 48, No. 9, 2005, pp. 3141-3152.

[6] T. Tsuruo, H. Iida, M. Nojiri, S. Tsukagoshi and Y. Sakurai, "Circumvention of Vincristine and Adriamycin Resistance *in Vitro* and *in Vivo* by Calcium Blockers," *Cancer Research*, Vol. 43, No. 6, 1983, pp. 2905-2910.

[7] B. B. Snider and Z. Shi, "Biometric Synthesis of (.+-.) Crambines A, B, C1 and C2. Revision of the Structure of Carbmines B and C1)," *The Journal of Organic Chemistry*, Vol. 58, No. 15, 1993, pp. 3828-3839.

[8] G. C. Rovnyak, S. D. Kimball, B. Beyer, G. Cucinotta, J. D. DiMarco, J. Gougoutas, A. Hedberg, M. Malley, J. P. McCarthy, R. Zhang and S. Moreland, "Calcium Entry Blockers and Activators: Conformational and Structural Determinants of Dihydro-Pyrimidine Calcium Channel Modulators," *Journal of Medicinal Chemistry*, Vol. 38, No. 1, 1995, pp. 119-129.

[9] A. Dondoni, A. Massi and S. Sabbatini, "Improved Synthesis and Preparative Scale Resolution for Racemic Monastrol," *Tetrahedron Letters*, Vol. 43, No. 34, 2002, pp. 5913-5916.

[10] R. Laville, O. P. Thomas, F. Berrué, D. Marquez, J. Vacelet and Ph. Amade, "Bioactive Guanidine Alkaloids from Two Caribbean Marine Sponges," *Journal of Natural Products*, Vol. 72, No. 9, 2009, pp.1589-1594.

[11] P. Biginelli, "Derivati Aldeiduredici Degli Eteri Acetile Dossal-Acetico," *Gazzetta Chimica Italiana*, Vol. 23, 1893, pp. 360-416.

[12] A. S. Paraskar, G. K. Dewkar and A. Sudalai, "$Cu(OTf)_2$ a Reusable Catalyst for High-Yield Synthesis of 3,4-Dihydropyrimidin-2(1*H*)-Ones," *Tetrahedron Letters*, Vol. 44, No. 16, 2003, pp. 3305-3308.

[13] G. Sabitha, K. B. Reddy, J. S. Yadav, D. Shailaja and K. S. Sivudu, "Ceria/Vinyl pyridine Polymer Nanocomposite: An Ecofriendly Catalyst for the Synthesis of 3,4-dihydropyrimidin-2(1*H*)-ones," *Tetrahedron Letters*, Vol. 46, No. 47, 2005, pp. 8221-8222.

[14] J. Azizian, A. A. Mohammadi, A. R. Karimi and M. R. Mohammadizadeh, "$KAl(SO_4)\cdot 12H_2O$ Supported on Silica Gel as a Novel Heterogeneous System Catalyzed Biginelli Reaction: One-Pot Synthesis of Dihydropyrimidinones Under Solvent-Free Conditions," *Applied Catalysis A: General*, Vol. 300, No. 1, 2006, pp. 85-88.

[15] K. U. Sadek, F. Al-Qalaf, M. M. Abdelkhalik and M. H. Elnagdi, "Cerium (IV) Ammonium Nitrate as an Efficient Lewis Acid for One-Pot Synthesis of 3,4-Dihydropyrimidin-2(1*H*)-ones and their Corresponding 2(1*H*)-thiones," *Journal of Heterocyclic Chemistry*, Vol. 47, No. 2, 2010, pp. 284-286.

[16] N. Shapiro and A. Vigalok, "Highly Efficient Organic Reactions 'on Water' and 'in Water' and Both," *Angewandte Chemie*, Vol. 120, No. 15, 2008, pp. 2891-2894. doi: 10.1002/ange. 200705347.

[17] S. D. Salim and K. G. Akamanchi, "Sulfated Tungstate: An Alternative, Eco-Friendly Catalyst for Biginelli Reaction," *Catalysis Communications*, Vol. 12, No. 12, 2011, pp. 1153-1156. doi: org/10.1016/j.catcom.2011.02.018.

[18] S. Chitra and K. Pandiarajan, "Calcium Fluoride: An Efficient and Reusable Catalyst for the Synthesis of 3,4-Dihydropyrimidin-2(1*H*)-ones and Their Corresponding 2-(1*H*)-thiones: An Improved High Yielding Protocol for Biginelli Reaction," *Tetrahedron Letters*, Vol. 50, No. 19, 2009, pp. 2222-2224.

[19] P. G. Mandhane, R. S. Joshi, D. R. Nagargoje and C. H. Gill, "An Efficient Synthesis of 3,4-Dihydropyrimidin-2(1*H*)-ones Catalyzed by Thiamine Hydrochloride in Water Under Ultrasound Irradiation," *Tetrahedron Letters*, Vol. 51, No. 23, 2010, pp. 3138-3140.

[20] K. K. Pasunooti, H. Chai, C. N. Jensen, B. K. Gorityala, S. Wang and X.-W. Liu, "A Microwave-Assisted, Copper Catalyzed three-component Synthesis of Dihydro-Pyrimidinones under Mild Conditions," *Tetrahedron Letters*, Vol. 52, No. 1, 2011, pp. 80-84.

[21] S. Patil, S. D. Jodhav and M. B. Deshmukh, "Natural Acid Catalyzed Multi-Component Reactions as a Green Approach," *Archives of Applied Science Research*, Vol. 3, No. 1, 2011, p. 203. www.scholarresearchlibirary.com

[22] A. Debache, M. Amimour, A. Belfaitah, S. Rhouati and B. Carboni, "A One-Pot Biginelli Synthesis of 3,4-Dihydropyrimidin-2(1*H*)-Ones/Thiones Catalyzed by Triphenylphosphine as Lewis Base," *Tetrahedron Letters*, Vol. 49, No. 42, 2008, pp. 6119-6121.

[23] Z.-L. Shen, X.-P. Xu and S.-J. Ji, "Bronsted Base-Catalyzed One-Pot Three Component Biginelli Type Reaction: An Efficient Synthesis of 4,5,6-Triaryl-3,4-Dihydropyri-

midin-2(1*H*)-one and Mechanistic Study," *The Journal of Organic Chemistry*, Vol. 75, No. 4, 2010, pp. 1162-1167.

[24] D. Ding and C.-G. Zhao, "Primary Amine Catalyzed Biginelli Reaction for the Enantioselective Synthesis of 3,4-Dihydropyrimidin-2(1*H*)-ones," *European Journal of Organic Chemistry*, Vol. 2010, No. 20, 2010, pp. 3802-3805.

[25] F. Tamaddon, Z. Razmi and A. A. Jafari, "Synthesis of 3,4-Dihydropyrimidin-2(1*H*)-ones and 1,4-Dihydropyridines Using Ammonium Carbonate in Water," *Tetrahedron Letters*, Vol. 51, No. 8, 2010, pp. 1187-1189.

[26] R. A. Mekheimer, A. M. Abdel Hameed and K. U. Sadek, "Solar Thermochemical Reactions: Four-Component Synthesis of Polyhydroquinoline Derivatives Induced by Solar Thermal Energy," *Green Chemistry*, Vol. 10, No. 5, 2008, pp. 592-593.

[27] K. U. Sadek, R. A. Mekheimer, A. M. Abdel Hameed, F. Elnahas and M. H. Elnagdi, "Green and Highly Efficient Synthesis of 2-Arylbenzothiazoles Using Glycerol without Catalyst at Ambient Temperature," *Molecules*, Vol. 17, No. 5, 2012, pp. 6011-6019.

[28] K. U. Sadek, R. M. Shaker, M. AbdElrady and M. H. Elnagdi, " A Novel Method for the Synthesis of Polysubstituted Diaminobenzonitrile Derivatives Using Controlled Microwave Heating," *Tetrahedron Letters*, Vol. 51, No. 48, 2010, pp. 6319-6321.

[29] C.-J. Liu and J.-D. Wang, "Copper (II) Sulfamate: An Efficient Catalyst for One-Pot Synthesis of 3,4-Dihydropyrimidin-2(1*H*)-ones and Thiones," *Molecules*, Vol. 14, No. 2, 2009, pp. 763-770.

[30] J. S. Yadav, B. V. S. Reddy, K. B. Reddy, K. S. Raj and A. R. Prasad, "Ultrasound-Accelerated Synthesis of 3,4-Dihydropyrimidin-2(1*H*)-ones with Ceric Ammonium Nitrate," *Journal of the Chemical Society*, *Perkin Transactions*, Vol. 1, No. 16, 2001, pp. 1939-1941.

[31] H. W. Zhan, J. X. Wang and X. T. Wang, "Solvent- and Catalyst-Free Synthesis of Dihydropyrimidinthiones in One-Pot under Focused Microwave Irradiation," *Chinese Chemical Letters*, Vol. 19, No. 10, 2008, pp. 1183-1185.

[32] A. D. Patil, N. V. Kumar, W. C. Kokke, M. F. Bean, A. J. Freyer, C. D. Brosse, S. Mai, A. Truneh, D. J. Faulkner, B. Carte, A. L. Breen, R. P. Hertzberg, R. K. Johnson, J. W. Westly and B. C. M. Potts, "Novel Alkaloids from the Sponge Batzellasp: Inhibitors of HIV gp120-Human CD4 Binding," *European Journal of Organic Chemistry*, Vol. 60, No. 7, 1995, pp.1182-1188.

[33] R. Ghosh, S. Maiti and A. Chakraborty, "In(OTf)$_3$-Catalyzed One-Pot Synthesis of 3,4-Dihydropyrimidin-2(1*H*)-ones," *Journal of Molecular Catalysis A: Chemical*, Vol. 217, No. 1, 2004, pp. 47-50.

A Green Protocol for Efficient Synthesis of 1,8-Dioxo-Octahydroxanthenes Using Ionic Liquid

Sangita Makone*, Shreyas Mahurkar
Chemical Sciences Research Laboratory, School of Chemical Sciences,
Swami Ramanand Teerth Marathwada University, Nanded, India

ABSTRACT

A clean, simple, efficient synthesis of 1,8-dioxo-octahydroxanthenes derivatives using ionic liquid 1-butyl, 3-methylimidazolium perchlorate ([bmim] ClO_4) has been developed. The method provides several advantages such as simple work up, environmental friendliness and excellent yields in short reaction time. The ionic liquid [bmim] ClO_4 used was recovered and reused for three times.

Keywords: 1,8-Dioxo-Octahydroxanthenes; Ionic Liquid; Aldehydes; Dimedone

1. Introduction

1,8-dioxo-octahydroxanthenes are important class of oxygen heterocycles in which a phenyl substituted pyran ring is fused on either side with two cyclohexanone rings. Presence of conjugated bis-dienone functionality makes these compounds sensitive to be attacked by nucleophiles and light energy. In the past decade, synthesis of xanthenes derivatives has been of considerable interest to organic chemists because they possess various biological and pharmaceuticals activities such as antiviral [1], antibacterial [2] and anti-inflammatory [3] properties. These are being utilized as antagonists for paralyzing action of zoxazolamine [4] and in photodynamic therapy [5]. Xanthenes and benzoxanthenes derivatives are the parent frame works found in a large number of naturally occurring as well as synthetic products possessing prominent position in medicinal chemistry [6,7]. Xanthenediones are likewise special structural units constituting various natural products [8] and being used as versatile synthons, because of inherent reactivity of their inbuilt pyran ring [9].

Apart from their use as valuable synthetic precursors for many organic compounds [10] and dyes [11], they are also used in laser technologies [12] and pH—sensitive fluorescent materials for the visualization of biomolecules assemblies [13].

There are a wide variety of methods for the preparation of the xanthenes having been reported and they are classified according to starting compounds which includes synthesis by cyclization of polycyclic aryltiflate esters [14], intra-molecular trapping of benzynes by phenols [15] and reaction of aryloxy magnesium halides with triethyl orthoformate [16] as well as cyclo-condensation of 2-hydroxyl aromatic aldehyde with 2-tetralone [17].

One of the commonly used methods reported for the synthesis of xanthenediones involves the condensation of aromatic aldehydes with 1, 3-cyclohexanedione or 5,5-dimethyl-1, 3-cyclohexanedione. The various catalysts have been reported for synthesis of 1,8-dioxo-octahydroxanthens: NSPVPC [18], $[Et_3NH][HSO_4]$ [19], $SiCl_4$ [20], $[Et_3NC_4SO_3H]$ $[HSO_4]/Al_2O_3$ [21], polyaniline p-toluenesulfonate [22], alumina-sulfuric acid [23], CAN [24], $[BMIm][BF_4]$-$Mg(BF_4)$ [25], I_2 [26], $[Et_3N\text{-}SO_3H]Cl$ [27], DABCO [28], $SmCl_3$ [29], cellulose sulfonic acid [30], heteropolyacid-MCM-41 *i.e.* ($H_3PW_{12}O_{40}$) [31], [bmim] HSO_4 [32], $Fe(HSO_4)_3$ [33], $[Hbim]BF_4$ [34], TBAHS [35] tetramethylguanidium trifluoroacetate [36], [DDPA] $[HSO_4]$ [37], [TMPSA] HSO_4 [38], [Hmim] TFA [39], trichloro isocyanuric acid [40], $SbCl_3/SiO_2$ [41], SiO_2-R-SO_3H [42], $InCl_3/P_2O_5$ [43], p-dodecyl benzenesulfonic acid [44], triethyl benzyl ammonium chloride (TEBA) [45], diammonium hydrogen phosphate [46], PPA-SiO_2 [47], montmorillonite K-10 [48], Fe^{3+}-Montmorillonite

*Corresponding author.

[49], trimethyl silyl chloride [50], $NaHSO_4$-SiO_2 [51], ZnO-Acetyl chloride [52], PMA-SiO_2 [53], $ZrOCl_2·8H_2O$ [54], Amberlyst-15 [55], Dowex-50W [56]. Most of these methods have some advantages and disadvantages, taking the weakness into consideration such as long reaction time, low yields, use of hazardous catalysts and solvents, tedious work up processes and difficulty in recovery and reusability of catalyst. To overcome these problems we have developed the green protocol for efficient synthesis of 1,8-dioxo-octahydroxanthenes using [bmim]ClO_4 ionic liquid (**Scheme 1**).

In recent years, ionic liquids have emerged as a useful alternative to conventional organic solvents and catalysts due to their particular properties, such as negligible vapor pressure, chemical stability, excellent solvent power for organic and inorganic compounds, as well as the ease of recovery [57]. Recently, much attention has been paid to organic reactions carried by using ionic liquids greater than ever importance in the context of green synthesis. Although ionic liquids were primarily introduced as an alternative green reaction medium, today they have progressed extremely beyond that, playing a significant role in controlling reactions as catalysts.

2. Experimental Procedure

All the chemicals were purchased from Sigma Aldrich and Merck chemicals. Melting points recorded on Veego melting point apparatus and were uncorrected. ^{1}H NMR spectra were recorded on a Brucker Avance II 400 MHz spectrometer with TMS as internal standard. IR spectra were determined on Perkin Elmer Spectrum RX FTIR spectrometer. The ionic liquid was synthesized according to ref. [58].

General Procedure for Synthesis of 1,8-Dioxo-Octahydroxanthens Derivatives

To a mixture of aldehyde (1 mmol) and 5,5-dimethyl-1,3-cyclohexanedione (2 mmol) and [bmim]ClO_4 (4 mmol) was added in 50 ml round bottom flask and was stirred at 100°C on heating magnetic stirrer (**Scheme 1**). The progress of the reaction was monitored by TLC after completion of the reaction the reaction mixture was cooled to room temperature and water (5 ml) was added, solid separated was filtered. The crude product was recrystallized from ethanol to give the pure product. The [bmim]ClO_4 was recovered by distillation and reused 3 times.

The comparison of various reported catalyst is summarized in (**Table 1**).

3,3,6,6-tetramethyl-9-(4-chloro-phenyl)-1,8-dioxo-octahydroxanthene (Entry 1) IR (cm^{-1}): 3030, 2963, 2952, 1679, 1661, 1469, 1361, 1198, 1166, 1003, 852. 1H NMR: 7.17 - 7.24 (dd, 4H), 4.71 (s, 1H), 2.46 (s, 4H), 2.18 (q, 4H), 1.10 (s, 6H), 0.98 (s, 6H). 13C NMR: 191.10, 157.17, 137.45, 126.79, 124.53, 122.97, 110.03, 45.45, 35.61, 26.96, 26.22, 24.03, 22.05 m/z: 385.2 (M + 1), 273.2.

3,3,6,6-tetramethyl-9-(4-methyl-phenyl)-1,8-dioxo-octahydroxanthene (Entry 2) IR (cm^{-1}): 3036, 2959, 2873, 1663, 1623, 1359, 1197, 1164, 1139, 1000, 842. 1H NMR: 7.17 (d, 2H), 7.01(d, 2H), 4.71(s, 1H), 2.45 (s, 4H), 2.24 (s, 3H), 2.18 (q, 4H), 1.09 (s, 6H), 0.99 (s, 6H). 13C NMR: 196.4, 162.0, 141.1, 135.7, 128.7, 128.2, 115.7, 50.7, 40.8, 32.2, 31.4, 29.2, 27.3, 21.0. m/z: 365.3 (M + 1), 273.2.

9,9'-(1,4-phenylene)bis(3,3,6,6-tetramethyl-3,4,5,6,7, 9-hexahydro-1H-xanthene-1,8 (2H)-dione (Entry 4) IR (cm^{-1}): 2957, 2871, 1665, 1620, 1365, 1201, 1167, 1144, 1005, 590. 1H NMR: 7.07 (s, 4H), 4.70 (s, 2H), 2.42 (s, 8H), 2.17 (s, 8H), 1.07 (s, 12H), 0.97 (s, 12H). m/z: 13C NMR: 196.4, 162.4, 141.7, 127.9,115.7, 50.8, 40.8, 32.2, 30.1, 28.9, 27.7. m/z: 645.4 (M + Na), 273.2.

3,3,6,6-tetramethyl-9-(4-hydroxy-3-methoxy-phenyl)-1,8-dioxo-octahydroxanthene (Entry 5) IR (cm^{-1}): 3412, 3029, 2954, 1666, 1622, 1514, 1431, 1359, 1278, 1229, 1197, 1135, 1028, 624, 571. 1H NMR: 7.01 (s, 1H), 6.73 (d, 1H), 6.56 (dd, 1H), 5.49 (bs, 1H, -OH), 4.66 (s, 1H), 3.89 (s, 3H), 2.45 (s, 4H), 2.20 (q, 4H), 1.00 (s, 12H). 13C NMR: 196.6, 162.0, 145.8, 144.0, 136.4, 120.0, 115.8, 113.9, 112.2, 55.8, 50.7, 40.8, 32.2, 31.3, 29.3, 27.2. m/z: 396.47, 419.3 (M + Na), 273.2.

3,3,6,6-tetramethyl-9-(4-methoxy-phenyl)-1,8-dioxo-octahydroxanthene (Entry 7) IR (cm^{-1}): 3059, 2958, 2876, 1665, 1626, 1511, 1462, 1357, 1260, 1193, 1109, 1031, 841, 569. 1H NMR: 7.18 (d, 2H), 6.75 (d, 2H), 4.69 (s, 1H), 3.73 (s, 3H), 2.45 (s, 4H),1.09 (s, 6H), 0.99 (s, 6H). 13C NMR: 196.5, 162.0, 157.9, 136.4, 129.3, 115.7, 113.4, 55.1, 50.7, 40.8, 32.2, 30.9, 29.2, 27.3. m/z: 381.2, 273.2.

3. Result and Discussion

To a mixture of 4-chloro benzaldehyde (1 mmol) and 5,5-dimethyl-1,3-cyclohexanedione (2 mmol) and [bmim] ClO_4 (4 mmol) was added in 50 ml round bottom flask and was stirred at 100°C on heating magnetic stirrer. The progress of the reaction was monitored by TLC After completion of the reaction the reaction mixture was cooled to room temperature and water (5 ml) was added, solid separated was filtered and product was characterized by IR, NMR, ^{13}C-NMR and mass.

The reaction between dimedone and 4-chloro benzaldehyde in the presence of [bmim]ClO_4 was studied as a model reaction. To optimize the reaction conditions of reaction temperature we have tried the study of above model reaction from temperature 60 - 110°C in the gap of 10°C. The best result in terms of time and yield for the reaction has been observed at 100°C (**Table 2**).

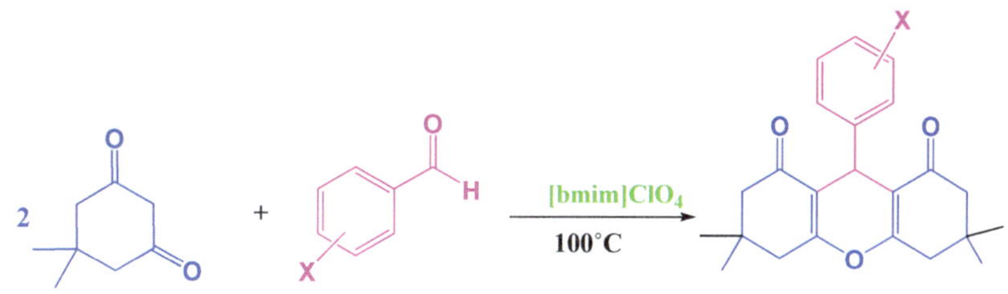

Scheme 1. Synthesis of 1,8-dioxo-octahydroxanthene using [bmim]ClO$_4$ ionic liquid.

Table 1. Comparison table with various reported catalysts.

Entry	Catalyst used	Time	Solvent	Temperature	Yield (%)	Ref.
1	Tetrachloro silane	3 hrs	CH_2Cl_2	60 - 70°C	92	[20]
2	$SmCl_3$	8 hrs	No solvent	120°C	98	[29]
3	Cellulose sulfonic acid	6 hrs	No solvent	110°C	95	[30]
4	Heteropoly acid supported MCM-41	5 hrs	Ethanol	90°C	90	[31]
5	[Bmim]HSO_4	3.5 hrs	No solvent	80°C	76 - 94	[32]
6	[Hbim]BF_4	45 min.	Methanol	Ultrasonic 25 - 30°C	85	[34]
7	Trimethyl silyl chloride	8 - 10 hrs	Acetonitrile	Reflux	72 - 84	[50]
8	[Bmim]ClO_4	40 - 90 min.	----	100°C	85 - 95	Present work

Table 2. Optimization of reaction temperature for synthesis of 1,8-dioxo-octahydroxanthenes.

Entry	Temperature (in °C)	Time (in hrs.)	Yields (in %)
1	60	8 hrs	55
2	70	6 hrs	70
3	80	5 hrs	75
4	90	2.5 hrs	85
5	100	1 hr	94
6	110	1 hr	94

To evaluate the scope of this reaction a range of 1, 8-dioxo-octahydroxanthenes were prepared by the reaction of dimedone and aromatic aldehydes under optimized reaction conditions. The results are summarized in the table (**Table 3**). Various aromatic aldehydes with electron donating substituent's reacted efficiently and quickly with dimedone to give cyclocondensation products in high yields over short reaction times. Similarly aldehydes with electron withdrawing substituent's reacted efficiently to yield the corresponding 1,8-dioxo-octahydroxanthenes.

Study of reuse of the ionic liquid can be easily done up to 3 times with excellent yields without loss in the reactivity (as shown in graph/**Figure 1**). After 3rd cycle we observed sudden decrease in the 4th reuse of ionic liquid thus the ionic liquid used was recycled for 3 times. The structure of isolated products was assigned based on their spectral analyses as well as by matching with their melting point with reported analogous.

Table 3. Green protocol for efficient synthesis of 1, 8-dioxoxanthenes using [bmim]ClO$_4$.[a]

Entry	Aldehyde	Product[b]	Time (min.)	Yield[c] (%)	Mp. (°C)
1	4-Chloro benzaldehyde	3a	60	94	215 - 217
2	4-Methyl benzaldehyde	3b	70	95	217 - 218
3	Benzaldehyde	3c	40	92	205 - 206
4	Terephthaldehyde	3d	90	88	>300
5	Vanillin	3e	50	85	225 - 227
6	4-Fluro benzaldehyde	3f	50	92	232 - 233
7	4-Methoxy benzaldehyde	3g	60	95	241 - 243
8	3-Chloro benzaldehyde	3h	60	90	183 - 184
9	4-Bromo benzaldehyde	3i	60	90	222 - 224
10	4-Hydroxy benzaldehyde	3j	70	87	228 - 230

[a]Reaction condition: aromatic aldehydes (2 mmol), 5,5-dimethyl-1-3cyclohexanedione (4 mmol), [bmim]ClO$_4$ (4 mmol) at 100°C. [b]All the product were characterized by IR spectral data and comparison of their melting point with those of the authentic samples. Also the structures of the some products were confirmed by 1H NMR spectral data. [c]Isolated yield.

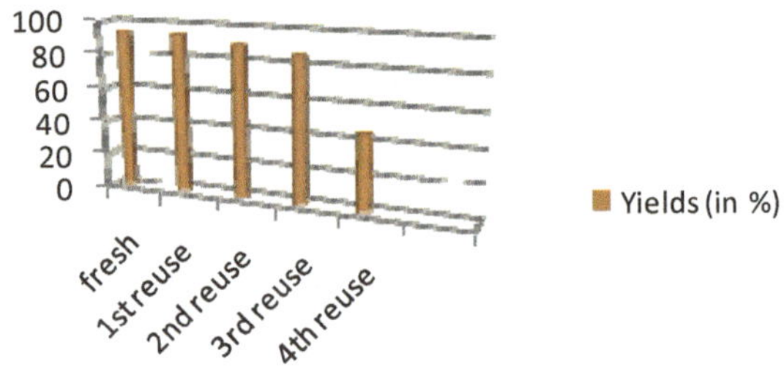

Figure 1. Study of reuse of ionic liquid.

4. Conclusion

In summary, we have developed a high-yielding simple, convenient, straight-forward and practical one-pot procedure for the synthesis of different types of 1,8-dioxo-octahydroxanthenes derivatives using [bmim]ClO_4. The ionic liquid used can be easily recovered and reused 3 times without significant decrease in the yield of the product.

REFERENCES

[1] R. W. Lamberk, J. A. Martin, J. H. Merrett, K. E. B. Parkes and G. J. Thomas, PCT Int. Appl. WO 9706178, 1997, *Chemical Abstract*, Vol. 126, 1997, p. 212377y.

[2] T. Hideu, JPn. Tokkyo Koho JP 56005480, 1981, *Chemical Abstract*, Vol. 95, 1981, 80922b.

[3] J. P. Poupelin, G. Saint-Rut, O. Fussard-Blanpin, G. Narciss, G. U. Ernouf and R. Lakroix, "Synthesis and Antiinflammatory Properties of Bis (2-Hydroxy, 1-Naphthyl) Methane Derivatives," *European Journal of Medicinal Chemistry*, Vol. 13, 1978, pp. 67-71.

[4] G. Saint-Ruf, A. De and H. T. Hieu, "The Effect of Dibenzoxanthenes on the Paralyzing Action of Zoxazolamine," *Naturwissenschaften*, Vol. 62, No. 12, 1975, pp. 584-585.

[5] R. M. Ion, D. Frackowiak, A. Planner and K. Wiktorowicz, "The Incorporation of Various Porphyrins into Blood Cells Measured via Flow Cytometry, Absorption and Emission Spectroscopy," *Acta Biochimica Polonica*, Vol. 45, No. 3, 1998, pp. 833-845.

[6] H. K. Wang, S. L. Morris-Natschke and K. H. Lee, "Recent Advances in the Discovery and Development of Topoisomerase Inhibitors as Antitumor Agents," *Medicinal Research Reviews*, Vol. 17, No. 4, 1997, pp. 367-425.

[7] A. V. Rukavishnikov, M. P. Smith, G. B. Birrell, J. F. W. Keana and O. H. Griffith, "Synthesis of a New Fluorogenic Substrate for the Assay of Phosphoinositide-Specific Phospholipase C," *Tetrahedron Letter*, Vol. 39, No. 37, 1998, pp. 6637-6640.

[8] S. Hatakeyma, N. Ochi, H. Numata and S. Takano, "A New Route to Substituted 3-Methoxycarbonyldihydropyrans; Enantioselective Synthesis of (–)-Methyl Elenolate," *Journal of the Chemical Society, Chemical Communications*, No. 17, 1988, pp. 1202-1204.

[9] C. N. O'Callaghan and T. B. H. McMurry, "Synthetic reactions of methyl XY-carbonyl-4H-1-benzopyran-4yl cyanoethanoate," *Journal of Chemical Research, Synopses*, 1995, pp. 214-218.

[10] Y. M. Shchekotikhin and T. G. Nikolaeva, "Transformations of *Sym*-Octahydroxanthene-1,8-Diones and 1,8-Dioxo-*Sym*-Octahydroxanthylium Salts in Recyclization under the Influence of Amines," *Chemistry Heterocyclic Compound*, Vol. 42, No. 1, 2006, pp. 28-33.

[11] S. A. Hilderbrand and R. Weissleder, "One-Pot Synthesis of New Symmetric and Asymmetric Xanthene Dyes," *Tetrahedron Letter*, Vol. 48, No. 25, 2007, pp. 4383-4385.

[12] G. Pohlers and J. C. Scaiano, "A Novel Photometric Method for the Determination of Photoacid Generation Efficiencies Using Benzothiazole and Xanthene Dyes as Acid Sensors," *Chemistry Materials*, Vol. 9, No. 12, 1997, pp. 3222-3230.

[13] C. G. Knight and T. Stephens, "Xanthene-Dye-Labelled Phosphatidylethanolamines as Probes of Interfacial pH. Studies in Phospholipid Vesicles," *Biochemical Journal*, Vol. 258, No. 3, 1989, pp. 683-687. http://www.biochemj.org/bj/258/bj2580683.htm

[14] J. Q. Wang and R. G. Harvey, "Synthesis of Polycyclic Xanthenes and Furans via Palladium-Catalyzed Cyclization of Polycyclic Aryltriflate Esters," *Tetrahedron*, Vol. 58, No. 29, 2002, pp. 5927-5931.

[15] D. W. Knight and P. B. Little, "The First High-Yielding Benzyne Cyclisation Using a Phenolic Nucleophile: A New Route to Xanthenes," *Synlett*, No. 10, 1998, pp. 1141-1143.

[16] G. Casiraghi, G. Casnati and M. Cornia, "Regiospecific Reactions of Phenol Salts: Reaction-Pathways of Alkylphenoxy-Magnesiumhalides with Triethylorthoformate," *Tetrahedron Letter*, Vol. 14, No. 9, 1973, pp. 679-682.

[17] A. Jha and J. Beal, "Convenient Synthesis of 12*H*-Benzo[*a*]xanthenes from 2-Tetralone," *Tetrahedron Letter*, Vol. 45, No. 29, 2004, pp. 8999-9001.

[18] F. Shirini, M. Abedini and R. Pourhasan, "*N*-Sulfonic Acid Poly(4-Vinylpyridinium) Chloride: A Novel Polymeric and Reusable Catalyst for the Preparation of Xanthenes Derivatives," *Dyes and Pigments*, Vol. 99 No. 1, 2013, pp. 250-255.

[19] Z. Zhou and X. Deng, "[Et_3NH][HSO_4] Catalyzed Efficient and Green Synthesis of 1,8-Dioxo-Octahydroxanthenes," *Journal of Molecular Catalysis A*: *Chemical*, Vol. 367, 2013, pp. 99-102.

[20] H. A. Soliman and T. A. Salama, "Silicon-Mediated Highly Efficient Synthesis of 1,8-Dioxo-Octahydroxanthenes and Their Transformation to Novel Functionalized Pyrano-

Tetrazolo[1,5-*a*] Azepine Derivatives," *Chinese Chemical Letters*, Vol. 24, No. 5, 2013, pp. 404-406.

[21] M. Khoshnevis, A. Davoodnia, A. Z. Bidaki and N. T. Hoseini, "Alumina Supported Acidic Ionic Liquid: Preparation, Characterization, and Its Application as Catalyst in the Synthesis of 1,8-Dioxo-Octahydroxanthenes," *Synthesis and Reactivity in Inorganic, Metal-organic and Nano-Metal Chemistry*, Vol. 43, No. 9, 2013, pp. 1154-1161.

[22] A. John, P. J. P. Yadav and S. J. Palaniappan, "Clean Synthesis of 1,8-Dioxo-Dodecahydroxanthene Derivatives Catalyzed by Polyaniline-*P*-Toluenesulfonate Salt in Aqueous Media," *Journal of Molecular Catalysis A: Chemical*, Vol. 248, No. 1-2, 2006, pp. 121-125.

[23] A. Pramanik and S. Bhar, "Alumina-Sulfuric Acid Catalyzed Eco-Friendly Synthesis of Xanthenediones," *Catalysis Communication*, Vol. 20, 2012, pp. 17-24.

[24] N. Mulakayala, G. Pavan Kumar, D. Rambabu, M. Aeluri, M. V. B. Rao and M. Pal, "A Greener Synthesis of 1,8-Dioxo-Octahydroxanthene Derivatives under Ultrasound," *Tetrahedron Letter*, Vol. 53, No. 51, 2013, pp. 6923-6926.

[25] K. Moghadam and S. C. Azimi, "$Mg(BF_4)_2$ Doped in [BMIm][BF_4]: A Homogeneous Ionic Liquid-Catalyst for Efficient Synthesis of 1,8-Dioxo-Octahydroxanthenes, Decahydroacridines and 14-Aryl-14*H*-Dibenzo[*a,j*]Xanthenes," *Journal Molecular Catalysis A: Chemical*, Vol. 363-364, 2012, pp. 465-469.

[26] N. Mulakayala, P. V. N. S. Murthy, D. Rambabu, M. Aeluri, R. Adepu, G. R. Krishna, C. M. Reddy, K. R. S. Prasad, M. Chaitanya, C. S. Kumar, M. V. Basaveswara Rao and M. Pal, "Catalysis by Molecular Iodine: A Rapid Synthesis of 1,8-Dioxo-Octahydroxanthenes and Their Evaluation as Potential Anticancer Agents," *Bioorganic & Medicinal Chemistry Letters*, Vol. 22, No. 6, 2012, pp. 2186-2191.

[27] A. Zare, A. R. Moosavi-Zare, M. Merajoddin, M. A. Zolfigol, T. Hekmat-Zadeh, A. Hasaninejad, A. Khazaei, M. Mokhlesi, V. Khakyzadeh, F. Derakhshan-Panah, M. H. Beyzavi, E. Rostami, A. Arghoon and R. Roohandeh, "Ionic Liquid Triethylamine-Bonded Sulfonic Acid {[Et_3N-SO_3H]Cl} as a Novel, Highly Efficient and Homogeneous Catalyst for the Synthesis of β-Acetamido Ketones, 1,8-Dioxo-Octahydroxanthenes and 14-Aryl-14*H*-Dibenzo[*a,j*] Xanthenes," *Journal of Molecular Liquids*, Vol. 167, 2012, pp. 69-77.

[28] P. Pradeep, J. Shrinivasa Rao and J. Shubha, "DABCO Promoted Multi-Component One-Pot Synthesis of Xanthenederivatives," *Research Journal of Chemical Sciences*, Vol. 2, No. 8, 2012, pp. 21-25. http://www.isca.in/rjcs/Archives/vol2/i8/4.ISCA-RJCS-2012-088.pdf

[29] A. Llangovan, S. Malayappasamy, S. Muralidharan and S. Maruthamuthu, "A Highly Efficient Green Synthesis of 1,8-Dioxo-Octahydroxanthenes," *Chemistry Central Journal*, Vol. 5, 2011.

[30] H. A. Oskooie, L. Tahershamsi, M. M. Heravi and B. Baghernejad, "Cellulose Sulfonic Acid: An Efficient Heterogeneous Catalyst for the Synthesis of 1, 8-Dioxo-Octahydroxanthenes," *E-Journal of Chemistry*, Vol. 7, No. 3, 2010, pp. 717-720.

[31] G. Karthikeyan and A. Pandurangan, "Heteropolyacid ($H_3PW_{12}O_{40}$) Supported MCM-41: An Efficient Solid Acid Catalyst for the Green Synthesis of Xanthenedione Derivatives," *Journal Molecular Catalysis A: Chemical*, Vol. 311, No. 1-2, 2009, pp. 36-45.

[32] K. Niknam and M. Damya, "1-Butyl-3-Methylimidazolium Hydrogen Sulfate [Bmim]HSO_4: An Efficient Reusable Acidic Ionic Liquid for the Synthesis of 1,8-Dioxo-Octahydroxanthenes," *Journal of the Chinese Chemical Society*, Vol. 56, No. 3, 2009, pp. 659-665.

[33] H. R. Shaterian, A. Hosseinian and M. Ghashang, "Ferric Hydrogen Sulfate as an Efficient Heterogeneous Catalyst for Environmentally Friendly Greener Synthesis of 1,8-Dioxo-Octahydroxanthenes," *Turkish Journal of Chemistry*, Vol. 33, No. 2, 2009, pp. 233-240.

[34] K. Vankatesan, S. S. Pujari, R. J. Lahoti and K. V. Srinivasan, "An Efficient Synthesis of 1,8-Dioxo-Octahydro-Xanthene Derivatives Promoted by a Room Temperature Ionic Liquid at Ambient Conditions under Ultrasound Irradiation," *Ultrasonic Sonochemistry*, Vol. 15, No. 4, 2008, pp. 548-553.

[35] H. N. Karade, M. Sathe and M. P. Kaushik, "An Efficient Synthesis of 1, 8-Dioxo-Octahydroxanthenes Using Tetrabutylammonium Hydrogen Sulfate," *ARKIVOC*, Vol. xiii, 2007, pp. 252-258.

[36] A. Rahmati, "A Rapid and Efficient Method for the Synthesis of 14*H*-Dibenzo[*a.j*]Xanthenes, Aryl-5*H*-Dibenzo [*b.i*]xanthene-5,7,12,14-(13*H*)-Tetraone and 1,8-Dioxo-Octahydroxanthenes by Acidic Ionic Liquid," *Chinese Chemical Letter*, Vol. 21, No. 7, 2010, pp. 761-764.

[37] D. Fang, J. Yang and Z. Liu, "Eco-Friendly Synthesis of 1,8-Dioxo-Octahydroxanthenes Catalyzed by Ionic Liquid in Aqueous Media," *Journal of Heterocyclic Chemistry*, Vol. 48, No. 2, 2011, pp. 468-472.

[38] D. Fang, K. Gong and Z. Liu, "Synthesis of 1,8-Dioxo-Octahydroxanthenes Catalyzed by Acidic Ionic Liquids in Aqueous Media," *Catalysis Letters*, Vol. 127, No. 3-4, 2009, pp. 291-295.

[39] M. Dabirir, M. Baghbanzadeh and E. Arzroomchilar, "1-Methylimidazolium triflouroacetate ([Hmim]TFA): An Efficient Reusable Acidic Ionic Liquid for the Synthesis of 1,8-Dioxo-Octahydroxanthenes and 1,8-Dioxo-Decahydroacridines," *Catalysis Communications*, Vol. 9, No. 5, 2008, pp. 939-942.

[40] M. A. Bigdeli, F. Nemati, G. H. Mahdavinia and H. Doo-

stmohammadi, "A Series of 1,8-Dioxooctahydroxanthenes Are Prepared Using Trichloroisocyanuric Acid," *Chinese Chemical Letters*, Vol. 20, No. 11, 2009, pp. 1275-1278.

[41] Z. H. Zhang and Y. H Liu, "Antimony Trichloride/SiO_2 Promoted Synthesis of 9-Ary-3,4,5,6,7,9-Hexahydroxanthene-1,8-Diones," *Catalysis Communications*, Vol. 9, No. 8, 2008, pp. 1715-1719.

[42] G. H. Mahdavinia, M. A. Bigdeli and Y. S. Hayeniaz, "Covalently Anchored Sulfonic Acid on Silica Gel (SiO_2-R-SO_3H) as an Efficient and Reusable Heterogeneous Catalyst for the One-Pot Synthesis of 1,8-Dioxo-Octahydroxanthenes Under Solvent-Free Conditions," *Chinese Chemical Letters*, Vol. 20, No. 5, 2009, pp. 539-541.

[43] G. K. Verma, K. Raghuvanshi, R. K. Verma, P. Dwivedi and M. S. Singh, "An Efficient One-Pot Solvent-Free Synthesis and Photophysical Properties of 9-Aryl/Alkyl-Octahydroxanthene-1,8-Diones," *Tetrahedron*, Vol. 67, No. 20, 2011, pp. 3698-3704.

[44] T. S. Jin, J. S. Zhang, J. C. Xiao, A. Q. Wang and T. S. Li, "Clean Synthesis of 1,8-Dioxo-octahydroxanthene Derivatives Catalyzed by *p*-Dodecylbenezenesulfonic Acid in Aqueous Media," *Synlett*, No. 5, 2004, pp. 866-870.

[45] X. S. Wang, D. Q. Shi, Y. L. Li, H. Chen, X. Y. Wei and Z. M. Zong, "A Clean Synthesis of 1 - Oxo - hexahydroxanthene Derivatives in Aqueous Media Catalyzed by TEBA," *Synthetic Communications*, Vol. 35, No. 1, 2005, pp. 97-104.

[46] F. Darvish, S. Balalaei, F. Chadegani and P. Salehi, "Diammonium Hydrogen Phosphate as a Neutral and Efficient Catalyst for Synthesis of 1, 8 - Dioxo - Octahydroxanthene Derivatives in Aqueous Media," *Synthetic Communications*, Vol. 37, No. 7, 2007, pp. 1059-1066.

[47] S. Kantevari, R. Bantu, L. Nagarapu, "$HClO_4$-SiO_2 and PPA-SiO_2 Catalyzed Efficient One-Pot Knoevenagel Condensation, Michael Addition and Cyclo-Dehydration of Dimedone and Aldehydes in Acetonitrile, Aqueous and solvent Free Conditions: Scope and Limitations," *Journal of Molecular Catalysis A: Chemical*, Vol. 269, No. 1-2, 2007, pp. 53-57.

[48] M. Dabiri, S. C. Azimi and A. Bazgir, "One-Pot Synthesis of Xanthene Derivatives under Solvent-Free Conditions," *Chemical Papers*, Vol. 62, No. 5, 2008, pp. 522-526.

[49] G. Y. Song, B. Wang, H. Luo and L. Yang, "Fe^{3+}-Montmorillonite as a Cost-Effective and Recyclable Solid Acidic Catalyst for the Synthesis of Xanthenediones," *Catalysis Communication*, Vol. 8, No. 4, 2007, pp. 673-676.

[50] S. Kantevari, R. Bantu and L. Nagarapu, ARKIVOC Part (xvi), 2006, pp. 136-148.

[51] B. Dashttp, P. Thirupathi, K. R. Reddy, B. Ravikanth and L. Nagarapu, "An Efficient Synthesis of 1,8-Dioxo-Octahydroxanthenes Using Heterogeneous Catalysts," *Catalysis Communications*, Vol. 8, No. 3, 2007, pp. 535-538.

[52] M. T. Maghsoodlou, S. M. Habibi-Khorassani, Z. Shahkarami, N. Maleki and M. Rostamizadeh, "An Efficient Synthesis of 2,2′-Arylmethylene Bis(3-Hydroxy-5,5-Dimethyl-2-Cyclohexene-1-One) and 1,8-Dioxo-Octahydroxanthenes Using ZnO and ZnO-Acetyl Chloride," *Chinese Chemical Letters*, Vol. 21, No. 6, 2010, pp. 686-689.

[53] P. Srihari, S. S. Mandal, J. S. S. Reddy and R. S. Rao and J. S. Yadav, "Synthesis of 1,8-Dioxo-Octahydroxanthenes Utilizing PMA-SiO_2 as an Efficient Reusable Catalyst," *Chinese Chemical Letters*, Vol. 19, No. 7, 2008, pp. 771-774.

[54] H. Y. Lü, J. J. Li and Z. H. Zhang, "$ZrOCl_2 \cdot 8H_2O$: A Highly Efficient Catalyst for the synthesis of 1,8-Dioxo-Octahydroxanthene Derivatives under Solvent-Free Conditions," *Applied Organometalic Chemistry*, Vol. 23, No. 4, 2009, pp. 165-169.

[55] B. Das, P. Thirupathi, I. Mahender, V. S. Reddy and Y. K. Rao, "Amberlyst-15: An Efficient Reusable Heterogeneous Catalyst for the Synthesis of 1,8-Dioxo-Octahydroxanthenes and 1,8-Dioxo-Decahydroacridinesao," *Journal of Molecular Catalysis A: Chemical*, Vol. 247, No. 1-2, 2006, pp. 233-239.

[56] G. I. Shakibaei, P. Mirzaei and A. Bazgir, "Dowex-50W Promoted Synthesis of 14-aryl-14*H*-Dibenzo[*a,j*]xanthene and 1,8-Dioxo-Octahydroxanthene Derivatives Under Solvent-Free Conditions," *Applied Catalysis A: General*, Vol. 325, No. 1, 2007, pp. 188-192.

[57] P. Wasserscheid and T. Welton, "In Ionic Liquids in Synthesis," Vols.1 and 2, 2nd Edition, Wiley-VCH, Weinheim, 2008.

[58] X.-D. Wang, W.-Y. Wu, G.-F. Tu and K.-X. Jiang, "Synthesis and Physico-Chemical Properties of New Green Electrolyte 1-Butyl-3-Methylimidazolium Perchlorate," *Transactions of Nonferrous Metals Society of China*, Vol. 20, No. 10, 2010, pp. 2032-2036.

Behaviour of Roofing Materials Facing to Micro-Organisms

Marielle Fassier[1], Céline Faugeron[2], Vincent Gloaguen[2], Céline Ducroquetz[1], Olivier Dupont[1]
[1]Centre Technique de Matériaux Naturels de Construction, Paris, France
[2]Laboratoire de Chimie des Substances Naturelles, Limoges, France

ABSTRACT

An accelerated water-streaming test was used to evaluate several roofing materials regarding their behavior to colonization by algae, by closely reproducing the phenomenon of natural biological soiling. A set of roofing materials with defined physical and chemical characteristics was thus investigated against the colonization by algae. Porosity, roughness and chemical composition showed to be factors of influence in the establishment of those micro-organisms.

Keywords: Building Materials; Fired Clay Products; Water-Streaming Test; Micro-Organisms; Colonization

1. Introduction

Algae and cyanobacteria are the first colonizers of the sheathing of the buildings in humid areas after bacteria which are omnipresent. In fact, algae and cyanobacteria colonize all the surfaces where humidity is high. So they develop in marine area but also on land substrates such as plants, rocks, grounds or even building materials [1]. These micro-organisms are present in the air in the form of spores, cells or filaments and have, in case of enough lights, humidity and nutriments, the capacity to adapt their morphology and their physiology to colonize the different types of products quoted above.

On buildings, the accumulation of micro-organisms is translated by the presence of greener black spots in continental area and by red spots in coastal area [2-5]. These spots appear most of the time near the outlet pipes, the architectural flaws, and the fallout of drainpipes or the ventilation of conditioning.

Beside an unaesthetic aspect, the presence of micro-organisms on the buildings causes physical and chemical damage to the materials. In fact, algae and cyanobacteria present on the surface of the material can cause for example the dissolution of the carbonate of calcium in the case of buildings made in stone or in concrete [5]. Except this case, the presence of micro-organisms on the surface of buildings promotes the penetration of the water in the porous network of the material which leads, first, to a inflating and the appearance of cracks and then to an increase of the humidity rate in the material which is propitious to the acceleration of the process of colonization [1].

At the present time, there are different means to fight against the micro-organisms such as the mechanical cleaning, the chemical cleaning or the hydrofugation. However, these treatments are, most of the time, toxic for the environment, not very effective and very expansive.

Algae and cyanobacteria form a series of varied species of unicellular and pluricellular plants which size is generally between 0.5 µm and 1 mm. These organisms are photo-autotrophe and need an environment vivid in light and humidity useful to their photosynthesis reaction.

Green algae, presented on building, are eukaryote organisms which the wall is composed of a rigid fibrillaire part and of a mucilagineuse part whereas cyanobacteria are prokaryote organisms composed of a wall formed by peptidoglycans and lipoproteins.

The colonization of the substrate by micro-organisms shows a dynamic aspect. Bacteria colonize quickly the surface of the material and they are followed by the establishment of algae and cyanobacteria. This establishment can be divided into five steps which are the followings: The development of an organic biofilm which absorbs ions and organic molecules at the surface of the material, the transport of micro-organisms on the surface and the formation of co-aggregates of micro-organisms, the reversible adhesion of micro-organisms and aggregates, the co-adhesion of micro-organisms each other and the irreversible adhesion of micro-organisms on the surface of material by the secretion of exo polymers [1]. This adhesion of micro-organisms is influenced by the topography, the chemical composition and the surface tension of the substrate.

Follow a seeding of the surface of building materials

due to the uprooting, from a colonized substrate, of spores, cells or fragments of filaments and their transport, algae and cyanobacteria develop on very various surfaces [6]. The biological development of algae and cyanobacteria is influenced by several environmental factors such as light, humidity, pluviometry, outside temperature, wind, atmospheric pollution and vegetal environment [1] but also by the physical and chemical characteristics of the colonized material such as porosity, roughness, surface tension and pH.

Researches on facades led during the twenty last years by Rindi and Gaylarde have allowed to list algae and cyanobacteria presented most of the time on buildings [1,7-9].

Concerning algae, Klebsormidium Flaccidum seems to be the most frequent with a presence rate of 55%. It follows by Trebouxiaspp, Stichococcus Bacillaris, Chlorosarcinopsis Minoe, Chlorella cf Mlr and Blls which have respectively a presence rate of 49%, 46%, 28% and 27%.

For the cyanobacteria, the most represented on facades is Cyanosarana Parthenonensis with a presence rate of 28% followed by Chroococcus Lithophilus, Calothrix Pulvinata, Leptolyngbya Foveolarum, Gloeocapsa Sanguinea and Phormidium Vorium which respectively have a presence rate of 21%, 21%, 18%, 15% and 13%.

As Klebsormidium Flaccidum is the specie of algae the most present on the surfaces of building materials, it has been withheld for this study. The physical, chemical and surface characteristics of the different roofing products have been determined. According to previous studies, these criteria could have an influence on the colonization of substrates by micro-organisms and that is why they have been retained. This study aims to confirm these observations and to establish correlations between the physical and chemical parameters of products and their ability to be colonized by micro-organisms.

2. Experimental

2.1. Klebsormidium Flaccidum

Klebsormidium Flaccidum is a filamentous algae hard-wearing to area strongly dehydrated even if it grows, most of the time, in area which have a high humidity.

This algae is composed of cells to barred-shaped, forming long filaments slightly constricted at cross-walls, one parietal and cup-shaped chloroplast, covering about 2/3 of the cells, with one clearly visible pyrenoid (L = 5 - 25 μm, w = 4 - 12 μm) [10,11].

The culture of Klebsormidium Flaccidum is realized in a modified Bold's basal medium composed of three mother solutions individually pressure-sealed, after their preparation, at 120°C for 20 minutes. 1 liter of liquid culture medium is composed of 10 mL of solution A, 10 mL of solution B and 1 mL of solution trace introduced in 1 liter of ultra-pure water. The medium is adjusted at pH 7 and then pressure-sealed before the introduction of the strain of algae [6].

Each Erlenmeyer flask is put on magnetic agitator, to oxygenate again the medium to have a better reproducetion and to homogenize the suspension, in anair-conditioned culture room at 20°C ± 0.5°C where the light intensity is controlled (white light) and the photoperiod is 16 h/8 h (light/darkness).

The dosage of algae contained in the 2 liters Erlenmeyer flasks is determined by the measurement of the dry mass of the algae. A known volume of algae suspension is filtered on membranes in micro fibers of glass which have a porosity of 1, 2 μm (Whatman GF/C). These membranes are dried at 105°C for 3 hours. Then, the dry algae mass is determined and attributed to the filtered volume.

2.2. Analysis

2.2.1. pH Measurement

Before pH measurement, materials are grinded to ob- tain powders which are dispersed in pure water and shaken with a magnetic agitator. This measurement is realized with a pHmeterCyberscan pH 1100 (Eutech instruments).

2.2.2. Roughness Measurement

The surface of a bright or not material is not a per- fectly polished surface. In fact, it presents a lot of micro or macro geometric irregularities which are defined by comparison with an average line and are classified into two categories: peak and hollow. All these surface flaws form the roughness.

The surface roughness can be determined by the measurement of some parameters such as:

- Rp: The maximal peak observed on the analyzed length;
- Rc: The maximal hollow observed on the analyzed length;
- Rt: The total roughness. It corresponds to the total of the maximal peak and the maximal hollow observed on the analyzed length;
- Ra: The average distances difference between successive peaks and hollows. It corresponds to the difference between the average distance and the average line.

When visual and tactile comparisons are not sufficient to distinguish a difference of roughness between two surfaces, it is necessary to use a roughness device. This device is composes of an induction transducer which has a diamond point. The variations in the generated tensions are located by a sensor and then converted in different roughness parameters by the electronic of the device.

This measurement has been realized with a roughness

device TR 110 TIMe (Phynix). The obtained value is the average of fifty measurements realized on a sample of the substrate.

2.2.3. Contact Angle Measurement

Among the techniques used to determine the surface tension of a liquid put on a solid the measurement of the contact angle of the solid is the technique the most used. This technique consists in putting a drop of a liquid on the surface of a solid and in reading the angle between this drop and the substrate. The measured value allows going back to the surface energy of the solid due to the Young equation:

$$\gamma_{SV} = \gamma_{SL} + \gamma_{LV}\cos\theta$$

where γ_{LV} is the superficial tension of the liquid in presence of vapour, γ_{SL}, the interfacial tension between the solid and the liquid, γ_{SV}, the superficial tension of the solid in presence of liquid and θ, the contact angle between the liquid and the solid surface.

The experimental device used is composed of a light source and of a digital camera allowing taking a photograph of the drop. The damping angle is measured with software from digitized pictures. The drop is put on the solid surface with a syringe. The horizontality of the substrate is precisely adjusted. The software used allows digitizing the outline of the drop by the analysis of the pictures taken by the digital camera. To conclude, this software determines the static contact angle, of the liquid on the surface, thanks to interpolation methods by polynomials or by arcs of the circle.

2.2.4. Porosity Measurement

The determination of the density with a helium pycnometer is realized like that: the device precisely measures the variation of gaseous volume caused by the introduction of the sample in a calibrated enclosure. The volume is calculated from the pressure transfer observed when the gas passes from the cell containing the sample (pressure P_2) to the reference cell (pressure P_1). This method is based on the Mariotte's law.

$$V_{ech} = V_{cell} - V_{exp} / \left(P_1 / P_2 - 1\right)$$

The volume of the cell, V_{cell}, and the expansion volume V_{exp} are constants. The maximal accuracy is ±0.1% on the volume and ±0.001 g on the weighing.

Then, the density is given by the relation:

$$\rho = \frac{m}{V}$$

where m is the sample mass and V, the sample volume.

This measure has been realized with a helium pycnometer ACCUPYC 1330 (Micromeritics). From the densities of the bulkyand the grinded sample, the opened porosity has been deduced to use the equation:

$$P = \left(\rho_{(grinded)} - \rho_{(bulky)}\right) / \rho_{(grinded)}$$

2.2.5. The Water-Streaming Test

To evaluate the behavior of building materials facing with the presence of micro-organisms, an experimental device has been developed. This device simulates speeding up the colonization, by algae, on the exposed surface of materials.

The principle of the water-streaming test consists in the dampening of the surface of the material by water streaming which is comparable to the rain flow on facades or roof tiles of buildings.

This device, developed by Boulon and Dubosc, is composed of a closed box where the humidity is relatively high, favorable to the development of micro-organisms [12]. In this box, the samples are put back to back on two supports in stainless steel at an angle of 45° to increase the colonization process. At regular time intervals, the samples are sprayed by the Klebsormidium-Flaccidum culture, kept at a constant temperature by a thermo-regulator, presented in the bottom of the box. This spraying is realized by a ramp pierced by holes which have a constant diameter at equal distance each other. Moreover, the box is installed in a room where the temperature and the relative humidity are controlled and have two footlights composed of two neon tubes which have a known intensity (1500 lux, 30 W) and according to a photo period controlled (12 h/12 h).

At regular time intervals, the biological development is evaluated by pictures analysis. The kinetic of colonization on the surface of materials is evaluated by pictures analysis. The surface of the samples installed in the water-streaming test is digitized at regular time intervals with an office scanner. Then, the digitized pictures are treated to digitize the signal as on the **Figure 1**.

Then, the digitized pictures are treated with a pictures

Figure 1. Example of a digitized and converted into 256 grey levels roof tile.

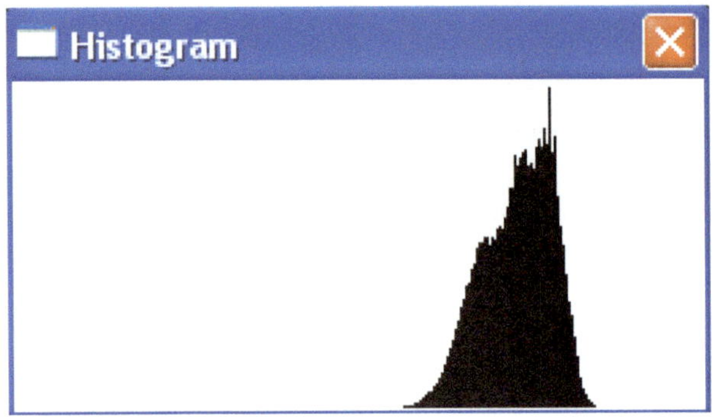

Figure 2. Histogram of the digitized picture representing the number of pixels in function of their intensity.

analysis program to establish ahistogram (**Figure 2**) representing the number of pixels in function of their intensity.

In this treatment, the intensity 0 is attributed to the black pixels and the white pixels correspond to the intensity 255. At the end, to make the area corresponding to the colonization of the micro-organisms more detectable, a threshold is realized for the maximal intensity of the peaks of the histogram. So, the micro-organisms seem to be white on the black support. Then, the percentage of colonization on the sample is obtained making the ratio between the pixels corresponding to the micro-organisms and the total number of pixels. To the end, a graph showing the evolution of the growing of micro-organisms in function of the exposure time in the water-streaming test is drawn for each sample series.

2.3. Materials

During this study, the materials used are the followings:

Slate: It is a metamorphic rock belonging to the family of clayey shales. It is generally composed of silica, alumina, iron oxide, magnesia, potash and soda. It is hard-wearing and its color can vary from the white to the black. Slate can be rectangular or scale-shaped. Its thickness varies between 3 and 9 mm.

Concrete: It is a powdery material, making with water or a saline solution, a binding agent able to agglomerate varied substances when it hardens. This hydraulic bonding agent hardens quickly and reaches in few days its maximal resistance. So, after hardening, this paste keeps its resistance and its tenacity even under water. A homogenous mixture is realized by humid, half-humid, half dry or dry way and then the mixture is introduced in a tubular rotary kiln, lightly sloped, at a temperature about 2000°C. A sudden and fast cooling of the mixture leads to the elaboration of the clinker. The composition of the concretes is a mixture of calcium silicates and calcium aluminates due to the combination of the lime with silica, alumina and iron oxide.

Glass: Glass is a hard, fragile, brittle and transparent to the visible radiation material, most of the time derived from siliceous sand. This is an amorphous material having an important structural disorder and the phenomenon of vitreous transition. It is generally composed of silica and flux. This material has good compatibility with most of chemical compounds exception to hydrofluoric acide.

Fibrocement: It is a building material composed of concrete reinforced by synthetic or natural fibers.

Clay roofing tiles: A roof tile is, most of the time, a fired clay product used as a roofing element. Fired clay products are made from fired clays at high temperature. So, the firing temperature is between 800°C and 1150°C to confer to the products a low porosity, a mechanical resistance, functionality and a surface aspect. The surface aspect depends of several parameters such as the choice of the raw materials, the drawing of the pattern and the process of fabrication. These products can be without coating, with a silicone coating or with a glazed coating.

3. Results and Discussion

3.1. Physical and Chemical Characteristics of the Roofing Products

The physical, chemical and surface characteristics of the different roofing products have been determined and summed up in the **Table 1**. These characteristics are the porosity, the contact angle, the roughness and the pH.

Thanks to the porosity of the roofing products and particularly to their porosity rate, it is possible to divide the roofing products into two categories. In fact, fired clay roofing tiles, which have or not a silicone coating, show a high porosity such as the concrete tiles. The noticed porosity is between 15% and 33%. Concerning the fibrocement sheets they have also a high porosity about 30%. Contrary, the glass tiles and the slates show a low porosity widely inferior to the previous products because it does not exceed 1%.

Concerning the contact angle of the surface of these different roofing products, it is very different from one

Table 1. Physical, chemical and surface characteristics of the roofing products.

Material	Porosity (%)	Contact angle (°)	Roughness (μm)	pH
Fired clay tile without coating	14.9	48.5	5.1	8.4
Fired clay tile with a silicone coating	29.4	67.5	4.4	8.2
Concrete tile	33.0	0.0	7.9	11.6
Glass tile	0.5	30.0	2.5	
Slate	0.7	20.0	3.8	
Fibrocement	29.9	0.0	6.8	9.9

product to another but stays however in the area of hydrophilicity. In fact, the values of the static contact angle of a drop of water deposited on the surface of the different materials does not exceed 68° which means that all the products cannot prevent water to stagnate at their surface provoking its streaming. So, these materials are propitious to their colonization by micro-organisms due to the fact of the presence of water at their surface. However, it seems that when a water drop is deposited on the surface of a concrete tile or on a fibrocement sheet, a state of total damping is observed. This lets to think that these materials will be more favorable to the colonization by micro-organisms then the others and so the colonization will be faster and higher.

The average roughness of the surface of the different products is spread between 2.5 and 7.9 μm. The difference between the values is not very important but the higher value observed for the concrete tile lets to think that this product will be easier colonized by the micro-organisms than the others and precisely than the glass tile which has the average roughness lower.

Concerning the pH values, they show that all the products have a basic surface which is propitious to their colonization by micro-organisms. In fact, algae are inclined to hang on substrates which have pH values between 9 and 11 [1]. These observations let to think that concrete tile, having a pH value about 11.6, such as the fibrocement sheet (pH $\approx$ 9.9), are more propitious to the colonization by micro-organisms than the others. So the colonization of their surface by algae will be faster and higher.

3.2. Behavior of the Roofing Materials Facing with Micro-Organisms

Concerning the roofing products, trials have been realized to compare their resistance to micro-organisms. The results are presented on **Figure 3**.

First, it is important to notice that whatever the coating

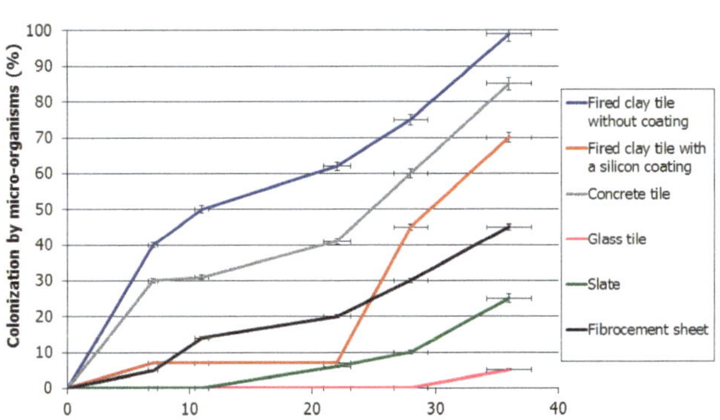

Figure 3. Kinetic of colonization and of growth of micro-organisms on the surface of three samples of several roofing materials.

or the material, the colonization of the surface by micro-organisms does not appear after the same duration passed in the time passed in the water-streaming test. For the samples of fired clay tile without coating, between the 1st and the 9th day, the colonization and the proliferation of micro-organisms on the surface of the samples is very high and very important. In fact, the surface of the samples is covered at 57% after nine days of exposure. This development is linear with regard to the exposure time during this period. Between the 9th and the 20th day, the proliferation of algae keeps accelerating becoming extremely important. After twenty days of exposure in the water-streaming test, the surface of the samples of fired clay tile is covered at 61%. Between the 20th and the 32nd day of exposure, the speed of colonization of micro-organisms strongly reduces to become independent of time. In fact, the rate of colonization becomes nearly constant to reach a total covering of the samples of 99% at the end of the experience. To conclude, the fired clay tile shows a fast and important greening.

For the samples of fired clay tile with a silicone coating, between the 1st and the 7th day, the percentage of colonization of the surface of the samples is about 7% which means that the colonization and the proliferation of micro-organisms is very slow. Micro-organisms have difficulties to hold on the substrate and to develop. Between the 7th and the 22nd day, the proliferation of micro-organisms stays constant which means that the proliferation of micro-organisms is stopped and that any new algae find hanging point at the surface of the samples. Then, between the 22nd and the 28th day, the proliferation of the micro-organisms strongly and quickly increases to reach a percentage of colonization of the surface about 45%. So, algae find many hanging points which allow their development. This can be explained by the deterioration of the silicone layer located at the surface of the fired clay tile. In fact, after about twenty days of exposure in the water-streaming test, the silicone layer is damaged by the streaming of water and the fired clay tile with a silicone coating retakes the aspect and the properties of a fired clay tile without coating. Moreover, these results are in harmony with the release tests and the contact angle measurements realized on this product which detects a degradation of the silicone layer after 15 to 20 days of soaking. Finally, between the 28th and the 36th day, the proliferation of micro-organisms continues to be very important and fast. In fact, after 36 days of exposure, the surface of the samples is colonized at 70%. This result is comparable to the result obtained with the fired clay tile without coating taking the delay of greening due to the silicone layer into account. To conclude, the colonization and the proliferation of micro-organisms on the surface of fired clay tile with a silicone coating is very slow and light during the first twenty two days of the experiment. The micro-organisms do not find hanging points, they cannot develop and they are washed by the successive watering. Silicone products, showing a hydrophobic surface, promote water streaming on their surface and so the slowdown in the colonization of surfaces by micro-organisms during this period. However, after twenty days of exposure, the silicone layer at the surface of samples is damaged by the successive water streaming and the behavior of the surface of the product becomes comparable to the behavior of a fired clay tile without coating. The colonization and the proliferation of micro-organisms become more important and the final covering rate of the samples reaches 70%.

For the concrete tile, their behavior facing the greening is comparable to the behavior of a fired clay tile without coating. In fact, between the 1st and the 7th day, the colonization and the proliferation of micro-organisms at the surface of the samples are very high and very important. The surface of the samples is covered at 30% after seven days of exposure. This development is linear with regard to time exposure during this period. Between the 7th and the 22nd day, the proliferation of algae stays notable but the speed decreases with regard to the first period of the experiment. So, after twenty two days of exposure in the water-streaming test, the surface of the samples on concrete tile is covered about 41%. Between the 22nd and the 36th day of exposure, the speed of colonization of micro-organisms strongly increases with regard to the exposure time. In fact, the colonization rate becomes very important and high to reach a final covering of the samples of 85% at the end of the experiment. To conclude, concrete tile shows a fast and important greening from the beginning of the experiment. Micro-organisms find hanging and development points easily and they are uniformly distributed at the surface of the samples. The covering rate is important and reaches a final value of 85% of the surface.

For the fibrocement sheet, the kinetic of colonization of micro-organisms at their surface is comparable to the kinetic observed on fired clay tile without coating but the rate and the speed of proliferation of micro-organisms stay lower. So, between the 1st and the 7th day, the colonization and the proliferation of micro-organisms at the surface of the samples are limited and low. In fact, the surface of the samples is covered at 5% after seven days of exposure. This development is linear with regard to the exposure time during this period. Between the 7th and the 11th day, the proliferation of algae is so important and the development of algae at the surface of samples becomes faster and more present. So, after eleven days of exposure in the water-streaming test, the surface of samples is covered at 14%. Between the 11th and the 20th day of exposure, the speed of colonization of micro-organisms seems to stagnate and become independent of the exposure time. Algae stay at the surface of the samples but do not grow. The colonization rate stays around 20% of the surface. Then, between the 22nd and the 36th day of exposure, the colonization and the proliferation of micro-organisms at the surface of the fibrocement sheet continue and are very strong. The development of algae is faster and more important than in the previous period of the trial. The final covering of the samples is about 45%. To conclude, the fibrocement sheet shows a more limited greening than the fired clay tile without coating at the beginning of the trial. However, the colonization of micro-organisms intensifies and becomes faster after about twenty days of exposure in the water-streaming test. So it seems that after a certain exposure time, the surface of the samples of fibrocement sheet is totally saturated with water and shows a biological layer and a nutritive medium propitious to the fixing and the development of micro-organisms. The final covering rate reaches 45% of the surface.

Concerning the slate, it is important to notice that the colonization by micro-organisms appears on these surfaces only after eleven days of exposure. Then, between the 11th and the 28th day, the colonization and the proliferation of micro-organisms at the surface of the samples are limited and low. In fact, the surface of the samples is covered at about 10% after twenty eight days of exposure. This development is linear with regard to the exposure time during this period. Between the 28th and the 36th day, the proliferation slightly increases but so significant and the development of algae at the surface of the samples becomes more important and present. So, after thirty six days of exposure in the water-streaming test, the surface of the samples is covered at about 25%. To conclude, the samples of slate show an important delay of greening with regard of fired clay tile without coating. In fact, the greening appears only after eleven days of exposure in the water-streaming test. At the end of this period, algae develop at the surface of the samples but slowly and it is limited because the final covering after thirty six days of exposure does not exceed 25%. So, these products show the property to delay so important the greening with regard to the fired clay product without coating thanks to, in majority, their physical, chemical and surface properties.

At the end, for the glass tile, it is important to notice that the colonization by micro-organisms appears on these surfaces only after twenty eight days of exposure. Then, between the 28th and the 36th day, the colonization and the proliferation of micro-organisms at the surface of the samples are limited and low. In fact, the surface of the samples is covered between 5% and 6% after thirty six days of exposure. This development is linear with regard to the exposure time during this period. To conclude, the samples of glass tile show an important delay of greening with regard to the fired clay tile without coating. In fact, the greening appears only after twenty eight days of exposure in the water-streaming test. Further to this period, algae develop at the surface of the samples but this development is slow and limited because the final covering after thirty six days of exposure dos not exceed 6%. So these products show the property to delay so important the greening with regard to the fired clay tile without coating thanks to, in majority, their physical, chemical and surface characteristics.

4. Conclusions

The analysis of physical, chemical and surface characteristics of roofing products allows putting above differences between materials and doing supposition with regard to their behavior facing micro-organisms. The analysis of curves representing the percentage of colonization of these surfaces by micro-organisms allows confirming these suppositions.

In fact, it appears that:

- More the porosity is high, more the colonization of substrates, by micro-organisms, is fast and important;
- More the contact angle of the surface of the products is low, that is to say more water is inclined to spread on the support, more the colonization of substrates is fast and important;
- More the average roughness of the surface of the substrates is high, more the colonization of supports is fast and important;
- For basic pH values, micro-organisms hold on the substrates and develop easier at the surface of substrates.

For the future, it could be interesting to observe the behavior of the building materials in front of micro-organisms when they are treated with commercial hydrofuge solutions. This future study can be interesting to evaluate the efficiency of these commercial products against micro-organisms.

REFERENCES

[1] H. Barbebrousse, B. Ruot, C. Yéprémian and G. Boulon, "An Assessment of Façade Coatings against Colonisation by Aerial and Cyanobacteria," *Building and Environment*, Vol. 42, No. 7, 2007, pp. 2555-2561.

[2] P. Berdahl, H. Akbari, R. Levinson and W. A. Miller, "Weathering of Roofing Materials—An Overview," *Construction and Building Materials*, Vol. 22, No. 4, 2008, pp. 423-433.

[3] D. M. John, "Algae Growths on Buildings: A General Review and Methods of Treatment," *Biodeterioration Abstracts*, Vol. 2, No. 67, 1988, pp. 81-102.

[4] R. J. Lombardo, H. Barberousse, G. Tell and A. Couté, "Factors Involved in the Colonization of Building Facades by Algae and Cyanobacteria in France," *Biofouling*, Vol. 22, No. 2, 2006, pp. 69-77.

[5] H. E. Schlichting, "Some Subaerial Algae from Ireland," *British Phycological Journal*, Vol. 10, No. 3, 1975, pp. 257-261.

[6] H. C. Bold, "Arbornealgae: Their Abundance and Heterogeneity," *Science*, Vol. 143, No. 3606, 1964, pp. 583-585.

[7] C. C. Gaylarde, P. M. Gaylarde, J. Copp and B. Neilan, "Polyphasic Detection of Cyanobacteria in Terrestrial Biofilms," *Biofouling*, Vol. 20, No. 2, 2004, pp. 71-79.

[8] F. Rindi and M. D. Guiry, "Composition and Distribution of Subarial Algal Assemblages in Galway City, Western Ireland," *Cryptogamie-Algologie*, Vol. 24, No. 3, 2003, pp. 245-267.

[9] F. Rindi and M. D. Guiry, "Composition and Spatial Variability of Terrestrial Algal Assemblages Occurring at the Bases of Urban Walls in Europe," *Phycologia*, Vol. 43, No. 3, 2004, pp. 225-235.

[10] M. A. Shirakawa, K. Loh, V. M. John, M. E. S. Silva and C. C. Gaylarde, "Biodeterioration of Painted Mortar Surfaces in Tropical Urban and Coastal Situations: Comparison of Four Paint Formulations," *International Biodeterioration & Biodegradation*, Vol. 65, No. 5, 2011, pp. 669-674.

[11] K. D. Stewart and K. R. Mattox, "Structural Evolution in the Flagellated Cells of Green Algae and Land Plants," *Biosystems*, Vol. 10, No. 1-2, 1978, pp. 145-152.

[12] A. Dubosc, G. Escadeillas and P. J. Blanc, "Characterization of Biological Stains on External Concrete Walls and Influence of Concrete as Underlying Material," *Cement and Concrete Research*, Vol. 31, No. 11, 2001, pp. 1613-1617.

Green Cycle: Sulfate Sorption from Natural Water on Anionic Clay Compound Obtained from Industry Wastewater

Genoveva Rosano-Ortega[1*], Carlos Vega Lebrún[1], Juan M. López Oglesby[1], Sonia Martínez Gallegos[2], Daniel Cruz-González[3], Beatriz Pérez Armendariz[1]

[1]Research Department, Universidad Popular Autónoma del Estado de Puebla (UPAEP), Puebla, México
[2]Division of Graduate Studies and Research, Instituto Tecnológico de Toluca, México City, México
[3]Chemical Engineering Faculty, Benemérita Universidad Autónoma de Puebla, Puebla, México

ABSTRACT

In this paper we are describing a green cycle process. The first step was a novel hydrotalcite-like compound (HTLC) synthesized by a co-precipitating method, under standard ambient temperature and pressure, using chemical industry wastewater rich in divalent and trivalent cations, activated by a thermal treatment and finally characterized by scanning electron microscopy (SEM), energy dispersive X-ray fluorescence (EDS) and thermogravimetric analysis (TGA). The second step was a series of batch sorption tests performed with this activated HTLC and untreated underground sulfurous water from the state of Puebla, México. The HTLC calcined at 500°C/3 h exhibited the best sorption ability for ions, demonstrated a decrease of the hardness and sulfate ions to below the regional legal standards for drinking water. Once inactive after being used in water treatment, the sorbed ions were removed by ion exchange in a carbonate-containing solution, resulting in an 80% recycling of the material which upon activation demonstrated a retained capacity for water treatment. This recyclability suggests the exciting possibility of this novel compound as an efficient "green" technology in water treatment processes.

Keywords: Hydrotalcite-Like Compound; Wastewater; Sulfurous Underground Water; Ion Sorption; Green Cycle

1. Introduction

Underground water in the State of Puebla, México is characterized by a slightly acidic pH (~6.5), high hardness (~1200 mg/L $CaCO_3$) and the presence of sulfurous chemical species mainly as sulfate ions such as SO_4^{2-} (~950 mg /L) and H_2S (~8.0mg/L), which give the water an unpleasant odor and flavor. This is particularly a problem in areas of volcanic activity where in many cases the drinking water contains traces of sulfurous compounds, even after standard purification. Purification treatments which address this lingering problem must be able to process large amounts of water at low cost to be effective [1].

Layered double hydroxides (LDH) or compounds are referred to in this paper as hydrotalcites (HT) or anionic clays. From the general formula for HT, $M(II)_{1-x}M(III)_x(OH)_2(A^{n-})_{x/n}mH_2O$ [2-5], the crystal structure of these LDH (**Figure 1**) consists of positively-charged brucite-type main sheets which forms an octahedron as a subunit of a layered structure of HT, where partial replacement of divalent M(II) with trivalent metal ion M(III) in the brucite-type structure results in an excess of positive charge, which is compensated in the HT structure by an n-valent anion (A^{n-}) and x can have values between 0.2 and 0.33. These anions such as carbonate, sulfate and hydroxide (which may also be hydrated) can be introduced between the layers of the crystal structure. The

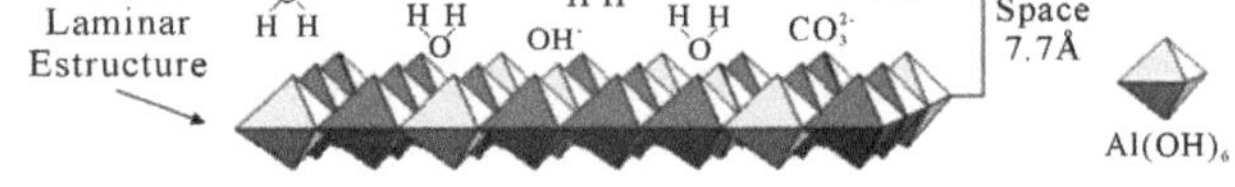

Figure 1. Interlayer hydroxides in an LDH with the general formula: $[M(II)_{1-x}M(III)x(OH)_2]^{+x}[A^{n-}{}_{x/n}zH_2O]^{-x}$.

*Corresponding author.

anions and water molecules introduced into interlayer spaces are only weakly bonded to the crystal layers. Therefore, HT can be used as an anion exchanger, polymer stabilizer, antacid anion scavenger and adsorbent [6-8].

The preparation of anionic clays has been described vastly in the literature [9-12]. Two important reviews of anionic clay chemistry were published in which the synthesis methods available for anionic clay synthesis have been summarized [13,14]. HT activation takes place through thermal decomposition by heating HT in air up to 500°C. This causes it to be calcined into periclase-like compounds (Mg-Al oxide) which possess a well-known "memory effect" or the ability to reconstruct themselves to their original structure when it is exposed to water [14].

In general, crystalline HT compounds are excellent adsorbents for a variety of anions in many different applications [15,16]. Nevertheless, they are relatively expensive due to the purity required of the reagents and complexity of the synthesis conditions. Special activation steps like calcination increase the intricacy and costs of the synthesis process even more [17]. An objective of the present work was to give an alternative called "green cycle" in which a novel method for the synthesis of an HTLC suitable for the adsorption of sulfate ions from sulfurous water was provided. A further requirement was that the processes not involve expensive reagents or complicated synthesis steps allowing the HTLC to be produced in a cost-efficient manner. This goal was achieved by taking advantage of a wastewater byproduct from a chemical industry, which has an excess of Mg^{2+} and Al^{3+} ions.

2. Experimental Details

2.1. Synthesis of Mg-Al-CO_3 HTLC

Commercially, bleaching earths are provided by extracting naturally-occurring clays, which contain a variety of different metals like Na, K, Ca, Mg, Al and Fe, with strong mineral acids. During the extraction process, some of these metals are dissolved from the clay mineral and form a cation-rich acid solution (wastewater) which needs treatment. The acid-treated clay may be used as a bleaching earth for the treatment of edible oils or as catalyst material. Due to the high relevance of bleaching earths in many industrial applications and their economic value, the cation-rich byproduct solution is highly available and inexpensive [18].

The HTLC with carbonate as the interlayer anion referred in this paper were prepared following a standard co-precipitation method [19]. The acid extract obtained in the production of bleaching earth from an industrial manufacturer in Puebla, Mexico, was selected as a byproduct such that it contained at least 6.5 g/L Al^{3+} ions and 4.5 g/L Fe^{3+} ions, both calculated as elemental Al and Fe. Generally, the clays used as the starting material in the production of bleaching earths also contain Mg^{2+} ions which are extracted during the treatment of the clay with strong mineral acids. Therefore, in most cases only a small amount of Mg^{2+} ions must be added to adjust the concentration of the acid extract to at least 6.8 g/L Mg^{2+}, allowing a range of 6.8 to 25 g/L for the final solution, calculated as elemental Mg. This corresponds to the difference in Mg concentration in the acid extract and the adsorbent to be synthesized [20].

The wastewater solution used for the laboratory synthesis of the HTLC corresponded to a typical bleaching earth extract: it contained 9.7 g/L of Al^{3+} and 20 g/L of Mg^{2+}, as determined by Inducted Couple Plasma (ICP, **Table 1**). In a three-necked flask equipped with a stirrer, 1 L of cation-rich wastewater was mixed with 20 mL of a 20 (wt%) solution of $Mg(OH)_2$ at 600 rpm, the pH value (pH 20 Conductronic electrode) was adjusted to 7 with a 20 (wt%) NaOH solution. After that, 800 mL of a 15 (wt%) Na_2CO_3 solution was added to reach a pH 10. Finally, the stirring speed was increased to 1800 rpm and the solution was stirred for 30 min. A solid precipitate was formed which was named HT3015D.

To optimize the previously-mentioned laboratory method and test the technical synthesis methodology for an expansion to the industrial level-production of HT3015D using the cation-rich wastewater, the composition shown in **Table 1** was used. The synthesis was carried out on a pilot scale as follows: The solution was heated to 60°C and then 98 g of $Mg(OH)_2$ (aqueous solution, 20 wt%) was added dropwise under high-shear agitation that supported an uniform product-forming process and prevented an early agglomeration-flocculation. During addition of the $Mg(OH)_2$ solution the temperature of the mixture was kept at 60°C during 10 min of agitation. Thereafter, for all the trials reported for further on in the

Table 1. Technical information by ICP of an industry wastewater from the bleaching earths process.

Analysis	Values
Acidity, $g \cdot L^{-1}$	35.5 - 45.0
pH	<1
Density at 20°C, $g \cdot cm^{-3}$	1.09
Solid content, %	<0.3
Composition, $g \cdot L^{-1}$	
Al^{3+}	9.7
Fe^{3+}	4
Ca^{2+}	2.3
Mg^{2+}	20
Colloidal Silica	0.3

paper, an alkaline solution was dropwise added, until the pH reached 7 by using NaOH solution (10 - 30 wt%). Though not used here, other solutions such as KOH, $Ca(OH)_2$ or $Mg(OH)_2$ can be used as a base to neutralize Al/Mg source mixture. During adjustment of the pH-value a precipitate containing metal ions was produced. To achieve complete precipitation of the metal ions, solution of Na_2CO_3 (20 wt%) was added a until the pH reached 10. Again, though not used here, other solutions such as $NaHCO_3$, K_2CO_3, $KHCO_3$, $Ca(HCO_3)_2$, $MgCO_3$ and $Mg(HCO_3)_2$ in pure form or as aqueous solutions with a solid content within a range of 10 to 25 wt% can be used to form the precipitate. The mixture was kept at 60°C for one hour while continuing high-shear agitation, then cooled to room temperature and then filtered through a Buchner funnel. To remove sulfate impurities from the H_2SO_4 bleaching earth process, the precipitates were washed 3 times with 500 mL tap water that did not exceed the regional legal standards for drinking water quality, then placed on a glass dish and dried at 120°C in an oven. The dried solid was milled and sieved through a 75 μm mesh to reach the desired particle size. This dehydrated solid precipitate was named HT3015O.

Finally, both solid samples synthesized HT3015D and HT3015O, with $x = (Al^{3+}/Al^{3+} + Mg^{2+}) = 3$, were calcined for activation by heating at 500°C/3h, obtaining HT3015DT and HT3015OT, respectively, where these periclase-like compounds or metallic oxides (Tan *et al.* 2011) were utilized for the sulfurous underground water treatment.

2.2. Sorption Test

Sorption batch tests were carried out with a ratio of 2:0.01, 2:0.02 and 2:0.03 L:Kg, using HT3015OT and sulfurous underground water with a quality described in the Introduction section. These tests were carried at room temperature and pressure conditions and stirred for 1 h at low shear, and then the adsorbent was separated by filtering through a paper filter. The water samples were analyzed for their SO_4^{2-} concentration by UV spectroscopy at 420 nm. Water hardness and pH measurements were done by the methods described by the American Public Health Association [21].

2.3. Characterization Techniques

Sorption batch tests X-ray diffraction analysis (XRD) was employed to identify and characterize the species present and their degree of crystallinity of the synthesized materials. The XRD patterns were recorded using a SIEMENS D500 powder diffractometer system with CuKα radiation (Kα = 0.1518 Å, 36 kV, 20 mA), a voltage of 35 kV, a current of 25 mA, and a goniometer speed of 3°/min). The samples were scanned for 2θ values ranging from 0° to 75°.

The thermal behavior data were obtained on TGA51 TA Instruments system in a nitrogen atmosphere at 35 mL/min and set at a heating rate of 10°C/min. SEM images and elemental contents were obtained using a high-vacuum scanning electron microscope, JEOL 5900LV, operating at an accelerating voltage of 15 kV, with an OXFORD probe for energy-dispersive X-ray emission analysis (EDS).

2.4. Recycling of the HT3015O

To regenerate the HT3015OT used in sorption tests as a HTLC carbonated for recycling, an ionic exchange of the sulfates ions trapped in the lamelar structure of HT was performed with an oversaturated-carbonates ionic solution of Na_2CO_3 at pH 7. Its structure was then evaluated by XRD, re-activated at 500°C/3 h and recycled for sorption tests [22,23].

3. Results & Discussion

3.1. Characterization

HT3015D and HT3015O samples show diffraction patterns which are typical for the LDH structure (**Figure 2**) with sharp and symmetric reflections of the basal (003), (006) planes and broad, referred by Miyata and Okada [24]. Thermally-treated sample HT3015OT corresponded to periclase, a typical calcined product of HTLC (JCPDS 45-0946). The calculated basal spacing and lattice parameters for HT3015D showed a basal spacing d003 of 7.68 Å and 7.36 Å for HT3015O, and are in agreement with the values reported in the literature indicating the formation of well-crystallized HTLC, the literature data have indicated that the basal d spacing of the (003) plane is dependent on the size of the guest anion in the inter

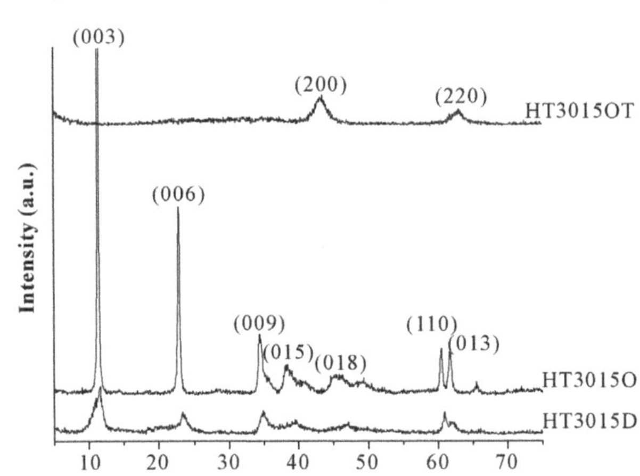

Figure 2. X-ray diffraction patterns of HTLC synthesized in the laboratory (HT3015D), optimized and synthesized at the industrial scale (HT3015O) and thermally treated for activation (HT3015OT).

layer. These values correspond well with the most intense peak at 8.64 Å, determined by Miyata and Okada [24], Heraldy [25] and Alexa [8]. The increment of d003 spacing suggested major crystallinity and no other diffraction peaks were detected in any of the patterns obtained. This indicated that the $MgAlCO_3$ -type HTLC was the only crystalline-precipitated product using the cation-rich byproduct solution.

Intensity and crystalline differences were observed in the XRD patterns between HT3015D synthesized directly and HT3015O from the optimized industrial-scale process (**Figure 2**).

Figure 3(a) shows that the original HT3015O powder particles presented a very singular morphology: they seemed to be hollow spheres, having diameter size distribution as an average of the three consecutive measurements between 0.6 µm and 2.5 µm. Additionally, the surface of these particles seemed to be corrugated, with seashell-like structures. In fact, this kind of morphology has been already reported for certain synthesized metal oxides [26]. Later, during the heat treatment, HTLC decomposed, producing gases (H_2O and CO_2). Additionally, the production and release of the gases, from the core of the particles, inhibited the densification of the surfaces and produced porous materials.

For HT3015O, the mixing speed was slower than for HT3015D at all stages, and the pH value for HT3015O was higher than HT3015D. These two changes in the synthesis process yielded a more crystalline product, and with more intense and better-defined peaks for HT3015O.

Thermally-treated sample HT3015OT showed parts of diffraction peaks disappearing, indicating that the structure of HT began to decompose. Furthermore, the non-existence of peak (006) also suggested the collapse of the layered structure.

From **Figure 3(b)**, the HT3015O elemental composition was almost entirely: Al, Mg, O and C. The x factor was determined with the following Equation (1):

$$x = \frac{M^{3+}}{M^{3+} + M^{2+}} \tag{1}$$

The obtained experimental concentration values are shown as an inset in the same **Figure 4**, and with the value of 0.30 + 0.19, which corresponded to that of x expected by its synthesis.

The elemental composition of HT3015OT that was exposed to sulfurous underground water for 24 h (**Figure 4(a)**) contained sulfates due to the sorption process. The main components remained the same as in the original HT3015O (**Figure 4**) but also included S, Ca, and some Si. Elemental composition of the gray crystals and white spherical particles in **Figure 4(b)** showed that the white ones were richer in Ca and were attached to surface of the gray crystals.

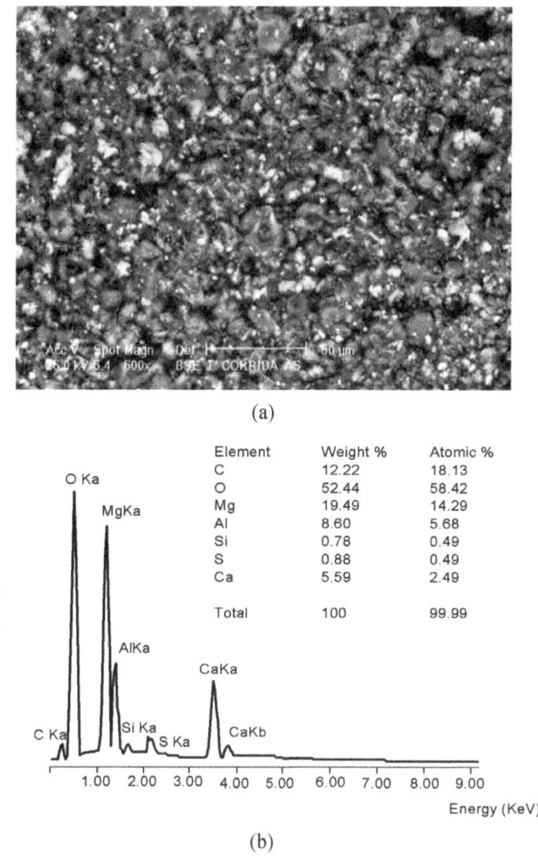

Figure 3. (a) HT3015O morphology as a seashell-like structure obtained from SEM; (b) EDS of HT3015O.

To complete the characterization study, TGA was applied to determine the high-temperature thermal stability of the samples prepared. **Figure 5** shows the TGA analysis results for the HT3015D and HT3015O samples. The diagrams for the two samples synthesized are very close to those of HTLCs reported elsewhere [27]. The thermal decomposition process could be described in two consecutives overlapped steps: 1. The mass-loss level of 9.63% was close to the theoretical value corresponding to the loss of crystal water, indicating that the main weight-loss phase involved in this step was the removal of interlayer water showing an endothermic peak corresponded to 25˚C to 200˚C [28,29].

3.2. Sorption Test

Figure 6 shows the sorption tests in 20 mL of sulfurous water for HT3015OT. This graph shows the uptake of the main problematic elements (sulfate ions, hardness, and alkalinity) considered in this underground water for different clay loads. To make this a viable large-scale application, minimizing the required quantities of HTLC for treatment is desired while achieving the goals set by official Mexican law of 400 mg/L for sulfate ions and 500 mg/L for hardness [30]. In the tested samples, the

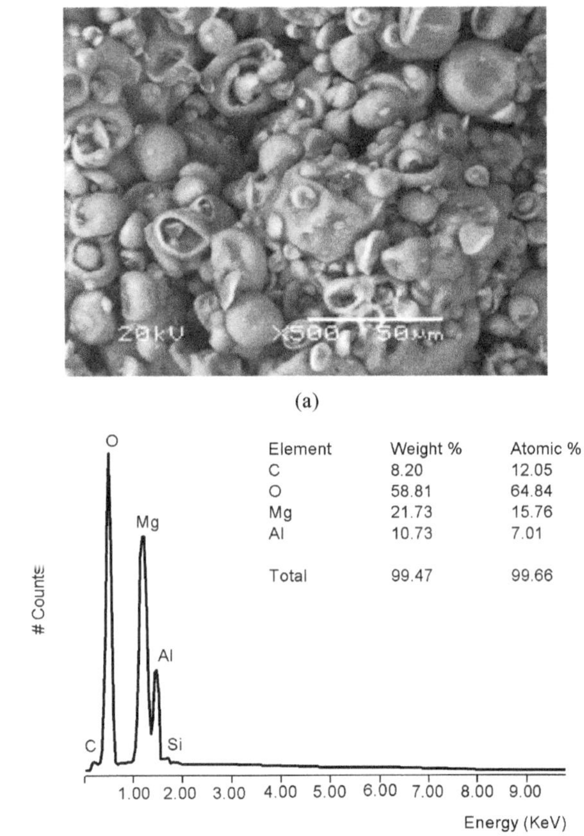

(b)

Figure 4. (a) Morphology of HT3015OT and exposed to sulfurous underground water for 24 h; (b) EDS of HT3015OT and exposed to sulfurous underground water for 24 h.

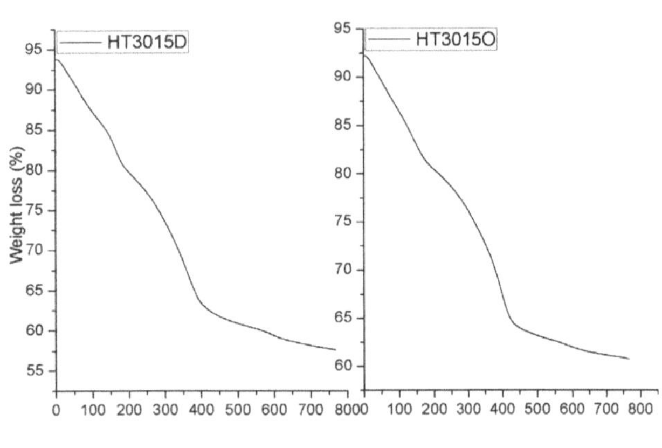

Figure 5. TGA analysis for HT3015D and HT3015O samples.

amount of HTLC employed successfully decreased the initial concentration of $CaCO_3$ by half. For higher quantities of HTLC, (200 mg/L of HT3015OT), the three water quality standards previously mentioned were met or exceeded. This larger concentration of clay removed up to 62% of the initial sulfate ions. In a similar fashion, alkalinity values were decreased below detectable levels

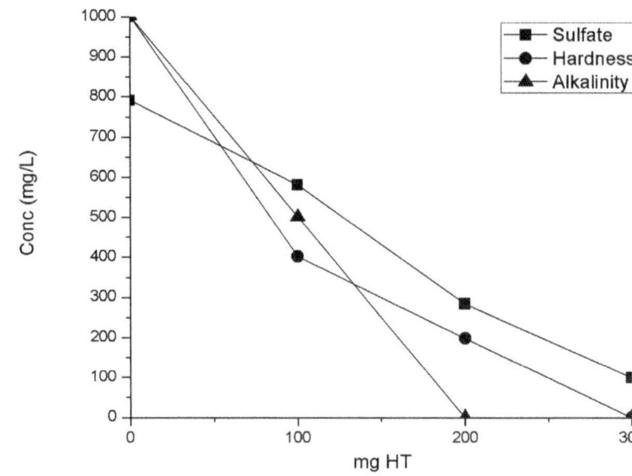

Figure 6. Sorption batch test on for sulfate, hardness and alkalinity using HT3015OT.

for the employed assays when using 200 mg/L of HT3015OT.

Adding HT3015OT to sulfate-contaminated water raised the initial pH 7 up to 10. It is believed this increase in pH helps to reduce hardness parameter. This behavior is important because the HTLCs were thought of only as means for anion exchange [31], but now we can see that it also precipitated cations. The mechanism proposed from these observations is that when HT3015OT comes in contact with the sulfurous water a catalytic activity of dissociation of the water molecules occurs ($H^+ + OH^-$). The mixed oxides of Mg and Al [Mg(Al)O] started to reconstruct its crystal structure of LDH trapping H+ and the hydroxides $\left[Mg(OH)_6^{4-}\right]$ and $\left[Al(OH)_6^{4-}\right]$ were formed. This was evidenced by the change in water's pH from 6 - 7 to 10 - 12. Once the mixed oxides of Al and Mg were formed, the structure was able to trap anions (An^-) present in the solution as shown by Equation (2):

$$2Mg_3AlO_4(OH) + 8H_2O + SO_4^{2-} \longrightarrow \left[Mg_6Al_2(OH)_{16}\right](SO_4) + 2OH^- \quad (2)$$

The removal mechanism of SO_4^{2-} from sulfurous water by HT3015OT, happened because of both the memory effect [14], during the regeneration of the lamellar structure of HTLC and the mechanism involved in diminishing hardness by the chemical precipitation of $CaCO_3$ and to a lesser degree $MgCO_3$ and $MgCa(CO_3)_2$. Because of the former it was also concluded that removal of CO_3^{2-} ions takes place through the two mechanisms mentioned, sorption and precipitation.

Sulfate sorption is due to the slow anion exchange ability of the thermally-treated HTLC, caused by reconstruction of tightly-held layers with SO_4^{2-} ions in the interlayer positions [13]. The sulfated HTLC could be explained by the memory effect because the regeneration of the HT structure is reversible if the calcination temperature does not exceed 500˚C - 600˚C. The HTLC can rehydrate and incorporate anions, sulfates in this case, in order to rebuild the initial HT lamellar structure. On the other hand, the ion exchange equilibrium constant for HTLCs is greater for divalent anions than for monovalent anions.

Table 2 indicates the real sorption capacity of HT3015OT for sulfate ions present in untreated sulfurous well water. After the three sequential treatment runs, the resultant water has been cleansed of the vast majority of the anions and cations present in the original sample, and the sulfate ions have been reduced to below the measurement threshold. The sorption capacity of this new anionic clay is very favorably compared to other reported conventional HT clay behaviors such as those reported by Parker et al. that report a ~3 meq/g HT sorption capacity when prepared in carbonate-free solutions, as well as a near-total sorption of sulfates present in the materials being treated. [32] As seen in **Table 2**, the anionic and cationic sorption exceeds the values reported by Parker *et al.* Currently, the only commercially-viable method for removing sulfates from groundwater is via reverse osmosis. [33] This series of tests shows great promise for a green method of sulfate removal that has yet to be adequately implemented on an industrial scale.

Table 2. Batch sorption of anions and cations present in sulfurous water and water recovered after treatment with the HT3015OT. A reading of N.D., represents no detectable ions with the given measurement techniques.

CONCENTRATION	Sulfurous Water $meq \cdot L^{-1}$	Treated Water $meq \cdot L^{-1}$ (% Reduction)
SO_4^{2-}	19.47	N.D. (>99)
Σ Anions	49.43	4.25 (91.4)
Σ Cations	39.26	6.04 (84.6)
SORPTION		
meq anions/g HT3015OT	-	4.45
meq cations/g HT3015OT	-	3.23
Maximum sorption capacity (in standard sulfate sol.): meq SO_4^{2-} /g HT3015OT		3.85
Real sorption capacity (in water): meq SO_4^{2-} /g HT3015OT		2.58

4. Conclusions

A novel HTLC synthesized by a co-precipitating method, under standard ambient temperature and pressure, using chemical industry wastewater rich in divalent and trivalent cations was performed. Batch sorption tests with this activated HTLC and untreated underground sulfurous water from the state of Puebla, México, showed to be suitable for the adsorption of sulfate ions.

The powder X-ray diffraction patterns suggest that HT3015O is comparable to HT clays made by conventional methods. The overall process presented in this work is very flexible, enabling a wide variety of materials to be prepared with standard industrial equipment, inexpensively and in an environmental-friendly manner. The resulting HTLC proved to have a sorption capacity similar to conventional HT compounds, but generated in an ecologically-sound green chemistry process that takes advantage of existing residue products.

The use of a wastewater from the bleaching earth's process allows a very cost-efficient synthesis of the adsorbent. The presence of metal ions does not have any detrimental effect on the formation of HT and the adsorption capacity of this adsorbent, and no heavy metal ions were leached out during the treatment of sulfur water by the sorption batch tests conducted in this work.

It was found that the HT3015OT obtained is suitable for removing sulfur-containing anions from sulfurous underground water, particularly sulfate ions. Additionally the adsorbent decreases the hardness of water by removing the carbonate and sulfate ions, thereby lowering the amount of hardly soluble salts. The HTLC clays synthesized as part of this work have performed in an excellent manner for diminishing the relevant parameters of originally sulfurous water, below the limits set forth for drinking water according to the reference norm.

In terms of the HT3015O regenerating or recycling, it was shown that the repeating cycle calcination-rehydration-anion exchange allowed the clay to continue performing at stable efficiency >75% of the original HTLC.

To date, little work has been done applying HT compounds to the treatment of groundwater. This work fulfilled the dual purpose of showing the effectiveness of the green chemistry proposed in generating the HTLC HT3015OT from waste products, as well as its novel application to the removal of sulfates from groundwater, avoiding the expense and complications associated with standard reverse osmosis methods by exploiting the capacity of HTLCs to be regenerated for multiple uses before losing their sorption efficiency. We believe this novel compound and application has an important implication to the future of industrial-scale water treatment options.

5. Acknowledgements

The authors thank ININ, ITT, UPAEP University, UNAM and BUAP for the technical work for the different characterization techniques.

REFERENCES

[1] G. Rosano-Ortega, "Evaluación de Hidrotalcita Industrial Para el Tratamiento del Agua Sulfatada del Valle de Puebla," Master Degree Dissertation, Universidad Autónoma del Estado de México (UAEM), Toluca, 2003.

[2] S. J. Palmer, L. M. Grand and R. L Frost, "Thermal Analysis of Hydrotalcite Synthesized from Aluminate Solutions," *Journal of Thermal Analysis and Calorimetry*, Vol. 103, No. 2, 2011, pp. 473-478.

[3] S. J. Palmer, R. L. Frost and T. Nguyen, "Thermal Decomposition of Hydrotalcite with Molybdate and Vandate Anions in the Interlayer," *Journal of Thermal Analysis and Calorimetry*, Vol. 92, No. 3, 2008, pp. 879-886.

[4] G. Fetter, E. Ramos, M. T. Olguín, P. Bosch, T. López, and S. Bulbulian, "Sorption of $^{131}I^-$ by Hydrotalcites," *Journal of Radioanalytical and Nuclear Chemistry*, Vol. 221, No. 1-2, 1997, pp. 63-66.

[5] J. S. Wu, Y. K. Xiao, J. Y. Wan and L. R. Wen, "The Growth Mechanism of Hydrotalcite Crystal," *Science China Technological Sciences*, Vol. 55, No. 4, 2012, pp. 872-878.

[6] S. P. Newman, W. Jones, P. O'Connor and D. N. Stamires, "Synthesis of the $3R_2$ Polytype of a Hydrotalcite-Like Mineral," *Journal of Materials Chemistry*, Vol. 12, No. 2, 2002, pp. 153-155.

[7] H. Khaled and S. Ezzeddine, "Using Kaolinitic Clay for Preparation of a Hydrotalcite-Like Compound," *Materials Sciences and Applications*, Vol. 2, No. 6, 2011, pp. 684-691.

[8] I. F. Alexa, R. F. Popovici and M. Ignat, "Non-Toxic Nanocomposite Containing Captopril Intercalated into Green Inorganic Carrier," *Digest Journal of Nanomaterials and Biostructures*, Vol. 6, No. 3, 2011, pp. 1091-1101.

[9] Z. Q. Yang, K.-M. Choi and N. Z. Jiang, "Microwave Synthesis of Hydrotalcite by Urea Hydrolysis," *Bulletin of the Korean Chemical Society*, Vol. 28, No. 11, 2007, pp. 2029-2033.

[10] V. Dávila, E. Lima, S. Bulbulian and P. Bosch, "Mixed Mg(Al)O Oxides Synthesized by the Combustion Method and Their Recrystallization to Hydrotalcites," *Microporous and Mesoporous Materials*, Vol. 107, No. 3, 2008, pp. 240-246.

[11] T. W. Kim, M. Sahimi and T. T. Tsotsis, "Preparation of Hydrotalcite Thin Films Using an Electrophoretic Technique," *Industrial & Engineering Chemistry Research*, Vol. 47, No. 23, 2008, pp. 9127-9132.

[12] M. R. Othman, Z. Helwani, F. Martunus and W. J. N. Fernando, "Synthetic Hydrotalcites from Different Routes and Their Application as Catalysts and Gas Adsorbents: A Review," *Applied Organometallic Chemistry*, Vol. 23, No. 9, 2009, pp. 335-346.

[13] R. Salomao, L. M. Milena, H. H. Wakamatsu and V. C. Pandolfelli, "Hydrotalcite Synthesis via Co-Precipitation Reactions Using MgO and $Al(OH)_3$ Precursors," *Ceramics International*, Vol. 37, No. 8, 2011, pp. 3063-3070.

[14] T. Qi, W. Z. Yin and L. Zhang, "Synthesis of Hydrotalcite Using Brucite as the Source of Magnesium," *Advanced Materials Research*, Vol. 158, 2011, pp. 241-247.

[15] Y. Liu, E. Lotero, J. G. Goodwin and X. Mo, "Transesterification of Poultry Fat with Methanol Using Mg-Al Hydrotalcite Derived Catalysts," *Applied Catalysis A: General*, Vol. 331, 2007, pp. 138-148.

[16] S.-L. Wang and X.-M. Song, "Synthesize and Application of Mg-Al Hydrotalcite in Flame Retardant Paper Preparation," *Advanced Materials Research*, Vol. 174, 2011, pp. 362-365.

[17] M. Eisgruber, J. Ladebeck, J. Koy, H. Schiessling, W. Buckl and H. Ebert, "Method for the Production of Hydrotalcites," United State Patent No. 7, 2011, pp. 1-11.

[18] A. W. Nursulihatimarsyila, K. Y. Cheah, T. G. Chuah, W. L. Siew and T. S. Y. Choong, "Deoiling and Regeneration Efficiencies of Spent Bleaching," *American Journal of Applied Sciences*, Vol. 7, No. 3, 2010, pp. 434-437.

[19] O. Meyer, F. Roessner, R. A. Rakoczy and R. W. Fischer, "Impact of Organic Interlayer Anions in Hydrotalcite Precursor on the Catalytic Activity of Hydrotalcite-Derived Mixed Oxides," *ChemCatChem*, Vol. 2, No. 3, 2010, pp. 314-321.

[20] Q. Zhang and F. Saito, "Novel Waste Processing by Means of Mechanochemical Treatment," *Materials Science Forum*, Vol. 561-565, 2007, pp. 1569-1573.

[21] American Public Health Association, "Standard Methods for the Examination of Water and Wastewater," 19th Edition, American Water Works Association, and Water Pollution Control Federation, American Public Health Association, Washington, 1995, p. 1050.

[22] A. Kawamoto, T. Suzuki, N. Kiba and T. Sato, "Regeneration and Reuse of Hydrotalcite-Like Anion Exchanger with High Selectivity to Phosphate Anion," *Journal of the Society of Inorganic Materials, Japan*, Vol. 10, 2003, pp. 167-172.

[23] K. František, E. Kovácsová and D. Koloušek, "Removal of Anions from Solution by Calcined Hydrotalcite and Regeneration of Used Sorbent in Repeated Calcination-Rehydration-Anion Exchange Processes," *Collection of Czechoslovak Chemical Communications*, Vol. 64, No. 9, 2009, pp. 1517-1528.

[24] S. Miyata and A. Okada, "Synthesis of Hydrotalcite Like Compounds and Their Physico-Chemical Properties—The Systems Mg^{2+}-Al^{3+}-SO_4^{2-} and Mg^{2+}-Al^{3+}-CrO_4^{2-}," *Clays and Clay Minerals*, Vol. 25, 1977, pp. 14-18.

[25] E. Heraldy, S. J. Santosa and K. Wijaya, "Synthesis of Mg/Al Hydrotalcite-Like from Brine Water and Its Application for Methyl Orange Removal: A Preliminary Study," *Makara Science Serie*, Vol. 15, 2011, pp. 9-14.

[26] G. Mondragón-Gutiérrez, D. Cruz, H. Pfeiffer and S. Bulbulian, "Low Temperature Synthesis of Li_2SiO_3: Effect on Its Morphological and Textural Properties," *Research*

Letters in Materials Science, Vol. 2008, 2008, pp. 1-4.

[27] V. Rives, "Layered Double Hydroxides: Present and Future," Nova Science Publishers, Inc., New York, 2001, p. 350.

[28] Z. Li, Y. Song, J. Wang, Q. Liu, P. Yang and M. Zhang, "Study of Structural Transformations and Phases Formation upon Calcination of Zn-Ni-Al Hydrotalcite Nanosheets," *Bulletin of Materials Science*, Vol. 34, No. 2, 2011, pp. 183-189.

[29] M. J. Kang, S. W. Rhee and H. Moon, "Sorption of MO^{4-} (M = Tc, Re) on Mg/Al Layered Double Hydroxide by Anion Exchange," *Radiochimica Acta*, Vol. 75, 1996, pp. 169-173.

[30] Federal Official Newspaper, "NOM-127-SSA1-1994," Health Secretariat, 1994. http://www.salud.gob.mx/unidades/cdi/nom/127ssa14.html

[31] M. S. Martínez-Gallegos, H. Pfeiffer, E. Lima, M. Espinosa, P. Bosch and S. Bulbulian, "Cr(VI) Immobilization in Mixed (Mg,Al) Oxides," *Microporous and Mesoporous Materials*, Vol. 94, No. 1-3, 2006, pp. 234-242.

[32] L. M. Parker, N. B. Milestone and R. H. Newman, "The Use of Hydrotalcite as an Anion Absorbent," *Industrial & Engineering Chemistry Research*, Vol. 34, No. 4, 1995, pp. 1196-1202.

[33] S. A. Exyco, "Geohydrological Aquifer Study of Puebla Valley," Management Studies and Projects of the National Water Commission and Ministry of Agriculture and Water Resources, Mexico, 1990.

9

The Best Available Technology of Water/Wastewater Treatment and Seawater Desalination: Simulation of the Open Sky Seawater Distillation

Djamel Ghernaout
Chemical Engineering Department, Saad Dahlab University of Blida, Blida, Algeria

ABSTRACT

This review suggests the concept of the best available technology of water/wastewater treatment and seawater desalination which is in fact a simulation of the seawater distillation at the open sky: coagulation in salty water aerated basin/coagulation using seawater as coagulant solution with distillation using stored solar energy followed by waterfall on a natural mountain. This natural, green, and technico-economical technology is composed of three steps: the first one is coagulation which may be achieved: 1) in salty water aerated basin (air stripping, AS; dissolved air flotation, DAF) where the raw water is "diluted" in seawater; or 2) in "conventional" coagulation using seawater as coagulant solution instead of alum/ferric salts. The first option seems to be more natural as it simulates river water dilution in seawater and the second one is more practical for "rapid" water consummation. For colloids and microorganisms' removal, double-layer compression and charge neutralisation, as main coagulation and disinfection mechanisms, would be involved in the first and second options, respectively. Aerated basin (AS/DAF) reproduces the natural aeration to simulate healthy natural water basin. Using stored solar energy, distillation as the best liquid-solid/liquid-liquid separation process provides the removal of dissolved pollutants. For well balanced calco-carbonic equilibrium, the last step of this green treatment is the waterfall on a natural mountain providing useful gases, dissolved oxygen and carbon dioxide, and mineral salts to the water.

Keywords: Air Stripping (AS); Seawater; Water Treatment; Coagulation; Dissolved Air Flotation (DAF); Distillation

1. Introduction

Although water is a renewable resource and we use little more than 10% of the total precipitation surplus for public water-supply, irrigation, and industrial processes, its availability is restricted through an uneven distribution, both in time and space. In this respect, there is no essential difference between ancient times and the present day; society has always experienced problems with water: too little, too much, too variable and too polluted. Over more than 6000 years mankind has tried to manage these water problems by intervening in its natural courses through redistribution, storage, and regulation, to accommodate their requirements for irrigation, drainage, flood protection, drinking water, sanitation, and power generation [1,2].

Health and aesthetics are the principal motivations for water treatment [3,4]. In the late 1800s and early 1900s, acute waterborne diseases, such as cholera and typhoid fever, spurred development and proliferation of filtration and chlorination plants. Subsequent identification in water supplies of additional disease agents (such as *Legionella, Cryptosporidium*, and *Giardia*) [5] and contaminants (such as cadmium and lead) resulted in more elaborate pretreatments to enhance filtration and disinfection [6]. Additionally, specialised processes such as granular activated carbon (GAC) adsorption and ion exchange were occasionally applied to water treatment [7] to control taste- and odour-causing compounds and to remove contaminants such as nitrates. In addition, water treatment (**Table 1**) can be used to protect and preserve the distribution system [8].

A variety of developments in the water quality field since the 1970s and an increasing understanding of health effects [4] have created an upheaval in the water treatment [7] field. With the identification in water of low levels of potentially harmful organic compounds, coliform-free and low-turbidity water is no longer suffi-

Table 1. Selection of unit processes for the removal of specific parameters [3].

Parameter	Water treatment process options
Algae	Powdered activated carbon adsorption, microscreens, rapid filtration
Colour	Activated carbon adsorption, C/F, filtration
Floating matter	Coarse screens
Hardness	Coagulation, filtration, lime softening
Coliforms	
>100 per 100 mL^{-1}	Pre-chlorination, coagulation, filtration, post-chlorination
>20 per 100 mL^{-1}	Coagulation, filtration, post-chlorination
<20 per 100 mL^{-1}	Post-chlorination
Hydrogen sulphide	Aeration
Fe and Mn	Pre-chlorination, aeration, coagulation, filtration
Odour and taste	Aeration, activated carbon adsorption
Suspended solids	Fine screens, microscreens
Trace organics	Activated carbon adsorption
Turbidity	Coagulation, sedimentation, post-chlorination

cient. New information regarding inorganic contaminants, such as lead, is forcing suppliers to tighten control of water quality within distribution systems. Increasing pressures on watersheds have resulted in a heavier incoming load of microorganisms to many treatment plants. Although a similarly intense reevaluation of the aesthetic aspects of water quality has not occurred, aesthetic quality is important. Problems, such as excessive minerals, fixture staining, and colour, do affect consumer acceptance of the water supply. However, significant advances in the identification of taste- and odour-causing organisms and their metabolites have occurred within the last two decades [8].

Coagulation/flocculation (C/F) may be broadly described as chemical and physical processes that mix coagulating chemicals [9,10] and flocculation aids with water [11]. The overall purpose is to form particles large enough to be removed by the subsequent settling or filtration processes [12-14]. Particles in source water that can be removed by C/F, sedimentation [15], and filtration [16] include colloids, suspended material, bacteria, and other organisms [17]. The size of these particles may vary by several orders of magnitude. Some dissolved material can also be removed through the formation of particles in the C/F processes. The importance of dissolved material removal has become much more critical in recent years with increased regulatory emphasis on disinfection by-products (DBPs) and total organic carbon (TOC) removal [18,19].

This review is divided in two main sections. The first one concerns green, such as air stripping (AS), dissolved air flotation (DAF) and distillation and greenable processes such as coagulation using seawater salts as coagulant. The second section discusses briefly simulation of seawater distillation.

2. Green and Greenable Processes

2.1. Air Stripping (AS) and Aeration

Several different types of AS and aeration systems (**Figure 1**) are widely used for a variety of water treatment applications (**Table 2**) [7,20]. The most common types are diffused-air, surface aerator, spray, and packed-tower systems [21]. Water treatment applications for these systems include the absorption of reactive gases for water stabilisation and disinfection, precipitation of inorganic contaminants, and AS of volatile organic compounds (VOCs) and nuisance-causing dissolved gases [22,23]. The diffused-aeration (or bubble) systems are primarily used for the absorption of reactive gases, such as oxygen (O_2), ozone (O_3), and chlorine (Cl_2). Oxygen is frequently used for the oxidation/precipitation of iron and manganese. Ozone is used for disinfection, colour removal, and oxidation of TOC. Chlorine is primarily used for disinfection and sometimes as a pre-oxidant for the oxidation of iron and manganese or for other purposes. Diffused-aeration systems have also been used for the stripping [23] of odour-causing compounds and VOCs. Surface-aeration systems are primarily used for VOCs removal. The packed-tower and spray nozzle systems are primarily used for the removal of NH_3, CO_2, H_2S, and VOCs. The packed-tower systems include counter-cur-

Table 2. Classification of adsorptive bubble separation processes [21].

I. Classification according to the technique used for generating fine gas bubbles:

1. Dissolved gas system (Example: DAF)
2. Dispersed gas system (Example: Dispersed air flotation)
3. Vaccum system (Example: Vaccum air flotation)
4. Electrolysis system (Example: Electroflotation)
5. Biological system (Example: Biological flotation)

II. Classification according to the technique used for separating impurities or pollutants:

1. Foam separation:
   - A. Foam fractionation
   - B. Froth flotation
     - B1. Precipitate flotation
     - B2. Ion flotation
     - B3. Molecular flotation
     - B4. Microflotation and colloid flotation
     - B5. Macroflotation and ore flotation
     - B6. Adsorption flotation
     - B7. Adsorbing colloid flotation
2. Nonfoaming adsorptive bubble separation
   - A. Bubble fractionation
   - B. Solvent sublation
   - C. Nonfoaming flotation
     - C1. Nonfoaming precipitate flotation
     - C2. Nonfoaming adsorption flotation
     - C3. Nonfoaming flotation thickening

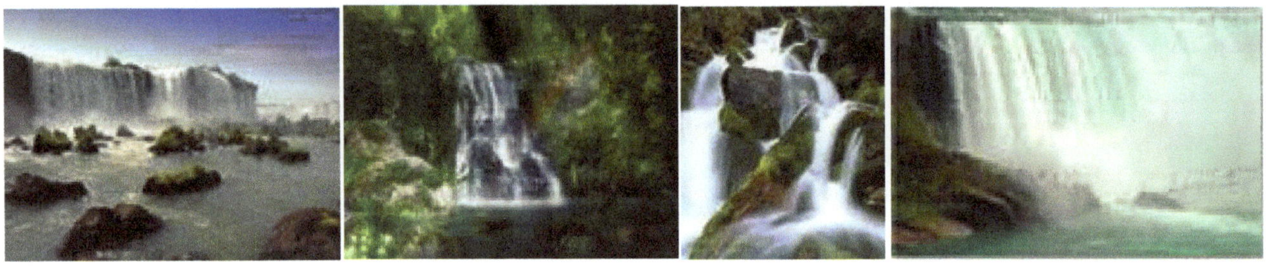

Figure 1. Waterfalls in the "virgin" Earth providing dissolved gas ($O_{2(g)}$, $CO_{2(g)}$) and salts (Ca^{2+}, Mg^{2+}) to water for well balanced calco-carbonic equilibrium.

rent flow, co-current flow, and cross-flow configurations. Spray nozzle systems can include tower and fountain-type configurations [24].

Hand *et al.* [24] discussed a fundamental understanding of the theory of gas transfer, followed by a description of the various unit operations, development of the design equations, and example design calculations. As unit operations, Hand *et al.* [24] presented packed towers, diffused or bubble aeration, surface aeration, spray aerators. In this review, we give here some generalities about the three last unit operations which are practically more convenient to actual (conventional) water treatment plants [7] since they can be added or introduced in sedimentation basin [15]. More theoretical details and practical examples may be found in this good reference [24].

2.1.1. Diffused or Bubble Aeration

The diffused or bubble aeration process consists of contacting gas bubbles with water for the purposes of transferring gas to the water (e.g., O_3, CO_2, O_2) or removing VOCs from the water by stripping [23]. The process can be carried out in a clear well or in special rectangular concrete tanks typically 2.74 to 4.57 m (9 to 15 ft) in depth [24]. **Figure 2** displays different types of diffused aeration systems. The most commonly used diffuser system consists of a matrix of perforated tubes (or membranes) or porous plates arranged near the bottom of the tank to provide maximum gas-to-water contact. Various types of diffusers and diffuser system layouts are presented in the Environmental Protection Agency's technology transfer design manual on fine-pore aeration systems (loc.cit.,EPA/625/1-89/023) [24]. Jet aerator devices are also used to provide good air-to-water contact. These aerators consist of jets that discharge fine gas bubbles and provide enhanced mixing for increased absorption efficiency [24].

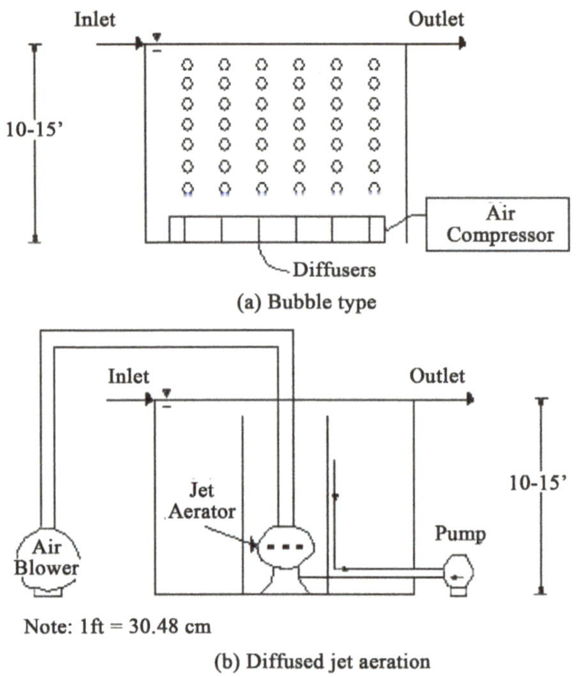

Figure 2. Schematic of various bubble aeration systems [24].

2.1.2. Surface Aeration

Surface aeration has been primarily used for oxygen absorption and the stripping of gases and volatile contaminants [23] when the required removals are less than about 90 percent. Surface aeration devices consist of the brush type or turbine type, as shown in **Figure 3**. The brush-type aerator consists of several brushes attached to a rotary drum, which is half-submerged in water in the centre of the tank. As the drum rotates, it disperses the water into the surrounding air providing reasonable contact between the air and water for mass transfer to take place. The turbine-type aerator consists of a submerged propeller system located in the centre of the tank and surrounded by draft tubs. As the submerged propeller rotates it draws water from outside the draft tubs through the inner section and into the air creating contact between the air and water. These types of systems have been extensively used in the aeration of wastewater [25] and their design and operation have been well documented [24].

2.1.3. Spray Aerators

Spray aerators have been used in water treatment for many years to oxygenate groundwater [26] for the purpose of iron and manganese removal and the AS of gases (*i.e.*, CO_2, H_2S) and VOCs [23]. Effective iron oxidation

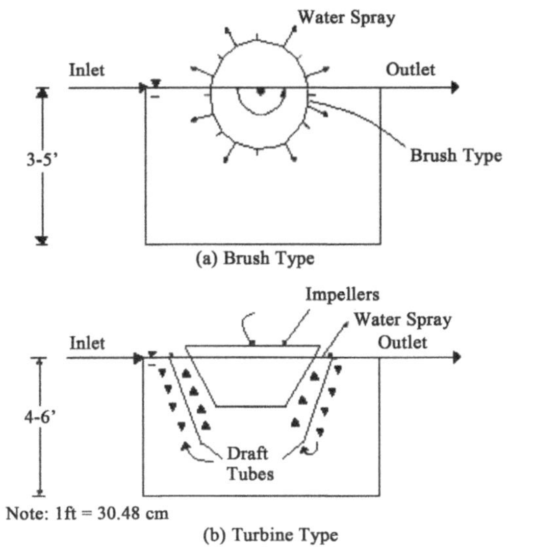

Figure 3. Schematic of various surface aeration devices [24].

by aeration usually requires at least 1 h of retention time after aeration. Manganese oxidation by aeration is very slow and not practical for waters with pH values below 9.5. Manganese removal usually requires a stronger oxidant. CO_2 and H_2S removals have ranged from 50 to 90 percent depending upon the pH of the water. VOCs removals have been as high as 90 percent depending upon the Henry's law constant [22,24,27]. **Figure 4** displays a schematic of a single fountain-type spray aerator.

Spray aerator systems consist of a series of fixed nozzles on a pipe grid. The grids can be placed in towers, commonly known as spray towers (or fountains), that spray onto the surface [28] of raw water reservoirs. Pressurised nozzles disperse fine water droplets into the surrounding air, creating a large air-water surface for mass transfer. Two types of pressurised spray nozzles, hollow- and full-cone, are commonly used in water treatment [7]. Full-cone nozzles deliver a uniform spray pattern of droplets. The hollow-cone nozzle delivers a circular spray pattern with most of the droplets concentrated at the circumference. The hollow-cone nozzle is generally preferred over the full-cone type because it provides smaller droplets for better mass transfer even though it has a larger pressure drop requirement. Hollow-cone spray droplets are around 5 mm and are prone to plugging. It is recommended that in-line strainers be installed in the spray nozzle manifold to prevent plugging [24].

The fountain-type spray aerators have been more widely used in water treatment because they can be easily adapted to existing water treatment systems. The design approach and application of the fountain type is presented in [24].

2.2. Coagulation

Coagulation [18,29,30] is a process for increasing the

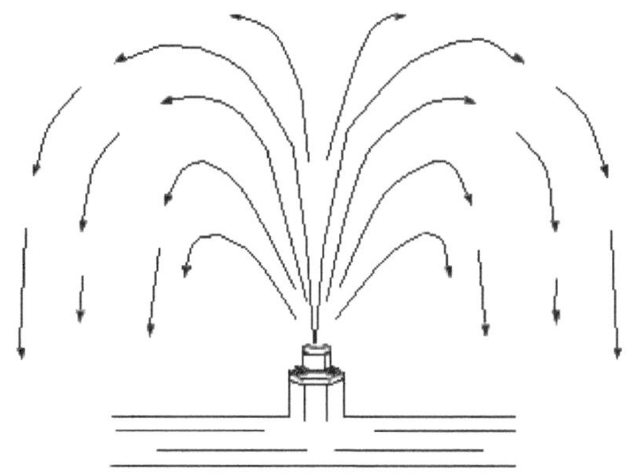

Figure 4. Schematic of a single-fountain spray aerator [24].

tendency of small particles in an aqueous suspension to attach to one another and to attach to surfaces such as the grains in a filter bed (**Table 3**). It is also used to effect the removal of certain soluble materials by adsorption or precipitation [31]. The coagulation process typically includes promoting the interaction of particles to form larger aggregates [18]. It is an essential component of conventional water treatment systems [7] in which the processes of coagulation, sedimentation [15], filtration, and disinfection are combined to clarify the water and remove and inactivate microbiological contaminants such as viruses, bacteria, and the cysts and oocysts of pathogenic protozoa [32]. Although the removal of microbiological contaminants continues to be an important reason for using coagulation, a newer objective, the removal of natural organic material (NOM) to reduce the formation of DBPs, is growing in importance [19,33-35].

Aluminium and ferric iron salts have long been used to remove colour caused by NOM [36]. These organic substances are present in all surface waters and in many groundwaters [26]. They can be leached from soil, diffused from wetland sediments [37], and released by plankton and bacteria. NOM adsorbs on natural particles and acts as a particle-stabilising agent in surface water [35]. It may be associated with toxic metals and synthetic organic chemicals (SOCs) [38]. NOM includes precursor compounds that form health-related by-products [4] when chlorine and other chemical disinfectants are used for disinfection and oxidation. For these reasons, considerable attention is being directed at the removal of NOM by coagulation in water treatment, even when colour removal is not the principle objective [7]. A treatment technique requirement in the U.S. Environmental Protection Agency's (USEPA's) Stage 1 Disinfection By-Products Rule requires NOM removal in conventional treatment systems by the practice of *enhanced coagulation* [11, 18,33,39].

Coagulation has been an important component of high-rate filtration plants in the United States since the 1880s. Alum and iron (III) salts have been employed as coagu-

Table 3. Terms used in coagulation literature [18].

Term	Definition
Coagulation	The process in which chemicals are added to water, causing a reduction of the forces tending to keep particles apart. Particles in source water are in a stable condition. The purpose of coagulation is to destabilise particles and enable them to become attached to other particles so that they may be removed in subsequent processes. Particulates in source waters that contribute to colour and turbidity are mainly clays, silts, viruses, bacteria, fulvic and humic acids, minerals (including asbestos, silicates, silica, and radioactive particles), and organic particulates. At pH > 4.0, particles or molecules are generally negatively charged. The coagulation process physically occurs in a rapid mixing process.
Mixing	Commonly referred to as *flash mixing*, *rapid mixing*, or *initial mixing*. The purpose of rapid mixing is to provide a uniform dispersion of coagulant chemical throughout the influent water.
Enhanced coagulation	A phrase used by the US Environmental Protection Agency (USEPA) in the Disinfectants and Disinfection By-products Rule. The rule requires that the coagulation process of some water supplies be operated to remove a specified percentage of organic material (OM) from the source water, as measured by total organic carbon (TOC). Enhanced coagulation (removal of TOC) can be achieved in most cases by either increasing coagulant chemical dosage or adjusting the pH during the coagulation reaction.
Coagulant chemicals	Inorganic or organic chemicals that, when added to water at an optimum dosage, cause particle destabilisation. Most coagulants are cationic when dissolved in water and include chemicals such as alum, ferric salts, lime, and cationic organic polymers.
Flocculation	The agglomeration of small particles and colloids to form settleable or filterable particles (flocs). Flocculation begins immediately after destabilisation in the zone of decaying mixing energy following rapid mixing, or as a result of the turbulence of transporting flow. In some instances, this incidental flocculation may be an adequate flocculation process. A separate flocculation process is most often included in the treatment train to enhance contact of destabilised particles and to build floc particles of optimum size, density, and strength.
Flocculation aids	Chemicals used to assist in forming larger, denser particles that can be more easily removed by sedimentation or filtration. Cationic, anionic, or non-ionic polymers are most often used in dosages of less than 1.0 mg L^{-1}.
Direct filtration	A treatment train that includes C/F, and filtration, but excludes a separate sedimentation process. With direct filtration, all suspended solids are removed by filtration. In the process sometimes called in-line filtration, flocculation occurs in the conduit between the rapid mixing stage and the filter, in the volume above the filter media, and within the filter media.
Solids contact clarifiers	Proprietary devices that combine rapid mixing, flocculation, and sedimentation in one unit. These units provide separate C/F zones and are designed to cause contact between newly formed floc and settled solids.

lant chemicals [9] since the beginning, with alum having the most widespread use. In the 1930s, Baylis perfected activated silica as a "coagulant aid". This material, formed on site, is an anionic polymer or a small, negatively charged colloid [40]. Synthetic organic polymers [41] were introduced in the 1960s, with cationic polymers having the greatest use. Natural starches were employed before the synthetic compounds. Polymers have helped change pretreatment and filtration practice, including the use of multimedia filters and filters with deep, uniform grain-size media, high-rate filtration, direct filtration (rapid mixing, flocculation, and filtration, but no sedimentation), and in-line filtration (rapid mixing and filtration only) [23,33].

Coagulants are also being used to enhance the performance of membrane microfiltration systems [42] and in pretreatment that prolongs the bed life of GAC contactors [43]. The development of new chemicals, advances in floc [44] removal process and filter design, and particle removal performance standards and goals have stimulated substantial diversity in the design and operation of the coagulation process, and change can be expected to continue into the future [45]. In evaluating high-rate filtration plants that were producing high-quality filtered water, Cleasby *et al.* [46] concluded, "Chemical pretreatment prior to filtration is more critical to success than the physical facilities at the plant." Their report recommends that plant staff use a well-defined coagulant chemical control strategy that considers variable rawwater quality. There is no question that high-rate (rapid sand) filtration plants are coagulant-based systems that work only as well as the coagulants that are used [32,33].

2.2.1. Properties of Colloidal Systems

Colloids are very small particles that have extremely large surface area [47]. Colloidal particles [48] are larger than atoms and ions but are small enough that they are usually not visible to the naked eye. They range in size from 0.001 to 10 μm resulting in a very small ratio of mass to surface area. The consequence of this smallness in size and mass and largeness in surface area is that in colloidal suspensions [49]: a) gravitational effects are negligible, and b) surface phenomena predominate.

Because of their tremendous surface, colloidal particles have the tendency to adsorb various ions from the surrounding medium that impart to the colloids an electrostatic charge relative to the bulk of surrounding water [40,50]. The developed electrostatic repulsive forces prevent the colloids from coming together and, consequently, contribute to their dispersion and stability.

1) Electrokinetic Properties

The electrokinetic properties of colloids can be attributed to the following three processes [33,49]:

a) Ionisation of groups within the surface of particles.

b) Adsorption of ions from water surrounding the particles.

c) Ionic deficit or replacement within the structure of particles.

Organic substances and bacteria acquire their surface

charges [40] as a result of the ionisation of the amino and carboxyl groups as shown below:

$$R\text{-}NH_3^+ \rightarrow R\text{-}NH_2 + H^+ \quad (1)$$

$$R\text{-}COOH \rightarrow R\text{-}COO^- + H^+ \quad (2)$$

The resulting charge on the surface of such particles is a function of the pH [40]. At high pH values or low hydrogen ion concentrations, the above reactions shift to the right and the colloid is negatively charged. At a low pH, the reactions shift to the left, the carboxyl group is not ionised, and the particle is positively charged due to the ionised amino group. When the pH is at the isoelectric point (IEP), the particle is neutral, *i.e.*, neither negatively nor positively charged. Proteinaceous material, containing various combinations of both amino and carboxyl groups, are usually negatively charged at pH values above 4 [47].

Oil droplets adsorb negative ions, preferably hydroxides (OH^-), from solution and, consequently, they develop a negative charge [40,50]. Some other neutral particles adsorb selected ions from their surrounding medium such as calcium (Ca^{2+}) or phosphate $\left(PO_4^{3-}\right)$ ions rendering them either positively or negatively charged, respectively.

Clays and other colloidal minerals may acquire a charge as a result of a deficit or imperfection in their internal structure [40]. This is known as isomorphic replacement [47]. Clays consist of a lattice formed of cross-linked layers of silica and alumina. In some clays there are fewer metallic atoms than nonmetallic ones within the mineral lattice producing a negative charge [40]. In others, higher valency cations may be replaced by lower valency cations during the formation of the mineral lattice that renders the clay particles negatively charged [40]. Examples of such imperfection include: a) the substitution of an aluminium ion (Al^{3+}) by either Mg^{2+} or Fe^{2+}; and b) the replacement of Si^{4+} cation by Al^{3+}. According to Letterman *et al.* [33], the type and strength of the charge resulting from this imperfection in the clay structure are independent of the surrounding water properties and pH. This is in contrast to the first two processes discussed above, in which both pH and ionic makeup of the surrounding solution play a big role in determining the sign and magnitude of the acquired charge on colloidal particles [47].

2) Hydration

Water molecules may also be sorbed on the surface of colloids, in addition to or in place of, other molecules or ions. The extent of this hydration depends on the affinity of particles for water. Colloidal particles that have water-soluble groups on their surface such as hydroxyl, carboxyl, amino, and sulfonic exhibit high affinity for hydration and cause a water film to surround the particles. Such colloids are classified as hydrophilic (water loving) particles. On the other hand, colloids that do not show affinity for water and do not have bound water films are classified as hydrophobic (water hating) [47].

3) Brownian Movement

Colloids exhibit a continuous random movement caused by bombardment by the water molecules in the dispersion medium [51]. This action, called *Brownian movement*, imparts kinetic energy to the particles that tends to cause an increase in the frequency of collisions [44], thus promoting coagulation. Elevated temperature increases molecular velocity resulting in more kinetic energy and more intense Brownian movement [44,47].

4) Tyndall Effect

Because colloidal particles have an index of refraction different from water, light passing through the dispersion medium and hitting the particles will be reflected. The turbid appearance due to this interference with the passage of light is termed the *Tyndall effect*. However, it should be noted that this might not always be the case. Water-loving, hydrophilic, colloids may produce just a diffuse Tyndall cone or none at all. The reason for this behaviour can be attributed to the bound water layer surrounding colloids. These particles will have an index of refraction not very different from that of the surrounding water. Hence, the dispersed phase and the dispersion medium behave in a similar fashion toward the passage of light [47].

2.2.2. Destabilisation of Colloids

Destabilisation [44] of colloidal particles is accomplished by coagulation through the addition of hydrolysing electrolytes such as metal salts and/or synthetic organic polymers [52]. Upon being added to the water, the action of the metal salt is complex [33,47]. It undergoes dissolution, the formation of complex highly charged hydrolysed metal coagulants (hydroxyoxides of metals), interparticle bridging [11,53], and the enmeshment of particles into flocs [44]. Polymers work either on the basis of particle destabilisation or bridging between the particles [53].

The destabilisation process is achieved by the following four mechanisms of coagulation [11,53-55]:

a) Double-layer compression (DLC).

b) Adsorption and charge neutralisation (CN).

c) Entrapment of particles in precipitate (*sweep coagulation, SC*).

d) Adsorption and bridging between particles.

1) Double-Layer Compression

When high concentrations of simple electrolytes are introduced into a stabilised colloidal dispersion, the added counter-ions penetrate into the diffuse double layer [51] surrounding the particles rendering it denser and hence thinner and smaller in volume. The addition of counter-ions with higher charges [40], such as divalent and trivalent ions, will result in even steeper electrostatic potential gradients and more rapid decrease in charge with distance from the surface of the particles. The net repulsive energy (see **Figure 5**) would become smaller or even would be completely eliminated, allowing the particles to approach each other and agglomerate [47].

A mathematical model that describes this coagulation mechanism is explained in detail in [56]. The prediction of this model is in agreement with what is known as the *Schultze-Hardly rule*. This rule states that the coagulation of colloidal particles is achieved by ions of added electrolytes, which carry opposite charge to that of the colloids, and that the destabilisation capability [44] of the ions rises sharply with ion charge. **Table 4** [49] illustrates the relative effectiveness of various electrolytes in the coagulation of negatively and positively charged colloids. For example, the relative power of Al^{3+}, Mg^{2+}, and Na^{+} for the coagulation of negative colloids is shown to vary in the ratio of 1000:30:1. A similar ratio is observed for the relative capability of PO_4^{3-}, SO_4^{2-}, and Cl^{-} for the coagulation of positively charged colloids.

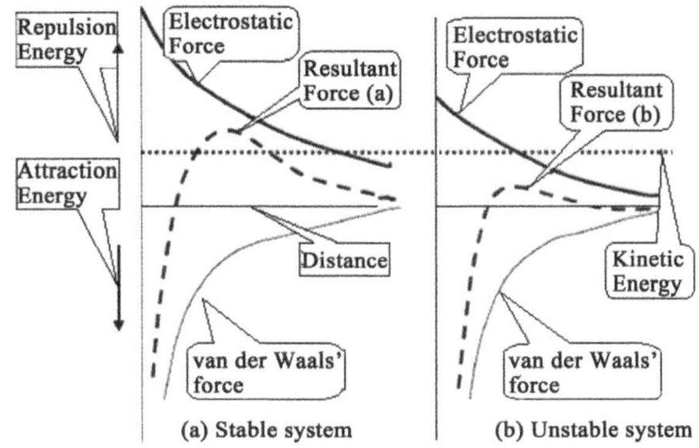

Figure 5. Effect of interparticle forces on the stability of a colloidal system [47].

Table 4. Relative coagulation power of electrolytes [47].

Electrolyte	Relative power of coagulation	
	Positive colloid	Negative colloid
NaCl	1	1
Na_2SO_4	30	1
Na_3PO_4	1000	1
$BaCl_2$	1	30
$MgSO_4$	30	30
$AlCl_3$	1	1000
$Al_2(SO_4)_3$	30	>1000
$FeCl_3$	1	1000
$Fe_2(SO_4)_3$	30	>1000

2) Adsorption and Charge Neutralisation (CN)

For all practical purposes, the ability of a chemical substance to destabilise and coagulate colloidal particles is the result of a combination of several mechanisms. Long-chained organic amines are often mentioned as being typical coagulants that function by adsorption and electrostatic neutralisation [33,47,49]. The positively charged organic amine molecules ($R\text{-}NH_3^+$) are easily and quickly attached to negatively charged colloidal particles. The charge on the particles gets neutralised and the electrostatic repulsion is decreased or eliminated resulting in the destabilisation of the colloids and hence their agglomeration [40]. The organic amines are hydrophobic because there is a lack of interaction between the CH_2 groups in their R-chain and the surrounding water. As a result, these positively charged ions are driven out of the water and get adsorbed on the particulate interface. An overdose of $R\text{-}NH_3^+$ counter-ions, however, can lead to charge reversal from negative to positive and the restabilisation of the dispersion system.

When coagulants such as metal salts are added to water, they dissociate yielding metallic ions, which undergo hydrolysis and form positively charged metallic hydroxyoxide complexes [47]. The commonly used coagulants, trivalent salts of aluminium and iron, produce numerous species because the hydrolysis products themselves tend to polymerise to give polynuclear metallic hydroxides [47]. Examples of aluminium salt polymers are $Al_6(OH)_{15}^{3+}$ and $Al_7(OH)_{17}^{4+}$ and of iron salt polymers are $Fe_2(OH)_2^{4+}$ and $Fe_3(OH)_4^{5+}$. When such polyvalent complexes possessing high positive charges get adsorbed on to the surface of the negatively charged colloids, the result is again a neutralisation of the charges, decrease in the repulsion energy, and destabilisation of the colloids. In a similar fashion to what occurs with the organic amines, an overdose of metallic salts could reverse the colloidal charge and restabilise the particles.

3) Entrapment of Particles in Precipitate

When the coagulants alum [$Al_2(SO_4)_3$] or ferric chloride ($FeCl_3$) are added in high enough concentration, they will react with hydroxides (OH^-) to form metal hydroxide precipitates, $Al(OH)_{3(s)}$ or $Fe(OH)_{3(s)}$ respectively. The colloidal particles get entrapped in the precipitates either during the precipitate formation or just after. This type of coagulation by enmeshment of colloids in precipitates is commonly called *sweep coagulation* (*SC*) [33,47,55].

There are three elements that influence this coagulation mechanism [47]:

a) Oversaturation: The rate of precipitation is a function of oversaturation with the metal hydroxide. To obtain fast precipitation and efficient *SC*, high concentrations of $Al(OH)_{3(s)}$ or $Fe(OH)_{3(s)}$ are required.

b) Presence of anions: The rate of precipitation is improved by the presence of various anions in water. The most effective anions in this respect are the sulphate ions.

c) Concentration of colloids: The rate of precipitation is also improved with higher concentration of colloidal particles. The reason for this is that the colloids themselves could act as nuclei for the formation of precipitates. In this case, it can be concluded that lower rather than higher coagulant dosage will be required to coagulate water having higher colloidal particle concentration.

4) Adsorption and Bridging between Particles

Polymers destabilise colloidal particles through the formation of bridges that extend between them [47]. The polymers have reactive groups that bind to specific sites on the surface of the colloidal particles. When a group on a polymer molecule attaches to a colloid, the remainder of the long-chain molecule extends away into the water. Once the extended portion of the polymer gets attached to another colloidal particle, the two particles become tied together or bridged by the polymer. If no other particle is available or if there is an overdose of polymer, the free extended portions of the polymer molecule would wrap around the same original particle, which could effectively bring about the restabilisation of the colloid. Restabilisation can also occur due to aggressive mixing or extended agitation, which may break the interparticle bridging and allow the folding back of the freed polymer portions around the same original particle (see **Figure 6**).

2.2.3. Phenomena of Coagulation

For Eckenfelder [29], coagulation results from two basic phenomena:

- Perikinetic (or electrokinetic) coagulation, in which the zeta potential (ZP) is reduced by ions or colloids of opposite charge to a level below the van der Waals attractive forces [57-59], and
- Orthokinetic coagulation—fluid motion [44,60]—in which the micelles aggregate and form clumps that agglomerate the colloidal particles.

The addition of high-valence cations depresses the particle charge and the effective distance of the double layer, thereby reducing the ZP [57]. As the coagulant dissolves, the cations serve to neutralise the negative charge on the colloids. This occurs before visible floc formation [44], and rapid mixing which "coats" the colloid is effective in this phase. Microflocs are then formed which retain a positive charge in the acid range because of the adsorption of H^+. These microflocs also serve to neutralise and coat the colloidal particle. Flocculation agglomerates the colloids with a hydrous oxide floc. In this phase, surface adsorption is also active. Colloids not initially adsorbed are removed by enmeshment in the floc [23,29].

A desired sequence of operation for effective coagulation was outlined [29]. If necessary, alkalinity should first be added (bicarbonate has the advantage of providing alkalinity without raising the pH). Alum or ferric salts are added next; they coat the colloid with Al^{3+} or Fe^{3+} and positively charged microflocs. Coagulant aids such as activated silica and/or polyelectrolyte [41] for floc build-up [45,61] and ZP control, are added last. After addition of alkali and coagulant, a rapid mixing of 1 to 3 min is recommended, followed by flocculation [23], with addition of coagulant aid, for 20 to 30 min. Destabilisation can also be accomplished by the addition of cationic polymers, which can bring the system to the IEP without a change in pH. Although polymers [3,62] are 10 to 15 times as effective as alum as a coagulant they are considerably more expensive (**Tables 5** and **6**). The mechanism of the coagulation process is shown in **Figure 7**.

However, DLC and CN may be classified as electrokinetic coagulation and *SC* and bridging between particles [63] may be attributed to orthokinetic coagulation [60] (**Figure 8**).

2.2.4. Polymer Toxicity

The normally used anionic and nonionic polymers are of low toxicity generally, but cationic types are more toxic, especially to aquatic organisms. Concerns about contaminants have led Japan and Switzerland not to permit the use of polyelectrolytes in drinking water treatment, whilst Germany and France have set stringent limits. The monomers are more toxic than the polymers [64]. Limits on the level of monomer are strictly controlled, especially with acrylamide products, where as a general rule the maximum allowable content of free acrylamide is 0.025%, and the residue in drinking water is limited to 0.5 $\mu g \cdot L^{-1}$ [41].

2.3. Distillation

Distillation is the most commonly used method for the separation of homogeneous fluid mixtures [27,65]. Separation exploits differences in boiling point, or volatility, between the components in the mixture [66,67]. Repeated vaporisation and condensation of the mixture allows virtually complete separation of most homogeneous fluid mixtures [68]. The vaporisation requires the input of energy [69,70]. This is the principal disadvantage of distillation: its high energy usage [71,72]. However, distillation has three principle advantages over alternative methods for the separation of homogeneous fluid mixtures [73]:

1) The ability to handle a wide range of feed flow rates. Many of the alternative processes for the separation of

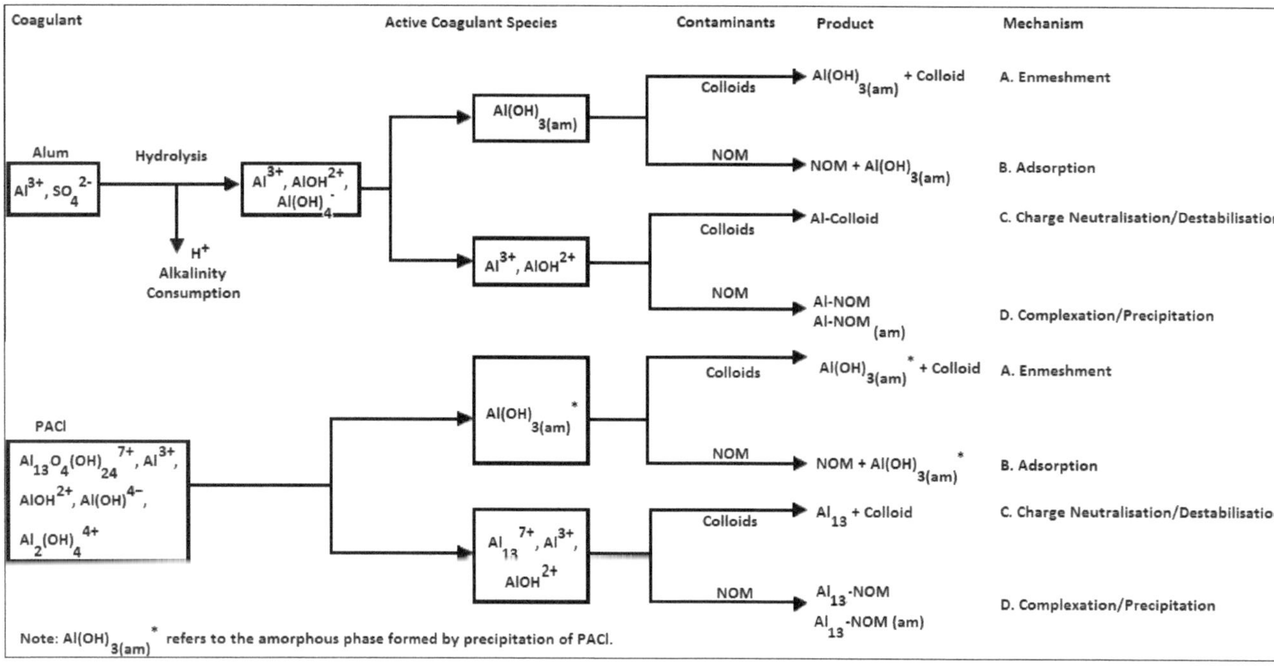

Figure 6. Conceptual view of coagulation reactions [54].

Table 5. Coagulants and flocculants, adjustment of conditions or addition of specific chemicals achieves required increase in particle size [62].

Metal salts, especially of aluminium or ferric iron ($Al_2(SO_4)_3 \cdot 18H_2O$; $Fe_2(SO_4)_3 \cdot 9H_2O$)
Natural flocculants
Starch
Gums
Tannin
Alginic acid
Sugar/sugar acid polymers
Polyglucosamine (chitosan)
Synthetic flocculants
Polyacrylamides
Polyamines/imines
Cellulose derivatives (e.g. carboxymethyl cellulose)
Polydiallydimethyl ammonium chloride
Chilling temperatures below 20°C, particularly yeast cells
pH adjustment in range 3 - 6
Concentration-increases particle concentration, increasing collision frequency

fluid mixtures can only handle low flow rates, whereas distillation can be designed for the separation of extremely high or extremely low flow rates.

2) The ability to separate feeds with a wide range of feed concentrations. Many of the alternatives to distillation can only separate feeds that are already relatively pure.

3) The ability to produce high product purity. Many of the alternatives to distillation only carry out a partial separation and cannot produce pure products.

It is no accident that distillation is the most common method for the separation of homogeneous mixtures [74]. It is a versatile, robust and well-understood technique [73].

2.3.1. Process Description

Distillation is a physical process for separating a liquid mixture into its constituents [75]. When such a mixture is partially vaporised, the vapour normally has a composition different from that of the residual liquid [15]. Implied in the method is the condensation of the vapour to form a product liquid, called the *distillate* [68]. The residual liquid product is often called the *bottoms.*

Distillation in crude form was practiced over 2000 years ago, usually for the concentration of alcoholic spirits. The first formalised documentation of distillation appears to be the treatise by Brunschwig in 1500 [75]. Distillation has since emerged as the key method for separating liquid mixtures in chemical processing and related industries because of its versatility, simplicity, economy, and many years of experience.

Early distillations were of the batch, takeover type, sometimes called *simple distillation* or *differential distillation.* A charge of liquid mixture is vaporised from a *still*, or *stillpot*, by heat addition, and the product vapour is condensed into one or more fractions [15]. Thus the term *fractional distillation*, or *fractionation*, has become associated with any distillation operation designed to obtain defined or specified constituent fractions [75].

Most distillations today are of the multistage rectification type, operated continuously or in the batch mode [75]. They are characterised by vertical vessels *(distillation columns*) with internal contacting devices (usually trays or packings) that provide intimate contacting of

Table 6. Properties of variety of coagulants [3].

Type of water	Alum	Ferric salts	Polymer
Type 1: high turbidity, high alkalinity (easiest to coagulate)	Effective over pH range 5 - 7. No need to add alkalinity or use coagulant aid	Effective over pH range 6 - 7. No need to add alkalinity or use coagulant aid	Cationic polymers usually very effective. Anionic and non-ionic may also work
Type 2: high turbidity, low alkalinity	Effective over pH range 5 - 7. May need to add alkalinity to control pH. Coagulant aid not needed	Effective over pH range 6 - 7. May need to add alkalinity to control pH. Coagulant aid not needed	Cationic polymers usually very effective. Anionic and non-ionic may also work
Type 3: low turbidity, high alkalinity	Relatively high dose needed to form sufficient floc, pH ~ 7. Coagulant aid may help	Relatively high dose needed to form sufficient floc. Coagulant aid may help	Will not work well alone due to low turbidity. Adding a clay to increase turbidity may be effective
Type 4: low turbidity, low alkalinity (most difficult to coagulate)	Relatively high dose needed to form sufficient floc, pH ~ 7. Alkalinity or clay needs to be dosed to produce type 2 or 3 water	Relatively high dose needed to form sufficient floc, pH ~ 7. Alkalinity or clay needs to be dosed to produce type 2 or 3 water	Will not work well alone due to low turbidity. Adding a clay to increase turbidity may be effective

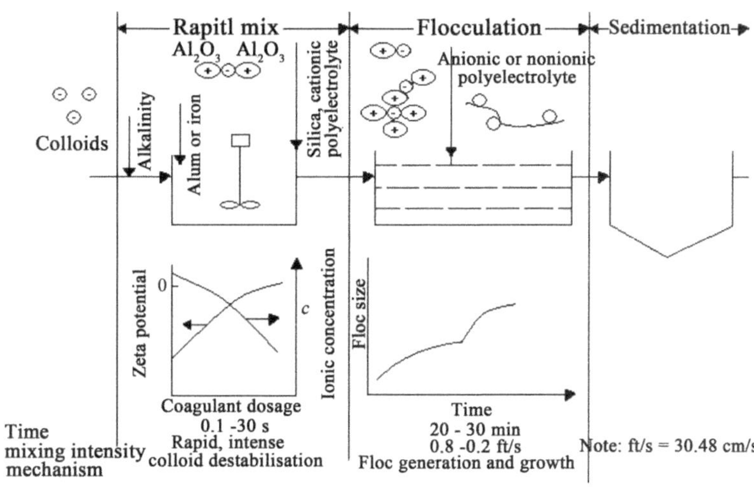

Figure 7. Phenomena of coagulation [29].

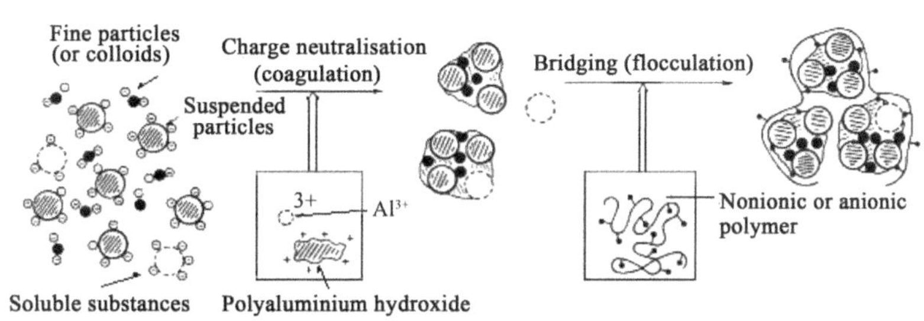

Figure 8. Schematic model of C/F process [63].

vapour and liquid. When operated continuously (the usual preferred mode), the towers are normally fed with a liquid mixture near the centre of the column, providing a *stripping zone* below the feed point and a *rectifying zone* (or "enriching zone") above the feed point. A diagram of a typical distillation column is shown in **Figure 9**. Heat is added at the base of the column by vapour from a *reboiler* and is removed at the top of the column in a *condenser* to provide the distillate product [15]. Part of the distillate is returned to the column as reflux liquid.

2.4. Dissolved Air Flotation (DAF)

DAF is a solid-liquid separation process for the removal of fine suspended material from an aqueous suspension [76,77]. The basic principle underlying DAF is Henry's law, which gives the solubility of air in water. According to Henry's law, the solubility of air in water is directly proportional to its partial pressure. A supersaturated solution of water is produced using high pressure in a saturator. The bubbles are generated by the pressure release

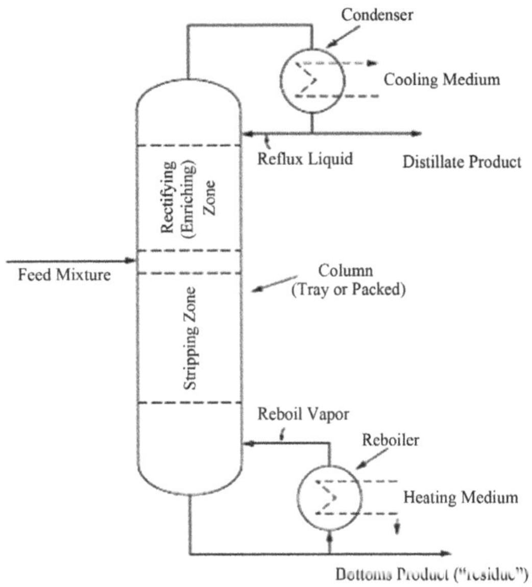

Figure 9. Distillation system operated in a continuous mode [75].

of this water stream [12]. These bubbles attach to suspended material present in the aqueous stream, causing them to float to the surface, where they are collected as *floc* [44,76].

DAF can be carried out by vacuum or pressurised methods [78]. In the vacuum flotation method [23] the water to be treated is saturated with air at atmospheric pressure. The bubbles are produced by applying a vacuum to the flotation tank, releasing the air as fine bubbles [15,79]. The vacuum flotation process has several disadvantages. These are 1) the amount of air available for flotation is limited by the vacuum achievable, 2) it is a batch process, and 3) it requires special equipment to produce and to maintain high vacuum. These disadvantages limit the application of vacuum flotation and it is only used in wastewater [80] sludge thickening [76].

The pressure flotation process is the most widely used DAF technique [78]. High pressure water is saturated with air. This pressurised water forms small bubbles when injected into water at atmospheric pressure. Three types of pressurisation processes can be used in DAF: full flow, partial flow and recycle flow pressurisation. The entire inlet stream is pressurised in full flow pressure DAF. It is commonly used when the wastewater [80] contains large amounts of suspended solids and the pressurisation process does not affect the treatment efficiency of the system. Partial flow pressurisation is used where the wastewater [81,82] contains moderate to low concentrations of suspended solids. In the recycle flow pressurisation system, 10% - 25% of the clarified effluent is recycled through a pressure vessel to the flotation tank. The flocculation process [23,83] in not disturbed in the recycle flow system because of intense mixing and pressurisation as clear water is pumped. A recycle flow system is cost-efficient because it pressurises only part of the water, thus requiring less compressor power. Recycle flow pressure flotation is the best-suited system for most DAF applications [76].

DAF is an effective alternative to sedimentation [15]. The advantages and disadvantage of DAF relative to sedimentation are presented in **Table 7** [76].

2.4.1. Process Description

A schematic diagram of a DAF process for wastewater treatment is shown in **Figure 10**. Its essential elements are a flocculation tank, a flotation tank, an air compressor, an air saturator, a recycling pump and a hydrosweep system. The wastewater [80,84,85] is pumped to the flocculation tank after being treated with coagulant/flocculent agents such as aluminium sulphate [23,83]. A portion of the clarified effluent is recycled for pressurisation. Compressed air is introduced into the discharge stream of the recycle pump, and the water is saturated with air at high pressure. The pressurised water stream is introduced to the flotation tank through nozzles, where fine bubbles (20 - 100 μm) in diameter are formed. The bubbles attach themselves to suspended solid particles, causing the agglomerates to float to the surface of the tank [78]. The float can be mechanically skimmed from the surface, and the clarified water is taken from the bottom of the flotation tank [76].

2.4.2. Principles of Dissolved Air Flotation (DAF)

DAF facilities are composed of the following four principal steps:

1) C/F prior to flotation.

2) Bubble generation.

3) Bubble-floc collision and attachment in the mixing zone [78].

4) Rising of the bubble-floc aggregates in a flotation tank [76].

3. Simulation of Seawater Distillation

This paper presents the concept of the best available

Table 7. The advantages and disadvantage of DAF relative to sedimentation [76].

Advantages
1. Clarification rates are higher in DAF, resulting in smaller flocculation tank volumes.
2. More concentrated sludge solids are produced in DAF than from sedimentation.
3. DAF uses lower amounts of coagulants and flocculent aids.
4. Oxygenation effects in DAF reduce odour problems.
5. DAF provides better removal of low density particles and algae, which can plug filters.
Disadvantage
1. DAF processes are more costly to operate and maintain than sedimentation processes.

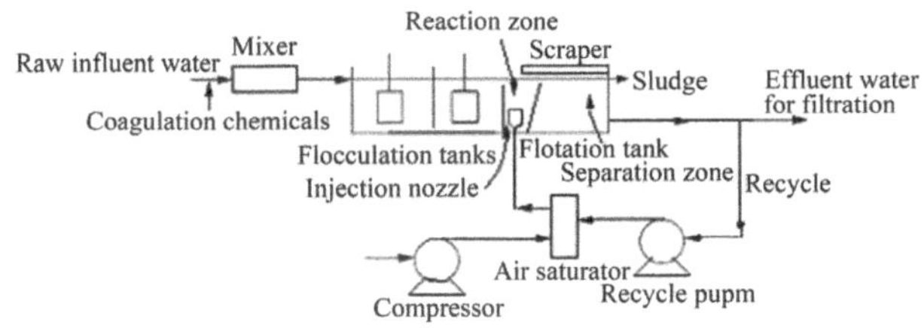

Figure 10. Schematic diagram of the DAF process for water treatment [76].

technology (BAT) of water/wastewater treatment and desalination which is in fact a simulation of seawater distillation at open sky: rapid mixing (coagulation) in salty water aerated basin (AS/DAF)/rapid mixing (coagulation) using seawater as coagulant solution with heating (distillation) using stored solar energy followed by waterfall (natural AS) on natural (artificial) mountain. This natural (*i.e.* non-"artificial" chemical) technology is composed of three steps: the first one is coagulation which may be achieved: 1) in salty water aerated basin (AS/DAF) where raw water is "diluted" in seawater; or 2) in "conventional" coagulation using seawater as coagulant solution instead of alum/ferric chloride. The first option is more natural as it simulates river water dilution in seawater and the second is more practical for "rapid" water consummation. For colloidal matters removal, *double-layer compression* (*DLC*) and *charge neutralisation* (*CN*), as main coagulation (and disinfection) mechanisms, would be involved in the first and second options, respectively. Aerated basin (AS/DAF) reproduces the natural aeration to simulate healthy natural water basin. Using stored solar energy, distillation as the best liquid-solid/liquid-liquid separation process provides the removal of dissolved pollutants. The last step is the waterfall (natural AS) on natural (or made of rocks) mountain providing dissolved gas ($O_{2(g)}$, $CO_{2(g)}$) and salts (Ca^{2+}, Mg^{2+}) to water for well balanced calco-carbonic equilibrium. This natural "three-therapy", needing technico-economical studies for its large application, will be helped by direct substitution of the actual convention water treatment plants as it uses rapid mixing basin for coagulation step, the flocculation, sedimentation, and filtration basins may be used for their initial aims, distillation basin may be added before the waterfall (natural AS) on natural (artificial) mountain.

Distillation [27,68] is a unit operation in which the components of a liquid solution are separated by vaporisation and condensation. Specially designed reactors are used to vaporise the water undergoing treatment, leaving behind waste brine that must be disposed of.

In desalination industry [86], distillation methods comprise the following modifications:

1) Multiple-effect evaporation (ME).
2) Multi-stage-flash evaporation (MSF).
3) Vapour-compression methods (VC).
4) Solar distillation method (SD).

Distillation (**Figure 11**) is the most developed process of removing water from a saline solution. It is applied up to very large capacities with various types of evaporators and accounts for about 59.4% of the total world plant capacity. The latent heat of changing phase is an important factor in the overall process economics, but the degree of salinity of the raw water is of no importance. MSF distillation and ME evaporation are reducing considerably the economic effect of the latent heat of vaporisation [86].

3.1. Decomposition of Organic Matter (OM) in Seawater

Organic matter (OM) is represented on this planet by living (autotrophic and heterotrophic) organisms and their excretory products and after-death remains, but may also be "inert" or non-living (refractory). The latter is found in large accumulations (fuel deposits, soils), as well as in the dispersed state in most mountain rocks and ocean waters [87]. Many authors believe that "inert" OM is of biogenic origin and that in reservoirs it dominates "living" OM [88].

In seas and oceans [87] refractory OM is mainly of autochthonous origin, the intake from land being comparatively small. Its primary source is phytoplankton. Heterotrophic organisms use phytoplankton as food for growth and for replacing expended energy. Excretory products of organisms and their remains are consumed by bacteria. As a result, the primary produced OM undergoes various conversions caused mainly by the activity of enzymes [89]. But despite all transformations, the OM of phytoplankton is not completely converted into

Figure 11. The first historically known solar distillation equipment, according to Giovanni Batista De La Porta; the sun evaporates the water inside the glass vessels and distilled water is collected beneath the vessels [86].

the mineral components which initially served as nutrients. Experiments testify that the remaining part is not large, and the rate of its biochemical conversion is slow. Krogh [90] considered the dissolved OM of oceans to be the result of the total OM turnover. One way of studying the processes of OM decomposition is to perform a series of long-term experiments on decomposition of OM from dead algae [88,90].

3.1.1. Water Humus

The analysis of the results of the experiments on decomposition of OM in dead plankton presents a view of the time-related changes of particulate and dissolved organic C, N and P, and some components of OM. The processes occurring are responsible not only for oxidation of the initial OM, but also in soils [91] for the polymerisation (condensation of the more biochemically resistant dissolved and particulate fractions of OM) [88].

Synchronically, new forms of OM are synthesised by bacteria. The combined processes lead to the formation of the refractory organic substance-water humus-in particulate and dissolved state. This surely occurred initially in the Precambrian period with blue-green algae and bacteria in water reservoirs under anoxic conditions [92-94]. As shown by experiments, the degree of decomposition of the OM of dead hydrobionts under such conditions was less. That is why more organic residues settled on the bottom of the reservoir. Naturally this influenced the further transformation of OM in sediments [37] and the accumulation of oil precursors [88,90].

Berzelius [94] appears to have pioneered a serious study of OM in natural waters. He discovered the dissolved organic compounds with acidic properties in the mineral spring Porla (Sweden). These compounds were called crenic and apocrenic acids. Berzelius [94] assumed that their salts were washed out from soil humus "intact". Oden [95] found the common term for these acids—fulvic acids (fulvus = yellow). Aschan [96] attributed them to the water humus group, being typical for peat, river and lake waters [88].

Thus, fulvic acids may originate from continental plants, and when carried by rivers and streams to the reservoirs, seas and oceans [87], they constitute the major part of the allochthonous water humus. Kalle (1966) [97] suggested they were only partly responsible for the "yellow substance" *(Gelbstoffe)* present in oceanic waters, the concentration of which in the water decreases with the increase of salinity. He considered that a part of the "yellow substance" was of autochthonous origin. It seems probable, that its existence in water is due to transformed products of excretion of dissolved and thinly dispersed OM which was discovered in littoral algae during studies conducted by some researchers [88]. According to Birge and Juday [98], the refractory OM appears in the process of decomposition of dead plankton in lakes. This is the water humus of autochthonous origin. Krogh [90] considered that in the deep waters of the ocean OM consists partly of "humus", which is resistant even for bacteria. Waksman [99], studying the question of the formation of OM in reservoirs, distinguished three types of humus: river, lake and sea humus. Besides this natural water humus, there may enter and be formed a humus from sewage of domestic origin. It was Odum [100] who considered humic substance (humus), the most stable product of OM decomposition in nature, to be a necessary component of the ecosystem.

Thus, it appears that water humus is the indispensable resistant product resulting from transformation, decomposition and synthesis of OM of excretions and dead remains of plankton in the ocean [87]: its final structure is largely determined by the activity of microorganisms [88].

Table 8 [101-103] summarises data to allow an approximate budget of water humus in the ocean. It is assumed that the annual decrease as a result of oxidation, aggregation and adsorption on the surface of particles with their subsequent sedimentation, equals its input [88].

The main differences observed between the calculated elements in a budget are typical in the amounts of organic C from the annual input reaching the ocean bottom. The bottom deposits also contain the aggregation products of dissolved OM. Scientists [88] estimated it to be $n \times 10^{15}$ g·C, where $n < 5$. River inputs are relatively low in comparison to the total organic production. The input of OM from atmospheric precipitation has been neglected, since a similar amount of OM seems to be carried out from the ocean surface by strong winds. Duce and Duursma [104] draw special attention to the works devoted to studies of the release of marine OM into the atmosphere and the input of river OM into the oceans. Recent data obtained give some estimation of OM content in the atmosphere. Researchers [88] investigated the surface-active OM fraction collected on platinum wires or glass fibre filters. As a result of analyses of the chloroform extracts, carried out with a gas chromatograph, fatty acids were found (C_{14} - C_{18}). The concentration ratios of single fatty acids in the sea aerosols were the same as those in the surface layer of oceanic waters [87]. The contents of methylesters formed from lipids in aerosols collected off the coast of Hawaii constituted 5%, and in the samples taken from the equatorial Pacific Ocean were 4% of the total OM isolated from sea aerosols. According to other researchers [88], the total content of organic C in the investigated sea aerosols ranged from 0.2 to 1.0 $\mu g\ m^{-3}$. In the aerosols, collected on the coast of the Bermuda Islands, the ratio of organic C to the total content of included salts ranged from 0.010 to 0.19, with an

Table 8. Elements of the annual balance of OM in the water of the world ocean (in carbon) [88].

Content, input and loss of OM	Williams [101]	Skopintsev [103][1]	Williams (1975) [102]
Average concentration of C (mg L^{-1}) (dissolved and particulate)	*From* 0 *to* 300 m need to add 1.0 coagulant aid. *From* 300 *to* 3000 m 0.5	≤1.5	0.70
Total C ($\times 10^{18}$ g) Phytoplankton production ($\times 10^{16}$ g)[2] Phytoplankton excretions of OM (10% of the annual production of OM) ($\times 10^{15}$ g) Input from atmospheric precipitations ($\times 10^{14}$ g)[3]	0.76 3.6 (100) 2.2 (1.0)	2.0 3.8 (120)	1.03 3.6 (100) 3.6 2.2 (1.0)
Input by rivers ($\times 10^{14}$ g)[3] OM of phytoplankton production used by organisms ($\times 10^{16}$ g)[4]	0.31 (2.0)	1.8 (5.0) 3.5 (92)	1.8 (5.0)
Input in water of the resistant soluble OM of planktonic origin ($\times 10^{15}$ g)[4] Sedimentation of particulate organic remains ($\times 10^{15}$ g)[4]	~0.1 (0.3)	1.1 (3) 1.9 (5)	1.8 (5) ~0.1 (0.3)

[1]For world oceans without adjacent seas; [2]Figures in parentheses: calculated as $g \cdot m^{-2} \cdot yr^{-1}$; [3]Figures in parentheses: in $mg \cdot L^{-1}$; [4]Figures in parentheses: in percentage of the annual production of phytoplankton.

average of 0.051.

The annual amount of the terrigenous water humus entering rivers might have been overestimated as a result of: 1) its partial coagulation by the salts of the sea water, and 2) its extensive oxidation in the presence of decomposing remains of plankton. The latter phenomenon has been reported by many authors [88]. According to Sholkovitz [105], only 3% - 11% of dissolved OM from waters of four Scottish rivers when mixed with sea water flocculated within half an hour. The humic acid (HA) content (colorimetrically determined) of river waters was 4% - 20% of the total OM. Fulvic acid and possibly non-humic compounds amounted to 80% - 90%. Fulvic acids form water-soluble compounds with Ca^{2+} and Mg^{2+}; a large proportion of the HA would have aggregated in the mixing zones between river and sea water. According to calculations, approximately 60% of the total HA content coagulated in these experiments.

3.1.2. Natural Coagulation of River Water Humus in Seawater

Colloidal properties are characteristic of humic compounds of river waters. Owing to their aggregation, the colour of the filtered water decreased only by 10% - 25% of its initial value, which indicates only a partial removal of OM of river water on mixing with sea water. At the same time, the concentrations of the elements Fe, Mn, Al and P in the water were markedly decreased (relative to their initial content). Similar results were obtained by Sholkovitz *et al.* [106] during experiments carried out in the waters of two rivers, the Luce (Scotland) and the Amazon (Brazil). Rapid flocculation [23] was characteristic for the high molecular-weight fractions of HA. This resulted in an aggregation of 60% - 80% of HA within half an hour (this amount represented 34 % of the initial dissolved OM).

The results of these works are principally in agreement with earlier published data [88]. In these experiments filtered marsh water was mixed with sea water in various proportions. Whilst carrying out these experiments, the water was kept in the dark at room temperature. After 30 days the colour of the filtered (size = 1.5 μm) water with a salinity of 0.2‰ was 98% of the initial value. At a salinity of 28.9‰ the colour was equal to 70% of the initial value of 62° (colour grade). In a similar experiment after 200 days the colour of the filtered water with a salinity of 0.2‰ and 32.5‰ was 50% of the initial value (44°); in such a continuous experiment a partial oxidation of the water humus might have taken place. The greatest rate of coagulation was observed on the first day. In some experiments with peat water, the coagulation effect was practically zero when a number of salts ($CaCl_2$, $MgCl_2$, $MgSO_4$, and HCl) were introduced. Equal effects might be found by mixing of peat water with sea water. Another effect is caused by the load of suspended particles. As has been shown by experiments, coagulation (or aggregation) of suspended particles in sea water [88] is less effective at a low particulate load even when humic compounds are present.

Bordovsky [107] is right in his belief that HA of marine deposits is of autochthonous and allochthonous origin, the latter playing a minor role. Based on chemical investigations and isotope determinations of C in HA isolated from ocean sediments, Nissenbaum and Kaplan [108] do not agree with this point of view. The allochthonous HA [87] in the bottom sediments of the ocean is evidently at the limits of detection of the analytical methods employed. According to **Table 9** the total annual input of water humus to the ocean constitutes 1.5×10^{15} g C, provided the distribution is proportional, this amounts to 1 μg $C \cdot L^{-1}$.

3.2. Why Simulation of Seawater Distillation?

It is difficult to imagine anything more important to the

Table 9. Elements of the annual balance of OM in the water of the world ocean (in carbon) [88].

Input	Output
1) From atmospheric precipitates: 2.2×10^{14} g (1.0)	1) From surface water to the atmosphere?
2) From rivers: 1.8×10^{14} g (5.0) 3) The relative resistant organic matter of planktonic origin: 1.1×10^{15} g (~3% of annual productivity equals 3.8×10^{16} g)	2) Biochemical oxidation (mainly by bacteria), sorption onto particles, aggregation (followed by partial use of aggregates by deep-water organisms) and sedimentation at bottom; a) humus of terrigenous origin: 1.8×10^{14} g; b) humus of planktonic
$\sum = 1.5 \times 10^{15}$ g	origin: 1.1×10^{15} g, $\sum = 1.3 \times 10^{15}$ g

Number in brackets: in $mg{\cdot}L^{-1}$; annual input of water humus in the oceans equals < 0.1% of its total content in ocean water.

human population than safe drinking water [109]. Lack of clean drinking water is still the major cause of illness and death in young children in developing countries. In more fortunate communities, where water treatment is practiced [7], the primary aim of water authorities is to provide water that is free from pathogens and toxins. A secondary and very important objective is to provide water that is clear, colourless, and pleasant to taste. These latter objectives, while admirable, are often very difficult to achieve. Most countries now have water quality regulations, or guidelines, which are driving water authorities to produce purer water, with the minimum of contamination from natural or manmade origin. At the same time consumers are demanding that chemicals [9] added during the treatment of drinking water be kept to a minimum. As a consequence, conventional clarification methods are being challenged to comply with the new regulations and restrictions and our understanding of the mechanisms involved is being tested as never before [110].

Drinking water is produced from sources such as lakes, rivers, reservoirs (surface waters) or is abstracted from below the surface (groundwater) [26]. In all cases the water is affected by the external environment. Surface and ground waters [26] contain dissolved, colloidal and large particulate materials, which are composed of minerals or NOM from the terrestrial and aqueous environments. Also present are small organisms such as bacteria, algae, protozoa and diatoms [90,93]. In many instances micro-contaminants from human origin are present. These include industrial and agricultural chemicals [9], pharmaceuticals and personal care products. It is inevitable that in the next few years our raw water supplies will be further contaminated by the products of the burgeoning nano-technological industries that are generating advanced materials for drug delivery, catalysis, energy storage devices and sensors, amongst others. Whilst it is the ability to engineer these materials at the molecular level which bestows their unique properties and functions, it is this reduced size and vastly enhanced surface area that will also challenge our existing treatment technologies [110].

The result of all these natural and anthropogenic activities is that our typical raw water is a complex "soup" that often requires a range of treatment processes for the achievement of water quality targets. Once those objectives have been met at the water treatment plant [7], the water is distributed through a series of pipes of various types and sizes to the consumer tap. Almost every step of the journey from source to consumer tap involves some sort of interaction at a surface or an interface. Therefore an understanding of the processes taking place at interfaces through the series of complex processes is vital for the provision of safe and palatable, drinking water [110].

Moreover, it is interesting to trace the history of water quality monitoring from the earliest times when colour (NOM, manganese) and turbidity (particulates) were the sole criteria, to more recent times when taste and odour (algal metabolites) have become important, to today when health [4] is the major determinant. Thus the water industry is now concerned with inorganics such as arsenic, organics such as endocrine disruptors and most recently microbiological contaminants, a good example of which is *Cryptosporidium* [5,110].

On the other hand, if we are invited to select two major challenges in water treatment industry [7] we will not hesitate to design: 1) using chemicals [9] in drinking water treatment processes and 2) NOM coagulability. An analysis of chemicals (Al, Cl_2 and polyelectrolytes) addition to the raw water during its treatment in order to produce drinking water is well presented in the literature such as the cited handbooks in this review (e.g., [21, 111,112]). Apart from human actions, some basic truths of chemistry [113] must be kept in mind: 1) a chemical reaction is rarely complete and may have sec- ondary reactions producing undesirable by-products (*i.e.*, all chemical disinfectants produce inorganic and/or or- ganic DBPs that may be of concern); 2) chemicals by their selves are toxic [9,114-119] and the reaction prod- ucts are also toxic even at low levels which are always diminished by the updated standards [19]. Further, chemicals need always appropriate mixing conditions [44] and hydraulic retention times, e.g., C/F to assure their distribution, conversion, and selectivity which are technically and practically difficult to realise considering the complicated microbial and physico-chemical composition of the raw water. It is well known that:

1) OM in water is very toxic than supposed (it was well known since 1970s that OM can produce DBPs but recent studies have proven that humic substances (HS) are toxic by their selves as they can be responsible of *Blackfoot Disease* [120-123];

2) NOM present in the natural environment is of a

complex nature (it would be a mistake just to focus entirely on one fraction of NOM and to assume that removal of this fraction will solve the problem of haloform formation) [10];

3) It has been shown that coagulant doses are controlled by NOM concentration rather than by turbidity [53];

4) The removal of OM using C/F is very difficult especially in its dissolved form which can encourage microbial growth [124].

Hence, water treatment using chemicals [7] is really and practically difficult to optimise technically and economically to safe drinking water. In this sense, Clark [125] and Ghernaout *et al.* [126] presented the concept of the greening of chemistry.

However, drinking water treatment is not from yesterday (**Figure 12**). God, The Great and The Perfect, uses distillation at the open sky to give perfect drinking water, satisfying all the qualities, to the world since at least The Adam and Eve's First Day until the Last Day. Distillation process would be presented as the future water treatment technology since it does not use chemicals (so, it has not the residual chemicals and their by-products problems) and its efficiency had been proven since the humankind dawn at the great scale. Furthermore, distillation technology is well known and developed in seawater desalination [68].

On the other hand, we may organise water treatment technologies (**Table 10**) into three general areas [12]:

- Physical methods,
- Chemical methods, and
- Energy intensive methods (such as distillation [15]).

All three of these technology groups can be combined in water treatment, or they may be used in select combinations depending upon the objectives of water treatment [12]. A good review of water treatment technologies is presented by Cheremisinoff [12].

Table 10. Water treatment technologies [12].

Water treatment technologies
Physical methods
Represent a body of technologies that we refer largely to as solid-liquid separations techniques, of which filtration plays a dominant role. Filtration technology can be broken into two general categories - conventional and non-conventional. This technology is an integral component of drinking water and wastewater treatment applications. It is, however, but one unit process within a modern water treatment plant scheme, whereby there are a multitude of equipment and technology options to select from depending upon the ultimate goals of treatment. To understand the role of filtration, it is important to make distinctions not only with the other technologies employed in the cleaning and purification of industrial and municipal waters, but also with the objectives of different unit processes.
Chemical methods
Rely upon the chemical interactions of the contaminants we wish to remove from water, and the application of chemicals that either aid in the separation of contaminants from water, or assist in the destruction or neutralisation of harmful effects associated with contaminants. Chemical treatment methods are applied both as stand-alone technologies, and as an integral part of the treatment process with physical methods.
Intensive technologies
Among them, thermal methods have a dual role in water treatment applications. They can be applied as a means of sterilisation, thus providing high quality drinking water, and/or these technologies can be applied to the processing of the solid wastes or sludge, generated from water treatment applications. In the latter cases, thermal methods can be applied in essentially the same manner as they are applied to conditioning water, namely to sterilise sludge contaminated with organic contaminants, and/or these technologies can be applied to volume reduction. Volume reduction is a key step in water treatment operations, because ultimately there is a trade-off between polluted water and hazardous solid waste. Energy intensive technologies include electrochemical techniques, which by and large are applied to drinking water applications. They represent both sterilisation and conditioning of water to achieve a palatable quality.

3.3. Models of Oceanic Plankton Aggregation

Two of the most fundamental properties of any particle, inert or living, are its length and its mass. These two properties determine how a particle interacts with planktonic organisms as food or habitat, how it affects light, and how fast it sinks. Because organisms are discrete entities, particle processes affect them as well as nonliving material [127].

Life in the ocean coexists with two competing physical processes favouring surface and bottom of the ocean: light from above provides the energy to fuel the system; gravity from below collects essential materials encapsulated in particles. Coagulation is the formation of single, larger, particles by the collision and union of two smaller particles; very large particles can be made from smaller particles by multiple collisions. Coagulation makes bigger particles, enhances sinking rates, and accelerates the removal of photosynthate. One result is that coagulation can limit the maximum phytoplankton concentration in the euphotic zone [127].

Particle size distributions have been measured since the advent of the Coulter Counter in the early 1970s, when researchers [127] reported on size distributions predominantly from surface waters around the world. They reported values for particles ostensibly between 1 and 1000 μm, although sampling and instrumental consideration suggest that the range was significantly smaller [127].

3.4. The Effects of Dissolved Salts

Any substance dissolved in a liquid has the effect of increasing the density of that liquid [128]. The greater the amount dissolved, the greater the effect. Water is no exception. The density of freshwater is close to 1.00×10^3 $kg{\cdot}m^{-3}$, while the average density of seawater [129] is about 1.03×10^3 $kg{\cdot}m^{-3}$.

Another important effect of dissolved substances is to depress the freezing point of liquids. For example, the

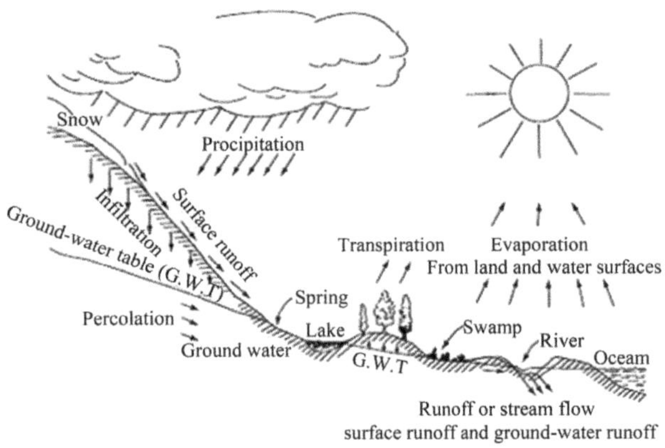

Figure 12. The hydrologic cycle [7].

addition of common salt (sodium chloride, NaCl) lowers the freezing point of water-which is why salt is spread on frozen roads. It also lowers the temperature at which water reaches its maximum density. That is because dissolved salts inhibit the tendency of water molecules to form ordered groups, so that density is controlled only by the thermal expansion effect. The oceans have high salinity, about 35 $g \cdot kg^{-1}$ on average (of which about 30 $g\ kg^{-1}$ are contributed by dissolved sodium ions (Na^+, ~11 g) and chloride ions (Cl^-, ~19 g)). Therefore, the density of seawater [129] increases with falling temperature right down to the freezing point. This is a crucial distinction between freshwater and seawater and it has a profound effect on the formation of sea-ice and on oceanic circulation processes [128].

4. Conclusions

This review concerns the concept of the best available technology of water/wastewater treatment and seawater desalination which is in fact a simulation of the seawater distillation at the open sky: coagulation in salty water aerated basin/coagulation using seawater as coagulant solution with distillation using stored solar energy followed by waterfall on a natural mountain. This natural, green, and technico-economical technology is composed of three steps: the first one is coagulation which may be achieved:

1) In salty water aerated basin (AS and DAF) where the raw water is "diluted" in seawater or;

2) In "conventional" coagulation using seawater as coagulant solution instead of alum/ferric salts.

The first option seems to be more natural as it simulates river water dilution in seawater and the second one is more practical for "rapid" water consummation. For colloids and microorganisms' removal, double-layer compression and charge neutralisation, as main coagulation and disinfection mechanisms, would be involved in the first and second options, respectively. Aerated basin (AS/DAF) reproduces the natural aeration to simulate healthy natural water basin. Using stored solar energy, distillation as the best liquid-solid/liquid-liquid separation process provides the removal of dissolved pollutants. For well balanced calco-carbonic equilibrium, the last step of this green treatment is the waterfall on a natural mountain providing useful gases, dissolved oxygen and carbon dioxide, and mineral salts to the water.

REFERENCES

[1] J. J. de Vries, "History of Groundwater Hydrology," In: J. W. Delleur, Ed., *The Handbook of Groundwater Engineering*, Ch. 1, 2nd Edition, CRC Press, Taylor & Francis Group, Boca Raton, 2007.

[2] I. R. Falconer, "Cyanobacterial Toxins of Drinking Water Supplies, Cylindrospermopsins and Microcystins," CRC Press, Taylor & Francis e-Library, Boca Raton, 2005.

[3] N. F. Gray, "Water Technology, an Introduction for Environmental Scientists and Engineers," 2nd Edition, Elsevier Science & Technology Books, London, 2005.

[4] J. A. Salvato Jr., "Environmental Health," In: J. R. Pfafflin and E. N. Ziegler, Eds., *Encyclopedia of Environmental Science and Engineering*, 5th Edition, CRC Press, Taylor & Francis Group, Boca Raton, 2006, p. 334.

[5] R. Fayer, "General Biology," In: R. Fayer and L. Xiao, Eds., *Cryptosporidium and Cryptosporidiosis*, Ch. 1, 2nd Edition, CRC Press, Taylor & Francis Group, Boca Raton, 2008.

[6] A. D. Russell and P. J. Ditchett, "Disinfection," In: J. R. Pfafflin and E. N. Ziegler, Eds., *Encyclopedia of Environmental Science and Engineering*, 5th Edition, CRC Press, Taylor & Francis Group, Boca Raton, 2006, p. 224.

[7] P. H. Jones and M. A. Tompeck, "Water Treatment," In: J. R. Pfafflin and E. N. Ziegler, *Encyclopedia of Environmental Science and Engineering*, 5th Edtion, CRC Press, Taylor & Francis Group, Boca Raton, 2006, p. 1311.

[8] P. D. Cohn, M. Cox and P. S. Berger, "Health and Aesthetic Aspects of Water Quality," In: R. D. Letterman, Tech. Ed., *Water Quality and Treatment, a Handbook of Community Water Supplies*, Ch. 2, 5th Edition, McGraw-Hill, Inc., New York, 1999.

[9] P. Baham and J. R. Pfafflin, "Effects of Chemicals," In: J. R. Pfafflin and E. N. Ziegler, Eds., *Encyclopedia of Environmental Science and Engineering*, 5th Edition, CRC Press, Taylor & Francis Group, Boca Raton, 2006, p. 271.

[10] D. Ghernaout, B. Ghernaout and A. Kellil, "NOM Removal and Enhanced Coagulation as a Link between Coagulation and Electrocoagulation," *Desalination and Water Treatment*, Vol. 2, No. 1-3, 2009, pp. 203-222.

[11] B. A. Dempsey, "Coagulant Characteristics and Reactions," In: G. Newcombe and D. Dixon, Eds., *Interface Science in Drinking Water Treatment, Theory and Applications*, Ch. 2. In: A. Hubbard, Series Ed., *Interface Science and Technology*, Vol. 10, Academic Press, Elsevier, Amsterdam, 2006.

[12] N. P. Cheremisinoff, "Handbook of Water and Wastewater Treatment Technologies," Butterworth-Heinemann, A

Member of the Reed Elsevier Group, Boston, 2002.

[13] C. Tien, "Introduction to Cake Filtration, Analyses, Experiments and Applications," Elsevier, Amsterdam, 2006.

[14] K. Sutherland, "Filters and Filtration Handbook," 5th Edtion, Butterworth-Heinemann, Elsevier, Oxford, 2008.

[15] J. F. Richardson, J. H. Harker and J. R. Backhurst, "Coulson and Richardson's Chemical Engineering," In: R. K. Sinnott, *Particle Technology and Separation Processes*, Vol. 2, 5th Edition, Butterworth-Heinemann, an Imprint of Elsevier Science, Oxford, 2002.

[16] D. A. Nield, "Modelling Fluid Flow in Saturated Porous Media and at Interfaces," In: D. B. Ingham and I. Pop, Eds., *Transport Phenomena in Porous Media II*, Ch. 1, Pergamon, an Imprint of Elsevier Science, Amsterdam, 2002.

[17] F. N. Kemmer (Ed.), "The NALCO Water Handbook," 2nd Edtion, McGraw-Hill Book Company, New York, 1988.

[18] P. J. Delphos and G. M. Wesner, "Mixing, Coagulation, and Flocculation," In: E. E. Baruth, Tech. Ed., *Water Treatment Plant Design, American Water Works Association, American Society of Civil Engineers*, Ch. 6, 4th Edition, McGraw-Hill, Inc., New York, 2005.

[19] D. Ghernaout, M. W. Naceur and A. Aouabed, "On the Dependence of Chlorine by-Products Generated Species Formation of the Electrode Material and Applied Charge during Electrochemical Water Treatment," *Desalination*, Vol. 270, No. 1-3, 2011, pp. 9-22.

[20] F. R. Spellman, "Water Treatment Operations and Unit Processes," In: F. R. Spellman, *Handbook of Water and Wastewater Treatment Plant Operations*, Ch. 17, CRC Press, Boca Raton, 2003.

[21] J. De Zuane, "Handbook of Drinking Water Quality," 2nd Edition, John Wiley & Sons, Inc., New York, 1997.

[22] J. E. Dyksen, "Aeration and Air Stripping," In: E. E. Baruth, Tech. Ed., *Water Treatment Plant Design, American Water Works Association, American Society of Civil Engineers*, Ch. 5, 4th Edition, McGraw-Hill, Inc., New York, 2005.

[23] A. Anzalone, J. K. Bewtra and H. I. Ali, "Physical and Chemical Treatment of Wastewaters," In: J. R. Pfafflin and E. N. Ziegler, Eds., *Encyclopedia of Environmental Science and Engineering*, 5th Edition, CRC Press, Taylor & Francis Group, Boca Raton, 2006, p. 972.

[24] D. W. Hand, D. R. Hokanson and J. C. Crittenden, "Air Stripping and Aeration," In: R. D. Letterman, Tech. Ed., *Water Quality and Treatment, a Handbook of Community Water Supplies, American Water Works Association*, Ch. 5, 5th Edition, McGraw-Hill, Inc., New York, 1999.

[25] A. Stoddard, J. B. Harcum, J. T. Simpson, J. R. Pagenkopf and R. K. Bastian, "Municipal Wastewater Treatment: Evaluating Improvements in National Water Quality," John Wiley & Sons, Inc., New York, 2002.

[26] P. Chan, Y. Ding and J. R. Schuring Jr., "Groundwater Resources," In: J. R. Pfafflin and E. N. Ziegler, Eds., *Encyclopedia of Environmental Science and Engineering*, 5th Edition, CRC Press, Taylor & Francis Group, Boca Raton, 2006, p. 439.

[27] T. Asano, F. L. Burton, H. L. Leverenz, R. Tsuchihashi and G. Tchobanoglous, "Water Reuse: Issues, Technologies, and Applications," Metcalf & Eddy, Inc., The McGraw-Hill companies, New York, 2007.

[28] H. F. Hemond and E. J. Fechner-Levy, "Chemical Fate and Transport in the Environment," 2nd Edition, Academic Press, San Diego, 2008.

[29] W. W. Eckenfelder Jr., "Industrial Water Pollution Control," 3rd Edition, McGraw-Hill Companies, Inc., New York, 2000.

[30] G. Tchobanoglous, F. L. Burton and H. D. Stensel, "Wastewater Engineering, Treatment and Reuse," 4th Edition, Metcalf & Eddy, Inc., McGraw-Hill Companies, Inc., New York, 2003.

[31] D. Ghernaout and M. W. Naceur, "Ferrate(VI): *In Situ* Generation and Water Treatment–A Review," *Desalination and Water Treatment*, Vol. 30, No. 1-3, 2011, pp. 319- 332.

[32] A. Rushton, A. S. Ward and R. G. Holdich, "Solid-Liquid Filtration and Separation Technology," VCH Verlagsgesellschaft Mbh, Weinheim, 1996.

[33] R. D. Letterman, A. Amirtharajah and C. R. O'Melia, "Coagulation and Flocculation," In: R. D. Letterman, Tech. Ed., *Water Quality and Treatment, a Handbook of Community Water Supplies, American Water Works Association*, Ch. 6, 5th Edition, McGraw-Hill, Inc., New York, 1999.

[34] R. Beckett and J. Ranville, "Natural Organic Matter," In: G. Newcombe and D. Dixon, Eds., Interface Science in Drinking Water Treatment, Theory and Applications, Ch. 17. In: A. Hubbard, Series Ed., *Interface Science and Technology*, Vol. 10, Academic Press, Elsevier, Amsterdam, 2006.

[35] A. I. Schäfer, "Natural Organics Removal Using Membranes, Principles, Performance and Cost," Technomic Publishing Company, Inc., Lancaster, Pennsylvania, 2001.

[36] J. E. Smith Jr., R. C. Renner, B. A. Hegg and J. H. Bender, "Upgrading Existing or Designing New Drinking Water Treatment Facilities," Noyes Data Corporation, Park Ridge, 1991.

[37] W. Lick, "Sediment and Contaminant Transport in Surface Waters," CRC press, Taylor & Francis Group, Boca Raton, 2009.

[38] J. G. Farmer and M. C. Graham, "Freshwaters," In: R. M. Harrison, Ed., *Understanding Our Environment, an Introduction to Environmental Chemistry and Pollution*, Ch. 3, 3rd Edition, Royal Society of Chemistry, Cambridge, 1999.

[39] H. L. Shorney, "Disinfection by-Product Precursor Removal by Enhanced Softening and Coagulation," Ph.D. Thesis, University of Kansas, Lawrence, 1998.

[40] G. Lagaly, "From Clay Mineral Crystals to Colloidal Clay Mineral Dispersions," In: H. Stechemesser and B. Dobiáš, Eds., *Coagulation and flocculation*, Ch. 8, 2nd Edition, Taylor & Francis Group, Boca Raton, 2005.

[41] B. A. Bolto, "Coagulation and Flocculation with Organic Polyelectrolytes," In: G. Newcombe and D. Dixon, Eds.,

Interface Science in Drinking Water Treatment, Theory and Applications, Ch. 5. In: A. Hubbard, Series Ed., *Interface Science and Technology*, Vol. 10, Academic Press, Elsevier, Amsterdam, 2006.

[42] M. R. Wiesner, M. M. Clark and J. Mallevialle, "Membrane Filtration of Coagulated Suspensions," *Journal of Environmental Engineering*, Vol. 115, No. 1, 1989, pp. 20-40.

[43] K. O. Nowack and F. S. Cannon, "Ferric Chloride plus GAC for Removing TOC," *Proceedings of the AWWA Water Quality Technology Conference*, Boston, 17-21 November 1996, pp. 65-78.

[44] J. Gregory, "Floc Formation and Floc Structure," In: G. Newcombe and D. Dixon, Eds., *Interface science in drinking water treatment, theory and applications*, Ch. 3. In: A. Hubbard, Series Ed., Interface Science and Technology, Vol. 10, Academic Press, Elsevier, Amsterdam, 2006.

[45] J. F. Atkinson, R. K. Chakraborti and J. E. VanBenschoten, "Effects of Floc Size and Shape in Particle Aggregation," In: I. G. Droppo, G. G. Leppard, S. N. Liss and T. G. Milligan, Eds., *Flocculation in Natural and Engineered Environmental Systems*, Ch. 5, CRC Press, Boca Raton, 2005.

[46] J. L. Cleasby, A. H. Dharmarajah, G. L. Sindt and E. R. Baumann, "Design and Operations Guidelines for Optimization of the High-Rate Filtration Process: Plant Survey Results," Final Project Report, AWWA and AWWA Research Foundation, Denver, 1989.

[47] N. K. Shammas, "Coagulation and Flocculation," In: L. K. Wang, Y.-T. Hung and N. K. Shammas, Eds., *Handbook of Environmental Engineering, Physicochemical Treatment Processes*, Vol. 3, Ch. 4, The Humana Press Inc., Totowa, 2005.

[48] F. Woodard, "Industrial Waste Treatment Handbook," Butterworth–Heinemann, Boston, 2001.

[49] C. N. Sawyer, P. L. McCarty and G. E. Parkin, "Chemistry for Environmental Engineering," 4th Edition, McGraw-Hill, New York, 1994.

[50] T. D. Reynolds, "Unit Operations and Processes in Environmental Engineering," Brooks/Cole Engineering Division, Monterey, 1982.

[51] J. H. Masliyah and S. Bhattacharjee, "Electrokinetic and Colloid Transport Phenomena," John Wiley & Sons, Inc., Hoboken, 2006.

[52] H. S. Weinberg, V. J. Pereira and Z. Ye, "Drugs in Drinking Water Treatment Options," In: D. S. Aga, Ed., *Fate of Pharmaceuticals in the Environment and in Water Treatment Systems*, Ch. 10, CRC Press, Taylor & Francis Group, Boca Raton, 2008.

[53] K. J. Wilkinson and A. Reinhardt, "Contrasting Roles of Natural Organic Matter on Colloidal Stabilization and Flocculation in Freshwaters," In: I. G. Droppo, G. G. Leppard, S. N. Liss and T. G. Milligan, *Flocculation in Natural and Engineered Environmental Systems*, Ch. 7, CRC Press, Boca Raton, 2005.

[54] D. J. Pernitsky, "Coagulation 101," P:\OFFICE\Conferences\2003TechTrans\Dave P\DAVE_paper.doc (consulted 25/07/05).

[55] D. Ghernaout and B. Ghernaout, "Sweep Flocculation as a Second Form of Charge Neutralisation—A Review," *Desalination and Water Treatment*, Vol. 44, No. 1-3, 2012, pp. 15-28.

[56] E. J. W. Verwey and J. Th. G. Overbeek, "Theory of the Stability of Lyophobic Colloids," Elsevier, Amsterdam, 1948.

[57] C. R. O'Melia, "Fundamentals of Particle Stability," In: G. Newcombe and D. Dixon, Eds., *Interface Science in Drinking Water Treatment, Theory and Applications*, Ch. 18. In: A. Hubbard, Series Ed., *Interface Science and Technology*, Vol. 10, Academic Press, Elsevier, Amsterdam, 2006.

[58] D. J. Shaw, "Introduction to Colloid and Surface Chemistry," 4th Edition, Butterworth-Heinemann, an Imprint of Elsevier Science, Oxford, 1992.

[59] R. M. Pashley and M. E. Karaman, "Applied Colloid and Surface Chemistry," John Wiley & Sons, Ltd., Chichester, 2004.

[60] K. J. Ives, "Coagulation and Flocculation, Part II-Orthokinetic Flocculation," In: L. Svarovsky, Ed., *Solid-Liquid Separation*, 4th Edition, Butterworth-Heine-mann, Oxford, 2000.

[61] S. N. Liss, T. G. Milligan, I. G. Droppo and G. G. Leppard, "Methods for Analyzing Floc Properties," In: I. G. Droppo, G. G. Leppard, S. N. Liss and T. G. Milligan, Eds., *Flocculation in Natural and Engineered Environmental Systems*, Ch. 1, CRC Press, Boca Raton, 2005.

[62] T. Beveridge, "Large-Scale Centrifugation," In: I. D. Wilson, E. R. Adlard, M. Cooke and C. F. Poole, *Encyclopedia of Separation Science*, Academic Press, London, 2000, pp. 320-342.

[63] S. Takahide, "Kurita Handbook of Water Treatment," 2nd Edition, Kurita Water Industries, Ltd., Tokyo, 1999.

[64] J. Criddle, "A Review of the Mammalian and Aquatic Toxicity of Polyelectrolytes," Report NR 2545, Foundation for Water Research, Medmenham, 1990.

[65] M. F. Doherty, Z. T. Fidkowski, M. F. Malon and R. Taylor, "Distillation (Section 13), Perry's Chemical Engineering Handbook," 8th Edition, McGraw-Hill Companies, Inc., New York, 2008.

[66] Z. Lei, B. Chen and Z. Ding, "Special Distillation Processes," Elsevier, Amsterdam, 2005.

[67] R. K. Sinnott, "Separation Columns (Distillation, Absorption and Extraction)," In: R. K. Sinnott, *Coulson and Richardson's Chemical Engineering*, Vol. 6, Ch. 11, 4th Edition, Elsevier Butterworth-Heinemann, Oxford, 2005.

[68] H. T. El-Dessouky and H. M. Ettouney, "Fundamentals of Salt Water Desalination," Elsevier, Amsterdam, 2002.

[69] G. Towler and R. Sinnott, "Chemical Engineering Design, Principles, Practice and Economics of Plant and Process Design," Elsevier Inc., Amsterdam, 2008.

[70] F.-W. Möllenkamp, "Energy Sources—Alternatives," In: J. R. Pfafflin and E. N. Ziegler, Eds., *Encyclopedia of Environmental Science and Engineering*, 5th Edition, CRC Press, Taylor & Francis Group, Boca Raton, 2006, p.

295.

[71] H. Silla, "Chemical Process Engineering, Design and Economics," Marcel Dekker, Inc., Taylor & Francis Group, New York, 2003.

[72] P. S. Buckley, W. L. Luyben and J. P. Shunta, "Design of Distillation Column Control Systems," Creative Services Inc., New York, 1985.

[73] R. Smith, M. Jobson, "Distillation," In: I. D. Wilson, E. R. Adlard, M. Cooke, C. F. Poole, Eds., *Encyclopedia of Separation Science*, Academic Press, London, 2000.

[74] R. Smith, "Chemical Process Design," McGraw-Hill, Inc., New York, 1995.

[75] J. R. Fair, "Distillation," In: L. F. Albright, Ed., *Albright's Chemical Engineering Handbook*, Ch. 12, CRC Press, Taylor & Francis Group, Boca Raton, 2009.

[76] D. Shekhawat and P. Srivastava, "Dissolved Air Flotation," In: I. D. Wilson, E. R. Adlard, M. Cooke and C. F. Poole, Eds., *Encyclopedia of Separation Science*, Academic Press, London, 2000.

[77] K. A. Matis and A. I. Zouboulis, "An Overview of the Process, In: K. A. Matis, Ed., *Flotation Science and Engineering*, Ch. 1, Marcel Dekker, Inc., New York, 1995.

[78] J. K. Edzwald, "Dissolved Air Flotation in Drinking Water Treatment," In: G. Newcombe and D. Dixon, Eds., *Interface Science in Drinking Water Treatment, Theory and Applications*, Ch. 5. In: A. Hubbard, Series Ed., *Interface Science and Technology*, Vol. 10, Academic Press, Elsevier, Amsterdam, 2006.

[79] R. W. Crites, J. Middlebrooks and S. C. Reed, "Natural Wastewater Treatment Systems," Taylor & Francis Group, Boca Raton, 2006.

[80] J. R. Pfafflin, "Municipal Wastewater," In: J. R. Pfafflin and E. N. Ziegler, *Encyclopedia of Environmental Science and Engineering*, 5th Edition, CRC Press, Taylor & Francis Group, Boca Raton, 2006, p. 727.

[81] G. Bitton, "Wastewater Microbiology," 3rd Edition, John Wiley & Sons, Inc., New York, 2005.

[82] V. Vinceviciene, "Wastewater Regulation," In: P. Quevauviller, O. Thomas and A. van der Beken, Eds., *Wastewater Quality Monitoring and Treatment*, Ch. 1.1, John Wiley & Sons, Ltd., Chichester, 2006.

[83] R. E. Weiner and R. A. Matthews, "Environmental Engineering," 4th Edition, Butterworth-Heinemann, Elsevier Science (USA), Amsterdam, 2003.

[84] M. B. Hocking, "Handbook of Chemical Technology and Pollution Control," 3rd Edition, Academic Press, an Imprint of Elsevier, Amsterdam, 2005.

[85] A. R. Jantrania and M. A. Gross, "Advanced Onsite Wastewater Systems Technologies," CRC Press, Taylor & Francis Group, Boca Raton, 2006.

[86] E. Delyannis and V. Belessiotis, "Desalination," In: J. R. Pfafflin and E. N. Ziegler, Eds., *Encyclopedia of Environmental Science and Engineering*, 5th Edition, CRC Press, Taylor & Francis Group, Boca Raton, 2006, p. 194.

[87] L. Thomsen, "Organic Rich Aggregates in the Ocean: Formation, Transport Behavior, and Biochemical Composition," In: I. G. Droppo, G. G. Leppard, S. N. Liss and T. G. Milligan, Eds., *Flocculation in Natural and Engineered Environmental Systems*, Ch. 11, CRC Press, Boca Raton, 2005.

[88] B. A. Skopintsev, "Decomposition of Organic Matter of Plankton, Humification and Hydrolysis," In: E. K. Duursma and R. Dawson, Eds., *Marine Organic Chemistry, Evolution, Composition, Interactions and Chemistry of Organic Matter in Seawater*, Ch. 6. In: Elsevier Oceanography Series, Vol. 31, Elsevier Scientific Publishing Company, Amsterdam, 1981, pp. 125-177.

[89] E. Kreps, "Organic Catalysts or Enzymes in Sea Water," In: A. E. Daniel, Ed., *James Johnston Memorial*, Liverpool University Press, Liverpool, 1934, pp. 193-202.

[90] A. Krogh, "Conditions of Life at Great Depth in the Ocean," *Ecological Monographs*, Vol. 4, No. 4, 1934, pp. 430-439.

[91] M. M. Kononova, "Organic Matter of Soil," Izvestija. Akademija Nauk SSSR, Moscow, 1963.

[92] M. G. Rutten, "The Origin of the Life by Natural Causes," Elsevier, Amsterdam, 1971.

[93] B. Ghernaout, D. Ghernaout and A. Saiba, "Algae and Cyanotoxins Removal by Coagulation/Flocculation: A Review," *Desalination and Water Treatment*, Vol. 20, No. 1-3, 2010, pp. 133-143.

[94] D. Ghernaout and B. Ghernaout, "On the Concept of the Future Drinking Water Treatment Plant: Algae Harvesting from the Algal Biomass for Biodiesel Production—A Review," *Desalination and Water Treatment*, Vol. 49, No. 1-3, 2012, pp. 1-18.

[95] S. Oden, "Die Huminsiiuren," *Kolloidchemische Beihefte*, Vol. 11, No. 3-9, 1919, pp. 75-260.

[96] O. Aschan, "Om Wattenhumus Och Dess Medverkan Vid Sjomalms Bildingen," *Arch. Kemi, Mineral. Geol*, Vol. 10A, 1932, pp. 1-143.

[97] K. Kalle, "The Problem of the Gelbstoff in the Sea," *Oceanography and Marine Biology. An Annual Review*, Vol. 4, 1966, pp. 91-104.

[98] B. A. Birge and C. Juday, "Particulate and Dissolved Organic Matter in Inland Waters," *Ecological Monographs*, Vol. 4, No. 4, 1934, pp. 440-474.

[99] S. A. Waksman, "Humus," Bailliere, Tindall and Cox, London, 1936.

[100] E. Odum, "The Essentials of Ecology," Nauka, Moscow, 1975.

[101] P. M. Williams, "The Distribution and Cycling of Organic Matter in the Ocean," In: S. D. Faust and I. P. Hunter, Eds., *Organic Compounds in Aquatic Environment*, Marcel Dekker, New York, 1971, pp. 145-163,.

[102] P. J. le B. Williams, "Biological and Chemical Aspects of Dissolved Organic Material in Sea Water," In: J. P. Riley and G. Skirrow, Eds., *Chemical Pceanography*, Vol. 2, 2nd Edition, Academic Press, London, 1975, pp. 301-363.

[103] B. A. Skopintsev, "Recent Progress in Study of Organic Matter in Ocean Water," *Okeanologiya*, Vol. 11, 1971, pp. 939-956.

[104] R. A. Duce and E. K. Duursma, "Inputs of Organic Matter to the Ocean," *Marine Chemistry*, Vol. 5, No. 4-6, 1977, pp. 319-339.

[105] E. R. Sholkovitz, "Flocculation of Dissolved Organic and Inorganic Matter during the Mixing of River Water and Sea Water," *Geochimica et Cosmochimica Acta*, Vol. 40, No. 7, 1976, pp. 831-845.

[106] E. R. Sholkovitz, E. A. Boyle and N. B. Price, "The Removal of Dissolved Humic Acids and Iron during Estuarine Mixing," *Earth and Planetary Science Letters*, Vol. 40, No. 1, 1978, pp. 130-136.

[107] O. K. Bordovsky, "Organic Matter of Sea and Ocean Sediments in the Stage of Early Diagenesis," Nauka, Moscow, 1974.

[108] A. Nissenbaum and J. Kaplan, "Chemical and Isotopic Evidence of *in Situ* Origin of Marine Humic Substances," *Limnology and Oceanography*, Vol. 17, No. 4, 1972, pp. 570-582.

[109] G. Newcombe and D. Dixon, "Preface," In: G. Newcombe and D. Dixon, Eds., *Interface Science in Drinking Water Treatment, Theory and Applications*, In: A. Hubbard, Series Ed., *Interface Science and Technology*, Vol. 10, Academic Press, Elsevier, Amsterdam, 2006.

[110] G. Newcombe and D. Dixon, "Introduction," In: G. Newcombe and D. Dixon, Eds., *Interface Science in Drinking Water Treatment, Theory and Applications*, Ch. 1. In: A. Hubbard, Series Ed., *Interface Science and Technology*, Vol. 10, Academic Press, Elsevier, Amsterdam, 2006.

[111] E. R. Alley, "Water Quality Control Handbook," 2nd Edition, McGraw-Hill Companies, Inc., New York, 2007.

[112] J. T. O'Connor, T. O'Connor and R. Twait, "Water Treatment Plant Performance Evaluations and Operations," John Wiley & Sons, Inc., New York, 2009.

[113] W. Stumm, M. Forsberg and S. Gherini, "Water Chemistry," In: J. R. Pfafflin and E.N. Ziegler, Eds., *Encyclopedia of Environmental Science and Engineering*, 5th Edition, CRC Press, Taylor & Francis Group, Boca Raton, 2006, p. 1256.

[114] A. Becaria, D. K. Lahiri, S. C. Bondy, D. Chen, A. Hamadeh, H. Li, R. Taylor and A. Campbell, "Aluminum and Copper in Drinking Water Enhance Inflammatory or Oxidative Events Specifically in the Brain," *Journal of Neuroimmunology*, Vol. 176, No. 1, 2006, pp. 16-23.

[115] J. R. Walton, "A Longitudinal Study of Rats Chronically Exposed to Aluminium at Human Dietary Levels," *Neuroscience Letters*, Vol. 412, No. 1, 2007, pp. 29-33.

[116] C. Gourier-Fréry and N. Fréry, Aluminium, *EMC-Toxicologie-Pathologie*, Vol. 1, No. 3, 2004, pp. 79-95.

[117] T. P. Flaten, "Aluminium as a Risk Factor in Alzheimer's Disease, with Emphasis on Drinking Water," *Brain Research Bulletin*, Vol. 55, No. 2, 2001, pp. 187-196.

[118] L. Lévesque, C. A. Mizzen, D. R. McLachlan and P. E. Fraser, "Ligand Specific Effects on Aluminum Incorporation and Toxicity in Neurons and Astrocytes," *Brain Research*, Vol. 877, No. 2, 2000, pp. 191-202.

[119] E. Gauthier, I. Fortier, F. Courchesne, P. Pepin, J. Mortimer and D. Gauvreau, "Aluminum Forms in Drinking Water and Risk of Alzheimer's Disease," *Environmental Research*, Vol. 84, No. 3, 2000, pp. 234-246.

[120] M.-L. Cheng, H.-Y. Ho, D. T.-Y. Chiu, F.-J. Lu, "HA-Mediated Oxidative Damages to Human Erythrocytes: A Possible Mechanism Leading to Anemia in Blackfoot Disease," *Free Radical Biology Medicine*, Vol. 27, No. 3-4, 1999, pp. 470-477.

[121] C.-H. Tseng, "Blackfoot Disease and Arsenic: A Never-Ending Story," *Journal of Environmental Science and Health*, Vol. 23, 2005, pp. 55-74.

[122] R.-J. Gau, H.-L. Yang, J.-L. Suen and F.-J. Lu, "Induction of Oxidative Stress by HA through Increasing Intracellular Iron: A Possible Mechanism Leading to Atherothrombotic Vascular Disorder in Blackfoot Disease," *Biochemical and Biophysical Research Communications*, Vol. 283, No. 4, 2001, pp.743-749.

[123] Y.-C. Hseu, H.-W. Huang, S.-Y. Wang, H.-Y. Chen, F.-J. Lu, R.-J. Gau and H.-L. Yang, "HA Induces Apoptosis in Human Endothelial Cells," *Toxicology and Applied Pharmacology*, Vol. 182, No. 1, 2002, pp. 34-43.

[124] J. E. Van Benschoten and J. K. Edzwald, "Chemical Aspects of Coagulation Using Aluminium Salts-II. Coagulation of Fulvic Acid Using Alum and Polyaluminum Chloride," *Water Research*, Vol. 24, No. 12, 1990, pp. 1527-1535.

[125] J. H. Clark, "Introduction," In: J. Clark and D. Macquarrie, Eds., *Handbook of Green Chemistry and Technology*, Ch. 1, Blackwell Science Ltd, Oxford, 2002.

[126] D. Ghernaout, B. Ghernaout and M. W. Naceur, Embodying the Chemical Water Treatment in the Green Chemistry—A Review," *Desalination*, Vol. 271, No. 1-3, 2011, pp. 1-10.

[127] G. A. Jackson, "Coagulation Theory and Models of Oceanic Plankton Aggregation," In: I. G. Droppo, G. G. Leppard, S. N. Liss and T. G. Milligan, Eds., *Flocculation in Natural and Engineered Environmental Systems*, Ch. 13, CRC Press, Boca Raton, 2005.

[128] E. Brown, A. Colling, D. Park, J. Phillips, D. Rothery and J. Wright, "Seawater: Its Composition, Properties and Behaviour," 2nd Edition, Butterworth-Heinemann, Oxford, 2004.

[129] K. S. Venkateswarlu, "Water Chemistry, Industrial and Power Station Water Treatment," New Age International (P), Ltd., New Delhi, 1996.

10

Solvent Free One-Pot Synthesis of 1,2,4,5-Tetrasubstituted Imidazoles Catalyzed by Secondary Amine Based Ionic Liquid and Defective Keggin Heteropoly Acid

Pranab Jyoti Das[1*], Jupitara Das[1], Munni Ghosh[1], Sabera Sultana[2]
[1]Department of Chemistry, Gauhati University, Guwahati, India
[2]Indian Institute of Technology, Guwahati, India

ABSTRACT

Secondary amine based ionic liquid and defective Keggin type heteropoly acid (HPA) are separately used for efficient one-pot four-component synthesis of 1,2,4,5-tetrasubstituted imidazoles assisted by microwave (MW). Eco-friendly solvent free procedure, short reaction time, high yield of products and reusability of catalysts are important features of the synthesis. A comparative study on the efficiency of the two catalysts is reported. This work further demonstrates the alternate use of urea, instead of often used ammonium acetate, as source of nitrogen.

Keywords: Ionic Liquids; Keggin Heteropolyacid; Microwave; Tetrasubstituted Imidazole; Solvent Free Synthesis; Multicomponent Reaction

1. Introduction

Tetrasubstituted imidazoles are important heterocycles with wide biological applications. Their synthesis and reactions form a significant part of the study of heterocycles with medicinal use [1]. Tetrasubstituted imidazole scaffold is the most active constituent in many biological systems and drug molecules such as olmesartan medoxomil, losartan, eprosartan and trifenagrel [2-4] as well as other natural products of pharmaceutical importance [5-7]. Multicomponent reaction (MCR) is a powerful tool in generating molecules with diverse functionality in a single synthetic protocol [8-10] and forms the basis of the synthesis of 1,2,4,5-tetrasubstituted imidazoles. Important synthetic methods include one-pot thiazolium catalyzed addition of an aldehyde to an acyl imine followed by ring closure to the imidazole [11], condensation of arylglyoxal, 1°-amine, carboxylic acids and isocyanates on Wang resin followed by cyclization in the presence of acetic acid [12], by a hetero-Cope rearrangement [13], a direct synthesis from alkenes involving two step ketoiodonation/cyclization protocol [14], Domino reaction of 2-azido acrylates and nitrones [15]. Other synthetic methods include the condensation of benzoin and benzoin acetate with aldehyde, 1°-amine, ammonia in the presence of copper acetate [16], one-pot synthesis using BF_3-SiO_2 as catalyst [17], cyclization of sulphonamide with mesoionic 1,3-oxazolinium-5-olates [18], condensation of β-carbonyl-N-acyl-N-alkylamines with NH_4OAc in refluxing acetic acid [19,20], conversion of N-2-oxoamides with ammonium triflate [21] besides others [22-25]. Several catalysts have been explored for performing the four components' reaction to give the target product notably the use of L-proline [26], zeolite HY and silica gel [27], Keggin type heteropolyacid [28], $LaCl_3$ catalyzed synthesis using urea as a source of ammonia [29]. Shaterian *et al.* reported the synthesis of tetrasubstituted imidazole used Brønsted acidic ionic liquid, N-methyl-2-pyrrolidonium hydrogen sulphate, as catalyst, but long reaction time, high amount of catalyst used and the time-consuming procedure for the preparation of the IL (12 hrs) are some of the disadvantages [30]. Most of the procedures reported are associated with one or more disadvantages such as use of expensive reagents and catalysts, mineral acids and requirement of large amount of catalysts which eventually results in generation of toxic wastes, longer reaction time and tedious

*Corresponding author.

work-up.

2. Results and Discussion

In a broad programme of developing efficient, selective and eco friendly synthetic methods for pharmacologically important heterocycles, we explored the applicability of green promoters namely the simple acidic ionic liquids (IL) as well as the defective Keggin type heteropolyacids (HPA) in the synthesis of 1,2,4,5 tetrasubstituted imidazoles using microwave (MW) heating in both procedures. Combination of two green techniques is reported to give better results in terms of short reaction time and high yield of products. Catalytic efficiencies of both the ionic liquid and heteropolyacid are well documented. The unique thermodynamic properties of both are increasingly enticing chemists to explore their use as media as well as promoters in organic synthesis [31-33], however, high cost of the often used imidazole based ILs and the heteropoly acids deter their use in industrial processes. Herein, we explored the possibility of using simple and cost effective secondary amine based IL namely di-*n*-propylammonium hydrogensulphate in the synthesis of imidazoles in a solvent free multicomponent reaction (MCR) using MW heating. In a related study, the efficiency of this IL *vis a vis* a heteroploy acid as an alternative catalyst for the synthesis was examined under identical experimental conditions. Initially the IL and the HPA used were prepared by adopting a simple atom economy procedure using cheap and easily available substrates. The IL used herein was prepared by the action of conc. H_2SO_4 on di-*n*-propylamine [34] and the lacunary and defective Keggin type HPA namely $H_6PAlMo_{11}O_{40}$ was prepared by a reported procedure [35] albeit with minor modification. Microwave heating was used while preparing the HPA instead of conventional heating which resulted in less reaction time and high yield of the HPA.

For the purpose of optimization of the amount of catalyst and the reaction time, representative reaction using benzil, benzaldehyde, benzylamine and NH_4OAc as the substrates was performed by varying the concentration of the catalyst from 5 mol% to 10 mol% and concomitant variation of reaction time from 2 mins to 10 mins. In case of IL as the catalyst, yield of the desired product **4a** improved by 20% when mol% was increased from 5% to 10% and in case of HPA 5 mol% of catalyst and 10 min reaction time gave the best results. No improvement in yield was apparent with increase in mol% of catalyst and/or the reaction time. Results summarized in **Table 1** indicated best results with 10 mol% of IL as catalysts and in case of HPA it was 5 mol%.

These two catalysts were used in turn for performing a one-pot four component reaction involving 1,2-diketone, aromatic aldehyde, 1°-amine and NH_4OAc to give the target 1,2,4,5-tetrasubstituted imidazole under solvent free condition mediated by MW. In a typical procedure, the reactants and the catalyst in the form of either HPA or IL, are thoroughly mixed and exposed to MW irradiation. In order to establish the generality of the synthetic procedure, several diversified examples were studied by varying the 1°-amine and the aldehyde. The yields obtained by using both HPA and the IL are good to excellent with very little or no side products which is a significant improved over other reported procedures. Further good results have been obtained with both the aliphatic as well as the aromatic amines. Reactions with the IL as promoter required 2 - 3 min for completion and reactions catalyzed by HPA required 10 - 13 min. These results are indicative of better efficiency of IL over HPA in this synthesis. The reaction is shown in **Scheme 1** and the results summarized in **Table 2**. Work up was accomplished by extraction of the reaction mixture with CH_2Cl_2 to separate out the target product. The CH_2Cl_2 solution was washed with water (3x 50 mL) to remove any trace of the impurity. Since both the catalysts are insoluble in CH_2Cl_2 the catalysts were precipitated out after addition of CH_2Cl_2 to the reaction mixture and recovered by simple filtration. The recovered IL was stored in desiccator for reuse and HPA was dried at 85°C for 24 hours and reused. Both the catalysts retained their activity after two subsequent reuses.

In order to preclude other possibilities, the reaction was examined using related catalyst namely tetra-*n*-propylammonium bromide and $NaHSO_4$ under MW irradiation. With tetra-*n*-propylammonium bromide, the reaction did not proceed at all whereas with $NaHSO_4$ about

Table 1. Effect of amount of catalyst for the formation of 4a via the 4-MCR of 1, 2, 3 and NH_4OAc.

Entry	mol%	Ionic liquid		Heteropoly acid	
		Time (min)	Yield (%)[a]	Time (min)	Yield (%)[a]
1	5	2	65	2	trace
2	5	10	69	10	89
3	10	2	85	2	trace
4	10	10	86	10	89

a. Yields refer to the pure isolated product.

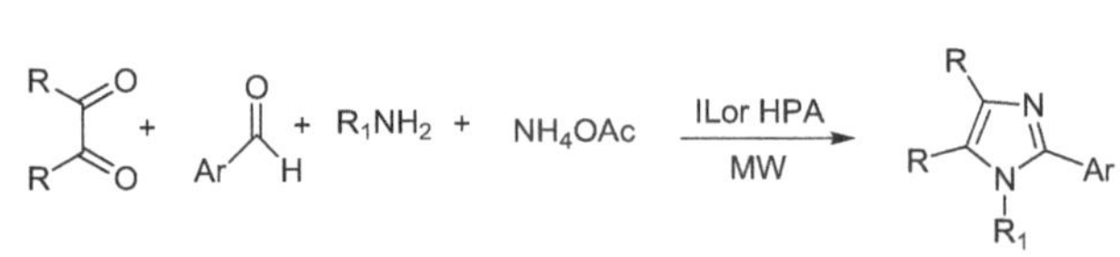

1 2 3 4 (a-s)

R= Phenyl, p-tolyl;
R1= C_6H_5-, $C_6H_5CH_2$-, 4-$NO_2C_6H_4$-, C_2H_5-, CH_3-
IL= [n-Pr_2NH_2][HSO_4], HPA = $H_6PAlMo_{11}O_{40}$

Scheme 1. Synthesis of imidazole derivatives mediated by IL/HPA.

Table 2. Solvent free MW induced one-pot synthesis of 1, 2, 4, 5-tetrasubstituted imidazole promoted by [n-Pr_2NH_2][HSO_4] or HPA using ammonium acetate as nitrogen source.

Product	R	Ar	R_1	Reaction with IL		Reaction with HPA	
				Time (min)	Yield (%)[a]	Time (min)	Yield (%)[a]
4a	Ph	C_6H_5	$C_6H_5CH_2$	2	85	10	89
4b	Ph	4-$CH_3C_6H_4$	$C_6H_5CH_2$	2	87	10	92
4c	Ph	2-$CH_3OC_6H_4$	$C_6H_5CH_2$	2	87	10	94
4d	Ph	3,4-$(OCH_3)_2C_6H_3$	$C_6H_5CH_2$	2	87	10	93
4e	Ph	4-ClC_6H_4	$C_6H_5CH_2$	2	85	13	90
4f[b]	Ph	4-EtC_6H_4	$C_6H_5CH_2$	2	87	10	92
4g	Ph	3-BrC_6H_4	$C_6H_5CH_2$	2	85	10	89
4h	Ph	4-$NO_2C_6H_4$	$C_6H_5CH_2$	2	85	10	88
4i[b]	Ph	2,5-$(CH_3)_2C_6H_4$	C_6H_5	2	84	10	90
4j	Ph	C_6H_5	C_6H_5	2	83	10	87
4k[b]	Ph	4-ClC_6H_4	CH_3	3	87	10	89
4l[b]	Ph	4-EtC_6H_4	CH_3	3	86	10	89
4m[b]	Ph	2-$CH_3OC_6H_4$	CH_3CH_2	3	90	10	92
4n	Ph	4-ClC_6H_4	CH_3CH_2	3	90	10	92
4o[b]	Ph	3,4-$(OCH_3)_2C_6H_3$	CH_3CH_2	3	90	10	92
4p[b]	4-CH_3Ph	2,6-$Cl_2C_6H_3$	CH_3CH_2	3	92	10	92
4q[b]	Ph	4-ClC_6H_4	4-$NO_2C_6H_4$	3	85	13	87
4r[b]	Ph	3,4-$(OCH_3)_2C_6H_3$	4-$NO_2C_6H_4$	3	85	13	88
4s	Ph	C_6H_5	4-$NO_2C_6H_4$	3	83	13	85

a. Yields refers to the pure isolated products; b. New compound.

40% yield of desired product was obtained indicating the essentiality of di-n-propylammonium hydrogensulphate as the appropriate catalyst for the reaction. The results obtained with HPA, [n-Pr_2NH_2][HSO_4], $NaHSO_4$, n-Pr_4NBr and in absence of catalyst for a model reaction are summarized in **Table 3**. In this study benzil, 2-anisaldehyde, benzyl amine and ammonium acetate were taken as the reference reactants and 10 mol% amounts of catalysts were used.

Investigations were also carried out using urea as the source of nitrogen in the reaction. It was observed that the reaction proceeded equally well with urea in the presence of both IL and HPA. Here efficiency was examined in some model reactions in presence of either IL or HPA. The results obtained by using urea instead of NH_4OAc are summarized in **Table 4**.

All products obtained were characterized by spectroscopic method such as IR, ^{1}H NMR, ^{13}C NMR, Mass spectrometry and by comparing their melting points with those reported in literature. Some new compounds are also reported. Single crystal X-ray analysis of one of the new products namely 1-ethyl-2-(2', 6'-dichlorophenyl)-4, 5-di(4'-methylphenyl)-imidazole (**4p**) confirms the structure (**Figure 1**). Crystals were obtained from ethanol in which case the results indicate the presence of residual solvent in the crystal. Attempt to remove the solvent destroyed crystallinity of the product and no worthwhile data could be obtained. Crystallographic data are given in **Table 5**. CCDC reference number for the compound is **CCDC 892443**.

Table 3. Effect of catalyst on the one-pot four component synthesis of tetra substituted imidazole 4c with the usual reactants and NH_4OAc.

Entry	Catalyst[b]	Time (min)	Yield (%)[a]
1	Catalyst free	10	-
2	HPA	10	95
3	[n-Pr_2NH_2][HSO_4]	10	88
4	$NaHSO_4$	10	40
5	n-Pr_4NBr	10	-

a. Yields refer to the pure isolated product; b. 10 mol% amount of catalysts were used.

3. Conclusion

In conclusion, a one-pot solvent free MCR promoted by the combined use of microwave and IL or HPA offer easy access to tetrasubstituted imidazoles in excellent yield. Procedural simplicity, short reaction time, solvent

Table 4. The effect of ammonium acetate and urea as the nitrogen source on the 4-MCR to form 4a, 4d and 4m.

Entry	Nitrogen source	Product	IL		HPA	
			Time (min)	Yield (%)[a]	Time (min)	Yield (%)[a]
1	Ammonium acetate	**4a**	2	86	10	89
2	Ammonium acetate	**4d**	2	86	10	95
3	Ammonium acetate	**4m**	3	90	10	90
4	Urea	**4a**	2	86	10	89
5	Urea	**4d**	2	86	10	95
6	Urea	**4m**	3	90	10	90

a. Yields refer to the pure isolated products.

Table 5. Crystallographic parameter of compound 4p.

empirical formula	$C_{25}H_{22}Cl_2N_2{\cdot}C_2H_5OH$
formula weight	467.43
crystal system	Monoclinic
space group	*P* 2(1)/n
T (K)	298
a/Å	11.7711(9)
b/Å	15.5380(12)
c/Å	13.9410(12)
α/deg	90.00
β/deg	95.289(5)
γ/deg	90.00
V / Å^3	2538.9(4)
D_{calcd} (g·cm^{-3})	1.223
μ (mm^{-1})	0.271
Z	4
reflns collected	25872
unique reflns	7416
Observed reflns	3022
$R_1[I > 2\sigma(I)]$	0.0837
wR2 (all)	0.1546
goodness-of-fit	1.287
diffractometer	SMART Bruker Apex-II

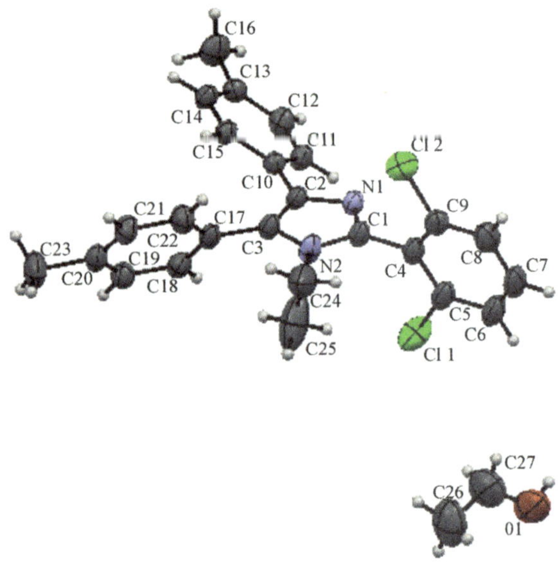

Figure 1. ORTEP diagram of 4p.

free condition and the use of non toxic and environmentally benign promoters are some of the important features of this new methodology. An added advantage is the simple recovery and efficient reusability of the catalyst. The reaction proceeds equally well with urea as a source of nitrogen instead of NH_4OAc which result in cost reduction of the transformation.

4. Experimental Section

4.1. General

Melting points were recorded in a VMP-D model melting point apparatus and are uncorrected. ^1H NMR and ^{13}C NMR spectra were recorded on a Bruker advance digital 300 MHz spectrometer in $CDCl_3$. TMS was used as internal standard. IR spectra were obtained on a Perkin Elmer FT-IR 1600 spectrophotometer using KBr pallets. Mass spectra were recorded on a Waters Micromass ZQ™ 400 Mass spectrometer system. The XRF analysis was carried out in Axios XRF Spectrometer. Microwave irradiation of the reaction mixture was performed in a reactor procured from Catalyst™ (India) reactor. The single crystal X-ray diffraction data for compound 1-ethyl-2-(2', 6'-dichlorophenyl)-4, 5-di(4'-methylphenyl) imidazole (4p) was collected on a SMART Bruker Apex-II diffractometer. The structure was solved by direct method and refined by full-matrix least squares based on F^2 using the SHELXTL 5.1 software package. The hydrogen atoms residing in the carbon atoms were located geometrically. All non-hydrogen atoms were refined anisotropically.

4.2. Synthetic Procedure for the Preparation of the $H_6PAlMo_{11}O_{40}$

A stoichiometric mixture of H_3PO_4 85% (0.58 g, 0.01 mol), $Al_2O_3{\cdot}6H_2O$ (1.21 g, 0.005 mol) and MoO_3 (14.4 g,

0.10 mol) was suspended in 150 ml of distilled water and irradiated with microwave for 10 minutes (560 W). On cooling the mixture to room temperature the insoluble MoO_3 precipitated and recovered by filtered. Reduced pressure removal of water and drying at 85°C for 24 hours gave dark green crystals of composition $H_6PAlMo_{11}O_{40}$ identical to that obtained earlier [35]. The metal composition was established by XRF analysis. IR (KBR): *v* 1074, 982, 875, 765, 362 and 345 cm^{-1}.

4.3. General Procedure for the Synthesis of Tetrasubstituted Imidazole

A mixture of 1, 2-diketone (1 mmol), aldehydes (1 mmol), primary amines (1 mmol), ammonium acetate (or urea) (1.5 mmol) and IL (10 mol%)/HPA (5 mol%) was irradiated with microwave (560 W) with stirring for 1 - 3/10 - 13 min as mentioned in **Table 2**. On completion of the reaction monitored by TLC, the reaction mixture was extracted with CH_2Cl_2 (5 mL × 3), washed with water, dried with anhydrous Na_2SO_4 and solvent removed under reduced pressure. The crude products were purified by recrystallization from ethanol.

1-*Benzyl*-2,4,5-*triphenylimidazole* (**4*a***): Mp 159 - 161°C (EtOH); IR(KBr): *v* 2890 (CH), 1590 (CN), 1480 cm^{-1}; ^1H NMR (300 MHz, $CDCl_3$): δ_H ppm 7.58 - 6.82 (m, 20H, Ar), 5.12 (s, 2H, CH_2); ^{13}C NMR (75 MHz, $CDCl_3$): δ ppm 148.08, 138.07, 137.56, 134.46, 131.08, 131.02, 130.94, 130.05, 129.08, 128.92, 128.81, 128.61, 128.59, 128.09, 127.36, 126.78, 126.37, 126.02, 48.28.

1-*Benzyl*-2-(4'-*methylphenyl*)-4,5-*diphenylimidazole* (**4*b***): Mp 164 - 166°C (EtOH); IR(KBr): *v* 2885 (CH), 1580 (CN), 1482 cm^{-1}; ^1H NMR (300 MHz, $CDCl_3$): δ_H ppm 7.58 - 6.81 (m, 19H, Ar), 5.11 (s, 2H, CH_2), 2.38 (s, 3H, CH_3); ^{13}C NMR (75 MHz, $CDCl_3$): δ ppm 148.21, 138.84, 137.67, 134.52, 131.07, 129.28, 128.94, 128.75, 128.55, 128.06, 126.78, 125.99, 48.24, 21.36.

1-*Benzyl*-2-(2'-*methoxyphenyl*)-4,5-*diphenylimidazole* (**4*c***): Mp 188 - 190°C (EtOH); IR(KBr): *v* 2823 (CH), 1593 (CN), 1460 (CC), 1250 cm^{-1}; ^1H NMR (300 MHz, $CDCl_3$): δ_H ppm 7.59 - 6.82 (m, 19H, Ar), 5.09 (s, 2H, CH_2), 3.83 (s, 3H, OCH_3); ^{13}C NMR (75 MHz, $CDCl_3$): δ ppm 160.04, 147.94, 137.61, 131.02, 130.37, 128.72, 128.53, 128.02, 127.27, 126.72, 126.25, 125.92, 123.29, 113.95, 55.27, 48.16; Anal. Calcd for $C_{29}H_{24}N_2O$: C, 83.63; H, 5.81; N, 6.73, O, 3.84. Found: C, 83.65; H, 5.83; N, 6.72; O, 3.84; HRMS (ESI): *m/z* = 416.1889 ($[M]^+$), calcd for $C_{29}H_{24}N_2O$: 416.1889.

1-*Benzyl*-2-(3',4'-*dimethoxyphenyl*)-4,5-*diphenylimidazole* (**4*d***): Mp 163 - 165°C (EtOH); IR(KBr): *v* 2835 (CH), 1600 (CN) cm^{-1}; ^1H NMR (300 MHz, $CDCl_3$): δ_H ppm 7.64 - 6.89 (m, 18H, Ar), 5.10 (s, 2H, CH_2), 3.89 (s, 3H, OCH_3), 3.69 (s, 3H, OCH_3); ^{13}C NMR (75 MHz, $CDCl_3$): δ ppm 149.63, 148.80, 137.99, 134.52, 131.10, 128.89, 128.75, 128.16, 127.42, 126.90, 126.42, 125.96, 121.65, 112.24, 110.99, 55.96, 55.69, 48.28.

1-*Benzyl*-2-(4'-*chlorophenyl*)-4,5-*diphenylimidazole* (**4*e***): Mp 163 - 165°C (EtOH); IR(KBr): *v* 2832 (CH), 1598 (CN) cm^{-1}; ^1H NMR (300 MHz, $CDCl_3$): δ_H ppm 8.00 - 6.82 (m, 19H, Ar), 5.09 (s, 2H, CH_2); ^{13}C NMR (75 MHz, $CDCl_3$): δ ppm 146.92, 137.37, 135.06, 131.08, 130.31, 129.98, 129.11, 128.92, 128.79, 128.20, 126.85, 125.92, 48.34.

1-*Benzyl*-2-(4'-*ethylphenyl*)-4,5-*diphenylimidazole* (**4*f***): Mp 138 - 140°C (EtOH); IR(KBr): *v* 2958 (CH), 1597 (CN), 1492 (CC), 1446, 1388 cm^{-1}; ^1H NMR (300 MHz, $CDCl_3$): δ_H ppm 7.59 - 6.83 (m, 19H, Ar), 5.11 (s, 2H, CH_2), 2.67 (q, 2H, J = 6 Hz, CH_2), 1.24 (t, 3H, J = 6 Hz, CH_3); ^{13}C NMR (75 MHz, $CDCl_3$): δ ppm 148.20, 145.13, 137.88, 137.63, 131.04, 128.97, 128.70, 128.51, 128.08, 128.02, 126.73, 126.26, 125.98, 48.22, 28.66, 15.41. Anal. Calcd for $C_{30}H_{26}N_2$: C, 86.92; H, 6.32; N, 6.76. Found: C, 86.90; H, 6.33; N, 6.78. HRMS (ESI): *m/z* = 414.2098 ($[M]^+$), calcd for $C_{30}H_{26}N_2$: 414.2096. 1-*Benzyl*-2-(3'-*bromophenyl*)-4,5-*diphenylimidazole* (**4*g***): Mp 148 - 150°C (EtOH); IR(KBr): *v* 2949 (CH), 1601(CN) cm^{-1}; ^1H NMR (300 MHz, $CDCl_3$): δ_H ppm 7.66 - 6.82 (m, 19H, Ar), 5.11 (s, 2H, CH_2); ^{13}C NMR (75 MHz, $CDCl_3$): δ ppm 132.17, 131.86, 130.99, 130.00,129.04, 128.86, 128.77, 128.69, 128.57, 128.11, 127.51, 127.22, 126.75, 126.51, 125.98, 125.90, 48.31.

1-*Benzyl*-2-(4'-*nitrophenyl*)-4,5-*diphenylimidazole* (**4*h***): Mp 170 - 172°C (EtOH); IR(KBr): *v* 2963 (CH), 1608 (CN) cm^{-1} ; ^1H NMR (300MHz, $CDCl_3$): δ_H ppm 8.12 - 6.73 (m, 19H, Ar), 4.88 (s, 2H, CH_2); ^{13}C NMR ($CDCl_3$): δ ppm 148.92, 143.00, 136.57, 133.23, 131.10, 128.93, 128.79, 128.38, 128.05, 126.64, 126.48, 126.41, 124.65, 48.27.

2-(2',5'-*Dimethylphenyl*)-1,4,5-*triphenylimidazole* (**4*i***): Mp 173 - 175°C (EtOH); IR(KBr): *v* 2942 (CH), 1598 (CN), 1492 (CC) cm^{-1}; ^1H NMR (300 MHz, $CDCl_3$): δ_H ppm 7.61 - 6.87 (m, 18H, ArH), 2.23 (s, 3H, CH_3), 2.10 (s, 3H, CH_3); ^{13}C NMR (75 MHz, $CDCl_3$): δ ppm 147.59, 137.64, 136.50, 134.74, 134.64, 134.51, 131.79, 130.94, 130.78, 130.40, 129.83, 129.73, 129.07, 128.46, 128.37, 128.05, 127.78, 127.71, 127.49, 127.41, 126.44, 20.73, 19.67; Anal. Calcd for $C_{29}H_{24}N_2$: C, 86.97; H, 6.04; N, 6.99. Found: C, 86.90; H, 6.04; N, 6.89. HRMS(ESI): *m/z* = 400.1939 ($[M]^+$), calcd for $C_{29}H_{24}N_2$: 400.1939.

1-(4'-*nitrophenyl*)-2-(4'-*chlorophenyl*)-4,5-*diphenylimidazole* (**4*q***): Mp 122 - 124°C (EtOH); IR(KBr): *v* 2928 (CH), 1590 (CN), 1489 (CC), 1351 (C-NO2), 739 (CCl) cm^{-1}; ^1H NMR (300 MHz, $CDCl_3$): δ_H ppm 8.05 - 7.03 (18H, m, ArH); ^{13}C NMR (75 MHz, $CDCl_3$): δ ppm 194.66, 152.50, 134.94, 132.87, 129.88, 129.02, 128.60, 127.77, 127.54, 126.47, 126.33, 113.31; Anal. Calcd for $C_{27}H_{18}N_3O_2Cl$: C, 71.76; H, 4.01; N, 9.30; O, 7.08; Cl, 7.85. Found: C, 71.68; H, 4.04; N, 9.32; O, 7.09; Cl, 7.83. HRMS (ESI): *m/z* = 451.1089 ($[M]^+$), calcd for

$C_{27}H_{18}N_3O_2Cl$: 451.1088.

1-(4'-*Nitrophenyl*)-2-(3',4'-*dimethoxyphenyl*)-4,5-*diphenylimidazole* (**4r**): Mp 125 - 126°C (EtOH); IR(KBr): ν 2939 (CH), 1597 (CN), 1489 (CC), 1346 (C-NO2), 1246 (CO) cm^{-1}; ^1H NMR (300 MHz, $CDCl_3$): δ_H ppm 8.14 - 8.11 (m, 2H, ArH), 7.59 - 7.56 (m, 2H, ArH), 7.30 - 7.12 (m, 12H, ArH), 6.70 (s, 1H, ArH), 3.86 (s, 3H, OCH_3), 3.78 (s, 3H, OCH_3); ^{13}C NMR (75 MHz, $CDCl_3$): δ ppm 149.60, 133.72, 131.02, 129.90, 129.18, 128.78, 128.56, 128.24, 127.40, 126.99, 124.36, 121.91, 112.28, 110.63, 55.83; Anal. Calcd for $C_{29}H_{23}N_3O_4$: C, 72.94; H, 4.85; N, 8.80; O, 13.40. Found: C, 72.96; H, 4.83; N, 8.82; O, 13.49. HRMS (ESI): m/z = 477.1689 ($[M]^+$), calcd for $C_{29}H_{23}N_3O_4$: 477.1689.

4.4. Synthesis of Compounds 4(k-p)

Carried as per the general procedure with 0.5 mL of aliphatic amines (methyl amine and ethyl amine).

1-*methyl*-2-(4'-*chlororphenyl*)-4,5-*diphenylimidazole* (**4k**): Mp 192 - 194°C (EtOH); IR(KBr): ν 2925 (CH), 1604 (CN), 1483 (CC), 735 (CCl) cm^{-1}; ^1H NMR (300 MHz, $CDCl_3$): δ_H ppm 7.71 - 7.18 (m, 14H, ArH), 3.50 (s, 3H, NCH_3); ^{13}C NMR (75 MHz, $CDCl_3$): δ ppm 146.79, 137.98, 134.90, 134.44,130.97, 130.89, 130.84, 130.32, 129.38, 129.15, 129.05, 128.92, 128.77, 128.20, 127.01, 126.54, 33.29; Anal. Calcd for $C_{22}H_{17}N_2Cl$: C, 76.63; H, 4.97; N, 8.12; Cl, 10.28. Found: C, 76.65; H, 4.99; N, 8.19; Cl, 10.31; HRMS(ESI): m/z = 344.1082 ($[M]^+$), calcd for $C_{22}H_{17}N_2Cl$: 344.1080.

1-*methyl*-2-(4'-*ethylphenyl*)-4, 5-*diphenylimidazole* (**4l**): Mp 127 - 130°C (EtOH); IR(KBr): ν 2920 (CH), 1597 (CN), 1480 (CC) cm^{-1}; ^1H NMR(300 MHz,$CDCl_3$): δ_H ppm 7.67 - 7.19 (m, 14H, ArH), 3.51 (s, 3H, NCH_3), 2.73 (q, 2H, J = 6 Hz, CH_2), 1.29 (t, 3H, J = 6 Hz, CH_3); ^{13}C NMR (75 MHz, $CDCl_3$): δ ppm 134.84, 130.79, 129.84, 128.95, 128.43, 128.00, 127.97, 126.87, 33.06, 28.65, 15.40; Anal. Calcd for $C_{24}H_{22}N_2$: C, 85.17; H, 6.55; N, 8.28. Found: C, 85.19; H, 6.58; N, 8.29. HRMS(ESI): m/z = 338.1786 ($[M]^+$), calcd for $C_{24}H_{22}N_2$: 338.1783.

1-*Ethyl*-2-(2'-*methoxyphenyl*)-4,5-*diphenylimidazole* (**4m**):Mp 123 - 125°C (EtOH); IR(KBr): ν 2924 (CH), 1600 (CN), 1462 (CC), 1260 (CO) cm^{-1}; ^1H NMR (300 MHz, $CDCl_3$): δ_H ppm 7.57 - 7.09 (14H, m, ArH), 3.85 (3H, s, OCH_3), 3.77 (2H, q, J = 6 Hz ,CH_2), 0.90 (3H, t, J = 6 Hz, CH_3); ^{13}C NMR (75 MHz, $CDCl_3$): δ ppm 157.53, 144.69, 137.57, 134.76, 132.66, 131.77, 130.92, 130.81, 128.93, 128.68, 128.40, 127.88, 126.72, 125.91, 120.88, 120.66, 110.85, 55.47, 39.47, 15.85; Anal. Calcd for $C_{24}H_{22}N_2O$: C, 81.33; H, 6.26; N, 7.90; O, 4.51. Found: C, 81.41; H, 6.28; N, 7.95; O, 4.56. HRMS(ESI): m/z = 354.1736 ($[M]^+$), calcd for $C_{24}H_{22}N_2O$: 354.1732.

1-*Ethyl*-2(4'-*chlorophenyl*)-4,5-*diphenylimidazole* (**4n**): Mp 311 - 313°C (EtOH); IR(KBr): ν 2930 (CH), 1596 (CN)cm^{-1}; ^1H NMR (300 MHz, $CDCl_3$): δ_H ppm 7.68 - 7.16 (14H, m, ArH), 3.94 (2H, q, J = 6 Hz, CH_2), 1.03 (3H, t, J = 6 Hz, CH_3); ^{13}C NMR (75 MHz, $CDCl_3$): δ ppm 194.57, 146.04, 137.95, 134.90, 134.33, 132.91, 131.26, 130.98, 130.34, 129.89, 129.82, 129.66, 129.08, 129.01, 128.87, 128.76, 128.05, 126.70, 126.30, 39.65, 16.22.

1-*Ethyl*-2-(3',4'-*dimethoxyphenyl*)-4,5-*diphenylimidazole* (**4o**): Mp 173 - 175°C (EtOH); IR(KBr): ν 2928 (CH), 1608 (CN), 1482(CC), 1258 (CO) cm^{-1}; ^1H NMR (300 MHz, $CDCl_3$): δ_H ppm 7.54 - 6.98 (13H, m, ArH), 3.96 - 3.90 (8H, m, $2OCH_3$ and CH_2), 1.02 (3H, t, J = 6 Hz, CH_3); ^{13}C NMR (75 MHz, $CDCl_3$): δ ppm 149.64, 149.04, 137.54, 134.63, 131.66, 131.08, 129.21, 129.08, 128.66, 128.04, 126.77, 126.18, 124.11, 121.51, 112.72, 110.98, 56.04, 55.99, 39.65, 16.28; Anal. Calcd for $C_{25}H_{24}N_2O_2$: C, 78.10; H, 6.29; N, 7.29; O, 8.32. Found: C, 78.14; H, 6.27; N, 7.32; O, 8.35. HRMS (ESI): m/z = 384.1839 ($[M]^+$), calcd for $C_{24}H_{22}N_2O$: 384.1838.

1-*Ethyl*-2-(2',6'-*dichlorophenyl*)-4,5-*di*(4'-*methylphenyl*)*imidazole* (**4p**): Mp 135 - 137°C (EtOH); IR(KBr): ν 2924 (CH), 1605 (CN), 1450(CC), 738 (CCl) cm^{-1}; ^1H NMR (300 MHz, $CDCl_3$): δ_H ppm 7.46 - 7.00 (11H, m, ArH), 3.66 (2H, q, J = 6 Hz, NCH_2), 2.45 (3H, s, CH_3), 2.28 (3H, s, CH_3), 1.00 (3H, t, J = 6 Hz, CH_3); ^{13}C NMR (75 MHz, $CDCl_3$): δ ppm 141.10, 138.43, 137.28, 135.59, 131.78, 131.22, 130.86, 129.72, 128.65, 128.22, 128.08, 126.54, 39.32, 21.40, 21.10, 15.88; Anal. Calcd for $C_{25}H_{22}N_2Cl_2$: C, 71.26; H, 5.26; N, 6.65; Cl, 16.83. Found: C, 71.29; H, 5.32; N, 6.64; Cl, 16.84. HRMS (ESI): m/z = 420.1160 ($[M]^+$), calcd for $C_{25}H_{22}N_2Cl_2$: 420.1160.

5. Acknowledgements

J. D. is grateful to CSIR, New Delhi for award of SRF. The authors are also grateful to SAIF, Gauhati University for recording the single crystal XRD data and Mr S.J. Bharali for analyzing the data.

REFERENCES

[1] M. R. Grimmet and A. R. Katritzky, "In Comprehensive Heterocyclic Chemistry," In: C. W. Rees, Ed., Pergamon Press, London, Vol. 5, 1984, p. 374.

[2] D. Norwood, E. Branch, B. Smith and M. Honeywell, "Olmesartan Medoxomil for Hypertension: A Clinical Review," *Drug Forecast*, Vol. 27, No. 12, 2002, pp. 611-618.

[3] S. L. Abrahams, R. J. Hazen, A. G. Batson and A. P. Phillips, "Trifenagrel: A Chemically Novel Platelet Aggregation Inhibitor," *Journal of Pharmacology and Experimental Therapeutics*, Vol. 249, No. 2, 1989, pp. 359-365.

[4] S. E. Wolkenberg, D. D. Wisnoski, W. H. Leister, Y. Wang, Z. Zhao and C. W. Lindsley, "Efficient Synthesis

of Imidazoles from Aldehydes and 1,2-Diketones Using Microwave Irrediation," *Organic Letters*, Vol. 6, No. 9, 2004, pp. 1453-1456.

[5] J. G. Lombardino and E. H. Wiseman, "Preparation and Antiinflammatory Activity of Some Nonacidic Trisubstituted Imidazoles," *Journal of Medicinal Chemistry*, Vol. 17, No. 11, 1974, pp. 1182-1188.

[6] J. G. Lombardino, "Pharmaceutical Imidazoles," *Ger. Offen*, DE 2155558 A 19720629, 1972.

[7] A. P. Phillips, H. L. White and S. Rosen, "Antithrombotic Triphenylimidazoles," European Patent, EP 58890 A1 19820901, 1982.

[8] J. Zhu, "Recent Developments in the Isonitrile-Based Multicomponent Synthesis of Heterocycles," *European Journal of Organic Chemistry*, Vol. 2003, No. 7, 2003, pp. 1133-1144.

[9] R. V. A. Orru and M. de Greef, "Recent Advances in Solution-Phase Multicomponent Methodology for the Synthesis of Heterocyclic Compounds," *Synthesis*, Vol. 2003, No. 10, 2003, pp. 1471-1499.

[10] D. J. Ramon and M. Yus, "Asymmetric Multicomponent Reactions (AMCRs): The New Frontier," *Angewandte Chemie International Edition*, Vol. 44, No. 11, 2005, pp. 1602-1634.

[11] D. E. Frantz, L. Morency, A. Soheili, J. A. Murry, E. J. J. Grabowski and R. D. Tillyer, "Synthesis of Substituted Imidazole via Organocatalysis," *Organic Letters*, Vol. 6, No. 5, 2004, pp. 843-846.

[12] C. Zhang, E. J. Moran, T. F. Woiwade, K. M. Short and A. M. M. Mjalli, "Synthesis of Tetrasubstituted Imidazoles via α-(N-acyl-N-alkylamino)-β-ketoamides on Wang Resin," *Tetrahedron Letters*, Vol. 37, No. 6, 1996, pp. 751-754.

[13] I. Lantos, W-Y. Zhang, X. Shui and D. S. Eggleston, "Synthesis of Imidazole via Hetero-Cope Rearrange ment," *The Journal of Organic Chemistry*, Vol. 58, No. 25, 1993, pp. 7092-7095.

[14] T. J. Donohoe, M. A. Kabeshov, A. H. Rathi and I. E. D. Smith, "Direct Preparation of Thiazoles, Imidazoles, Imidazopyridines and Thiazolidines from Alkenes," *Organic & Biomolecular Chemistry*, Vol. 10, No. 5, 2012, pp. 1093-1101.

[15] B. Hu, Z. Wang, N. Ai, J. Zheng, X-H. Liu, S. Shan and Z. Wang, "Catalyst-Free Preparation of 1,2,4,5-Tetrasubstituted Imidazoles from a Novel Unexpected Domino Reaction of 2-Azido Acrylates and Nitrones," *Organic Letters*, Vol. 13 , No. 24, 2011, pp. 6362-6365.

[16] B. H. Lipshutz and M. C. Morey, "An Approach to the Cyclopeptide Alkaloids (Phencyclopeptines) via Heterocyclic Diamide/Dipeptide Equivalents. Preparation and N-Alkylation Studies of 2,4(5)-Disubstituted Imidazoles," *The Journal of Organic Chemistry*, Vol. 48 , No. 21, 1983, pp. 3745-3750.

[17] B. Sadeghi, B. B. F. Mirjalili and M. M. Hashemi, "$BF_3 \cdot SiO_2$: An Efficient Reagent System for the One-Pot Synthesis of 1,2,4,5-Tetrasubstituted Imidazoles," *Tetrahedron Letters*, Vol. 49, No. 16, 2008, pp. 2575-2577.

[18] R. Consonni, P. D. Croce, R. Ferraccioli and C. L. Rosa, "A New Approach to Imidazole Derivatives," *Journal of Chemical Research* (*Synopses*), No. 7, 1991, pp. 188-189.

[19] D. A. Evans and K. M. Lundy, "Synthesis of Diphthamide: The Target of Diphtheria Toxin Catalyzed ADP-Ribosylation in Protein Synthesis Elongation Factor 2," *Journal of the American Chemical Society*, Vol. 114, No. 4, 1992, pp. 1495-1496.

[20] P. Schneiders, J. Heinze and H. Baumgartel, "Synthesis of 4,4',5,5'-Tetrasubstituted Di-2-imidazolyl Derivatives, Starting Materials for the Synthesis of 1,4,5,8-Tetraazafulvalenes," *Chemische Berichte*, Vol. 106, No. 7, 1973, pp. 2415-2417.

[21] C. F. Claiborne, N. J. Liverton and K. T. Nguyen, "An Efficient Synthesis of Tetrasubstituted Imidazole from N-(2-Oxo)-Amides," *Tetrahedron Letters*, Vol. 39, No. 49, 1998, pp. 8939-8942.

[22] S. Narayana Murthy, B. Madhav and Y. V. D. Nageswar, "DABCO as a Mild and Efficient Catalytic System for the Synthesis of Highly Substituted Imidazoles via Multi-Component Condensation Strategy," *Tetrahedron Letters*, Vol. 51, No. 40, 2010, pp. 5252-5257.

[23] S. Kantevari, S. V. N. Vuppalapati, D. O. Biradar and L. Nagarapu, "Highly Efficient, One-Pot, Solvent-Free Synthesis of Tetrasubstituted Imidazoles Using $HClO_4$-SiO_2 as Novel Heterogeneous Catalyst," *Journal of Molecular Catalysis A*: *Chemical*, Vol. 266, No. 1-2, 2007, pp. 109-113.

[24] M. Kidwai and P. Mothsra, "A One-Pot Synthesis of 1,2, 4,5-Tetraarylimidazole Using Molecular Iodine as an Efficient Catalyst," *Tetrahedron Letters*, Vol. 47, No. 29, 2006, pp. 5029-5031.

[25] S. Balalaie, M. M. Hashemi and M. Akhbari, "A Novel One-Pot Synthesis of Tetrasubstituted Imidazoles under Solvent-Free Conditions and Microwave Irradiation," *Tetrahedron Letters*, Vol. 44, No. 8, 2003, pp. 1709-1711.

[26] S. Samai, G. C. Nandi, P. Singh and M. S. Singh, "L-Proline: An Efficient Catalyst for the One-Pot Synthesis of 2,4,5-Trisubstituted and 1,2,4,5-Tetrasubstituted Imidazoles," *Tetrahedron*, Vol. 65, No. 49, 2009, pp. 10155-10161.

[27] S. Balalaie and A. Arabanian, "One-Pot Synthesis of Tetrasubstituted Imidazoles Catalyzed by Zeolite HY and Silica Gel under Microwave Irradiation," *Green Chemistry*, Vol. 2, No. 6, 2000, pp. 274-276.

[28] M. M. Heravi, F. Derikvand and F. F. Bamoharramb, "Highly Efficient, Four-Component One-Pot Synthesis of

Tetrasubstituted Imidazoles Using Keggin-Type Heteropolyacids as Green and Reusable Catalysts," *Journal of Molecular Catalysis A: Chemical*, Vol. 263, No. 1-2, 2007, pp. 112-114.

[29] M. R. Manafi, P. Manafi and M. R. Kalaee, "Versatile Microwave-Assisted and Lanthanum Chloride Catalysed Synthesis of Polysubstituted Imidazoles Using Urea/Thiourea as Benign Source of Ammonia," *E-Journal of Chemistry*, Vol. 9, No. 4, 2012, pp. 1773-1777.

[30] H. R. Shaterian and M. Ranjbar, "An Environmental Friendly Approach for the Synthesis of Highly Substituted Imidazoles Using Brønsted Acidic Ionic Liquid, N-Methyl-2-Pyrrolidonium Hydrogen Sulphate, as Reusable Catalyst," *Journal of Molecular Liquids*, Vol. 160, No. 1, 2011, pp. 40-49.

[31] P. Wasserscheid and T. Welton, "Ionic Liquids in Synthesis," 2nd Edition, Wiley-VCH, Weinheim, 2007.

[32] Y. Izumi, K. Urabe and M. Onaka, "Zeolite, Clay and Heteropolyacid in Organic Reactions," Kodansha/VCH, Tokyo, 1992.

[33] T. Okuhara, N. Mizuno and M. Misono, "Catalytic Che mistry of Heteropoly Compounds," *Advances in Catalysis*, Vol. 41, 1996, pp. 113-252.

[34] P. J. Das and J. Das, "Synthesis of Aryl/alkyl(2,2'-bis-3-methylindolyl)methanes and Aryl(3,3'-bisindolyl)Methanes Promoted by Secondary Amine Based Ionic Liquids and Microwave Irradiation," *Tetrahedron Letters*, Vol. 53, No. 35, 2012, pp. 4718-4720.

[35] I. V. Kozhevnikov, "Catalysis by Heteropoly Acids and Multicomponent Polyoxometalates in Liquid-Phase Reactions," *Chemical Review*, Vol. 98, No. 1, 1998, pp. 171-198.

11

Use of Solid-Supported Reagents towards Synthesis of 2-Arylbenzoxazole, 3,5-Diarylisoxazole and 1,3,5-Triarylpyrazole

Sneha Naik, Vidya Desai*
Department of Chemistry, Dnyanprassarak Mandal's College of Arts, Science and Commerce, Mapusa, India

ABSTRACT

Herein we report a convenient and efficient synthesis of 2-arylbenzoxazole from the Schiff's bases of 2-aminophenol, 3,5-Diarylisoxazole from α, β-unsaturated ketoxime and 1,3,5-triarylpyrazole from 2-pyrazoline and N-Arylhydrazone by using milder, less toxic and less expensive-NBS-SiO_2, $KMnO_4$-Al_2O_3, PCC-SiO_2 and ACC-Al_2O_3 as solid-supported oxidizing agents at room temperature. Within the framework of Green Chemistry, the use of solid supported reagents has many advantages such as 1) they are easy to remove from reactions by filtration 2) excess reagents can be used to drive the reaction without introducing any difficulties in purification 3) such solid-supported reagents react differently, mostly more selectively, than their unbound counterparts and 4) toxic, noxious and explosive chemicals are handled more safely when contained on solid support.

Keywords: Schiff's Bases; 3,5-Diarylisoxazoles; 2-Arylbenzoxazoles; 1,3,5-Triarylpyrazoles; Solid-Supported Reagents

1. Introduction

In the recent years heterocyclic compounds are playing important role in the synthetic organic chemistry. Among these heterocycles the bicyclic heterocycles-benzoxazoles [1-3] and five membered heterocycles pyrazoles [4,5] and isoxazoles [6-8] are known to exhibit various bioactive properties. Arylbenzoxazole possesses the important biaryl pharmacophore thus exhibiting a variety of biological and pharmacological activities such as antitumor [9], anti-viral [10] and many more. Isoxazoles act as intermediates in the synthesis of natural products and as building blocks for construction of new molecular systems [11]. It exhibits wide range of medicinal properties [12]. On the other hand 1,3,5-triarylpyrazole occupy an important position in agrochemicals and pharmaceutical industries due to its wide range of bioactive properties such as anti-inflammatory, anti-arrythmic etc. [13,14].

Lot of research has been carried out and various methods have been reported towards the synthesis of 2-arylbenzoxazoles, 3,5-diarylisoxazole and 1,3,5-triarylpyrazoles in the recent years. One of such synthesis of 2-arylbenzoxazole includes oxidative cyclization of phenolic Schiff's bases (derived from 2-aminophenols and aromatic aldehydes) using various oxidizing agents like $PhI(OAc)_2$ [15], $Mn(OAc)_3$ [16], $Ba(MnO_4)_2$ [17], NiO_2 [18], $Pb(OAc)_4$ [19] and DDQ [20]. Despite of these intensive works, the development of more effective and convenient method still remains challenging, since many of the known methods require severe reaction conditions and workup procedures. There are several routes by which 3,5-diaryl isoxazoles can be synthesized. The [5 + 0] route involving oxidative cyclization of α, β-unsaturated ketoximes using oxidizing agents like DDQ [21], MnO_2 [22], I_2/KI in $NaHCO_3$ [23], are some of the few known. Our method depicts use of less toxic reagent as well as good yields of product. On the other side, to carry out aromatization of 2-pyrazoline and oxidative cyclization of phenylhydrazone, limited reports are there in the literature which include reagents like $Pb(OAc)_4$ [24], $PhI(OAc)_2$ [25], Pd/C [26] and $Zr(NO_3)_4$ [27]. Our method stands out to be more feasible compared to literature known methods due to its versatility in variety of substrates being subjected to oxidation with good yields of the product.

In our work we report a simple, convenient and efficient procedure for the synthesis of 2-arylbenzoxazoles by oxidative cyclization of phenolic Schiff's bases using NBS-SiO_2, synthesis of 3,5-diarylisoxazole by oxidative

*Corresponding author.

cyclization of α, β-unsaturated ketoxime using ACC-Al_2O_3 and synthesis of 1,3,5-triarylpyrazoles by oxidation of 2-pyrazolines using $KMnO_4$-Al_2O_3, PCC-SiO_2 and oxidative cyclization of phenylhydrazones using ACC-Al_2O_3 and $KMnO_4$-SiO_2 as solid supported oxidizing reagents.

With our growing interest in developing greener routes towards the synthesis of these heterocycles we thought of using older and stronger oxidizing agents but adsorbed on a solid support. These solid supported oxidizing agents are chosen to be as greener reagents due to the fact that they are mild, have good selectivity, easy to handle, easily removed after completion of reaction and reduced toxicity of the reagents.

2. Experimental Details

2.1. Typical Procedure for the Synthesis of 2-Arylbenzoxazole

All the reactions were carried out in 25 mL round bottom flask and the stirring was carried out using a magnetic stirrer. In a typical procedure the phenolic Schiff's basse 1a (0.15 g, 0.8 mmol) was dissolved in dichloromethane (5 mL) and kept for stirring at room temperature. To this stirring mixture, NBS-Silica (0.27 g, 1.6 mmol) was added and further stirred for 30 minutes. The reaction was continuously monitored on TLC. After the completion of reaction, the mixture was filtered and the filtrate was evaporated on hot water bath. The residue that remained after evaporation was purified by column chromatogramphy over silica gel. Yields indicated are obtained after purification.

Selected product: 2-phenyl benzoxazole (2a), white solid, melting point: 98°C - 100°C, yield = 84% H^1NMR (300 MHz, $CDCl_3$): δ 8.20 (dd, 2H), 7.75 (m, 1H), 7.48 - 7.52 (m, 4H), 7.34 - 7.388 (m, 2H). $C^{13}NMR$ (75 MHz, $CDCl_3$): δ 164.0, 152.7, 143.87, 131.29, 130.89, 127.59, 126.79, 122.06, 112.6. IR (cm^{-1}): 1050, 1250, 1450, 1600, 3050.

Selected product: 2-(4'-nitrophenyl) benzoxazole (2d), pale yellow solid, melting point: 270°C, yield = 64% ^1HNMR (300 MHz, $CDCl_3$): δ 8.4 (dd, 2H), 7.82 (m, 1H), 7.43 - 7.62 (m, 4H), 7.39 - 7.42 (m, 2H). $^{13}CNMR$ (75 MHz, $CDCl_3$): δ 190.24, 167.0, 151.0, 141.96, 132.84, 128.43, 126.24, 125.2, 120.26, 110.6. IR (cm^{-1}): 1050, 1250, 1450, 1630, 3050.

Selected product: 2-(2'hydroxynaphthyl) benzoxazole (2i), white solid, melting point: 152°C, yield = 84% ^1HNMR (300 MHz, $CDCl_3$): δ 9.08(s, 1H), 7.88 - 7.95 (m, 4H), 7.76 (s, 1H), 7.34-7.388 (m, 6H). $^{13}CNMR$ (75 MHz, $CDCl_3$): δ 164.0, 152.7, 143.87, 131.29, 130.89, 127.59, 126.79, 122.06, 112.6. IR (cm^{-1}): 1050, 1250, 1450, 1630, 3400.

2.2. Typical Procedure for the Synthesis 3,5-Diarylisoxazole

All the reactions were carried out in 25 mL round bottom flask and the stirring was carried out using a magnetic stirrer. Ketoxime 7a (1 mmol), ACC-Alumina (1 mmol) and dichloromethane were placed in a 25 mL round bottom flask. The reaction was stirred for three hours and frequently monitored using T.L.C. After completion of the reaction, the reaction mixture was filtered and filtrate obtained was then evaporated to give the product 8a.

Selected product: 3,5-diphenyl isoxazole (8a) White solid, melting point = 140°C H^1NMR (300 MHz, $CDCl_3$): -δ 6.8 (s, 1H), 7.21 - 8.05 (m, 10H). IR (cm^{-1}): 1250, 1435, 3010.

Selected product: 3-(4'-bromophenyl)-5-(4"-methoxyphenyl) isoxazole (8c) White solid, melting point = 130°C H^1NMR (300 MHz, $CDCl_3$): δ 5.3 (s, 3H), 6.83(s, 1H), δ 7.34 - 7.86 (m, 10H). IR (cm^{-1}): 690, 1250, 1500, 1600, 3010.

Selected product: 3-(4'-bromophenyl)-5-phenyl isoxazole (8d) White solid, melting point = 178°C ^1H NMR (300 MHz, $CDCl_3$): -δ 6.814 (s, 1H), 7.19 - 7.25 (m, 5H), 7.45 - 8.07(m, 4H). ^{13}C NMR (75 MHz, $CDCl_3$): 142, 135.2, 134.7, 134.09, 133.644, 133.05, 132.17, 130.157, 128, 127.646, 116.9, 95.26. IR (cm^{-1}): 685, 1250, 1510, 1600, 3010.

2.3. Typical Procedure for the Synthesis of 1,3,5-Trisubstituted Pyrazoles Using

2.3.1. $KMnO_4$-Alumina

All the reactions were carried out in 25 mL round bottom flask and the stirring was carried out using a magnetic stirrer. In a typical procedure the 2-pyrazoline 3a (0.15 g, 0.4 mmol) was placed in round bottom flask containing magnetic needle/pellet. To this mixture dichloromethane (5 mL) was added and kept for stirring at room temperature. To this stirring mixture, $KMnO_4$-Alumina (0.206 g, 0.8 mmol) was added and further stirred for four hours. The progress of the reaction was continuously monitored on TLC. After the completion of reaction, the mixture was filtered and the filtrate was evaporated on hot water bath. After evaporation, a pale yellow solid pyrazole is obtained which was purified by recrystallization using ethanol as solvent.

Selected product: 1,3,5-triphenylpyrazole (4a), pale yellow solid, melting point: 144°C, yield = 66% ^1H NMR (300 MHz, $CDCl_3$): δ 6.8 (s, 1H), 7.21 - 7.45 (m, 13H), 7.92 (d, 2H) IR (cm^{-1}): 1495, 1500, 1600, 3010.

2.3.2. PCC-Silica

All the reactions were carried out in 25 mL round bottom flask and the stirring was carried out using a magnetic stirrer. In a typical procedure the 2-pyrazoline 5b (0.15 g,

0.4 mmol) was placed in round bottom flask containing magnetic needle/pellet. To this mixture dichloromethane (5 mL) was added and kept for stirring at room temperature. To this stirring mixture, PCC-Silica (0.546 g, 2 mmol) was added and further stirred for five hours. The progress of the reaction was continuously monitored on TLC. After the completion of reaction, the mixture was filtered and the filtrate was evaporated on hot water bath. After evaporation, a pale yellow solid pyrazole is obtained which was purified by recrystallization using ethanol as solvent.

Selected product: 5-(4'-chlorophenyl)-1,3-diphenyl pyrazole (6b), pale yellow solid, melting point: 116°C, yield = 51% ^{1}H NMR (300 MHz, $CDCl_3$): -δ 6.83(s, 1H), 7.20 - 7.29 (m, 5H), 7.31 - 7.42 (m, 5H), 7.92 (d, 2H), 7.45 (d, 2H) IR (cm^{-1}): 670, 1490, 1600, 3015.

2.3.3. $KMnO_4$-Silica

All the reactions were carried out in 25 mL round bottom flask and refluxing was carried out using a heating mantle. In a typical procedure the parent phenylhydrazone 9a (0.15 g, 0.33 mmol) was placed in round bottom flask. To this mixture dichloromethane (5 mL) was added and kept for stirring at room temperature. To this stirring mixture, $KMnO_4$-Silica (0.546 g, 0.33 mmol) was added and further refluxed for three hours. The progress of the reaction was continuously monitored on TLC. After the completion of reaction, the mixture was filtered and the filtrate was evaporated on hot water bath. After evaporation, a pale yellow solid pyrazole is obtained which was purified by recrystallization using ethanol as solvent.

Selected product: 5-(4'-chlorophenyl)-1,3-diphenyl pyrazole (6b), pale yellow solid, melting point: 116°C, yield = 63% ^{1}HNMR (300 MHz, $CDCl_3$): -δ 6.83 (s, 1H), 7.212 - 7.45 (m, 12H), 7.31 - 7.42 (m, 5H), 7.92 (d, 2H), IR (cm^{-1}): 670, 1500, 1577, 3010.

2.3.4. ACC-Alumina

All the reactions were carried out in 25 mL round bottom flask and refluxing was carried out using a heating mantle. In a typical procedure the 2,4-dinitro phenylhydrazone 11b (0.15 g, 0.26 mmol) was placed in a round bottom flask. To this mixture dichloromethane (5 mL) was added and kept for stirring at room temperature. To this stirring mixture, ACC-Alumina (0.546 g, 0.78 mmol) was added and further refluxed for three hours. The progress of the reaction was continuously monitored on TLC. After the completion of reaction, the mixture was filtered and the filtrate was evaporated on hot water bath. After evaporation, a pale yellow solid pyrazole is obtained which was purified by recrystallization using ethanol as solvent.

Selected product:

3-(p-Bromophenyl)-5-phenyl-1-(2',4'-dinitrophenyl)-pyrazole (12a), light yellow solid mp 142°C; yield = 65% IR (KBr): 3100, 1620, 1550, 1500, 680 cm^{-1}. ^{1}H NMR (300 MHz, $CDCl_3$): δ 6.7 (s, 1H, 4-H), 7.1 - 7.4 (m, 5H), 7.76 - 7.83 (m, 4H), 7.88 (d, 1H), 8.75 (d, 1H), 9.056 (d, 1H).

^{13}C NMR (75 MHz, $CDCl_3$): δ 116, 117, 126,127, 127.6, 128.3, 128.4, 128.7, 129.05, 129.11, 129.45, 129.46, 129.53, 129.6, 129.91, 130.0, 130.19, 130.39, 132.61, 137.469, 141.3, 156.1, 157.5.

Selected product:

5-(p-methoxyphenyl)-3-(p-nitrophenyl)-1-(2', 4'-dinitrophenyl) pyrazole (12b), yellow solid, melting point: 166°C, yield = 73% IR (cm^{-1}): 1248, 1290, 1490, 1600, 3050 ^{1}H NMR (300 MHz, $CDCl_3$): δ 3.88 (s, 3H), 6.90 (s, 1H), 7.25 - 7.44 (m, 4H), 7.78 (d, 2H), 7.35 (d, 2H) 7.79 (d, 2H), 8.74 (d, 1H).

Selected product:

5-(p-chlorophenyl)-3-(p-methoxyphenyl)-1-(2', 4'-dinitrophenyl) pyrazole (12c), yellow solid, melting point: 132°C; yield = 50% IR (cm^{-1}): 1255, 1280, 1520, 1620, 3070 ^{1}H NMR (300 MHz, $CDCl_3$): δ 3.85 (s, 3H), 6.87 (s, 1H), 7.22 - 7.48 (m, 4H), 7.70 (d, 2H), 7.40 (d, 2H) 7.82 (d, 2H), 8.80 (d, 1H).

3. Results and Discussion

Initially, we prepared nine phenolic Schiff's bases by reacting aromatic aldehydes with 2-aminophenol. Once the Schiff's bases were prepared, we carried out oxidative cyclization on it. Here the parent Schiff's base, prepared from benzaldehyde and 2-aminophenol, was dissolved in dichloromethane to which NBS-silica was added and kept for stirring at room temperature for 30 minutes. After that the reaction mixture was filtered so as to remove the solid support. The filtrate was concentrated and purified using column chromatography. The product, 2-phenyl benzoxazole was obtained in 84% yield after recrystallization. To check the feasibility of this reaction the remaining Schiff's bases (those prepared earlier) were also tried for similar oxidative cyclization using NBS-silica as the oxidizing agent. In all nine 2-arylbenzoxazoles were prepared in moderate-good yields (60% - 84%) after chromatographic separation and purification (**Scheme 1**, **Table 1**).

Next task was to synthesize 1,3,5-trisubstituted pyrazoles from the corresponding 2-pyrazolines. In this case the parent pyrazoline, which was synthesized from phenyl hydrazine and chalcone was dissolved in dichloromethane and to this solution $KMnO_4$-Alumina was added. The reaction mixture was stirred for four hours at room temperature. After the completion of reaction, the mixture was filtered and the filtrate was evaporated to give a pale yellow solid pyrazole as the product which was purified by recrystallization using ethanol. In all three 1,3,5-trisubstituted pyrazoles were prepared in good yields

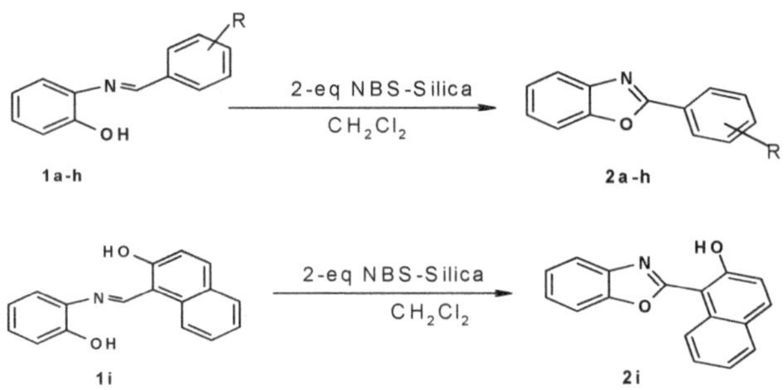

Scheme 1. Synthesis of 2-arylbenzoxazoles 1a: R = H, 1b: R = 2-OH, 1c: R = 4-Cl, 1d: R = 4-NO_2, 1e: R = 2-NO_2, 1f: R = 2-Cl, 1g: R = 4-OMe, 1h: R = 3-NO_2.

Table 1. Synthesis of 2-arylbenzoxazoles.

Entry	Phenolic Schiff's base	Product	Isolated %yield	M.p.°C (lit)
1	1a	2a	84	102 (100 - 102) [20]
2	1b	2b	77	122 - 124 (121 - 124) [28]
3	1c	2c	68	140 - 142 (141 - 142) [20]
4	1d	2d	67	270 (268) [29]
5	1e	2e	64	102 (103)[34]
6	1f	2f	62	69 - 70 (68 - 70) [36]
7	1g	2g	60	100 - 102 (100 - 102) [20]
8	1h	2h	70	211 (211 - 212) [34]
9	1i	2i	61	152

after purification (**Scheme 2**, **Table 2**).

We also used another reagent *i.e.* PCC-Silica for the oxidation of 2-pyrazolines to the corresponding 1,3,5-trisubstituted pyrazoles. Here the parent 2-pyrazoline was dissolved in dichloromethane and then subjected to oxidation by adding PCC-Silica. The reaction mixture was stirred for five to six hours at room temperature. After the completion of reaction, the mixture was filtered and the filtrate was evaporated to give a pale yellow solid pyrazole as the product which was purified by recrystallization using ethanol. In all three 1,3,5-trisubstituted pyrazoles were prepared in good yields after purification (**Scheme 3**, **Table 3**).

In addition, we also prepared 3, 5-diarylisoxazole from the corresponding α, β-unsaturated ketoxime. The α, β-unsaturated ketoxime was initially prepared by reacting α, β-unsaturated chalcones with hydroxylamine hydrochloride under weakly basic conditions. Once the α, β-unsaturated ketoxime were prepared, we carried out oxidative cyclization on it. Here the parent α, β-unsaturated ketoxime was dissolved in dichloromethane to which ACC-Alumina was added and kept for stirring at room temperature for three hours. After that the reaction mixture was filtered so as to remove the solid support. The filtrate was concentrated and purified by recrystallization using absolute ethanol as the solvent. To check the feasibility of this reaction the remaining α, β-unsaturated ketoxime were also tried for similar oxidative cyclization using ACC-Alumina as the oxidizing agent. In all five 3, 5-diarylisoxazole were prepared in moderate-good yields after purification (**Scheme 4**, **Table 4**).

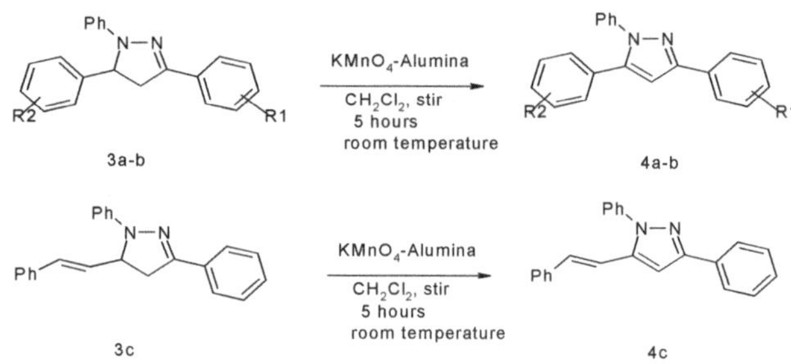

Scheme 2. Synthesis of 1,3,5-trisubstituted pyrazoles using $KMnO_4$-Alumina 3a: R1 = H, R2 = H; 3b: R1 = 4-OMe, R2 = 4-NO_2.

Table 2. Synthesis of 1,3,5-trisubstituted pyrazoles using $KMnO_4$-Alumina.

Entry	2-pyrazoline	Product	Isolated % yield	M.p. °C (lit)
1	3a	4a	66	138 (140) [31]
2	3b	4b	68	175 (176 - 180) [36]
3	3c	4c	62	145 (141 - 143) [37]

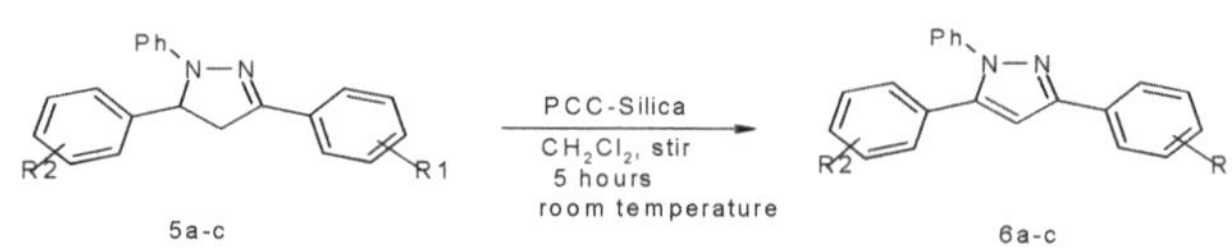

Scheme 3. Synthesis of 1,3,5-trisubstituted pyrazoles using PCC-Silica 5a: R1 = H, R2 = H; 5b: R1 = 4-OMe, R2 = 4-NO_2; 5c: R1 = H, R2 = 4-Cl.

Table 3. Synthesis of 1,3,5-trisubstituted pyrazoles using PCC-Silica.

Entry	2-pyrazoline	Product	Isolated % yield	M.p. °C (lit)
1	5a	6a	70	138 (140 - 141) [31]
2	5b	6b	72	180 (178) [36]
3	5c	6c	51	116 (114 - 116) [32]

Since we were also successful in synthesizing 1,3,5-

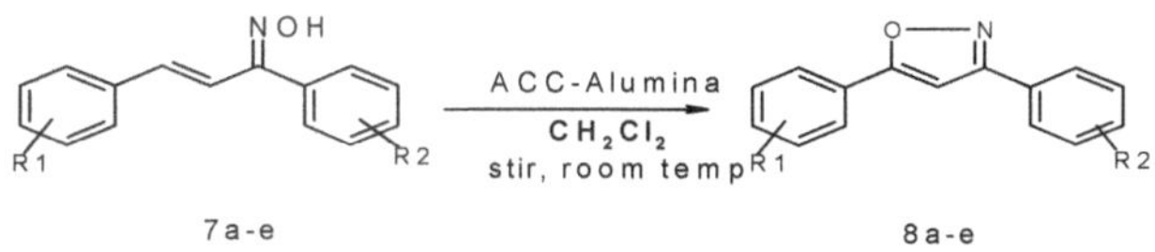

Scheme 4. Synthesis of 3,5-diarylisoxazole using ACC-Alumina. 7a: R1 = H, R2 = H; 7b: R1 = 4-Cl R2 = 4-OMe; 7c: R1 = 4-OMe R2 = Br; 7d: R1 = H, R2 = Br, 7e: R1 = H, R2 = 4-NO_2

Table 4. Synthesis of 3,5-diarylisoxazole using ACC-Alumina.

Entry	Product	Isolated % yield	M.p. °C (lit)
1	8a	54	145 (142) [21,22]
2	8b	77	185 (185 - 186) [29]
3	8c	55	130 (130) [22]
4	8d	51	178 (178 - 180) [30]
5	8e	50	224 (226 - 228) [22]

trisubstituted pyrazoles from 2-pyrazolines, we thought of preparing 1,3,5-trisubstituted pyrazoles from the corresponding N-Arylhydarzone and 2,4-dinitrophenylhydrazone via oxidative cyclization. For which we used $KMnO_4$-silica and ACC-alumina as the oxidizing agents. Here the parent phenyl hydrazone was initially synthesized from phenyl hydrazine and chalcone under acidic conditions. This phenyl hydrazone was dissolved in dichloromethane and then subjected to oxidative cyclization by adding $KMnO_4$-silica. The reaction mixture was refluxed for three hours. After the completion of reaction, the mixture was filtered and the filtrate was evaporated to give a pale yellow solid pyrazole as the product which was purified by recrystallization. In all three 1,3,5-trisubstituted pyrazoles were prepared in good yields after purification (**Scheme 5**, **Table 5**).

Similarly the 2,4-dinitrophenylhydrazone was dissolved in dichloromethane and then subjected to oxidative cyclization by adding ACC-alumina. The reaction mixture was refluxed for three hours. After the completion of reaction, the mixture was filtered and the filtrate was evaporated to give a pale yellow solid pyrazole as the product which was purified by recrystallization. In all three 1,3,5-trisubstituted pyrazoles were prepared in good yields after purification (**Scheme 6**, **Table 6**).

4. Conclusion

In conclusion, we have been successful in developing a convenient, efficient and greener method for the synthesis of 2-arylbenzoxazoles from phenolic Schiff's bases, 3,5-diarylisoxazole from α, β-unsaturated ke-

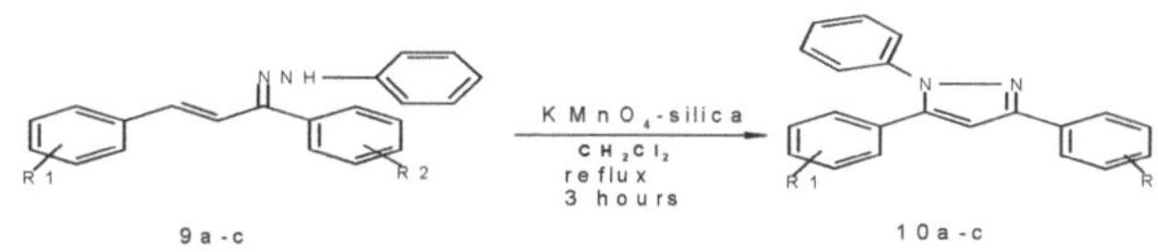

Scheme 5. Synthesis of 1,3,5-trisubstituted pyrazoles using $KMnO_4$-silica 9a: R1 = H, R2 = H; 9b: R1 = 4-OMe, R2 = 4-NO_2; 9c: R1 = H, R2 = 4-Cl.

Table 5. Synthesis of 1,3,5-trisubstituted pyrazoles using $KMnO_4$-Silica.

Entry	Product	Isolated % yield	M.p. °C (lit)
1	10a	52	140 (142) [31]
2	10b	80	179 (176 - 179) [36]
3	10c	63	116 [33]

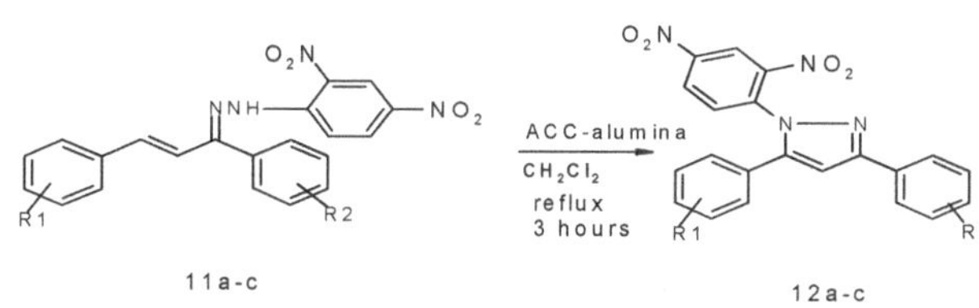

Scheme 6. Synthesis of 1,3,5-trisubstituted pyrazoles using ACC-alumina. 11a: R1 = H, R2 = Br; 11b: R1 = 4-OMe R2 = 4-NO_2; 11c: R1 = 4-Cl R2 = 4-OMe.

Table 6. Synthesis of 1,3,5-trisubstituted pyrazoles using ACC-alumina.

Entry	Product	Isolated % yield	M.p.°C
1	12a	65	142
2	12b	73	166
3	12c	50	132

toxime and 1,3,5-trisubstituted pyrazoles from the 2-pyrazolines and N-Arylhydrazones respectively. All the reactions were carried out at room temperature and the conditions used were quite mild. The product yields The products are obtained in moderate-good yield, which is compatible with literature known methods and moreover our methodology is eco-friendly. This methodology also shows greater versatility of using solid-supported reagents towards synthesis of variety of heterocycles. Use of solid supported reagents in our reaction has reduced the toxicity of the reagent which makes it much safer and easier to handle.

5. Acknowledgements

The authors are grateful to the Department of Science and Technology-Government of Goa for the financial support. Also, grateful to Professor Santosh G. Tilve,

Goa University for constant encouragement.

REFERENCES

[1] R. J. Perry, B. D. Wilson and R. J. Miller, "Synthesis of 2-Arylbenzoxazoles via the Palladium Catalysed Carbonylation and Condensation of Aromatic Halides and O-Aminophenols," *Journal of Organic Chemistry*, Vol. 57, No. 10, 1992, pp. 2883-2887.

[2] T. Kondo, S. Yang, K.-T. Huh, M. Kobayashi, S. Kotachi and Y. Watanabe, "Ruthenium Complex-Catalysed Facile Synthesis of 2-Substituted Benzoxazoles," *Chemistry Letters*, Vol. 20, No. 7, 1991, pp. 1275-1278.

[3] M. I. E. Sheikh, A. J. Marks and E. R. Biehl, "Investigation of the Synthesis of Benzoxazoles via Aryne Reaction," *Journal of Organic Chemistry*, Vol. 46, No. 16, 1981, pp. 3256-3259.

[4] T. L. Gilchrist, "Heterocyclic Chemistry," 3rd Edition, Addison-Wesley Longman, Ltd., London, 1998.

[5] D. Azarifar and M. Shaebanzadeh, "Synthesis and Characterisation of New 3,5-Dinaphthyl Substituted 2-Pyrazolines and Study of Their Anti-Microbial Properties," *Molecules*, Vol. 7, No. 12, 2002, pp. 885-895.

[6] A. Barco, S. Benitti, G. P. Pollini and D. Simoni, "Synthesis of Natural Products via Isoxazoles," *Synthesis*, Vol. 1987, No. 10, 1987, pp. 857-869.

[7] Y. Kondo, D. Uchiyama, T. Sakamoto and H. Yamanaka, "Synthesis and Reactions of 5-(Tributylstannyl)isoxazoles," *Tetrahedron Letters*, Vol. 30, No. 32, 1989, pp. 4249-4250.

[8] O. Morriyo, H. Nakamura, T. Kageyama and Y. Urata, "Synthesis of Isoxazolines and Isoxazoles from Aldoximes by the Use of Sodium Bromite with Organotin Halide," *Tetrahedron Letters*, Vol. 30, No. 30, 1989, pp. 3987-3990.

[9] O. Temiz, I. Rren, E. Sener, I. Yalcin and N. Ucartuk, "Synthesis and Microbiological Activity of Some Novel 5-and 6-Methyl-2-(2,4-disubstitutedphenyl) benzoxazole Derivatives," *Farmaco*, Vol. 53, No. 5, 1998, pp. 337-341.

[10] S. Aiello, G. Wells, E. L. Stone, H. Kadri, R. Bazzi, D. R. Bell, M. F. G. Stevens, C. S. Mathews, T. D. Bradshaw and A. D. Westwell, "Synthesis and Biological Properties of Benzothiazoles, Benzoxazoles and Chromene-4-One Analogues of the Potent Antitumor Agent 2-(3,4-Dimethoxyphenyl)-5-fluorobenzothiazole(PMX610, NSC 721648)," *Journal of Medicinal Chemistry*, Vol. 51, No. 16, 2008, pp. 5135-5139.

[11] E. Dominguez, E. Ibeus, E. M. de Margart, J. K. Palacius and E. San Marka, "A Convenient One-Pot Preparative Method for 4,5-Diaryl Isoxazoles involving Amine Exchange Reaction," *Journal of Organic Chemistry*, Vol. 61, No. 16, 1996, pp. 5435-5439.

[12] J. W. Wang, J. Jai, H. M. Li and C. Wang, "Synthesis of Novel Isoxazole-Contained Analogues of Losartan," *Chinese Chemical Letters*, Vol. 11, No. 11, 2000, pp. 961- 962.

[13] H. Katayama, T. Oshiyama, "Preparation and Bioactivity of Pyrazole Derivatives as Potential Cross-Linking Agents," *Canadian Journal of Chemistry*, Vol. 75, No. 6, 1997, pp. 913-919.

[14] H. M. Faid and H. M. Mokhtar, "Pyrazole Derivatives with Possible Hypoglycemic Activity," *Indian Journal of Chemistry*, Vol. 27B, No. 3, 1988, pp. 245-249.

[15] R. S. Varma, R. K. Saini and O. Prakash, "Hypervalent Iodine Oxidation of Phenolic Schiff's Bases: Synthesis of 2-Arylbenzoxazoles," *Tetrahedron Letters*, Vol. 38, No. 15, 1997, pp. 2621-2622.

[16] R. S. Varma and D. Kumar, "Manganese Triacetate Oxidation of Phenolic Schiff Bases: Synthesis of 2-Arylbenzoxazoles," *Journal of Heterocyclic Chemistry*, Vol. 35, No. 6, 1998, pp. 1539-1540.

[17] R. G. Shrivastava and P. S. Venkataramani, "Barium Manganate Oxidation in Organic Synthesis: Part III: Oxidation of Schiff's Bases to Benzimidazoles, Benzoxazoles and Benzthiazoles," *Synthetic Communications*, Vol. 18, No. 13, 1988, pp. 1537-1544.

[18] K. Nakagawa and H. Onoue, "Oxidation with Nickel Peroxide. IV. The Preparation of Benzoxazoles from Schiff's Bases," *Chemical and Pharmaceutical Bulletin*, Vol. 12, No. 10, 1964, pp. 1135-1138.

[19] F. F. Stephens and J. D. Bower, "The Preparation of Benzimidazoles and Benzoxazoles from Schiff's Bases. Part I," *Journal of Chemical Society*, Vol. 71, No. 9, 1949, pp. 2971-2972.

[20] J. Chang, K. Zhao and S. Pan, "Synthesis of 2-Arylbenzoxazoles via DDQ Promoted Oxidative Cyclization of Phenolic Schiff's Bases—A Solution Phase Strategy for Library Synthesis," *Tetrahedron Letters*, Vol. 43, No. 6, 2002, pp. 951-954.

[21] V. G. Desai and S. G. Tilve, "A Novel and Convenient Method for the Synthesis of 3,5-Diarylisoxazoles," *Synthetic Communications*, Vol. 29, No. 17, 1999, pp. 3017-3020.

[22] R. F. Kurangi, R. Kawthankar and V. G. Desai, "Convenient Synthesis of 3,5-Disubstituted Isoxazoles," *Synthetic Communications*, Vol. 37, No. 4, 2007, pp. 585-587.

[23] G. Buchi and J. C. Vedera, "Interchange of Functionality Compounds through Isoxazoles," *Journal of American Chemical Society*, Vol. 94, No. 26, 1972, pp. 9128-9131.

[24] S. P. Singh, D. Kumar, O. Prakash and R. P. Kapoor, "Hypervalent Iodine Oxidation of 1,3,5-Trisubstituted Pyrazolines: A Facile Synthesis of 1,3,5-Trisubstituted Pyrazoles," *Synthetic Communications*, Vol. 27, No. 15, 1997, pp. 2683-2689.

[25] I. Bhatnagar and M. V. George, "Oxidation with Metal Oxides-II: Oxidation of Chalcone Phenylhydrazones, Pyrazolines, O-Aminobenzylidine Anils and O-Hydroxy-

benzylidine Anils with Manganese Dioxide," *Tetrahedron*, Vol. 24, No. 3, 1968, pp. 1293-1298.

[26] W. A. F. Gladstone and R. O. C. Norman, "Reactions of Lead Tetra-Acetate Part VII: Some Reactions Leading to Pyrazoles," *Journal of Chemical Society* (*C*), 1966, pp. 1536-1540.

[27] G. Sabitha, G. S. K. Kumar Reddy, Ch. S. Reddy, N. Fatima and J. S. Yadav, "$Zr(NO_3)_4$: A Versatile Oxidizing Agent for Aromatization of Hantzsch 1,4-Dihydropyridines and 1,3,5-Trisubstituted Pyrazolines," *Synthesis*, No. 8, 2003, pp. 1267-1271.

[28] C. Wei-hua and Yi. Pang, "Efficient Synthesis of 2-(2'-Hydroxyphenyl)benzoxazole by Palladium(II)-Catalyzed Oxidative Cyclization," *Tetrahedron Letters*, Vol. 50, No. 48, 2009, pp. 6680-6683.

[29] K. H. Park, K. Jun, S. R. Shin and S. W. Oh, "2-Arylbenzoxazoles from Phenolic Schiffs Bases by Thianthrene Cation Radical," *Tetrahedron Letters*, Vol. 37, No. 49, 1996, pp. 8869-8870.

[30] M. Kidwai, S. Kukreja and R. Thakur, "K_2CO_3-Mediated Regioselective Synthesis of Isoxazoles and Pyrazolines," *Letters of Organic Chemistry*, Vol. 3, No. 2, 2006, pp. 135-139.

[31] S. Tang, J. He, Y. Sun and X. She, "Efficient and Regioselective One Pot Synthesis of 3-Substituted and 3,5-Disubstituted Isoxazoles," *Organic Letters*, Vol. 11, No. 17, 2009, pp. 3982-3985.

[32] S. Morrocohi, A. Ricca and A. Zanarotti, "New Features of the Reaction between Nitrilimines and Arylacetylenes," *Tetrahedron Letters*, Vol. 11, No. 37, 1970, pp. 3215-3218.

[33] D. Azarifar, M. A. Zulfigol and B. Maleki, "Silica-Supported 1,3-Dibromo-5,5-dimethylhydantoin (DBH) as a Useful Reagent for Microwave-Assisted Aromatization of 1,3,5-Trisubstituted Pyrazolines under Solvent-Free Conditions," *Synthesis*, No. 11, 2004, pp. 1744-1746.

[34] S. Tariq and S. Wakode, "Synthesis through Microwave Irradiation, Characterisation and Evaluation of Antimicrobial Activity of 2-Phenyl-1,3-benzoxazole Derivatives," *International Research Journal of Chemistry*, Vol. 3, No. 9, 2012, pp. 213-217.

[35] C. Praveen, A. Nandakumar, P. Dheenkumar, D. Murlidharan and P. T. Perumal, "Microwave Assisted One Pot Synthesis of Benzothiazole and Benzoxazole Libraries as Analgesics Agents," *Journal of Chemical Society*, Vol. 124, No. 3, 2012, pp. 609-624.

[36] B. C. Bishop, K. M. J. Brands, A. D. Gibbs and D. J. Kennedy, "Regioselective Synthesis of 1,3,5-Substituted Pyrazoles from Acetylenic Ketones and Hydrazines," *Synthesis*, No. 1, 2004, pp. 43-52.

[37] O. A. Ignatenko, A. N. Blandov and M. A. Kuznetsov, "Oxidative Addition of N-Aminophthalimide to Alkenyl-4,5-Dihydropyrazoles and Alkenylpyrazoles Synthesis of Aziridinylpyrazoles," *Russian Journal of Organic Chemistry*, Vol. 41, No. 12, 2005, pp. 1793-1801.

Carbonylative Coupling of 4,4'-Diiodobiphenyl Catalyzed by Pd(NHC) Complex

Do-Hun Lee[1*], Jung-Tai Hahn[2], Dai-Il Jung[1*]
[1]Department of Chemistry, Dong-A University, Busan, South Korea
[2]Department of Beautycare, Youngdong University, Youngdong, South Korea

ABSTRACT

To develop a luminescent material with high color purity, luminous efficiency, and stability, we synthesized diketone by carbonylative Suzuki coupling in the presence of Pd(NHC) complex as the catalyst. Carbonylative coupling of 4,4'-diiodobiphenyl and phenylboronic acid was investigated to study in detail the catalytic ability of the Pd(NHC) complex. Reactions were carried out using both CO and metal carbonyls. Bis-(1,3-dihydro-1,3-dimethyl-2H-imidazol-2-ylidene) diiodo palladium was used as the catalytic complex. Reaction products biphenyl-4,4'-diylbis (phenyl-methanone) 3 and (4'-iodobiphenyl-4-yl)(phenyl) methanone 4 were obtained as a result of CO insertion into the palladium(II)-aryl bond. However, when pyridine-4-yl boronic acid was used in place of phenylboronic acid as the starting reagent, synthetic reaction yielding 3 and 4 were found not to occur.

Keywords: Carbonylative Coupling; Metal Carbonyl; Pd(NHC) Complex; 4,4'-Diiodobiphenyl; Phenylboronic Acid

1. Introduction

Aryl ketones and flavanoids are common scaffolds in many natural products and biologically active small molecules [1-7]. A carbonylative coupling method for the synthesis of aryl compounds with CO was pioneered by Heck [8-11]. This method is one of the most efficient and direct routes to synthesize aryl ketones as it forms tow carbon-carbon bonds in a single step, in contrast to the conventional method of introducing ketone functional group in a stepwise fashion. Carbonylative coupling has since been further developed to synthesize a range of carbon nucleophiles [12], including those of tin [13-17], copper [18-22], boron [23-25], zinc [26], aluminum [27], magnesium [28], and silicon [29-31]. Our purpose is to synthesize a new distyryl biphenyl arylene (DBA) derivative as a blue-emitting material. To develop such a luminescent material with high color purity, luminous efficiency, and stability, first of all, we synthesized diketone with Pd(NHC) complex as a catalyst under a balloon of CO or metal carbonyl.

During the course of an on-going synthetic project for preparing aryl ketones, we decided to evaluate the applicability of *N*-heterocyclic carbene (NHC) ligands. NHC ligands have gained popularity in metal-catalyzed cross-coupling reactions for several reasons [32-35]: 1) the steric bulk that they introduce around the metal center facilitates reductive elimination; 2) their strong σ-donating character enables facile oxidative addition; and 3) their greater stability at elevated temperatures relative to phosphineligands enables their use under a broader range of reaction conditions. Carbonylative Suzuki coupling using the synthesized NHC-Pd complex was carried out under a balloon of CO or metal carbonyls. To study the scope of the process, the reaction conditions were optimizied for the cross-coupling of 4,4'-diiodobiphenyl and phenylboronic acid with *N*-heterocyclic carbene (NHC) ligand under a balloon (1 atm) of CO or metal carbonyls. 4,4'-diiodobiphenyl 1) and phenylboronic acid 2) were reacted under CO (1 bar, a balloon) atmosphere in the presence of the Pd(NHC) complex catalyst formed in situ [36,37].

*Corresponding authors.

2. Experimental

2.1. Carbonylative Coupling Reaction under Carbon Monoxide

In a typical reaction, Pd(NHC) complex (2×10^{-3} g, 5×10^{-2} mol) was dissolved in 15 mL anisole under N_2 gas. After the formation of a pale brown homogeneous solution, phenylboronic acid (0.112 g, 1×10^{-3} mol), 4,4'-diiodobiphenyl (0.203 g, 5×10^{-4} mol), and potassium carbonate (0.425 g, 1.5×10^{-3} mol) were added. The atmosphere was changed to carbon monoxide and the reaction mixture was kept at 80°C for 24 h. After elimination of Pd(NHC) complex by filteration, the reaction mixture was diluted water (10 mL) and CH_2Cl_2 (20 mL). The neutralized solution was extracted with CH_2Cl_2. The organic layer was dried (Na_2SO_4), filtered, and concentrated. The reaction mixture was analyzed immediately by GC-MS. The residue was chromatographed on a silica gel (n-hexane:ethylacetate = 20:1, v/v) yield 3 (0.154 g, 42.6%) and 4 (4.9×10^{-2} g, 12.7%).

2.2. Carbonylative Coupling Reaction under Metal Carbonyl

The mixture of 4,4'-diiodobiphenyl (0.203 g, 5×10^{-4} mol), phenylboronic acid (0.112 g, 1×10^{-3} mol), K_2CO_3 (0.425 g, 1.5×10^{-3} mol) and Di-(1,3-dihydro-1,3-dimethyl-2H-imidazol-2-ylidene)diiodopalladium (2×10^{-3} g, 5×10^{-2} mol) and Molybdenum hexacarbonyl (9.2×10^{-2} mol, 0.7 eq) was stirred in 15 mL anisole under N_2. The reaction mixture was kept at 80°C for 24 h. After elimination of Pd(NHC) complex by filteration, the reaction mixture was diluted water (10 mL) and CH_2Cl_2 (20 mL). The neutralized solution was extracted with CH_2Cl_2. The organic layer was dried (Na_2SO_4), filtered, and concentrated. The reaction mixture was analyzed immediately by GC-MS. The residue was chromatographed on a silica gel (n-hexane:ethylacetate = 20:1, v/v) yield 3 (0.189 g, 52.3%) and 4 (5.2×10^{-2} g, 13.5%).

2.3. Synthesis of Bis-(1,3-dihydro-1,3-dimethyl-2H-imidazol-2-ylidene) diiodopalladium

The synthetic scheme for producing bis-(1,3-dihydro-1,3-dimethyl-2H-imidazol-2-ylidene) diiodopalladium catalyst is as follows:

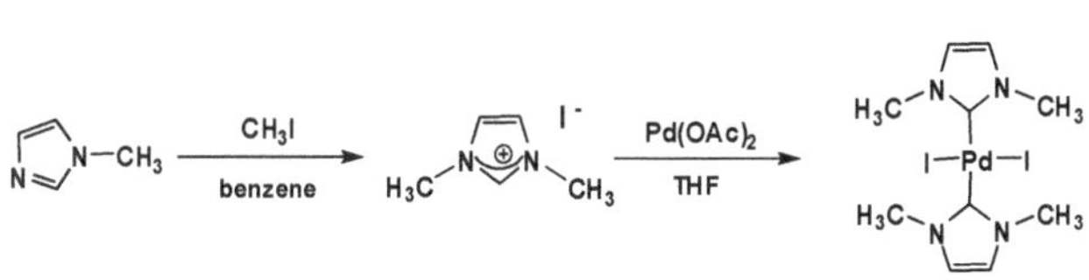

N,N'-dimethyl imidazolium iodide was obtained by the reaction of *N*-methylimidazole with methyl iodide. Following this, reaction of *N,N'*-dimethyl imidazolium iodide with palladium acetate resulted in NHC-Pd complex in good yield (72%).

3. Results and Discussion

The desired carbonylative products biphenyl-4,4'-diylbis (phenyl-methanone) 3 and (4'-iodobiphenyl-4-yl)(phenyl) methanone 4 were formed in all cases, irrespective of the reaction conditions.

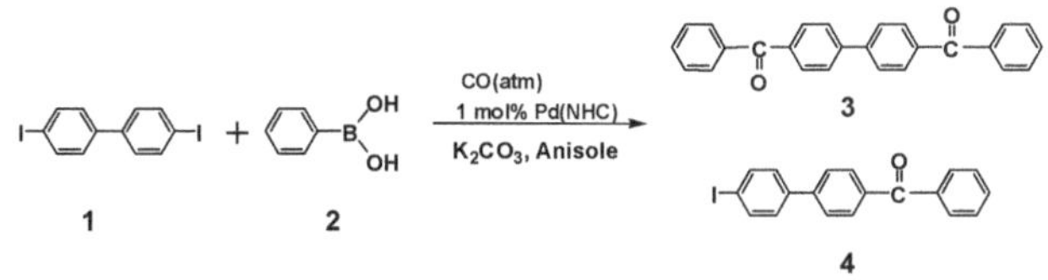

When metal carbonyl [for $Mo(CO)_6$: 3 = 42.6% and 4 = 12.7%; $Mn_2(CO)_{10}$: 3 = 6.6% and 4 = 33.1%; $Co_2(CO)_8$: 3 = 48.6% and 4 = 11.2%; $Fe_3(CO)_{12}$: 3 = 9.9% and 4 = 24.8%; $Fe(CO)_5$: 3 = 62.5% and 4 = 10.6%] was used in place of CO, we achieved the same reaction products.

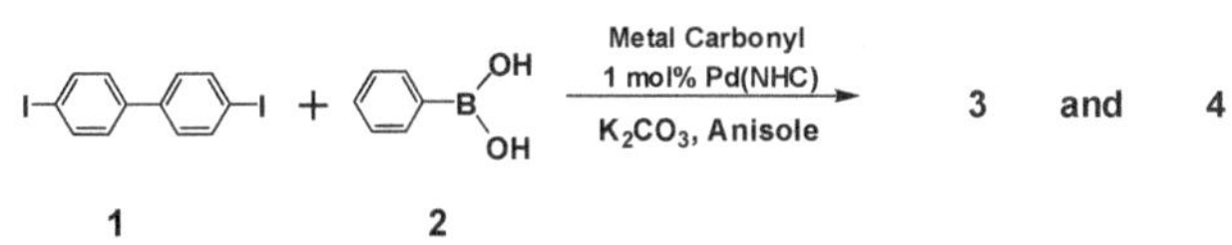

In reactions with $Mn_2(CO)_{10}$ and $Fe_3(CO)_{12}$ as metal carbonyals, yield of 4 was higher then that of 3. As seen in **Table 1**, various metal carbonyls were as effective as CO donors as CO itself. The plausible mechanism of diketone formation is assumed to be as shown in **Scheme 1**.

4. Conclusions

When metal carbonyl was used in place of CO, we achieved the same reaction products.

We assume that the two reactions needed to obtain 3 require a longer reaction time as 4 is formed as a reaction intermediate. When pyridine-4-ylboronic acid is used in place of phenylboronic acid,

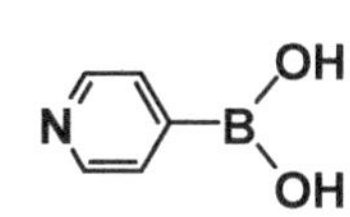

carbonylative Suzuki coupling under CO or metal carbonyls [$Mo(CO)_6$, $Mn_2(CO)_{10}$, $Co_2(CO)_8$, $Fe_3(CO)_{12}$, and $Fe(CO)_5$] is found not to occur. In future, we intend to

Table 1. Carbonylative Suzuki coupling with phenylboronic acid and 4,4'-diiodobiphenyl.

Run	Iodide	Boronic acid	CO	Reaction time (h)	Yield (%)	
					3	4
1	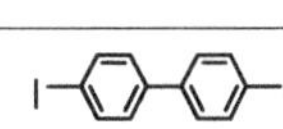	Ph-B(OH)$_2$	CO (gas)	24 h	42.6	12.7
2	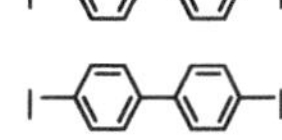	Ph-B(OH)$_2$	$Mo(CO)_6$	24 h	52.3	13.5
3	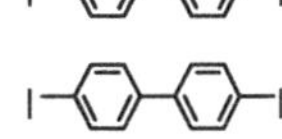	Ph-B(OH)$_2$	$Mn_2(CO)_{10}$	24 h	6.6	33.1
4	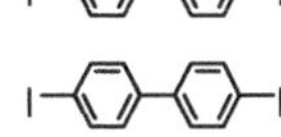	Ph-B(OH)$_2$	$Co_2(CO)_8$	24 h	48.6	11.2
5	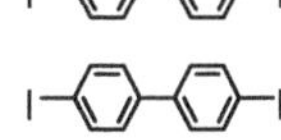	Ph-B(OH)$_2$	$Fe_3(CO)_{12}$	24 h	9.9	24.8
6	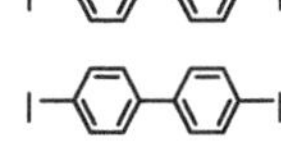	Ph-B(OH)$_2$	$Fe(CO)_5$	24 h	62.5	10.6

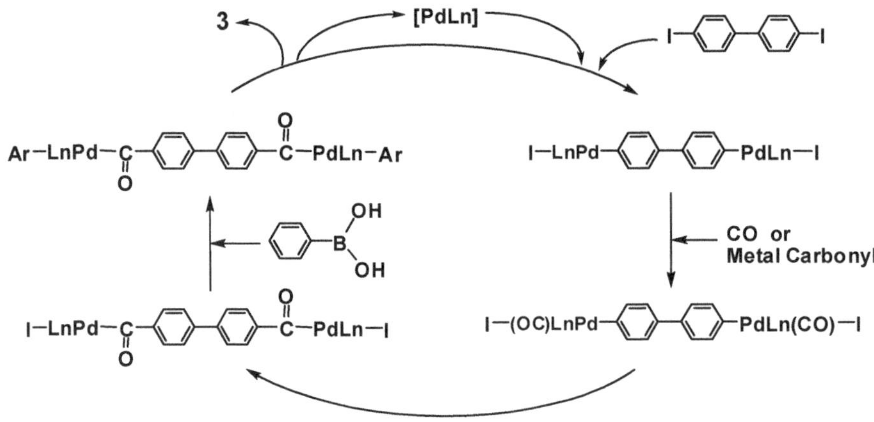

Scheme 1. Simplified catalytic cycle showing the formation of 3.

synthesize various diketones by using heteroaromatic boronic acid to develop a luminescent material.

5. Acknowledgements

The work was supported by a grant of Dong-A University (2012).

REFERENCES

[1] Y. Jiang and P. Tu, "Four New Phenones from the Cortexes of *Polygalatenuifolia*," *Chemical & Pharmaceutical Bulletin*, Vol. 53, No. 9, 2005, pp. 1164-1166.

[2] Nilar, L.-H. D. Nguyen, G. Venkatraman, K.-Y. Sim and L. J. Harrison, "Xanthones and Benzophenones from *Garciniagriffithii* and *Garcinia mangostana*," *Phytochemistry*, Vol. 66, No. 14, 2005, pp. 1718-1723.

[3] J. W. Lampe, C. K. Biggers, J. M. Defauw, R. J. Foglesong, S. E. Hall, J. M. Heerding, S. P. Hollinshead, H. Hu, P. F. Hughes, G. E. Jagdmann Jr., M. G. Johnson, Y.-S. Lai, C. T. Lowden, M. P. Lynch, J. S. Mendoza, M. M. Murphy, J. W. Wilson, L. M. Ballas, K. Carter, J. W. Darges, J. E. Davis, F. R. Hubbard and M. L. Stamper, "Synthesis and Protein Kinase Inhibitory Activity of Balanol Analogues with Modified Benzophenone Subunits," *Journal of Medicinal Chemistry*, Vol. 45, No. 12, 2002, pp. 2624-2643.

[4] S. Rancon, A. Chaboud, N. Darbour, G. Comte, C. Bayet, P.-N. Simon, J. Raymond, A. Di Pietro, P. Cabalion and D. Barron, "Natural and Synthetic Benzophenones: Interaction with the Cytosolic Binding Domain of P-Glycoprotein," *Phytochemistry*, Vol. 57, No. 4, 2001, pp. 553-557.

[5] H. Ito, E. Nishitani, T. Konoshima, M. Takasaki, M. Kozuka and T. Yoshida, "Flavonoid and Benzophenone Glycosides from *Coleogyne ramosissima*," *Phytochemistry*, Vol. 54, No. 7, 2000, pp. 695-700.

[6] J.-C. Li and T. Nohara, "Benzophenone C-Glucosides from Polygala Telephioides," *Chemical & Pharmaceutical Bulletin*, Vol. 48, No. 9, 2000, pp. 1354-1355.

[7] D. M. O'Keefe, N. Simmons and S. F. Martin, "Carbonylative Cross-Coupling of *ortho*-Disubstituted Aryl Io dides. Convenient Synthesis of Sterically Hindered Aryl Ketones," *Organic Letters*, Vol. 10, No. 22, 2008, pp. 5301-5304.

[8] R. F. Heck, "A Synthesis of Diaryl Ketones from Arylmercuric Salts," *Journal of the American Chemical Society*, Vol. 90, No. 20, 1968, pp. 5546-5548.

[9] A. Schoenberg, I. Bartoletti and R. F. Heck, "Palladium-Catalyzed Carboalkoxylation of Aryl, Benzyl, and Vinylic Halides," *Journal of Organic Chemistry*, Vol. 39, No. 23, 1974, pp. 3318-3326.

[10] A. Schoenberg and R. F. Heck, "Palladium-Catalyzed Amidation of Aryl, Heterocyclic, and Vinylic Halides," *Journal of Organic Chemistry*, Vol. 39, No. 23, 1974, pp. 3327-3331.

[11] J.-J. Brunetand R. Chauvin, "Synthesis of Diarylketones through Carbonylative Coupling," *Chemical Society Reviews*, Vol. 24, No. 2, 1995, p. 89.

[12] Y. Tamaru and M. Kimura, "Reactions of Acylpalladium Derivatives with Organometals and Related Carbon Nucleophiles," *Handbook of Organopalladium Chemistry for Organic Synthesis*, Vol. 2, 2002, pp. 2425-2454.

[13] V. Farina, V. Krishnamurthy and W. J. Scott, "Stille Reaction," *Organic Reactions*, Vol. 50, No. 1, 1997, pp. 1-652.

[14] S.-K. Kang, T. Yamaguchi, T.-H. Kim and P.-S. Ho, "Copper-Catalyzed Cross-Coupling and Carbonylative Cross-Coupling of Organostannanes and Organoboranes with Hypervalent Iodine Compounds," *Journal of Organic Chemistry*, Vol. 61, No. 26, 1996, pp. 9082-9083.

[15] A. M. Echavarren and J. K. Stille, "Palladium-Catalyzed Coupling of Vinyl Epoxides with Organostannanes," *Journal of the American Chemical Society*, Vol. 110, No. 12, 1988, pp. 4039-4041.

[16] J. K. Stille, "The Palladium-Catalyzed Cross-Coupling Reaction of Organotin Reagents with Organic Electro-

philes," *Angewandte Chemie International Edition English*, Vol. 25, No. 6, 1986, pp. 508-524.

[17] M. Tanaka, "Unsymmetrical Ketone Synthesis from Organic Halides, Carbon Monoxide, and Organotin Compounds Catalyzed by a Palladium Complex," *Tetrahdron Letters*, Vol. 20, No. 28, 1979, pp. 2601-2602.

[18] V. Sans, A. M. Trzeciak, S. Luis and J. J. Ziolkowski, "PdCl2(P(OPh)(3))(2) Catalyzed Coupling and Carbonylative Coupling of Phenylacetylenes with Aryl Iodides in Organic Solvents and in Ionic Liquids," *Catalysis Letters*, Vol. 109, No. 1-2, 2006, pp. 37-41.

[19] P. J. Tambade, Y. P. Patil, N. S. Nandurkar and B. M. Bhanage, "Copper-Catalyzed, Palladium-Free Carbonylative Sonogashira Coupling Reaction of Aliphatic and Aromatic Alkynes with Iodoaryls," *Synle*, No. 6, 2008, pp. 886-888.

[20] N. Haddad, J. Tan and V. Farina, "Convergent Synthesis of the Quinolone Substructure of BILN 2061 via Carbonylative Sonogashira Coupling/Cyclization," *Journal of Organic Chemistry*, Vol. 71, No. 13, 2006, pp. 5031-5034.

[21] M. S. M. Ahmed and A. Mori, "Carbonylative Sonogashira Coupling of Terminal Alkynes with Aqueous Ammonia," *Organic Letters*, Vol. 5, No. 17, 2003, pp. 3057-3060.

[22] S. Torii, H. Okomoto, L. H. Xu, M. Sadakane, M. V. Shostakovsky, A. B Ponomaryov and V. N. Kalinin, "Syntheses of Chromones and Quinolones via Pd-Catalyzed Carbonylation of O-Iodophenols and Anilines in the Presence of Acetylenes," *Tetrahdron*, Vol. 49, No. 31, 1993, pp. 6773-6784.

[23] T. Ohe, K. Ohe, S. Uemura and N. Sugita, "Palladium-(0)-Catalyzed Carbonylation of Alkenyl- and Arylborates and Boronic Acids with Carbon Monoxide," *Journal of Organometallic Chemistry*, Vol. 344, No. 1, 1988, pp. c5-c7.

[24] T. Ishiyama, H. Kizaki, T. Hayashi, A. Suzuki and N. Miyaura, "Palladium-Catalyzed Carbonylative Cross-Coupling Reaction of Arylboronic Acids with Aryl Electrophiles: Synthesis of Biaryl Ketones," *Journal of Organic Chemistry*, Vol. 63, No. 14, 1998, pp. 4726-4731.

[25] M. B. Andrus, Y. Ma, Y. Zang and C. Song, "Palladium-Imidazolium-Catalyzed Carbonylative Coupling of Aryl Diazonium Ions and Aryl Boronic Acids," *Tetrahedron Letters*, Vol. 43, No. 50, 2002, pp. 9137-9140.

[26] Q. Wang and C. Chen, "Nickel-Catalyzed Carbonylative Negishi Cross-Coupling Reactions," *Tetrahedron Letters*, Vol. 49, No. 18, 2008, pp. 2916-2921.

[27] N. A. Bumagin, A. B. Ponomaryov and I. P. Beletskya, "Ketone Synthesis via Palladium-Catalyzed Carbonylation of Organoaluminium Compounds," *Tetrahedron Letters*, Vol. 26, No. 39, 1985, pp. 4819-4822.

[28] T. Yamamoto, T. Kohara and A. Yamamoto, "Selective Formation of Ketone, Diketone and Aldehyde by the Co Insertion into Nickel-Alkyl Bonds of Dialkylnickel Complexes. A Novel Nickel-Catalyzed Syntheses of Ketones and Tertiary Alcohols from Grignard Reagents, Aryl Halides, and Carbon Monoxide," *Chemistry Letters*, Vol. 5, No. 11, 1976, pp. 1217-1220.

[29] Y. Hatanaka, S. Fukushima and T. Hiyama, "Carbonylative Coupling Reaction of Organofluorosilanes with Organic Halides Promoted by Fluoride Ion and Palldium Catalyst," *Tetrahedron*, Vol. 48, No. 11, 1992, pp. 2113-2126.

[30] Y. Hatanaka and T. Hiyama, "Highly Selective Cross-Coupling Reactions of Organosilicon Compounds Mediated by Fluoride Ion and a Palladium Catalyst," *Synlett*, Vol. 1991, No. 12, 1991, pp. 845-853.

[31] Y. Hatanaka and T. Hiyama, "Palladium-Catalyzed Carbonylative Coupling of Arylfluorosilanes with Aryl Iodides. A Convenient Synthesis of Diaryl Ketones," *Chemistry Letters*, Vol. 18, No. 11, 1989, pp. 2049-2052.

[32] E. A. B. Kantchev, C. J. O'Brien and M. G. Organ, "Palladium Complexes of N-Heterocyclic Carbenes as Catalysts for Cross-Coupling Reactions-A Synthetic Chemist's Perspective," *Angewandte Chemie International Edition*, Vol. 46, No. 16, 2007, pp. 2768-2813.

[33] K. J. Covell and D. S. McGuinness, "Redox Processes Involving Hydrocarbylmetal (*N*-Heterocyclic Carbene) Complexes and Associated Imidazolium Salts: Ramifications for Catalysis," *Coordination Chemistry Reviews*, Vol. 248, No. 7-8, 2004, pp. 671-681.

[34] W. A. Herrmann, "N-Heterocyclic Carbenes: A New Concept in Organometallic Catalysis," *Angewandte Chemie International Edition*, Vol. 41, No. 8, 2002, pp. 1290-1309.

[35] A. C. Hillier, G. A. Grasa, M. S. Viciu, H. M. Lee, C. Yang and S. P. Nolan, "Catalytic Cross-Coupling Reactions Mediated by Palladium/Nucleophilic Carbene Systems," *Journal of Organometallic Chemistry*, Vol. 653, No. 1-2, 2002, pp. 69-82.

[36] P. Andrea, P. Gabor, P. Zoltan and K. Laszlo, "Carbonylative and Direct Suzuki-Miyaura Cross-Coupling Reactions with 1-Iodo-Cyclohexene," *Journal of Molecular Catalysis*, Vol. 255, No. 1-2, 2006, pp. 97-102.

[37] L. M. Daniela, M. A. Heiddyand and C. S. A. Lucia, "Microwave-Assisted Suzuki Reaction Catalyzed by Pd(0)-PVP Nanoparticles," *Tetrahedron Letters*, Vol. 51, No. 52, 2010, pp. 6814-6817.

$AlCl_3$ Supported Catalysts for the Isomerization of *Endo*-Tetrahydrodicyclopentadiene

Min Ji[1*], Yiyong Zhou[1], Qunxing Luo[1], Sang-Eon Park[2], Tianxi Cai[1]
[1]School of Chemistry, Dalian University of Technology, Dalian, China
[2]Laboratory of Nano-Green Catalysis and Nano Center for Fine Chemicals Fusion Technology, Department of Chemistry, Inha University, Incheon, South Korea

ABSTRACT

Two of zeolites such as Hβ and HY, and mesoporous HMCM-41 were used as supports to immobilize $AlCl_3$ for the investigation of the isomerization of *endo*-tetrahydrodicyclopentadiene (*endo*-TCD). The dependences of porous structure and surface acidity of the $AlCl_3$ immobilized catalysts on the activity and selectivity were studied. Based on characterization studies, both the large pore diameter and strong acid sites of the supported catalysts contributed to the high activities in the *endo*-TCD conversion. And specifically the stronger Lewis acid sites seems to be responsible for the higher selectivity onto *exo*-TCD, and the stronger Brönsted acid sites were crucial to the formation of adamantane through the further isomerization of *exo*-TCD.

Keywords: Tetrahydrodicyclopentadiene; Isomerization; Zeolite; Immobilization; Aluminium Chloride

1. Introduction

Endo-tetrahydrodicyclopentadiene (*endo*-TCD) has been typically formed through the dimerization followed by hydrogenation from dicyclopentadiene, and its isomers, both *exo*-tetrahydrodicyclopentadiene(*exo*-TCD) and adamantane(ADH), are very useful fine chemical products[1-3]. In industry, *exo*-TCD and ADH are produced through the rearrangement of *endo*-TCD by acid catalysis [4], as shown in **Scheme 1**. Due to the advantages of low reaction temperature and strong resistance to ward coke, $AlCl_3$ has been used as the most widely used catalyst [5], but it causes environmental problems such as severe corrosion and difficulty in separation from the reaction mixture. In the 1980s, liquid type superacids, such as CF_3SO_3H and SbF_5-graphite, were used to catalyze *endo*-TCD to *exo*-TCD [6]. However, the problems of separation and corrosion were still remained. Since the 2000s, ionic liquids [7] were used to isomerize the *endo*-TCD, which resulted in pretty good result, but the expensive price restricts its industrial application. Hence, there is a strong motivation to develop new catalyst system for the isomerization that has minimum environmental impact. On the other hand, during the past decades, zeolites have been drawn much attentions in the rearrangement of *endo*-TCD [8,9]. However, the low activity of zeolites limits its industrial application.

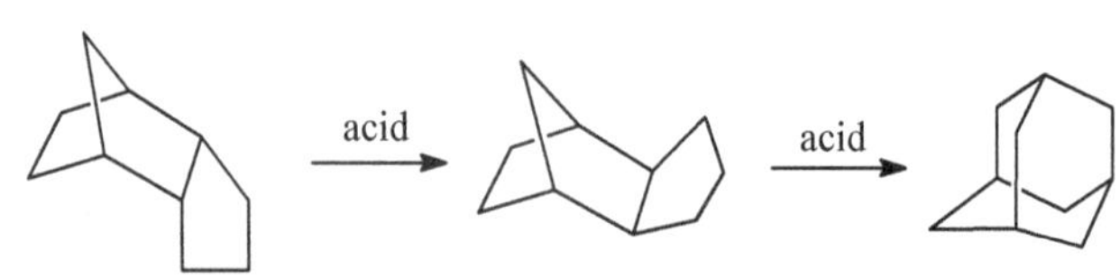

Scheme 1. *Exo*-TCD and ADH are produced through the rearrangement of *endo*-TCD by acid catalysis.

Since then, the immobilization of $AlCl_3$ on inorganic solid supports has been proposed [10-12]. In our previous study [13,14], we developed the catalysts of $AlCl_3$ immobilized on SiO_2, Al_2O_3 and MCM-41, and found that it could keep the excellent reactivities of traditional $AlCl_3$ catalyst at mild reaction conditions for *endo*-TCD isomerization, and the properties of supports strongly affected to the product selectivity. It is known that the reaction mechanism for the *endo*-TCD isomerization to *exo*-TCD and ADH is a carbenium ion skeleton rearrangement [4]. So, the pore structure and acid type of the catalyst seem to be important parameters in the conversion and product distribution. In the present report, Hβ and HY zeolites, and HMCM-41mesoporous material were used as catalyst supports to immobilize $AlCl_3$ for the purpose of dif-

*Corresponding author.

ferentiating both pore structures and acidities. Based on the property changes of supports after $AlCl_3$ immobilization, the influences of porous property, the type of acidities, and acidic strength of catalysts on both *endo*-TCD conversion and product selectivities in the *endo*-TCD rearrangement reaction were investigated.

2. Experiment

2.1. Catalyst Preparation

The Hβ and HY zeolites were purchased from Nankai Catalyst Co., China. HMCM-41 was prepared by the tetraethyl silicate (TEOS) in basic solution condensation with cetyltrimethylammonium bromide (CTAB) as a supramolecular template, and by ion-exchange with HNO_3 [15]. The immobilized $AlCl_3$ catalyst was prepared by a two-step method as described in our previous publications [13].

2.2. Catalyst Characterization

The BET surface areas and the pore properties of the catalysts were measured by nitrogen physisorption by using a Micromeritics Gemini 2395 instrument at 77 K. In-situ Fourier Transform Infrared Spectra (FT-IR) of chemically adsorbed pyridine were collected on a Nicolet 360 FT-IR spectrophotometer. All spectra were taken at a resolution of 2 cm^{-1} for 32 scans. Catalysts were activated in an IR cell at 550°C under vacuum for 2 h, cooled down to room temperature, followed by exposure to pyridine at this temperature for 30 min. The samples were heated up to 150°C or 250°C and evacuated under vacuum (10^{-5} Torr) for 1 h to remove physically adsorbed pyridine. Finally, the spectra were recorded at the desired temperatures.

2.3. Catalyst Activity Test

The isomerization rearrangement reaction was carried out in a batch suspension reactor. The catalyst was charged into 20 ml n-heptane solution. The reaction temperature was controlled by an oil bath equipped with a thermostat and an electric stirrer. The reaction pressure of 1.0 MPa was set by the replacement of air by N_2 to keep the solvent in liquid phase. The reaction time was 3 h. The products were analyzed by an Agilent 6890 GC equipped with an HP-5 capillary column and FID detector of HP 6890/5973 GC-MS.

3. Results and Discussion

The physico-chemical properties of the catalysts are summarized in **Table 1**. The specific surface area and pore diameter of HMCM-41 were 1001 m^2/g and 2.33 nm, respectively, which were much higher than that of

Table 1. Physico-chemical properties of supports and $AlCl_3$ supported catalysts.

Catalyst	Surface Area (m^2/g)	Pore Volume (cm^3/g)	Pore Size (nm)	Mass Fraction of [-O-$AlCl_2$](%)
HY	579	0.25	0.59	-
Hβ	478	0.16	0.58	-
HMCM-41	1001	0.58	2.33	-
$AlCl_3$/HY	46	0.02	0.51	45.32
$AlCl_3$/Hβ	291	0.12	0.52	13.20
$AlCl_3$/ HMCM-41	710	0.36	2.00	20.08

both HY and Hβ due to the mesoporosity. After $AlCl_3$ was loaded, the decreasing surface area, pore volumes, and pore diameters were observed for both $AlCl_3$/zeolites and $AlCl_3$/HMCM-41 catalysts. This was primarily due to the penetration of $AlCl_3$ into the pores of supports and located mostly inside of pores. Specially, HY showed the most drastic decresse in its surface area (46 m^2/g) and pore volume (only 0.02 cm^3/g) after $AlCl_3$ immobilization, indicating that the pore channel of HY seemed to be greatly blocked by the large amounts of $AlCl_3$ loading of 45.3%. However, the mass fractions of [-O-$AlCl_2$] on the case of Hβ and HMCM-41 were 13.2% and 20.1%, respectively. So, they sustained pretty high pore volumes and surface areas even after immobilization of $AlCl_3$. The mechanism of $AlCl_3$ immobilization onto the supports were proposed via the surface reaction of $AlCl_3$ with surface hydroxyl groups of the supports accompanied with HCl release [16,17]. Over 90% of the chloroaluminium species on the surface of the immobilized supports had the composition of [-O-$AlCl_2$] [16]. So, the higher [-O-$AlCl_2$] loading on HMCM-41 could be attributed to its higher surface area and more abundant surface hydroxyl groups.

The types of acid sites (Lewis and Brönsted) were studied by in-situ FT-IR adsorption of pyridine. The infrared spectra of the pyridine-adsorbed Hβ and HY (in **Figure 1**) revealed three peaks at ca. 1449 and 1540 cm^{-1} due to pyridine adsorbed on Lewis and Brönsted acid sites, respectively, and ~1480 cm^{-1} band corresponded to both Lewis and Brönsted (L+B) sites [18,19]. The infrared spectrum of the pyridine adsorbed HMCM-41 was somewhat different from those on HY and Hβ since it only showed Lewis acid sites at 1449 cm^{-1}. After immobilization of $AlCl_3$, the amounts of acid sites on catalyst surfaces increased remarkablely in all the cases. The infrared spectra of all the samples recorded after evacuation at 150°C and 250°C are indicating total acid sites and only strong acid sites, respectively, and the difference between them could be presumed to show weak acid sites of the catalysts. Based on the spectra in **Figure 1**, the adsorption capacities were the planimetered ac-

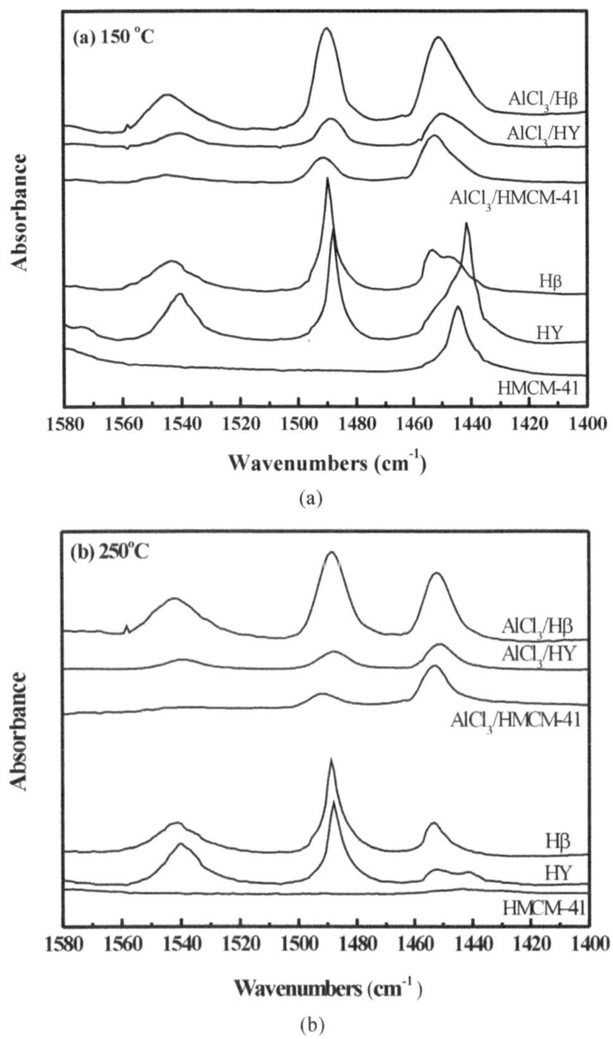

Figure 1. The infrared spectra of the pyridine-adsorbed over catalyst.

cording to the peak areas and normalized with both surface areas and wafer thickness of the catalysts [20], and the calculation results are listed in **Table 2**. It can be found that the pure HMCM-41 possessed only weak Lewis acid sites. After immobilization of $AlCl_3$, it still lacked of strong Brönsted acid sites but owned abundant strong Lewis acid sites, which was as high as 11.50 (a.u.)$\cdot$cm$^{-1}\cdot$cm$^2\cdot$mg^{-1}. HY showed fewer strong Brönsted acid sites and strong Lewis acid sites, but after immobilization of $AlCl_3$ on HY by forming $AlCl_3$/HY, both strong Brönsted acid sites and strong Lewis acid sites increased simultaneously. For H*β*, it exhibited both the strongest Brönsted and Lewis acidities among the three supports, and those properties were also enhanced after $AlCl_3$ immobilized on H*β* to form $AlCl_3$/H*β*. It is interesting that the amount of $AlCl_3$ loading on HY was much more than that on both HMCM-41 and H*β*, as shown in Table 1, but the number of total acid sites on $AlCl_3$/HY was lower than that on $AlCl_3$/H*β*, and $AlCl_3$/HMCM-41. It is mainly because of huge pore block in the case of $AlCl_3$/HY.

Since the isomerization reaction of *endo*-TCD → *exo*-TCD → ADH is a consecutive reaction, the pore structure and surface acidity of solid acid catalyst were speculated to play important roles in the *endo*-TCD conversion and product selectivity. **Table 3** shows the catalytic properties of supports and supported $AlCl_3$ catalysts for the isomerization of *endo*-TCD at 150°C and 250°C. Three pure supports showed no activity at 150°C. When reaction temperature was risen up to 250°C, HY and H*β* showed a certain degree of activity increase, while HMCM-41 still kept no activity. The results of FT-IR of pyridine adsorption listed in **Table 2** shows that HMCM-41 had only weak Lewis acid sites, instead HY and H*β* owned both weak and strong acid sites. So, it can be suggested that the weak Lewis acid sites of solid acid catalyst might contribute insignificantly to the reaction.

For all the of $AlCl_3$ immobilized catalysts, the catalytic activities were greatly increased whether HY, H*β* and HMCM-41 as supports. Although the pores of $AlCl_3$/HY catalyst were largely blocked due to high $AlCl_3$ loading compared with the pure zeolites, it still allowed the reaction by the contribution of $AlCl_3$ on external surface.

The greatly enhanced activity over $AlCl_3$/H*β* could be ascribed to the increase in strong acidity of the catalyst as well as low blocking due to low loading of $AlCl_3$. The order of the strong acidity of the supported $AlCl_3$ catalysts are listed in **Table 2** ranked as $AlCl_3$/H*β* > $AlCl_3$/HMCM-41 ≈ $AlCl_3$/HY, and the highest *endo*-TCD conversion of 95.7% was observed over $AlCl_3$/H*β* catalyst. It is noteworthy that the total strong acid sites on $AlCl_3$/HY and $AlCl_3$/HMCM-41 was quite similar, but the *endo*-TCD conversion of $AlCl_3$/HMCM-41 was much higher than that of $AlCl_3$/HY, which might be caused by the the diffusional limitation of partial pore blocking in the $AlCl_3$/HY catalyst. Therefore, the mesoporous structure of $AlCl_3$/HMCM-41 catalyst seems to be quite beneficial to the reactant diffusion to overcome the small amounts of Lewis acidity for the reaction.

The data in **Tables 2** and **3** also show the dependence of surface acidity of catalyst on the product selectivity.

The ADH selectivities obtained at 250°C shows $AlCl_3$/H*β* > $AlCl_3$/HY ≈ H*β* > HY ≈ $AlCl_3$/HMCM-41, which seemed to be corresponding to the amounts of strong Brönsted acid sites of catalysts (see **Table 2**). $AlCl_3$/H*β* possessed the large number of strong Brönsted acid sites, and its ADH yield reached 25.6%, while the others were lower than 4%. The $AlCl_3$/HY catalyst and H*β* support itself provided nearly the same amounts of strong Brönsted acid sites, and their ADH selectivities were almost same, which were 10.9% and 9.8%, respectively.

In contrast, $AlCl_3$/HMCM-41 showed only strong Lewis acidic sites and had no strong Brönsted acidity, then it exhibited the highest *exo*-TCD selectivity but the lowest ADH selectivity. Similar relationship between ADH selectivity and the amount of strong Brönsted acid sites could be found at the reaction temperature of 150°C. These observations obviously indicate that strong Lewis acid sites of catalysts are attributed to the increases of selectivity of intermediate isomerization product, *exo*-TCD, and the strong Brönsted acid sites are responsible for the farther ADH formation.

Table 2. Absorption intensities of FT-IR of pyridine adsorbed on acid sites of the catalysts.

Catalysts	Integrated absorption intensity $[(a.u.) \times cm^{-1} \times cm^2 \times mg^{-1}]$								
	Total acid			Strong acid			Weak acid		
	Total	B acid	L acid	Total	B acid	L acid	Total	B acid	L acid
HY	7.61	2.45	5.16	2.72	1.81	0.91	4.89	0.64	4.25
Hβ	10.21	4.52	5.69	5.12	3.74	1.38	5.09	0.78	4.31
HMCM-41	4.49	0	4.49	0	0	0	4.49	0	4.49
$AlCl_3$/HY	17.65	4.86	12.79	11.12	3.43	7.69	6.53	1.43	5.10
$AlCl_3$/Hβ	52.35	18.84	33.51	33.25	14.86	18.39	19.10	3.98	15.12
$AlCl_3$/HMCM-41	18.48	2.13	16.35	11.50	0	11.50	6.98	2.13	4.85

Table 3. Comparison of the catalyst performances in the isomerization of endo-TCD.

Catalyst	Temperature (°C)	Conversion (%)	Selectivity (%)			Yield (%)	
			Exo-TCD	ADH	Others	*Exo*-TCD	ADH
HY	150	0	0	0	0	0	0
Hβ	150	0	0	0	0	0	0
HMCM-41	150	0	0	0	0	0	0
$AlCl_3$/HY	150	15.6	98.4	0.8	0.8	15.4	0.1
$AlCl_3$/Hβ	150	37.9	95.4	3.1	1.5	36.1	1.2
$AlCl_3$/HMCM-41	150	64.4	99.3	0.6	0.1	63.9	0.4
HY	250	14.8	94.2	2.1	3.7	14.0	0.3
Hβ	250	37.3	88.3	9.8	1.9	33.0	3.7
HMCM-41	250	0	0	0	0	0	0
$AlCl_3$/HY	250	36.5	85.7	10.9	3.4	31.2	4.0
$AlCl_3$/Hβ	250	95.7	71.0	26.8	2.2	67.9	25.6
$AlCl_3$/HMCM-41	250	76.3	97.3	2.1	0.6	74.3	1.6

The synthesis mechanism of *exo*-TCD and ADH goes through a carbenium ion rearrangement (**Scheme 2**). Endo-TCD is isomerized on a acid by two consecutive reactions: endo-TCD is first isomerized to form exo-TCD, and then ADH was formed through the isomerization of exo-TCD. The first isomerization of endo-TCD can be catalyzed by strong Lewis acids, while the second step of exo-TCD to ADH requires strong Bronsted acids.

4. Conclusion

We have prepared a series of $AlCl_3$ supported catalysts onto Hβ, HY, and HMCM-41 having different both porous structures and surface acidities. The dependence of the activity and selectivity in the conversion of *endo*-TCD to *exo*-TCD and ADH was studied by correlation with both porous property and surface acidity of the catalysts. The suitable pore diameter and strong acid sites of catalyst are requisite for the isomerization of *endo*-TCD. And the weak Lewis acid center of catalyst surface seemed to be no use to the catalytic activity for isomeri-

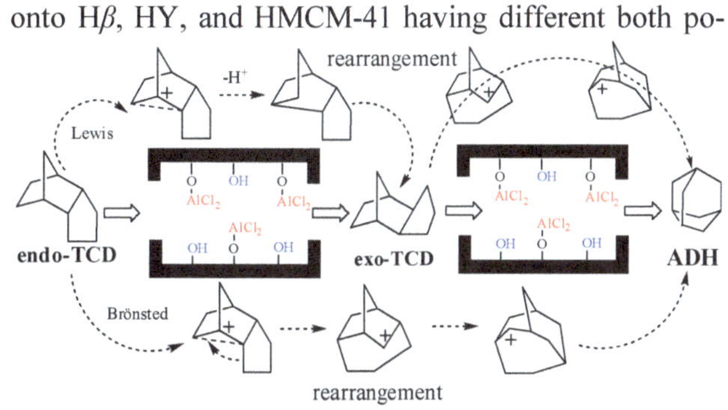

Scheme 2. The synthesis mechanism of *exo*-TCD and ADH.

zation even at the high reaction temperature of 250°C. However, the strong acid sites controlled the distribution of the products distinctly. Strong Lewis acid sites are responsible to get the intermediate isomerization product of *exo*-TCD, and the strong Brönsted acid sites of catalyst are crucial for the formation of ADH. The result found herein is believed to be potentially important for the selective formation of exo-TCD or ADH by controlling acidity and its strength.

5. Acknowledgements

Supports of this work by the National Natural Science Foundation of China (21076031) and the Fundamental Research Funds for the Central Universities (DUT12-ZD219) are gratefully acknowledged. Sang-Eon Park thanks Dalian University of Technology for a "Seasky" professorship

REFERENCES

[1] H. S. Chung, C. S. H. Chen, R. A. Kremer, J. R. Boulton and G. W. Burdette, "Recent Developments in High-Energy Density Liquid Hydrocarbon Fuels," *Energy and Fuels*, Vol. 13, No. 3, 1999, pp. 641-649.

[2] G. Lamoureux and G. Artavia, "Use of the Adamantane Structure in Medicinal Chemistry," *Current Medicinal Chemistry*, Vol. 17, No. 26, 2010, pp. 2967-2978.

[3] A. Masaaki, G. Akira and M. Shinji, EP Patent 2372799, 2011.

[4] E. M. Engler, M. Farcasiu, A. Sevin, J. M. Cense and P. V. R. Schleyer, "Mechanism of Adamantane Rearrangements," *Journal of the American Chemical Society*, Vol. 95, No. 17, 1973, pp. 5769-5771.

[5] "Société des Usines Chimiques Rhŏne-Poluenc," FR Patent 1431816, 1966.

[6] G. A. Olah and O. Farooq, "Chemistry in Superacids. 7. Superacid-Catalyzed Isomerization of Endo- to Exo-Trimethylenenorbornane (Tetrahydrodicyclopentadiene) and to Adamantine," *Journal of Organic Chemistry*, Vol. 51, No. 26, 1986, pp. 5410-5413.

[7] M. Y. Huang, J. C. Wu, F. S. Shieu and J. J. Lin, "Isomerization of Exo-Tetrahydrodicyclopentadiene to Adamantane Using an Acidity-Adjustable Chloroaluminate Ionic Liquid," *Catalysis Communications*, Vol. 10, No. 13, 2009, pp. 1747-1751.

[8] K. Honna, M. Sugimoto, N. Shimizu and K. Kurisaki, "Catalytic Rearrangement of Tetrahydrodicyclotadiene to Adamantine over Y-Zeolite," *Chemistry Letters*, Vol. 3, No. 3, 1986, pp. 315-318.

[9] M. Navrátilová and K. Sporka, "Synthesis of Adamantane on Commercially Available Zeolitic Catalysts," *Applied Catalysis A: General*, Vol. 203, No. 1, 2000, pp. 127-132.

[10] J. H. Clark, K. Martin, A. J. Teasdale and S. J. Barlow, "Environmentally Friendly Catalysis Using Supported Reagents: Evolution of a Highly Active Form of Immobilised Aluminium Chloride," *Journal of the Chemical Society, Chemical Communications*, Vol. 19, No. 19, 1995, pp. 2037-2040.

[11] C. DeCastro, E. Sauvage, M. H. Valkenberg and W. F. Hölderich, "Immobilised Ionic Liquids as Lewis Acid Catalysts for the Alkylation of Aromatic Compounds with Dodecene," *Journal of Catalysis*, Vol. 196, No. 1, 2000, pp. 86-94.

[12] Y. Takefumi, I. Tokuji, I. Shigeru, S. Michimasa, K. Yoshiyuki and T. Masanori, "Highly Active Supported Catalysts for Olefin Polymerization: Preparation and Characterization of the Catalyst," *Journal of Polymer Science, Part A: Polymer Chemistry*, Vol. 26, No. 2, 1988, pp. 477-489.

[13] L. M. Wu, M. Ji, M. He and T. X. Cai, "Synthesis of Adamantane Catalyzed by an Active Immobilized Aluminium Chloride Catalyst," *Chinese Journal of Catalysis*, Vol. 28, No. 7, 2007, pp. 585-587.

[14] L. L. Qi, M. Ji, M. He and T. X. Cai, "$AlCl_3$/MCM-41 as a Catalyst for Isomerization of Endo-Tricyclodecane," *Chinese Journal of Catalysis*, Vol. 31, No. 4, 2010, pp. 383-385.

[15] W. H. Zhang, J. L. Shi, L. Z. Wang and D. S. Yan, "Preparation and Characterization of ZnO Clusters inside Mesoporous Silica," *Chemistry of Materials*, Vol. 12, No. 5, 2000, pp. 1408-1413.

[16] R. S. Drago and E. E. Getty, "Preparation and Catalytic Activity of a New Solid Acid Catalyst," *Journal of the American Chemical Society*, Vol. 110, No. 10, 1988, pp. 3311-3312.

[17] T. Xu, N. Kob, R. S. Drago, J. B. Nicholas and J. F. Haw, "A Solid Acid Catalyst at the Threshold of Superacid Strength: NMR, Calorimetry, and Density Functional Theory Studies of Silica-Supported Aluminum Chloride," *Journal of American Chemical Society*, Vol. 119, No. 50, 1997, pp. 12231-12239.

[18] E. P. Parry, "An Infrared Study of Pyridine Adsorbed on Acidic Solids. Characterization of Surface Acidity," *Journal of Catalysis*, Vol. 2, No. 5, 1963, pp. 371-379.

[19] T. R. Hughes and H. M. White, "A Study of the Surface Structure of Decationized Y Zeolite by Quantitative Infrared Spectroscopy," *Journal of Physical Chemistry*, Vol. 71, No. 7, 1967, pp. 2192-2201.

[20] C. Defosse and P. Canesson, "Potentiality of Photoelectron Spectroscopy in the Characterization of Surface Acidity: Photoelectron and Infrared Spectroscopic Comparative Study of Pyridine Adsorption on NH_4-Y Zeolite Activated at Various Temperatures," *Journal of the Chemical Society, Faraday Transactions 1: Physical Chemistry in Condensed Phases*, Vol. 1, No. 72, 1976, pp. 2565-2576.

14

An Efficient $FeCl_3$ Catalyzed Synthesis of N,N'-Diarylformamidines

Pushkin Chakraborty, Subhas C. Roy
Department of Organic Chemistry, Indian Association for the Cultivation of Science, Jadavpur Kolkata, India

ABSTRACT

An efficient $FeCl_3$ catalyzed synthesis of N,N'-diarylformamidines using triethylorthoformate (1 equivalent) and primary aryl amines (2 equivalents) at ambient temperature has been described. This methodology provides an eco-friendly and simple procedure without using any hazardous and expensive chemicals.

Keywords: Fe(III) Chloride; Diarylformamidines; Aryl Amines; Triethylorthoformate

1. Introduction

Formamidines have structural similarity to the imidazole ring, a part of the histamine molecule, are supposed to possess enormous biological activities. The biochemical aims of formamidines include monoamine oxidase inhibitor [1,2], adrenergic, neurochemical receptors [3-8] and prostaglandin E2 synthesis [9]. Formamidines are also noted for their complexation with transition metals [10,11] and usage as auxiliaries in asymmetric synthesis [12,13], electrophiles [14]. The utility of formamidines as support linkers in solid phase synthesis [15] is now well established in the field of organic synthesis. Formamidines are now vastly used for the preparation of imidazolium salts which are the precursor for the synthesis of N-Heterocyclic carbenes [16]. Moreover, formamidines are useful subject of interest to the physical chemists for dynamic NMR study [17]. There have also been reported some cryoscopic molecular weight determination experiments utilizing the molecular association property of diarylformamidines in benzene solution [18].

2. Results and Discussion

There are only a few reports [16,19-26] in the literature for the synthesis of formamidines specially using triethyl orthoformate and amines. However, there is still scope for further improvement in this field since most of the reported methods suffer from long reaction times, elevated temperature or use of toxic and expensive reagents. Very recently, Sadek *et al.* [26] reported the synthesis of diarylformamidines using ceric ammonium nitrate (CAN) in water. But it is well known that CAN is a toxic and strong oxidizing reagent and especially in water it shows strong acidic property to affect many sensitive functional groups. So, a mild and efficient method is still desirable. We report herein an efficient $FeCl_3$ catalyzed synthesis of N,N'-diarylformamidines using triethylorthoformate (1 equivalent) and primary aryl amines (2 equivalents) at ambient temperature. Compared to other methods this method is much more environment friendly due to not using any toxic chemicals. In a preliminary experiment, a solution of aniline (1a) (2 mmol) and triethyl orthoformate (1 mmol) in the presence of a catalytic amount of $FeCl_3$ (10 mol%) in toluene (10 mL) was stirred for 3 h at room temperature. Solvent was removed and the solid mass obtained was purified by column chromatography over silica gel to afford pure formamidine 1b in excellent yield (**Scheme 1**).

Thus, a series of diarylformamidines have been synthesized using the reaction conditions and the results are summarized in **Table 1**. All the products were characterized by spectral and analytical studies and were compared with the reported data (10,13b,13d,13f-h). The probable mechanism of the formation of the product may be suggested with the line of the report by Sadek *et al.* [26] (**Scheme 2**). It is proposed that $FeCl_3$ as a Lewis acid activates ethoxy groups and enhances the C-O bond cleavage to generate a stable carbocation which facilitates the subsequent nucleophilic displacements by aromatic amines.

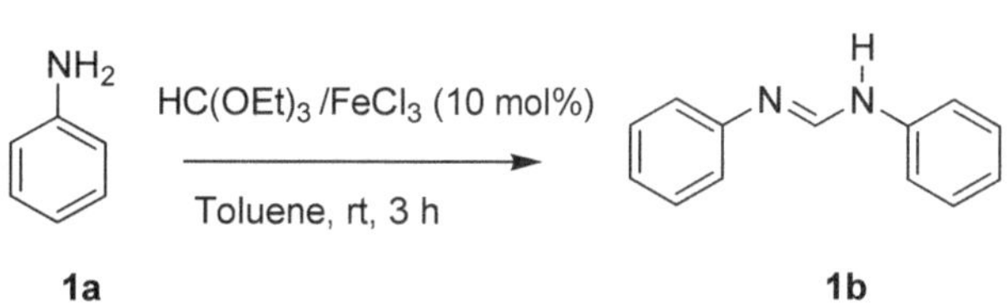

Scheme 1. Synthesis of diarylformamidines.

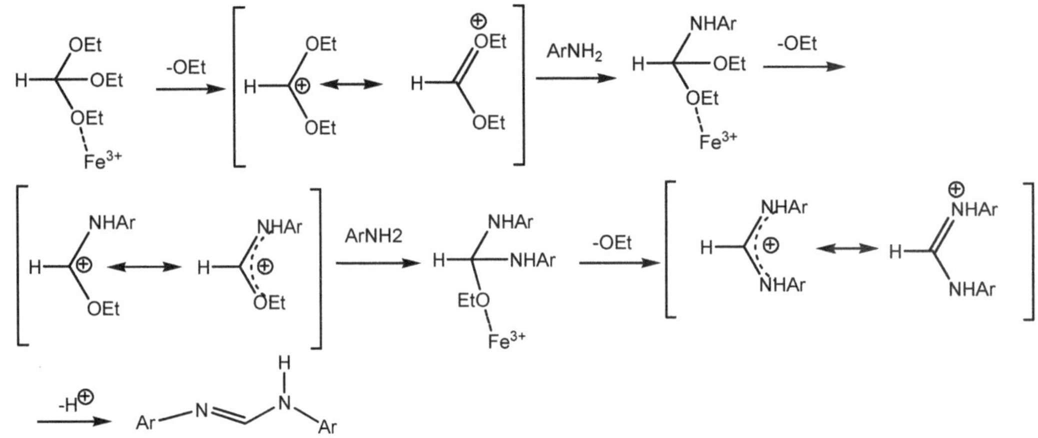

Scheme 2. Plausible mechanism for the formation of diarylformamidines.

Table 1. Synthesis of N,N'-diarylformamidines.

Entry	Amine	Product	Time	Yield (%)[a]	Ref.
1	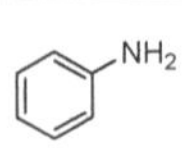 1a	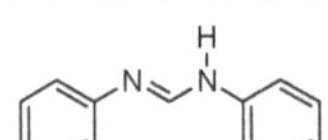 1b	3 h	87	10
2	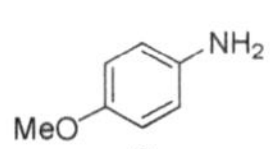 2a	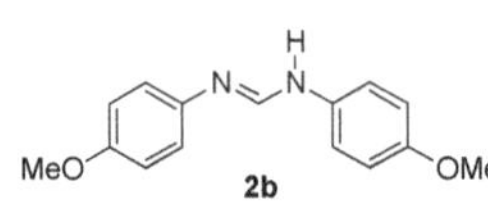 2b	3 h	82	13f
3	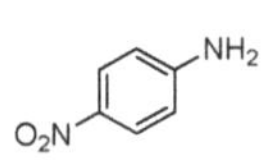 3a	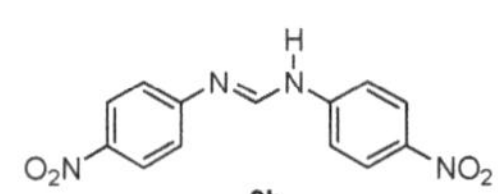 3b	3 hr[b]	55	13b
4	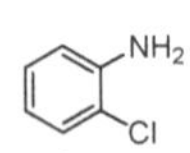 4a	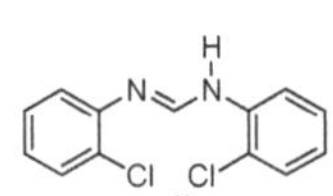 4b	3 h	76	13d
5	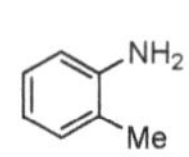 5a	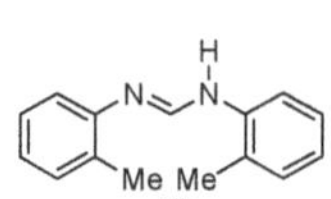 5b	3 h	79	10
6	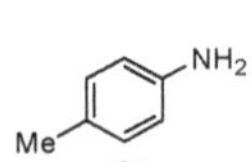 6a	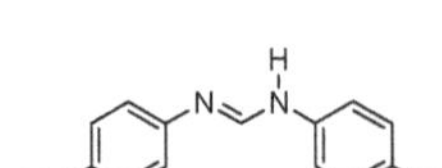 6b	3 h	88	13g
7	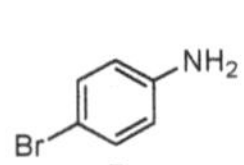 7a	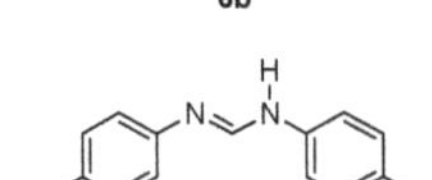 7b	3 h	89	13h
8	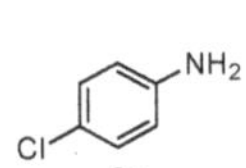 8a	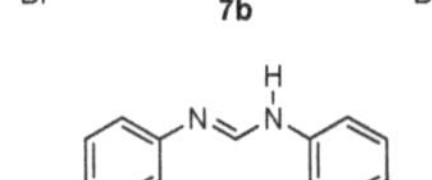 8b	3 h	88	13f
9	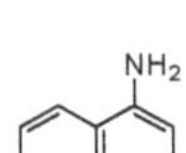 9a	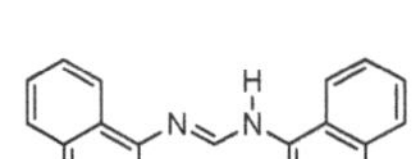 9b	3 h	91	10

[a]Yields refer to pure isolated products. [b]Refluxed in toluene.

3. Conclusion

In conclusion, we have developed a mild and efficient method for the direct conversion of primary aryl amine to N,N'-diarylformamidines using a catalytic amount of $FeCl_3$. This provides an eco-friendly and simple method without using any toxic and expensive reagents.

4. Experimental Section

4.1. General Procedures

All melting points were taken on a Gallenkamp melting point apparatus and are uncorrected. The ^1H and ^{13}C NMR were recorded in $CDCl_3$ using TMS as an internal standard on 300 and 75 MHz spectrometer (Bruker) respectively and IR were recorded using a Shimadzu FT IR-8300 instrument. High-resolution mass spectra were obtained using a Qt of Micro YA263 instrument. Toluene was dried over sodium. Chloroform was freshly distilled from phosphorus pentoxide. Petroleum ether of boiling range 60°C - 80°C and silica gel of 60 - 120 mesh were used for column chromatography.

4.2. Experimental Section

4.2.1. Representative Procedure for the Synthesis of Diarylformamidines

To a well stirred a solution of aniline (1a) (186 mg, 2 mmol) and triethyl orthoformate (148 mg, 1 mmol) in the presence of a catalytic amount of $FeCl_3$ (10 mol%) in toluene (10 mL) was stirred for 3 h at room temperature. Solvent was removed under reduced pressure and the solid mass was dissolved in chloroform (30 mL) and then filtered through a Whatmann filter paper. The solvent was removed under reduced pressure and the crude residue obtained was purified by column chromatography over silica gel (100 - 200 mesh) (40% ethyl acetate in petroleum ether) to afford N,N'-diphenylformamidine (1b) as a pale yellow solid; m.p.139°C -140°C. IR (KBr): 3356, 3037, 2875, 1685, 1600, 1498, 1442, 1313, 1174, 1028, 752, 692 cm^{-1}; ^1H NMR (300 MHz, $CDCl_3$): δ 7.04-7.18 (m, 6H), 7.29 - 7.34 (m, 4H), 8.19 (s, 1H); ^{13}C NMR (75 MHz, $CDCl_3$): δ 118.9, 120.2, 124.8, 125.3, 129.1, 129.8, 136.8, 137.0, 159.5 ; HRMS: calcd. for $C_{13}H_{13}N_2$ $[M+H]^+$: 197.1073; found: 197.1072.

4.2.2. N,N'-Bis(4-methoxyphenyl) formamidine (2b)

Colourless solid; m.p. 85°C - 86°C. IR (KBr): 2835, 1672, 1504, 1319, 1242, 1109, 1033, 825, 719, 580 cm^{-1}; ^1H NMR (300 MHz, $CDCl_3$): δ 3.81 (s, 6H), 6.89 - 6.92 (d, J = 8.8 Hz, 4H), 7.09 - 7.12 (d, J = 8.8 Hz, 4H), 8.06 (s, 1H); ^{13}C NMR (75 MHz, $CDCl_3$): δ 55.5, 55.6, 114.3, 115.0, 121.7, 121.9, 129.7, 130.1, 156.8, 157.7, 163.3; HRMS: calcd. for $C_{15}H_{17}N_2O_2$ $[M+H]^+$: 257.1285; found: 257.1283.

4.2.3. N,N'-Bis(4-nitrophenyl) formamidine (3b)

Pale yellow solid; m.p. 239°C - 240°C (decomposed). IR (KBr): 3086, 2885, 1687, 1562, 1500, 1411, 1330, 1271, 1111, 854, 752, 688, 540 cm^{-1}; ^1H NMR (300 MHz, DMSO-d_6): δ 7.82 - 7.97 (m, 5H), 8.22 - 8.25 (m, 3H), 8.42 (s, 1H); ^{13}C NMR (75 MHz, DMSO-d_6): δ 112.9, 119.5, 125.6, 126.9, 136.2, 143.1, 144.7, 156.2, 161.1; HRMS: calcd. for $C_{13}H_{10}N_4O_4$ $[M+H]^+$: 286.0775; found: 286.0776.

4.2.4. N,N'-Bis(2-chlorophenyl) formamidine (4b)

Pale yellow solid; m.p. 141°C -142°C. IR (KBr): 3010, 2981, 1685, 1503, 1363, 1091, 821 cm^{-1}; ^1H NMR (300 MHz, $CDCl_3$): δ 6.96 - 7.35 (m, 8H), 8.33 (s, 1H); ^{13}C NMR (75 MHz, $CDCl_3$): δ 115.9, 119.0, 119.3, 124.2, 127.6, 127.8, 129.4, 129.9, 143.0; HRMS: calcd. for $C_{13}H_{11}N_2Cl_2$ $[M+H]^+$: 265.0294; found: 265.0292.

4.2.5. N,N'-Di-O-tolylformamidine (5b)

Light brown solid; m.p. 152°C -153°C IR (KBr): 3010, 2806, 1662, 1580, 1480, 1465, 1310, 1210, 1187, 996, 780, 744, 723, 613 cm^{-1}; ^1H NMR (300 MHz, $CDCl_3$): δ 2.25 (s, 6H), 6.92 - 6.97 (m, 4H), 7.08 - 7.18 (m, 4H), 7.98 (s, 1H); ^{13}C NMR (75 MHz, $CDCl_3$): δ 18.0, 117.6, 123.8, 127.1,128.8, 130.5, 130.9, 147.8; HRMS: calcd. for $C_{15}H_{17}N_2$ $[M+H]^+$: 225.1386; found: 225.1386.

4.2.6. N,N'-Di-P-tolylformamidine (6b)

Brownish solid; m.p. 140°C - 141°C. IR (KBr): 3342, 2922, 1693, 1518, 1356, 1215, 1037, 817, 669, 509 cm^{-1}; ^1H NMR (300 MHz, DMSO-d_6): δ 2.24 (brs, 6H), 7.18 - 7.28 (m, 8H), 8.76 (brs, 1H); HRMS: calcd. for $C_{15}H_{17}N_2$ $[M+H]^+$: 225.1386; found: 225.1385.

4.2.7. N,N'-Bis(4-bromophenyl) formamidine (7b)

White solid; m.p. 19°C - 192°C. IR (KBr): 3003, 2976, 1697, 1491, 1352, 1076, 819, 634, 495 cm^{-1}; ^1H NMR (300 MHz, DMSO-d_6): δ 7.51 - 7.59 (m, 4H), 7.65 - 7.68 (m, 4H), 8.98 (brs, 1H); ^{13}C NMR (75 MHz, $CDCl_3$): δ 117.6, 118.4, 120.4, 121.6, 132.2, 132.9, 135.9, 136.0, 159.0; HRMS: calcd. for $C_{13}H_{11}N_2Br_2$ $[M+H]^+$: 352.9283; found: 352.9283.

4.2.8. N,N'-Bis(4-chlorophenyl) formamidine (8b)

Light brown solid; 179°C - 180°C. IR (KBr): 3013, 2974, 1691, 1501, 1367, 1100, 823, 656, 476 cm^{-1}; ^1H NMR (300 MHz, $CDCl_3$): δ 7.02 - 7.10 (m, 2H), 7.26 - 7.33 (m, 3H), 7.43 - 7.58 (m, 3H), 8.35 (s, 1H); ^{13}C NMR (75 MHz, $CDCl_3$): δ 120.2, 121.4, 129.2, 129.9, 130.9, 135.4, 135.5, 159.3; HRMS: calcd. for $C_{13}H_{11}Cl_2N_2$ [M+H]: 265.0294; found: 265.0295.

4.2.9. N,N'-Dinaphthalen-1-Yl-formamidine (9b)

Light purple solid; m.p. 200°C - 201°C. IR (KBr): 3047,

1660, 1573, 1394, 1300, 1263, 993, 788, 765 cm^{-1}; ^{1}H NMR (500 MHz, $CDCl_3$): δ 7.48 - 7.68 (m, 10H), 7.90 (s, 2H), 8.30 (s, 2H), 8.50 (brs, 1H); HRMS: calcd. for $C_{22}H_{16}N_2$ $[M+H]^+$: 297.1386; found: 297.1386.

5. Acknowledgements

The authors thank the Department of Science and Technology, New Delhi for financial assistance. P C thanks CSIR, New Delhi for awarding the research fellowships.

REFERENCES

[1] R. W. Beeman and F. Matsumura, "Chlordimeform: A Pesticide Acting upon Amine Regulatory Mechanisms," *Nature*, Vol. 242, No. 5395, 1973, pp. 273-274.

[2] S. A. Aziz and C. O. Knowles, "Inhibition of Monoamine Oxidase by the Pesticides Chlordimeform and Related Compounds," *Nature*, Vol. 242, No. 5397, 1973, pp. 417-418.

[3] V. K. S. Leung, T. Y. K. Chan and V. T. F. Yeung, "Ami-Traz Poisining in Humans," *Clinical Toxicology*, Vol. 37, No. 4, 1999, pp. 513-514.

[4] A. Nakayama, M. Sukekawa and Y. Eguchi, "Stereochemistry and Active Conformation of a Novel Insecticide, Acetamiprid," *Pesticide Science*, Vol. 51, No. 2, 1997, pp. 157-164.

[5] G. D. Baxter and S. C. Barker, "Isolation of a cDNA for an Octopamine-Like, G-Protein Coupled Receptor from the Cattle Tick, *Boophilus microplus*," *Insect Biochemistry and Molecular Biology*, Vol. 29, No. 5, 1999, pp. 461-467.

[6] M. Gall, J. M. McCall, R. E. TenBrink, P. F. Von-Voigtlander and J. S. Mohrland, "Arylformamidines with Antinociceptive Properties," *Journal of Medicinal Chemistry*, Vol. 31, No. 9, 1988, pp. 1816-1820.

[7] A. Donetti, E. Cereda, E. Bellora, A. Gallazzi, C. Bazzano, P. Vanoni, P. D. Soldato, R. Michelett, F. Pagani and A. Giachetti, "(Lmidazolylphenyl)Formamidines. A Structurally Novel Class of Potent Histamine H2 Receptor Antagonists," *Journal of Medicinal Chemistry*, Vol. 27, No. 3, 1984, pp. 380-386.

[8] T. Goto, H. Sakashita, K. Murakami, M. Sugiura, T. Kondo and C. Fukaya, "Novel Histamine H_3 Receptor Antagonists: Synthesis and Evaluation of Formamidine and S-Methylisothiourea Derivatives," *Chemical & Pharmaceutical Bulletin*, Vol. 45, No. 2, 1997, pp. 305-311.

[9] G. K. W. Yim, M. P. Holsapple, W. R. Pfister and R. M. Hollingworth, "Prostaglandin Synthesis Inhebited by Formamidine Pesticides," *Life Sciences*, Vol. 23, No. 25, 1978, pp. 2509-2515.

[10] D. I. Arnold, F. A. Cotton, J. H. Matonic and C. A. Murillo, "Bis(N,N'-diphenylformamidine)silver(I) Triflate: A Three-Coordinate Silver Formamidine Compound Stabilized by Intramolecular Hydrogen Bonds," *Polyhedron*, Vol. 16, No. 11, 1997, pp. 1837-1841.

[11] D. B. Mitzi and K. Liang, "Synthesis, Resistivity, and Thermal Properties of the Cubic Perovskite $NH_2CH=NH_2SnI_3$ and Related Systems," *Journal of Solid State Chemistry*, Vol. 134, No. 2, 1997, pp. 376-381.

[12] A. I. Meyers and R. Hutchings, "Asymmetric Dialkylation of Chiral 2-Benzazepine Formamidines," *Heterocycles* Vol. 42, No. 2, 1996, pp. 475-478.

[13] M. Matulenko and A. I. Meyers, "Total Synthesis of (-)-Tetrahydropalmatine via Chiral Formamidine Carbanions: Unexpected Behavior with Certain Ortho-Substituted Electrophiles," *The Journal of Organic Chemistry*, Vol. 61, No. 2, 1996, pp. 573-580.

[14] S. J. Benkovic, T. H. Barrows and P. R. Farina, "Studies on Models for Tetrahydrofolic Acid. IV. Reactions of Amines with Formamidinium Tetrahydroquinoxaline Analogs," *Journal of the American Chemical Society*, Vol. 95, No. 25, 1973, pp. 8414-8420.

[15] P. S. Furth, M. S. Reitman and A. F. Cook, "A Novel Formamidine Linker for Use in Soid-Phase Synthesis," *Tetrahedron Letters*, Vol. 38, No. 31, 1997, pp. 5403-5406.

[16] K. Hirano, S. Urban, C. Wang and F. Glorius, "A Modular Synthesis of Highly Substituted Imidazolium Salts," *Organic Letters*, Vol. 11, No. 4, 2009, pp. 1019-1022.

[17] L. Meschede and H-H Limbach, "Dynamic NMR Study of the Kinetic HH/HD/DD Isotope Effects on the Double Proton Transfer in Cyclic Bis(p-fluorophenyl)formamidine Dimers," *The Journal of Physical Chemistry*, Vol. 95, No. 25, 1991, pp. 10267-10280.

[18] R. M. Roberts, "The Molecular Association of Diarylformamidines. II. Effects of O- and P-Methyl Groups," *Journal of the American Chemical Society*, Vol. 78, No. 11, 1956, pp. 2606-2608.

[19] K. M. Kuhn and R. H. Grubbs, "A Facile Preparation of Imidazolinium Chlorides," *Organic Letters*, Vol. 10, No. 10, 2008, pp. 2075-2077.

[20] H. G. Mandel and A. J. Hill, "The Conversion of Formamides into Formamidines," *Journal of the American Chemical Society*, Vol. 76, No. 15, 1954, pp. 3978-3982.

[21] R. M. Roberts, R. H. DeWolfe and J. H. Ross, "Ortho Esters, Imidic Esters and Amidines. II. Disproportionation Reactions of Nitrophenyl-, Chlorophenyl- and Tolyl-substituted Formimidates and Formamidines," *Journal of the American Chemical Society*, Vol. 73, No. 5, 1951, pp. 2277-2281.

[22] C. D. Lewis, R. G. Krupp, H Tieckelmann and H. W. Post, "The Action of Ethyl Orthoformate on Aniline and Certain of Its Derivatives," *The Journal of Organic Che-

mistry, Vol. 12, No. 2, 1947, pp. 303-307.

[23] C. Lin, J. D. Protasiewicz, E. T. Smith and T. Ren, "Linear Free Energy Relationships in Dinuclear Compounds. 2. Inductive Redox Tuning via Remote Substituents in Quadruply Bonded Dimolybdenum Compounds," *Inorganic Chemistry*, Vol. 35, No. 22, 1996, pp. 6422-6428.

[24] M. L. Cole, P. C. Junk and L. M. Louis, "Synthesis and Structural Characterisation of Some Novel Lithium and Sodium N,N'-di(para-tolyl)formamidinate Complexes," *Journal of the Chemical Society, Dalton Transacions*, No. 20, 2002, pp. 3906-3914.

[25] L.-J. Han, "Redetermination of (E)-N,N'-bis(4-bromophenyl)formamidine," Acta Crystallographica section E, Vol. 67, No. 5, 2011, p. 1159.

[26] K. U. Sadek, A. Alnajjar, R. A. Mekheimer, N. K. Mohamed and H. A. Mohamed, "Cerium (IV) Ammonium Nitrate (CAN) Mediated Reactions IV. A Highly Efficient Synthesis of N,N'-Diarylsubstituted Formamidines in Water at Ambient Temperature," *Green and Sustainable Chemistry*, Vol. 1, No. 3, 2011, pp. 92-97.

15

Ionic Liquid as Useful Media for Dissolution, Derivatization, and Nanomaterial Processing of Chitin

Jun-Ichi Kadokawa
Graduate School of Science and Engineering, Kagoshima University, Kagoshima, Japan

ABSTRACT

This paper reviews studies on dissolution, derivatization, and nanomaterial processing of chitin using an ionic liquid as useful media. Because chitin is the second most abundant polysaccharide on the earth after cellulose, there is major interest in conversion of native chitin resources into various useful materials after proper dissolution in suitable solvents. For the derivatization and nanomaterial processing of chitin, the author has been focusing on ionic liquids because which have been found to be used as good solvents for cellulose in a past decade. The author found that an ionic liquid, 1-allyl-3-methylimidazolium bromide (AMIMBr), dissolved chitin in concentrations up to 4.8 wt% and mixtures of the higher amounts of chitin with AMIMBr gave ion gels. Acetylation, the simplest derivatization, of chitin using acetic anhydride was achieved in the AMIMBr solvent under mild conditions. Furthermore, the chitin nanofibers were fabricated by regeneration technique from the chitin ion gel with AMIMBr using methanol. Moreover, filtration of the chitin nanofiber dispersion with methanol was carried out to give a chitin nanofiber film. The chitin nanofiber-poly(vinyl alcohol) composite film was also prepared from the ion gel by co-regeneration method.

Keywords: Chitin; Ionic Liquid; Derivatization; Nanofiber; Composite Material

1. Introduction

Ionic liquids are low-melting-point molten salts, defined as that form liquids at room temperature or even at temperatures lower than the boiling point of water. The property is owing to that the liquid state is thermodynamically favorable due to the large size and conformational flexibility of the ions, in which these behaviors lead to small lattice enthalpies and large entropy changes that favor the liquid state [1,2]. In the past more than a decade, ionic liquids have attracted much attention due to their specific characteristics such as excellent thermal stabilities, a negligible vapor pressure, and controllable physical and chemical properties [3]. Beyond these traditional properties of ionic liquids, recently, interests and applications on ionic liquids have been extended to the researches related to biological macromolecules such as naturally occurring polysaccharides, because of specific good affinities of ionic liquids for them [4].

Polysaccharides are widely distributed in nature and have been regarded as structural materials and as suppliers of water and energy [5]. Of the many kinds of polysaccharides, cellulose and chitin are the most important biomass resources because they are the first and second most abundant natural polysaccharides on the earth, respectively [6-10]. Chitin is an aminopolysaccharide consisting of β (1→4)-linked *N*-acetyl-D-glucosamine residues, which has acetamido groups at the C-2 position in place of hydroxy groups of the glucose residues in cellulose (**Figure 1**). Despite its huge production in nature and easy accessibility, chitin still remains as an unutilized biomass resource primary because of its intractable bulk structure and insolubility in water and common organic solvents. Even in recent years, therefore, there is major interest in conversion into various useful materials after proper dissolution of chitin in suitable solvents.

Chitin occurs mainly in the exoskeletons of crustacean shells such as crab and shrimp shells in nature. Native chitin in crustacean shells is arranged as microfibrils embedded in a protein matrix [11,12]. The microfibril is composed of nanofibers with 2 nm - 5 nm diameters. Moreover, three types of crystalline forms, *i.e.*, α-, β-, and γ-chitins, are known depending on the arrangement of the polymeric chains. The most abundant form is α-chitin (e.g., from crab and shrimp shells) [13], where the polymeric chains are aligned in an antiparallel fashion. This arrangement is favorable for the formation of

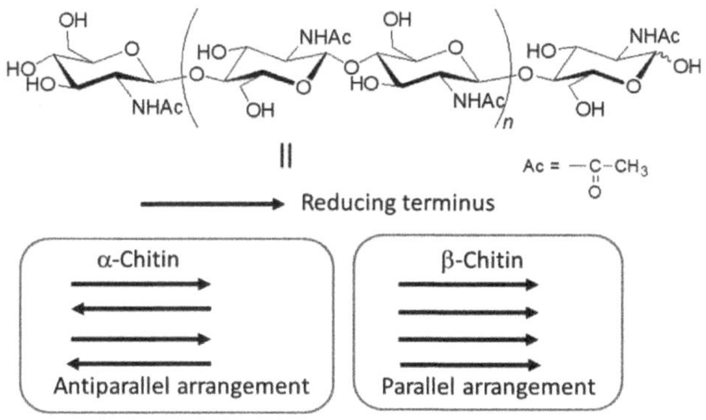

Figure 1. Chemical structure of chitin and schematic image of chitin crystalline structures.

strong intermolecular hydrogen bonding, leading to the most stable form in the three types of crystalline structures. In *β*-chitin (e.g., from squid pen), the polymeric chains are packed in a parallel arrangement [14], resulting in weaker intermolecular forces. Accordingly, *β*-chitin is considered to be less stable than *α*-chitin. *γ*-Chitin may be a mixture of *α*- and *β*-forms.

Since Rogers et al. comprehensively reported that an ionic liquid, 1-butyl-3-methylimidazolium chloride, dissolved cellulose in relatively high concentrations [15], ionic liquids have began to be used in the processing of cellulose, which mainly concern with the dissolution, homogeneous derivatization and modification, and regeneration [16-19]. Therefore, ionic liquids are considered as powerful solvents and are used for material processing of cellulose. On the other hand, only the limited investigations have been reported regarding the dissolution of chitin with ionic liquids including the author's study, in which the author found that an ionic liquid, 1-allyl-3-methylimidazolium bromide (AMIMBr), dissolved or swelled chitin to form weak gel-like materials (ion gels) (**Figure 2**) [20,21]. The other groups have also reported the dissolution of chitin with ionic liquids, 1-butyl- and 1-ethyl-3-methylimidazolium acetates (**Figure 2**) [22,23]. However, the dissolution of chitin with other ionic liquids having different cationic, anionic, and substituent structures has hardly been reported. The observation suggests that the aforementioned ionic liquids only having specific structures probably exhibit a potential for the dissolution of chitin [24-27].

Chemical derivatization of chitin is one of the efficient approaches to explore their potentials, giving the novel types of functional materials [28,29]. Derivatization of chitin, however, is generally difficult owing to a lack of solubility as aforementioned, and thus the reactions often take place under heterogeneous conditions, leading to low degree of substitution (DS). Even acetylation of chitin, a simplest derivatization, has been performed only under special conditions. The author has considered that ionic liquids, that dissolve chitin, act as efficient media for derivatizations of chitin with high DS, and thus investigated acetylation of chitin in AMIMBr [30].

On the other hand, the preparation of self-assembled fibrillar nanostructures, so-called nanofibers, is one of the most useful methods to practically utilize polymeric functional materials from native chitin sources [31-33]. Conventional approaches to the production of chitin nanofibers are mainly performed upon top-down procedures that break down the starting bulk materials from native chitin sources (**Figure 3**). As the representative techniques, acid hydrolysis [34-37], 2,2,6,6-tetramethylpiperidine-1-oxyl radical (TEMPO)-mediated oxidation [38,39], and grinding technique [40,41] have previously been performed to produce chitin nanofibers from native chitin sources. The other method accorded to self-assembling generative (bottom-up) route (**Figure 3**), in which fibrillar nanostructures were produced by regeneration from chitin solutions via appropriate process, as examples of the electrospinning equipment [42,43] and simple precipitation process [44,45]. To efficiently provide chitin nanofibers by the bottom-up technique, the author has focused on the ionic liquid, AMIMBr. Accordingly, the author successfully reported that chitin nanofiber films were facilely obtained by regeneration from the ion gels with AMIMBr using methanol, followed by filtration [46].

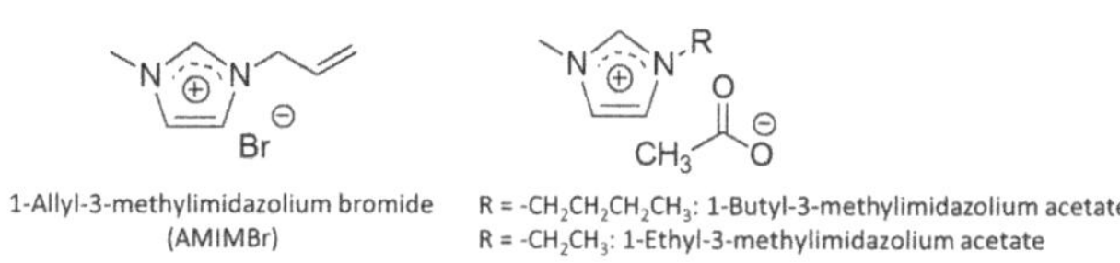

Figure 2. Representative ionic liquids that dissolve chitin.

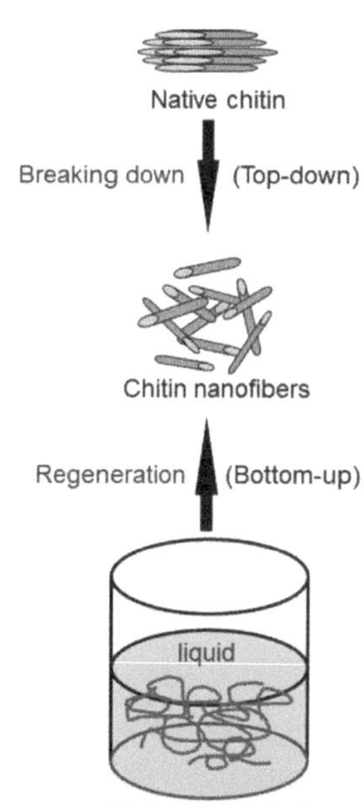

Figure 3. Top-down and bottom-up approaches for preparation of chitin nanofibers.

On the basis of the above backgrounds and viewpoints, this paper reviews the studies on the dissolution, acetylation, and nanofibrillation using AMIMBr. The series of the studies provide new approaches for material processing from native chitin sources.

2. Dissolution of Chitin with Ionic Liquid

For the dissolution study of chitin with ionic liquids, the author noted the previous study reporting that the imidazolium-type ionic liquids having a bromide counter anion were the good solvents for synthesis of polyamides and polyimides [47]. The result inspired to use the same kind of ionic liquids for dissolution of chitin, because chitin has the -N-C=O groups as same as polyamides and polyimides. Thus, three ionic liquids consisting of a bromide anion were prepared, which were AMIMBr, 1-methyl-3-propylimidazolium bromide, and 1-butyl-3-methylimidazolium bromide, and the dissolution experiments were conducted as follows [21].

A mixture of chitin with each ionic liquid was heated at 100˚C and the heating process was simply followed by a charge coupled device (CCD) camera on glass plate. When AMIMBr was used for the dissolution experiment, the clear solutions of chitin were formed in the concentrations up to ~4.8 wt% (**Figure 4(a)**), whereas the other ionic liquids did not show ability to dissolve chitin even in 1 wt% concentration. The disappearance of chitin powder in the solution of 4.8 wt% chitin with AMIMBr was confirmed further by the SEM measurement. The SEM image of the mixture of chitin with AMIMBr after heating it at 100˚C for 48 h did not show any solid of chitin (**Figure 4(b)**), suggesting that 4.8 wt% chitin was solvated with AMIMBr at least at the μm scale level.

It was confirmed as follows that degradation and decrease in the molecular weight of chitin did not frequently occur during dissolution with AMIMBr. The powder X-ray diffraction (XRD), thermal gravimetrical analysis (TGA), and IR results of the regenerated chitin from the solution with AMIMBr, which was prepared by addition of methanol into it, followed by filtration and dryness of the precipitate, were almost same as those of the original chitin. In addition, the hexanoyl derivative of the regenerated chitin, which was synthesized by the reaction of the regenerated material with hexanoyl chloride [48], showed the exact same ^{1}H NMR pattern as that of the hexanoylated sample from the original chitin and the GPC profiles of both the derivatives were not much different from each other.

(a) (b)

Figure 4. CCD camera view (a) and SEM image (b) of a mixture of chitin (4.8 wt%) with AMIMBr after dissolution experiment (100˚C, 48 h).

Interestingly, mixtures of higher amounts of chitin with AMIMBr gave more viscous materials, *i.e.*, gel-like materials when 6.5 wt% - 10.7 wt% of chitin were immersed in AMIMBr at room temperature for 24 h, followed by heating at 100˚C for 48 h and cooling to room temperature. Indeed, the obtained 6.5 wt% chitin with AMIMBr did not flow upon leaning a test tube, whereas the aforementioned 4.8 wt% chitin with AMIMBr started to flow upon leaning (**Figure 5**). The dynamic rheological measurements showed that both 4.8 wt% and 6.5 wt% chitins with AMIMBr behaved as the weak gels.

3. Acetylation of Chitin in Ionic Liquid

As a simplest and useful derivatization reaction of chitin, acetylation of chitin using acetic anhydride in an ionic liquid solvent, AMIMBr, was performed (**Figure 6**) [30]. Prior to acetylation, chitin was dissolved with AMIMBr (2.0 wt%) by heating the mixture at 100˚C for 24 h with stirring. Then, the 5 - 20 equiv. of acetic anhydride for a repeating unit was added to the solution and the mixture was heated at 60˚C - 100˚C for 24 h with stirring. The products were isolated as a fraction insoluble in methanol and the DS values of the products were estimated by comparison of the intensities of the two carbonyl absorptions due to amido I (ca. 1658 cm^{-1}) and ester (ca. 1745 cm^{-1})

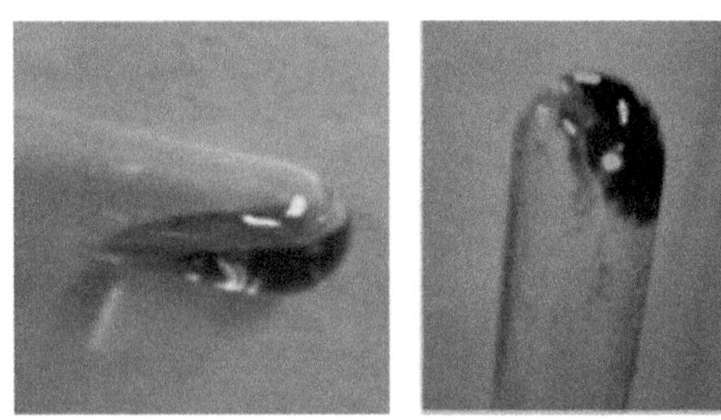

(a) (b)

Figure 5. Photographs of 4.8 wt% (a) and 6.5 wt% (b) chitin with AMIMBr.

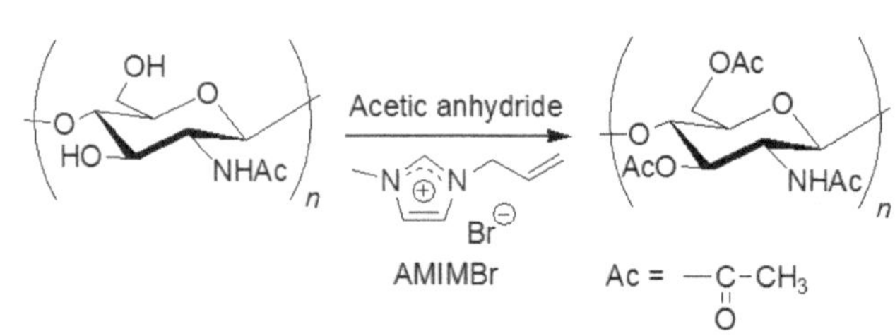

Figure 6. Acetylation of chitin using acetic anhydride in AMIMBr.

in the IR spectra (**Figure** 7). No acetylation took place under the conditions using 5 equiv. of acetic anhydride at 60°C. While the DS values increased under the conditions using larger equiv. of acetic anhydride at higher reaction temperatures. The highest DS value (1.86) was obtained when the reaction was carried out using 20 equiv. of acetic anhydride at 80°C.

Because the products with high DS, which were obtained by the reaction using 20 equiv. of acetic anhydride, were soluble in DMSO, the structures of these products were further confirmed by the ^{1}H NMR analysis measured in DMSO-d_6. On the basis of the integrated ratios of the signals due to the acetyl protons to the signal due to the anomeric protons, the DS values were calculated, which were in good agreement with those determined by the IR spectra. The all the above results indicated that AMIMBr was the powerful media for the efficient derivatization of chitin.

4. Preparation of Nanofiber Films from Chitin Ion Gel with Ionic Liquid

The author has reported that chitin nanofiber films were formed by regeneration from the aforementioned ion gels with AMIMBr using methanol, followed by filtration (**Figure 8**, Route **A**) [46]. First, chitin was swollen with AMIMBr according to the procedure as aforementioned to give the chitin ion gels with AMIMBr (9.1 wt% - 10.7 wt%). It was found that chitin dispersions were obtained when the gels were immersed in methanol at room temperature for 24 h to slowly regenerate chitin, followed by sonication. The resulting dispersion was diluted with methanol, which was subjected to the SEM measurement. The morphology of nanofibers with ca. 20 nm - 60 nm in width and several hundred nanometers in length was observed in the SEM image of the sample from the dispersion (**Figure 9(a)**), indicating the formation of the self-assembled chitin nanofibers by the above regeneration approach from the ion gel according to the bottom-up process. When the dispersion was filtered, the residue formed a film, which was further purified by Soxhlet extraction with methanol. The SEM image of the resulting film was also measured to confirm the nanoscaled morphology of chitin, which showed the pattern of highly entangled nanofibers (**Figure 9(b)**). This suggested that such entangled structure of the nanofibers contributed to formability of the film. The XRD pattern of the chitin nanofiber film mainly showed the diffraction peaks corresponding to crystalline structure of α-chitin and is in good agreement with that of an original chitin powder. The result indicated that the crystalline structure of α-chitin was reconstructed by the above regeneration procedure during the formation of the nanofibers.

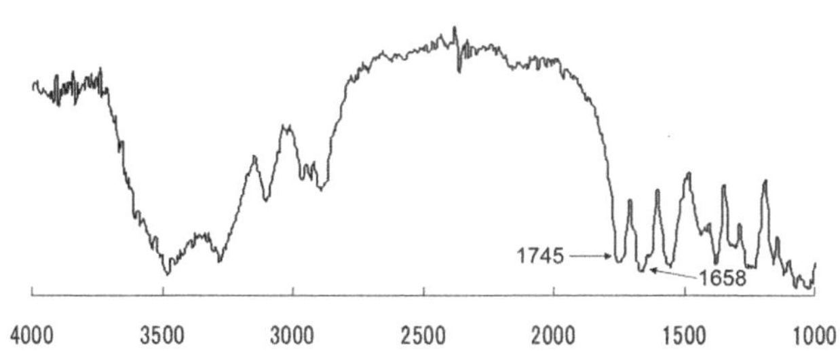

Figure 7. IR spectrum of acetylated chitin obtained using 20 equiv. of acetic anhydride at 60°C (DS value; 1.79).

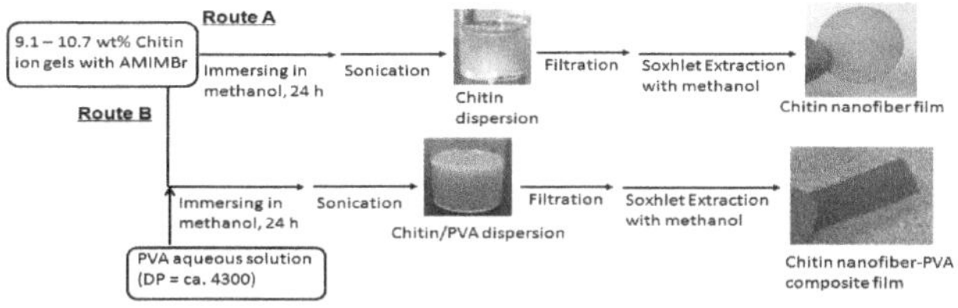

Figure 8. Procedures for the preparation of chitin dispersion/nanofiber film (Route A) and chitin nanofiber-PVA composite film (Route B).

As one of the possible applications of the present chitin nanofiber film, the preparation of the chitin nanofiber-poly(vinyl alcohol) (PVA) composite films was investigated (**Figure 8**, Route **B**) [46]. First, the 9.1 wt% chitin ion gel with AMIMBr was prepared according to the aforementioned procedure and a solution of PVA (DP = ca. 4300) in a small amount of hot water was added to the gel (the feed weight ratio of chitin to PVA = 1:0.30). Then, the co-regeneration using methanol was conducted, followed by filtration and Soxhlet extraction with methanol same as the aforementioned procedure to produce the chitin nanofiber-PVA composite film. It was confirmed that a small amount of AMIMBr was remained in the composite. The SEM image of the composite showed that the nanofiber-like morphology was maintained and PVA components probably filled in spaces among the fibers (**Figure 9(c)**), indicating relative immiscibility of chitin and PVA in the composite. Indeed, the DSC profile of the composite film exhibited an endothermic peak attributed to the melting point of PVA, suggesting some immiscibility of PVA toward chitin in the composite. However, the melting point peak of PVA in the composite film was broadened, indicating that the crystallinity of PVA decreased in the composite film. The DSC result suggested that chitin and PVA might be partially misci-

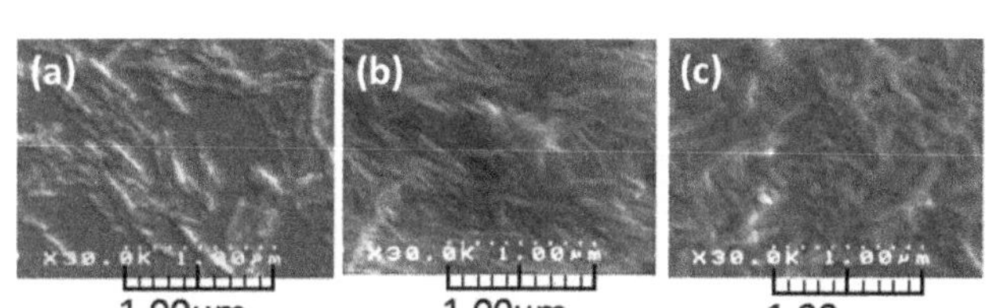

Figure 9. SEM images of chitin dispersion (a), chitin nanofiber film (b), and chitin nonofiber-PVA composite film (c).

ble at the interfacial area between the two polymers in the composite film by the formation of hydrogen bonding between them or by the presence of a small amount of AMIMBr.

5. Conclusion

This article reviewed the dissolution, derivatization, and nanomaterial processing of chitin using the ionic liquid, AMIMBr. Because the author found that AMIMBr dissolved chitin in the concentrations up to 4.8 wt% by heating and further the chitin ion gels were obtained from the mixture of the higher amounts of chitin with AMIMBr, acetylation of chitin using acetic anhydride was performed in the AMIMBr solvent to give the acetylated chitins with the high DS values. Chitin-based nanomaterials such as chitin nanofibers and nanocomposites with synthetic polymer were also fabricated using AMIMBr. Immersing the ion gel in methanol and subsequent sonication gave the chitin dispersion. The SEM image of the dispersion showed the formation of the self-assembled chitin nanofibers. Then, filtration of the dispersion was carried out to give the chitin nanofiber film. The composite film of the chitin nanofibers with PVA was prepared by a similar procedure as that for the chitin nanofiber film. The present methods using the ionic liquid, AMIMBr, provide new chitin-based functional and composite materials which have a highly potential to apply to practical material research filed in the future. Finally, the author would like to state that this review has especially presented to the readers how the ionic liquid media are useful for new approaches in the chitin research field, which will be increasingly developed with attracting much attention on the further technological interests in the efficient use of chitin as the abundant biomass resource.

6. Acknowledgements

The author is indebted to the co-workers, whose names are found in references from his papers, for their enthusiastic collaborations.

REFERENCES

[1] T. Welton, "Room-Temperature Ionic Liquids. Solvents for Synthesis and Catalysis," *Chemical Reviews*, Vol. 99, No. 8, 1999, pp. 2071-2084.

[2] P. Wasserscheid and W. Keim, "Ionic Liquids: New 'Solutions' for Transition Metal Catalysis," *Angewandte Chemie International Edition*, Vol. 39, No. 21, 2000, pp. 3772-3789.

[3] N. V. Plechkova and K. R. Seddon, "Applications of Ionic Liquids in the Chemical Industry," *Chemical Society Reviews*, Vol. 37, No. 1, 2008, pp. 123-150.

[4] O. A. E. Seoud, A. Koschella, L. C. Fidale, S. Dorn and T. Heinze, "Applications of Ionic Liquids in Carbohydrate Chemistry: A Windows of Opportunities," *Biomacromolecules*, Vol. 8, No. 9, 2007, pp. 2629-2648.

[5] C. Schuerch, "Polysaccharides," In: H. F. Mark, N. Bilkales and C. G. Overberger, Eds., *Encyclopedia of Polymer Science and Engineering*, 2nd Edition, John Wiley & Sons, New York, 1986, Vol. 13, pp. 87-162.

[6] D. Klemm, B. Heublein, H.-P. Fink and A. Bohn, "Cellulose: Fascinating Biopolymer and Sustainable Raw mMaterial," *Angewandte Chemie International Edition*, Vol. 44, No. 22, 2005, pp. 3358-3393.

[7] M. Rinaudo, "Chitin and Chitosan: Properties and Applications," *Progress in Polymer Science*, Vol. 31, No. 7, 2006, pp. 603-632.

[8] K. Kurita, "Chitin and Chitosan: Functional Biopolymers from Marine Crustaceans," *Marine Biotechnology*, Vol. 8, No. 3, 2006, pp. 203-226.

[9] C. K. S. Pillai, W. Paul and C. P. Sharma "Chitin and Chitosan Polymers: Chemistry, Solubility and Fiber Formation," *Progress in Polymer Science*, Vol. 34, No. 7, 2009, pp. 641-678.

[10] R. A. A. Muzzarelli, "Chitin Nanostructures in Living Organisms," In: S. N. Gupta, Ed., *Chitin Formation and Diagenesis*, Springer, New York, 2011.

[11] D. Raabe, P. Romano, C. Sachs, H. Fabritius, A. Al-Sawalmih, S. B. Yi, G. Servos and H. G. Hartwig, "Microstructure and Crystallographic Texture of the Chitin-Protein Network in the Biological Composite Material of the Exoskeleton of the Lobster Homarus Americanus," *Materials Science and Engineering A-Structural Materials Properties Microstructure and Processing*, Vol. 421, No. 1-2, 2006, pp. 143-153.

[12] P. Y. Chen, A. Y. M. Lin, J. McKittrick and M. A. Meyers, "Structure and Mechanical Properties of Crab Exoskeletons," *Acta Biomaterialia*, Vol. 4, No. 3, 2008, pp. 587-596.

[13] R. Minke and J. Blackwell, "The Structure of α-Chitin," *Journal of Molecular Biology*, Vol. 120, No. 2, 1978, pp. 167-181.

[14] K. H. Gardner and J. Blackwell, "Refinement of the Structure of β-Chitin," *Biopolymers*, Vol. 14, No. 8, 1975, pp. 1581-1595.

[15] R. P. Swatloski, S. K. Spear, J. D. Holbrey and R. D. Rogers, "Dissolution of Cellulose with Ionic Liquids," *Journal of the American Chemical Society*, Vol. 124, No. 18, 2002, pp. 4974-4975.

[16] T. Liebert and T. Heinze, "Interaction of Ionic Liquids with Polysaccharides 5. Solvents and Reaction Media for the Modification of Cellulose," *BioResources*, Vol. 3, No. 2, 2008, pp. 576-601.

[17] L. Feng and Z. I. Chen, "Research Progress on Dissolution and Functional Modification of Cellulose in Ionic Liquids," *Journal of Molecular Liquids*, Vol. 142, No. 1-3, 2008, pp. 1-5.

[18] A. Pinkert, K. N. Marsh, S. Pang and M. P. Staiger, "Ionic Liquids and their Interaction with Cellulose," *Chemical Reviews*, Vol. 109, No. 12, 2009, pp. 6712-6728.

[19] M. E. Zakrzewska, E. B. Lukasik and R. B. Lukasik, "Solubility of Carbohydrates in Ionic Liquids," *Energy & Fuels*, Vol. 24, No. 2, 2010, pp. 737-745.

[20] S. Yamazaki, A. Takegawa, Y. Kaneko, J. Kadokawa, M. Yamagata and M. Ishikawa, "An Acidic Cellulose-Chitin Hybrid Gel as Novel Electrolyte for an Electric Double Layer Capacitor," *Electrochemistry Communications*, Vol. 11, No. 1, 2009, pp. 68-70.

[21] K. Prasad, M. Murakami, Y. Kaneko, A. Takada, Y. Nakamura and J. Kadokawa, "Weak Gel of Chitin with Ionic Liquid, 1-Allyl-3-Methylimidazolium Bromide," *International Journal of Biological Macromolecules*, Vol. 45, No. 3, 2009, pp. 221-225.

[22] Y. Wu, T. Sasaki, S. Irie and K. Sakurai, "A Novel Biomass-Ionic Liquid Platform for the Utilization of Native Chitin," *Polymer*, Vol. 49, No. 9, 2008, pp. 2321-2327.

[23] Y. Qin, X. Lu, N. Sun and R. D. Rogers, "Dissolution or Extraction of Crustacean Shells Using Ionic Liquids to Obtain High Molecular Weight Purified Chitin and Direct Production of Chitin Films and Fibers," *Green Chemistry*, Vol. 12, No. 6, 2010, pp. 968-971.

[24] W. T. Wang, J. Zhu, X. L. Wang, Y. Huang and Y. Z. Wang, "Dissolution Behavior of Chitin in Ionic Liquids," *Journal of Macromolecular Science, Part B Physics*, Vol. 49, No. 3, 2010, pp. 528-541.

[25] R. A. A. Muzzarelli, "Biomedical Exploitation of Chitin and Chitosan via Mechano-Chemical Disassembly, Electrospinning, Dissolution in Imidazolium Ionic Liquids, and Supercritical Drying," *Marine Drugs*, Vol. 9, No. 9, 2011, pp. 1510-1533.

[26] M. M. Jaworska, T. Kozlecki and A. Gorak, "Review of the Application of Ionic Liquids as Solvents for Chitin," *Journal of Polymer Engineering*, Vol. 32, No. 2, 2012, pp. 67-69.

[27] A. M. Bochek, A. A. Muravev, N. P. Novoselov, M. Zaborski, N. M. Zabivalova, V. A. Petrova, E. N. Vlasova, B. Z. Volchek and V. K. Lavrentev, "Specific Features of Cellulose and Chitin Dissolution in Ionic Liquids of Varied Structure and the Structural Organization of Regenerated Polysaccharides," *Russian Journal of Applied Chemistry*, Vol. 85, No. 11, 2012, pp. 1718-1725.

[28] K. Kurita, "Controlled Functionalization of the Polysaccharide Chitin," *Progress in Polymer Science*, Vol. 26, No. 9, 2001, pp. 1921-1971.

[29] M. Morimoto, H. Saimoto and Y. Shigemasa, "Control of Functions of α-Chitin and Chitosan by Chemical Modification," *Trends in Glycoscience and Glycotechnology*, Vol. 14, No. 78, 2002, pp. 205-222.

[30] S. Mine, H. Izawa, Y. Kaneko and J. Kadokawa, "Acetylation of Chitin in Ionic Liquids," *Carbohydrate Reserach*, Vol. 344, No. 16, 2009, pp. 2263-2265.

[31] J. B. Zeng, Y. S. He, S. L. Li and Y. Z. Wang, "Chitin Whiskers: A Overview," *Biomacromolecules*, Vol. 13, No. 1, 2012, pp. 1-11.

[32] S. Ifuku and H. Saimoto, "Chitin Nanofibers: Preparations, Modifications, and Applications," *Nanoscale*, Vol. 4, No. 11, 2012, pp. 3308-3318.

[33] S. Ifuku, "Preparation of Chitin Nanofibers from Crab Shell and Their Applications," *Kobunshi Ronbunshu*, Vol. 69, No. 8, 2012, pp. 460-467.

[34] J. F. Revol and R. H. Marchessault, "*In-Vitro* Chiral Nematic Ordering of Chitin Crystallites," *International Journal of Biological Macromolecules*, Vol. 15, No. 6, 1993, pp. 329-335.

[35] J. Li, J. F. Revol, E. Naranjo and R. H. Marchessault, "Effect of Electrostatic Interaction on Phase Separation of Chitin Crystallite Suspensions," *International Journal of Biological Macromolecules*, Vol. 18, No. 3, 1996, pp. 177-187.

[36] J. Li, J. F. Revol and R. H. Marchessault, "Effect of Degree of Deacetylation of α-Chitin on the Properties of Chitin Crystallites," *Journal of Applied Polymer Science*, Vol. 65, No. 2, 1997, pp. 373-380.

[37] J. D. Goodrich and W. T. Winter, "α-Chitin Nanocrystals Prepared from Shrimp Shells and Their Specific Sur- face Area Measurement," *Biomacromolecules*, Vol. 8, No. 1, 2007, pp. 252-257.

[38] Y. Fan, T. Saito and A. Isogai, "Chitin Nanocrystals Prepared by TEMPO-Mediated Oxidation of α-Chitin," *Biomacromolecules*, Vol. 9, No. 1, 2008, pp. 192-198.

[39] Y. Fan, T. Saito and A. Isogai, "TEMPO-Mediated Oxidation of α-Chitin to Prepare Individual Nanofibers," *Carbohydrate Polymers*, Vol. 77, No. 4, 2009, pp. 832-838.

[40] S. Ifuku, M. Nogi, K. Abe, M. Yoshioka, M. Morimoto, H. Saimoto and H. Yano, "Preparation of Chitin Nanofibers with a Uniform Width as α-Chitin from Crab Shells," *Biomacromolecules*, Vol. 10, No. 6, 2009, pp. 1584-1588.

[41] S. Ifuku, M. Nogi, M. Yoshioka, M. Morimoto, H. Yano and H. Saimoto, "Fibrillation of Dried Chitin into 10-20 nm Nanofibers by a Simple Grinding Method under Acidic Conditions," *Carbohydrate Polymers*, Vol. 81, No. 1, 2010, pp. 134-139.

[42] J. D. Schiffman, L. A. Stulga and C. L. Schauer, "Chitin and Chitosan: Transformations Due to the Electrospin-

ning Process," *Polymer Engineering and Science*, Vol. 49, No. 10, 2009, pp. 1918-1928.

[43] R. Jayakumar, M. Prabaharan, S. V. Nair and H. Tamura, "Novel Chitin and Chitosan Nanofibers in Biomedical Applications," *Biotechnology Advances*, Vol. 28, No. 1, 2010, pp. 142-150.

[44] C. Zhong, A. Cooper, A. Kapetanovic, Z. Fang, M. Zhang and M. Rolandi, "A Facile Bottom-Up Route to Self-Assembled Biogenic Nanofibers," *Soft Matter*, Vol. 6, No. 21, 2010, pp. 5298-5301.

[45] C. Zhong, A. Kapetanovic, Y. Deng and M. Rolandi, "A Chitin Nanofiber Ink for Airbrushing, Replica Molding, and Microcontact Printing of Self-Assembled Macro-, Micro-, and Nanostructures," *Advanced Materials*, Vol. 23, No. 41, 2011, pp. 4776-4781.

[46] J. Kadokawa, A. Takegawa, S. Mine and K. Prasad, "Preparation of Chitin Nanowhiskers Using an Ionic Liquid and Their Composite Materials with Poly(Viny Alcohol)," *Carbohydrate Polymers*, Vol. 84, No. 4, 2011, pp. 1408-1412.

[47] Y. S. Vygodskii, E. L. Lozinskaya and A. S. Shaplov, "Ionic Liquids as Novel Reaction Media for the Synthesis of Condensation Polymers," *Macromolecular Rapid Communications*, Vol. 23, No. 12, 2002, pp. 676-680.

[48] K. Kaifu, N. Nishi and T. Komai, "Preparation of Hexanoyl, Decanoyl, and Dodecanoylchitin," *Journal of Polymer Science*: *Polymer Chemistry Edition*, Vol. 19, No. 9, 1981, pp. 2361-2363.

16

Ag, Hg and Cr Precipitation for Recycling Derived of Hazardous Liquid Waste

Silvia C. Gutiérrez-Gutiérrez[1*], Genoveva Rosano-Ortega[1], Carlos Vega-Lebrún[1], Juan Manuel López[1], Pedro Ávila-Pérez[2], Sonia M. Martínez Gallegos[3], Daniel Cruz González[4]

[1]Research Department, Universidad Popular Autónoma del Estado de Puebla (UPAEP), Puebla, México
[2]Technological Research División, Instituto Nacional de Investigaciones Nucleares, Ocoyoacac, México
[3]Division of Graduate Studies and Research, Instituto Tecnológico de Toluca, Toluca, México
[4]Chemical Engineering Faculty, Benemérita Universidad Autónoma de Puebla, Puebla, México

ABSTRACT

Production of different compounds can generate large amounts of hazardous wastes which are dangerous to the environment and human health. The disposal or treatment of hazardous liquid waste rich in heavy metals like silver (Ag), mercury (Hg) and chromium (Cr) is difficult due to the strong acidity and toxicity which usually present in these contaminants. For this study, several research works were reviewed in order to obtain an efficient and viable treatment in time and removal efficiency. A series of chemical precipitations were evaluated for efficiency in the reduction of heavy metals in liquid waste. The precipitation of all three metals lasted 30 minutes and after treatment the wastewater presented concentrations of 0.064 $mg \cdot L^{-1}$ Ag, 0.010 $mg \cdot L^{-1}$ Hg and 0.048 $mg \cdot L^{-1}$ Cr, with a standard pH (7.5 - 8.5); with removal efficiencies of 94.31% for Ag, 99.99% for Hg and 98.17% for Cr.

Keywords: Hazardous Liquid Waste; Treatment for Recycling; Silver; Mercury; Chromium

1. Introduction

Progress in the chemical industry has allowed the development of a wide variety of substances serving as products or raw materials for many goods consumed worldwide.

While these advances have significantly improved the standard of living for many people, they unfortunately exert an important pressure on human health and the environment [1]. The use of chemical substances and dangerous materials in industrial processes or in domestic applications can generate hazardous waste.

Hazardous waste is dangerous or potentially harmful to human and animal health or the environment, and can come in many forms: liquids, solids, gases, or sludges [2]. Hazardous waste can also be defined to possess one or more of the following characteristics: corrosivity, reactivity, explosivity, toxicity, flammability, or containing infectious agents [3]. Therefore, when a residue qualifies as a hazardous waste it must be safely handled and disposed. This can be achieved in different ways: minimization of their generation by recycling and reusing prior to treatment and disposal, treatment which reduces their danger and adequate confinement [1].

An important class of hazardous wastes is those which contain heavy metals. The term "heavy metal" has no strict scientific or chemical definition but it is generally accepted that these metals have a specific gravity more than about 5.0 which are considered as priority pollutants [4].

However, contact with a heavy metal does not necessarily mean that adverse health effects will result in. Health effects depend upon the amount, form and its method of exposure.

Silver (Ag) can be found as hazardous waste in the forms of $AgNO_3$, AgCl, Ag_2S and Ag_2O mixed with soil or water. Ag is used to make jewelry, silverware, electronic equipment, and dental fillings, and photographers use $AgNO_3$ or AgCl to reveal photographs. These materials are the major source of Ag that is released into the environment [5].

Associated health effects with Ag exposure are argyria, argyrosis, abdominal pain, respiratory irrigation, allergic

*Corresponding author.

response and others [6], while in animals it can cause brain damage, abnormal heart size and damage to reproductive tissues [5].

Mercury (Hg) is a metal element that occurs naturally in the environment where we can find it in compounds such as Hg_2Cl_2, $HgCl_2$, $Hg(O_2CCH_3)_2$, $HgSO_4$, CH_3HgCl, $(CH_3)_2Hg$ and $C_8H_8HgO_2$ [7]. Common Hg emissions are due to the burning of fossil fuels, smelting metal ores, Hg mining, industry, waste incinerators and crematoriums [8].

Some health effects from Hg toxicity are neurological damage, irritability, paralysis, blindness, insanity, chromosome breakage and birth defects [7].

Chromium (Cr) is another important heavy metal that is used for chrome plating, dyes and pigments, leather tanning in the form of $Cr_2(SO_4)_3$, and wood preserving in the form of $Cu_2Cr_2O_7$. The highest potential exposure occurs in the metallurgy and tanning industries, where workers may be exposed to high concentrations in air. Cr is known to be a human carcinogen and also can affect the immune, urinary and respiratory systems [9].

The purpose of this study was to test treatment protocols to easily, efficiently, and rapidly reduce the quantity of heavy metals, specifically Ag, Hg and Cr, present in industrial hazardous liquid waste down to levels define as safe by Mexican law. This was accomplished through a series of chemical precipitations that were tested in order to obtain an effluent without hazardous characteristics that would harm the environment or human/animal health, and could be easily disposed.

2. Experimental

Several research and methodologies that offer a treatment for Ag, Hg or Cr were reviewed in order to formulate an improved method that would treat this hazardous waste.

Theexperiments were conducted with a sample obtained from the mixture of the close-reflux COD (Chemical Oxygen Demand) waste and industry waste-water rich in heavy metals, both from the Green Chemistry laboratory at the Universidad Popular Autónoma del Estado de Puebla.

A 1 L sample was tested for baseline measurements. The analysis of the sample was performed by atomic absorption spectrophotometry and showed concentrations of 1.12 $mg \cdot L^{-1}$ Ag, 5171 $mg \cdot L^{-1}$ Hg and 2.72 $mg \cdot L^{-1}$ Cr. The pH values were below 0.01 at room temperature and pressure conditions.

Prior to the experiments, the wastewater sample was filtered to eliminate suspended solids that could interfere with the reactions. An apparatus for vacuum filtration was installed using a glass Büchner funnel (Kimble & Chase, no. 284003503, fine porosity) with glass microfiber filter paper (Whatman, 934-AH 24 mm∅, pore size 1.5μm) and a Kitasato flask. An evaporation of the 10% of the volume of the liquid was performed to concentrate the sample.

Due to the metals contained in the wastewater, the selected precipitation reactions were performed in the following order to increase the selectivity of the reactions and avoid interferences or unwanted reactions: Ag, Hg, and Cr [10].

2.1. Ag Precipitation, AgCl

2.00 $g \cdot L^{-1}$ of NaCl (J.T Baker CAS: 7647-14-5, 99.9% purity) was added to the sample, keeping it in constant agitation for 10 minutes using a 1.5 cm magnetic stirrer and a stirring hot plate (Thermolyne model SP131325) at medium speed. Two consecutive precipitations were required to precipitate the high quantity of Ag in the waste. The solution was filtered in vacuum using a Gooch crucible and glass microfiber filter paper (1.5 μm).

Forreactions 1 to 6 reported in this paper, a parenthesis nomenclature is used to indicate the reaction between different compounds with the wastewater and others assistant reactants. Reaction 1 describes the Ag precipitation into AgCl.

$$Ag_2SO_{4(aq)} + 2NaCl \rightarrow 2AgCl\downarrow + Na_2SO_{4(aq)} \quad (1)$$

2.2. Hg Precipitation, HgS

A sulphidation device was installed in a fume hood to avoid accidental exposure to the gases produced in the reaction.The remaining liquid from the previous cyclewas placed in a flask in a water bath with constant stirring using the Thermolyne plate. A second flask was set with 50 g of FeS (Golden Bell no. 28610, 99.9% purity) and 300 mL of HCl (37.3%, J.T Baker CAS: 764701-0). The two flasks were connected to a scrubber system, which was filled with distilled water to 1/3 of the volume.

The solutions were left in contact for 15 minutes and the solution from the first flask was processed with vacuum filtration as described for the AgCl precipitate. The precipitation steps are shown in Reactions 2, 3, and 4.

$$FeS + 2HCl_{(aq)} \rightarrow H_2S\uparrow + FeCl_{(aq)} \quad (2)$$

$$HgSO_{4(aq)} + H_2S_{(g)} \rightarrow HgS\downarrow + H_2SO_{4(aq)} \quad (3)$$

$$HgCl_{2(aq)} + H_2S_{(g)} \rightarrow HgS\downarrow + 2HCl_{(l)} \quad (4)$$

2.3. Cr Precipitation, $Cr(OH)_3$

The samplewas diluted to a concentration of 1:2 with distilled water. To increase the alkalinity, 30 $mg \cdot L^{-1}$ of Na_3PO_4 (Golden Bell no. 26497, 99.9% purity) was added to the solution remaining from the Hg precipitate process. The pH of this mixture was adjusted dropwise with a 50% solution of NaOH in distilled water (pellets,

J.T Baker CAS: 1310-73-2, 98.4% purity) to obtain a pH 8.5 ± 0.5, the optimum value for the formation of $Cr(OH)_3$.

$$Cr_2(SO_4)_{3(aq)} + 6NaOH_{(aq)} \rightarrow 2Cr(OH)_3 \downarrow + 3Na_2SO_{4(aq)} \quad (5)$$

The solution was centrifuged (IEC HN SII Centrifuge, Thermo Electron) for 60 minutes at 7500 rpm to separate the phases. The solid phase form was stabilized with H_2SO_4 (J.T Baker CAS: 7664-93-0, 97.9% purity), to obtain $Cr_2(SO_4)_3$.

$$2Cr(OH)_3 + 3H_2SO_{4(aq)} \rightarrow Cr_2(SO_4)_3 + 6H_2O_{(aq)} \quad (6)$$

2.4. Assays

Metal assays for all runs were performed by atomic absorption (AA) spectrophotometry according to the Mexican standard NOM-117-SSA1-1994 [11], which establishes the method for determining metals by atomic absorption spectrometry. The tests were performed in a certified laboratory by the Mexican Accreditation Entity, where five replicates were carried out to obtain an accurate and precise value with a maximum 5% deviation.

The degree of purity of the precipitated metal salts was determined by X-ray fluorescent (XRF) spectrometry. The fluorescent X-rays derived from the sample were detected with a solid-state lithium-drifted silicon detector of 20 mm^2 front area and cooled with liquid nitrogen. The energy resolution of the Si(Li) detector was 140 eV for Mn Kα and its Be window was 8 μm thick (Ital-Structures, 2000). The signal from the detector was processed in a counting chain (preamplifier, amplifier, analogue to digital converter) attached to the digital system. All samples were excited for 500 s.

With the purpose of confirming the chemical stability of the precipitated salts, a CRIT analysis was made in the National Institute of Nuclear Investigations (ININ from its acronym in Spanish); this procedure is composed of tests for corrosiveness, reactivity, flammability and toxicity (CRIT). CRIT analysis was performed under the Mexican standards NOM-052-SEMARNAT-2005 [12] and NOM-053-SEMARNAT-1993 [13] which establish the procedure to determinate if waste possesses any hazardous characteristics.

3. Results and Discussion

Table 1 shows a comparison of the heavy metal concentrations at the beginning and end of the treatment process with the contamination limits established in the relevant standards. The results indicate that the remaining liquid meets with the Mexican standards, NOM-001-SEMARNAT-1996 [14] and NOM-002-SEMARNAT-1996 [15], which establish the maximum permissible limits of pollutants in dischargesto sewer systems as well as in national waterways, in terms of pH and Ag, Hg and Cr concentrations.

These results (**Table 1**) indicates that the treatment process offers a removal efficiency of 94.31% for Ag, 99.99% for Hg and 98.17% for Cr, as well as acceptable pH, generating an effluent that would not need further treatment before being safely and legally dispose.

Comparison of the previous results shows that we obtained excellent outcomes precipitating over 94% of the metal all three metals tested. The duration of treatment for Ag, Hg and Cr was a total time of 30 minutes. Duration is an advantage over other treatments [17-19] that last more than 3 hours which made them not as economically viable for industries to apply. **Table 2** shows a comparison among several different treatments for Ag, Hg and Cr.

Ag precipitation with NaCl was employed by Mañunga *et al.*, Agudelo *et al.* and in this research, different quantities are used but NaCl is an excellent reagent to precipitate Ag in a short time with good removal efficiencies.

According to Mañunga *et al.* Hg precipitation is with 10 $g \cdot L^{-1}$ of FeS in constant agitation for 1h. Agudelo *et al.* propose Hg precipitation with NaOH with a contact time of 2 h while in the treatment procedure developed in this research it can be achieved with fewer reagents in 15

Table 1. Comparison between heavy metal concentrations at the beginning and end of the treatment process.

Parameter	Sample of wastewater	Sample of the remaining liquid[a]	MPL[b]
		mg/L	
Ag	1.124	0.064	5.0
Hg	5171	0.010	0.01
Total Cr	2.724	0.048	0.5
		pH units	
pH	0.01	8.5	$5.5 \leq pH \leq 10$

[a]Results obtained by atomic absorption with a 5% deviation; [b]Maximum Permissible Limits set forth by Mexican law, NOM-001-SEMARNAT-1996 and NOM-002-SEMARNAT-1996.

Table 2. Comparison of different treatments for the removal of Ag, Hg and Cr in liquids waste.

Author	Treatment time	Removal %		
		Ag	Hg	Cr
M. M. Isla [16]	30 min	94.31	99.99	98.08
Bittar *et al.* [17]	6 months 24 h. 15 min	99.78	99.23	99.98
Mañunga *et al.* [18]	3 h. 10 min	99.99	99.99	99.33
A. M. Agudelo [19]	3 h. 15 min	99.98	99.98	99.99

minutes. As presented in Reaction 3 in this paper, H_2S is produced in order to Hg precipitate in the form of HgS. H_2S is a flammable and colorless gas with a characteristic smell of rotten eggs and it is known as a pollutant that affects human health [20]. However, H_2S contamination during the treatment does not occur because H_2S is contained by the sulphidation device and is produced in necessary and sufficient quantities, according to stoichiometric calculation, to react with Hg. On a larger scale, further precise stoichiometric calculations will be required to secure the correct amount of H_2S that is produce agreeing with the amount of Hg in waste. Security measures can also be introduce in case of a H_2S leak by adapting a removal system of this gas, number of physicochemical and biological methods are available for remove H_2S from gas streams [21,22].

For Cr precipitation, Mañunga*et al.* made a dilution of the sample but at a concentration 1:4 and used a 50% solution of NaOH after they reduced Cr^{6+} with glucose in 1h. In the method proposed in our current research an excellent removal percentage was reached with a lower dilution to maintain a proper concentration of Cr, so the reaction with the 50% solution of NaOH could precipitate as much Cr possible. No preview reduction of Cr^{6+} with glucose were needed, reduction of Cr^{6+} to Cr^{3+} occurred during Hg precipitation by using H_2S [23].

To appreciate the efficiency of the precipitation method developed in this research, the salts we obtained were analyzed by X-ray fluorescence (XRF) spectrometry to determine their purity (**Table 3**).

The AgCl and HgS samples had a purity of 89.41% and 90.8%, respectively. The $Cr(OH)_3$ precipitate obtained had a high concentration of iron, sulfur, chlorine and sodium, which were presumed to be unreacted precipitant reagents due to the presence of other pollutant substances in the wastewater sample that interfered with the efficiency of the reaction. The precipitated salts presented a chemical stability based on their corrosivity, reactivity, flammability and toxicity characteristics, which facilitates their management and storage or disposal.

Table 3. Analysis results of the elements present in the precipitated solids of the reactions.

	Quantity of the elements present in each precipitated salt (g)		
Elements	**AgCl**	**HgS**	**$Cr(OH)_3$**
Ag	0.95	-	-
Cl	0.57	0.26	0.81
Hg	-	11.24	-
S	-	4.11	13.92
Fe	-	-	44.39
Na	-	-	12.95
Cr	-	-	1.63
Other elements	0.18	1.49	1.60

For each liter of treated wastewater we obtained approximately 0.5, 2.5 and 150 cm^3 of AgCl, HgS and $Cr(OH)_3$ respectively, therefore the volume of hazardous substances to be managed was reduced by approximately 85%.

During the precipitation of the metals we observed several changes of color; this was an important indicator because it signaled that the reactions were satisfactorily completed. In the precipitation of Ag, there was a change of color from orange-amber to light orange and the formation of a white precipitate, from 0.00112 gthe recovery of Ag was 1.7 g as AgCl. In the Hg precipitation we appreciated color changes in the solution from green to orange ending with a blue-green hue. The generated solid changed from white to black, indicating the end of the reaction and the recovery of Hg from 5.171 g resulted in 19.0 g in the form of HgS. The precipitation of Cr presented a color change from blue-green to orange-brown and ended with very dark green hue. A dark green gelatinous solid precipitate was formed during the procedure; the recovery of Cr from 0.00272 g was 75.3 g as $Cr(OH)_3$.

4. Conclusions

A simpler and economically-viable industrial wastewater treatment was achieved that removed heavy metal concentrations using common reactants, materials and equipment in a 30 minute sequential reaction. Ag, Hg and Cr present in wastewater were reduced to levels below those required by law [14,15] rendering safe discharge in sewers possible. A removal efficiency of 94.31% for Ag, 99.99% for Hg and 98.17% for Cr was achieved with a pH of 8.5. The effluent obtained from the treatment didn't exhibit any significant danger to human health or the environment with respect to corrosivity, reactivity, explosivity, toxicity, flammability or infectious agents. The volume of substances that must be stored and/or disposed was reduced by approximately 85%, and their purity and stability also means they could be reused by other processes.

This treatment method is been scaled up to an industrial process in the laboratory of Sistemas de Ingeniería Ambiental (SIA) S.A. de C.V. The method will treat the hazardous liquid waste obtained from the determination of Chemical Oxygen Demand parameter in the municipal wastewater treatment plant of the city Puebla, Pue, Mexico. This method will be used on this industrial scale to test the removal of Ag, Hg and Cr as hazardous waste and whether the precipitates can be reused as raw material in other industrial process.

It is noteworthy that the method development can be applied to other types of liquid waste containing Ag, Hg or Cr, as each one of the stages are sequential, but independent. Several preliminary trials were performed in other wastewater which contained Ag, Hg and/or Cr in order to prove that this treatment can apply to industrial wastewater, and the preliminary results showed the expected precipitates. We believe the green chemistry potential of this process is great, and warrants investment in future study.

5. Acknowledgements

The authors thank the Research Department for financial support and Integrated Technology University Services of UPAEP for the technical work of the different characterization techniques. Also thank Isla-López, M for her contribution in this work with her graduate thesis and to SIA S.A. de C.V. enterprise for allowed us to use their laboratory.

REFERENCES

[1] Secretariat of Environment and Natural Resources (SEMARNAT in Spanish), "Chapter 7. Residues," *Report of the State of the Environment in Mexico*, Mexico, 2009, pp. 326-357. http://www.semarnat.gob.mx/informacionambiental/Documents/pdf/cap_7_residuos.pdf

[2] US Environmental Protection Agency, "Hazardous Waste," 2012. http://www.epa.gov/osw/hazard/

[3] Secretariat of Environment and Natural Resources (SEMARNAT in Spanish), "General Law for the Prevention and Management of Waste," Federal Ecological Agenda, ISEF Editorial, Mexico, 2010.

[4] US: Environmental Protection Agency, "Priority Pollutants," 2012. http://water.epa.gov/scitech/methods/cwa/pollutants.cfm

[5] Agency for Toxic Substances and Disease Registry (ATSDR), "Public Health Statement Silver," Department of Health and Human Services, Atlanta, 1990.

[6] P. L. Drake and K. J. Hazelwood, "Exposure-Related Health Effects of Silver and Silver Compounds: A Review," Oxford University Press, Vol. 9, No. 7, 2005, pp. 575-585.

[7] Agency for Toxic Substances and Disease Registry (ATSDR), "Public Health Statement Mercury," Department of Health and Human Services, Atlanta, 1999. http://www.atsdr.cdc.gov/ToxProfiles/tp46-c1-b.pdf

[8] T. W. Clarkson, "Mercury: Major Issues in Environmental Health," *Environmental Health Perspectives*, Vol. 100, 1993, pp. 31-38.

[9] Agency for Toxic Substances and Disease (ATSDR), "Public Health Statement Chromium," Department of Health and Human Services, Atlanta, 2012. http://www.atsdr.cdc.gov/ToxProfiles/tp7-c1-b.pdf

[10] R. L. Rangel, "Fundamentals of Analytical Chemistry," 2nd Edition, Limusa, Mexico City, 1991, pp. 69-92.

[11] Ministry of Health (SSA in Spanish), "Test Method for the Determination of Cadmium, Arsenic, Lead, Tin, Copper, Iron, Zinc and Mercury in Food, Drinking Water and Purified Water by Atomic Absorption Spectrometry," Official Journal of the Federation, Mexico, 1994. http://200.77.231.100/work/normas/noms/1995/117-ssa1.pdf

[12] Secretariat of Environment and Natural Resources (SEMARNAT in Spanish), "Establishes the Characteristics of Hazardous Waste and the Listing of Them and the Threshold above Which Make a Waste Hazardous Because Its Toxicity to the Environment," Official Journal of the Federation, Mexico, 2005. http://www.aguascalientes.gob.mx/proespa/pdf/NOM-SEMARNAT-052%20RESIDUOS%20PELIGROSOS.pdf

[13] Secretariat of Environment and Natural Resources (SEMARNAT in Spanish), "Establishes the Procedure for Conducting the Extraction Test to Determine Constituents That Make a Waste Hazardous Because Its Toxicity to the Environment," Official Journal of the Federation, Mexico, 1993. http://www.spabc.gob.mx/views/files/tmp/NOM-052.pdf

[14] Secretariat of Environment and Natural Resources (SEMARNAT in Spanish), "Establishes the Maximum Permissible Limits of Pollutants in Wastewater Discharges into National Waters," Official Journal of the Federation, Mexico, 1996. http://www.hgm.salud.gob.mx/descargas/pdf/noticias/programa_mercurio/marco/norma_001.pdf%20

[15] Secretariat of Environment and Natural Resources (SEMARNAT in Spanish), "Establishes the Maximum Permissible Limits of Pollutants in Wastewater Discharges to Sewers," Official Journal of the Federation, Mexico, 1996. http://www.aguascalientes.gob.mx/proespa/pdf/NOM-SEMA-NAT-002%20DESCARGA%20DE%20AGUAS%20RESIDUALES%20AL%20ALCANTARILLADO.pdf

[16] M. M. I. López, "Stabilization and Disposal of Liquid Waste from the Determination of Chemical Oxygen Demand (in Spanish)," Bachelor Thesis, UPAEP University, Mexico, 2012.

[17] M. Bittar, S. Llamas M. C. Quiroga and L. E. Ferrer, "Treatment and Disposal of Liquid Waste Generated in the Chemical Oxygen Demand (in Spanish)," *3rd Iberoamerican Waste Engineering Seminar*, Castellón de la Plana, 2010.

[18] T. Mañunga, H. M. Gutiérrez, J. A. Rodríguez and A. VillarealDíaz, "COD Waste Treatment Generated in Environmental Laboratories Engineering and Research (in Spanish)," *Redalyc*, Vol. 30, No. 2, 2010, pp. 87-95.

[19] A. M. Agudelo, "Proposal for Treating Mercury, Chromium and Silver Metals Present in the Waste Generated by the Chemical Oxygen Demand," Bachelor Thesis, Pereira University, Colombia, 2010.

[20] Agency for Toxic Substances and Disease (ATSDR), "Hydrogen Sulfide," Department of Health and Human Services, Atlanta, 2006.

[21] T. J. Bandosz, "On the Adsorption/Oxidation of Hydrogen Sulfide on Activated Carbons at Ambient Temperatures," *Journal of Colloid and Interface Science*, Vol. 246, No. 1, 2002, pp.1-20.

[22] S. M. Zicari, "Removal of Hydrogen Sulfide from Biogas

Using Cow-Manure Compost," Master Thesis, Cornell University, New York, 2003.

[23] C. Kim, Q. Zhou, B. Deng, E. C. Thornton and H. Xu, "Chromium(IV) Reduction by Hydrogen Sulfide in Aqueous Media: Stoichiometry and Kinetics," *Environmental Science and Technology*, Vol. 35, No. 11, 2001, pp. 2219-2225.

17

Green Chemistry Approach to Fast and Highly Efficient One-Pot Synthesis of Bis-Isoxazolyl-1,2,5,6-Tetrahydro Pyridine-3-Carboxylates

Eligeti Rajanarendar*, Kammari Thirupathaiah, Saini Rama Krishna, Baireddy Kishore
Department of Chemistry, Kakatiya University, Warangal, India

ABSTRACT

The synthesis of the bis-isoxazolyl-1,2,5,6-tetrahydro pyridine-3-carboxylates is achieved in high yield without the production of toxic waste products by using room temperature ionic liquid (RTILs) triethyl ammonium acetate (TEAA). The RTIL TEAA played the dual role of efficient green solvent as well as recyclable catalyst.

Keywords: Bis-Isoxazolyl-1,2,5,6-Tetrahydro Pyridine-3-Carboxylates; Bis-Isoxazolyl-1,2,5,6-Tetrahydro Pyridine-3-Carboxylates; Green Synthesis

1. Introduction

Recently, there has been an increasing interest in the development of cleaner technologies to replace hazardous organic solvents with environmental benign solvents. Nowadays green chemistry emphasizes the optimization of synthetic methodologies to reduce pollution, cost and tedious work-ups. This new challenge has led to a growing interest in the field of organic synthesis. Among the several aspects of green chemistry, the reduction/replacement of volatile organic solvents from the reaction medium is of utmost importance. The use of a large excess of conventional volatile solvents required to conduct a chemical reaction creates ecological and economic concerns. The search for a nonvolatile and recyclable alternative solvent is thus holding a key role in this field of research. The use of fused organic salts, consisting of ions, is now emerging as a possible alternative. A proper choice of cations and anions is required to achieve ionic salts that are liquids at room temperature and are appropriately termed room temperature ionic liquids (RTILs). Common RTILs consists of *N,N'*-dialkylimidazolium, alkylammonium, alkylphosphonium or *N*-alkylimidazolium as cations [1]. These innovative fluids are organic salts, whose cation, anion and the alkyl chain attached to organic cation can be varied to change their chemical and physical properties as per the needs of the chemical process. A key feature of these liquids is their intrinsic ability to solvate a wide array of organic and inorganic substrates. They are blessed with qualities such as incredibly large liquidous range, negligible vapour pressure and recyclability [2]. In short, these ionic liquids have enlightened the way to environmentally benign procedures (3). The development of cleaner technologies is a major emphasis in green chemistry.

Several compounds consisting of reduced pyridine, like the 1,2,3,6-tetrahydropyridine **1** ring system is known to exhibit a variety of biological activities [4,5]. 1-Methyl-4-phenyl-1,2,3,6-tetrahydropyridine (MPTP) **2** was recently found to be neurotoxic, causing persistent parkinsonism in humans and other animal species [6-9]. Arecoline **3**, an alkaloid containing tetrahydropyridine ring system obtained from the betel nut (*Areca catechu*), the fruit of a palm tree, has been used previously as a leading centrally active muscarinic agent [10]. 3-Methoxy-5-methyl-4,5,6,7-tetrahydroisoxazolo[4,5-*c*]-pyridinium **4** [11-13] and 3-methoxy-5-methyl-6,7-di-hydro-4*H*-thiopyrano[3,4-*d*]isoxazol-5-ium **5** [14] salts are muscarinic agonists and partial agonists bio iso-sterically derived from arecoline (**Figure 1**). Biological activity of substituted isoxazoles has made them a focus of medicinal chemistry over the years. Spirocyclic isoxazoles such as aerothionin, aerophobin-1, and zamamistatin are biologically active alkaloids [15]. Isoxazoles are potent analgesic, anti-inflammatory [16], antimicrobial [17], COX-2

*Corresponding author.

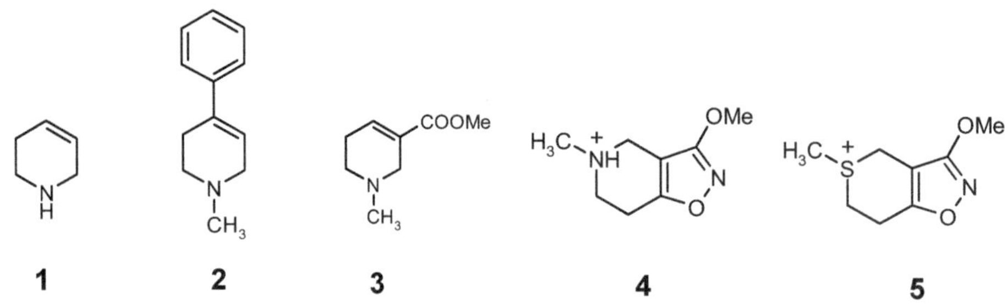

Figure 1. Biologically active tetrahydropyridines and arecoline analogs.

inhibitory [18], antitubercular [19], anticonvulsant [20], and anticancer agents [21]. In continuation of our studies on the synthesis of biologically important heterocyclic compounds [22-26], in the present work we have designed the synthesis of a novel tetrahydropyridine derivatives incorporated with isoxazole moiety utilizing an eco friendly, reusable and inexpensive ionic liquid as solvent as well as a catalyst.

2. Results and Discussion

Literature search on the synthesis of tetrahydropyridines provide only few previous reports which utilizes the primary amine, aromatic aldehydes and β-diketones [27-34] for their synthesis. Therefore, we have further investigated this molecule of interest using greener approach and in search of better alternative over reported methods. Triethyl ammonium acetate (TEAA) ionic liquid (IL) has been found to be effective in aza/thia Michael reactions [35]. We report for the first time eco-friendly method for the synthesis of bis-isoxazolyl-1,2,5,6-tetrahydro pyridine-3-carboxylates by using cost effective TEAA ionic liquid (**Scheme 1**). To optimize the reaction condition, initial studies were conducted on the reaction of the isoxazole amine **1**, aromatic aldehyde **2** and β-ketone ester **3** as a model reaction in the presence of different catalysts and solvents. The reaction afforded bis-isoxazolyl-1,2,5,6-tetrahydro pyridine-3-carboxy-lates **4** in 30% - 75% yields at 60°C (**Table 1**).

We further explored the reaction in the presence of TEAA ionic liquid. We started the five component reaction by examining the effect of TEAA ionic liquid on yield and time of the typical reaction of isoxazole amine **1**, aromatic aldehyde **2** and β-ketone ester **3** at room temperature. To our surprise, the use of TEAA ionic liquid decreased reaction time, temperature and resulted in improved yields. We also carried out the reaction in the presence of different ILs at room temperature. The results are summarized in **Table 2**. Among the ILs examined TEAA proved to be the most effective in terms of yields, reaction time, and more recyclability. Utilization of other ILs was found to be quite unsatisfactory.

Having established that, the best solvent and catalyst is TEAA for this transformation, the another advantage is,

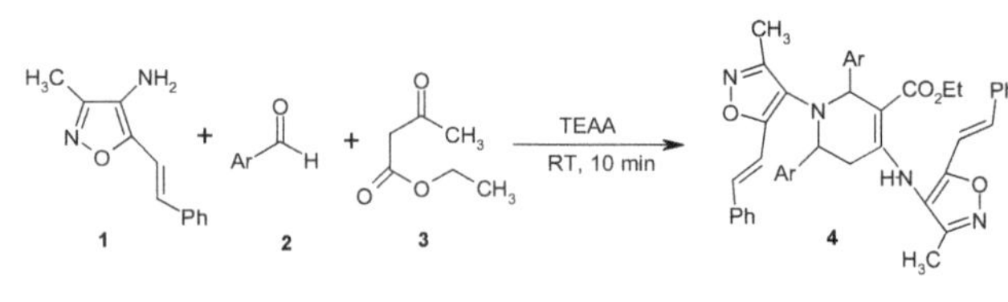

Scheme 1. Synthesis of bis-isoxazolyl-1,2,5,6-tetrahydro pyridine-3-carboxylates by using TEAA IL.

Table 1. Effect of various catalysts and solvents on the synthesis of bis-isoxazolyl-1,2,5,6-tetrahydro pyridine-3-carboxylates.[a]

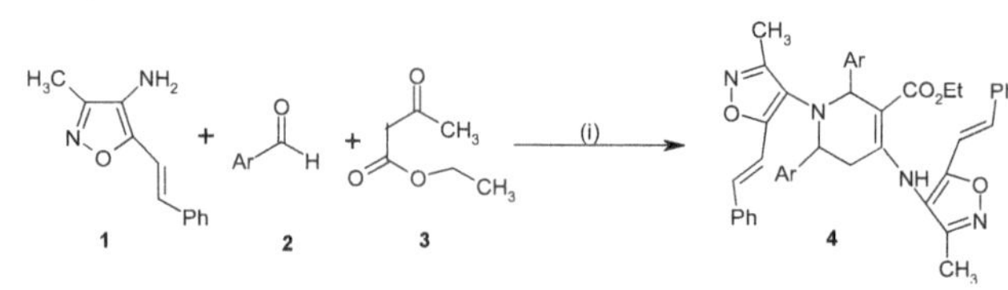

Entry	Catalyst (10 mol%)	Solvent	Time (min)	Yield[b] (%)
1	CAN	THF	30	75
2	Silica gel	EtOH	50	60
3	I_2	CH_3CN	46	50
4	$InCl_3$	CH_3CN	80	42
5	L-proline + TFA	CH_3CN	74	45
6	*p*TSA	DMSO	55	48
7	$ZnCl_2$	EtOH	45	30

[a]Reaction conditions: (i) isoxazole amine **1** (2 mmol), aldehyde **2** (2 mmol), EAA **3** (1 mmol) and catalyst 10 mol% in solvent (10 mL), 60°C. [b]Isolated yields.

Table 2. Optimization of reaction condition by using different ILs.[a]

Entry	IL	Time (min)	Yield[b] (%)
1	TEAA	10	98
2	[bmIm]BF_4	20	45
3	[bmIm]OH	45	20
4	[bmIm]Br	35	15
5	[HMIm]BF_4	60	65

[a]Reaction conditions: isoxazole amine 1 (2 mmol), aldehyde 2 (2 mmol), EAA 3 (1 mmol) and IL (5 mL), RT. [b]Isolated yields.

it can be easily recovered after the completion of the reaction and can be reused in subsequent runs (five runs) without loss of efficiency and with negligible loss of IL (**Table 3**). The products were easily separated by simple extraction with ether and characterized by IR, ^{1}H NMR, ^{13}C NMR and mass spectroscopy and also by analytical data.

To investigate the scope of TEAA IL under the optimized reaction conditions, a number of bis-isoxazolyl-1, 2,5,6-tetrahydro pyridine-3-carboxylate derivatives were synthesized and the results are summarized in **Table 4**. In all the cases, the products were obtained in excellent yields. The great advantages of the present procedure using TEAA in the synthesis of bis-isoxazolyl-1,2,5,6-tetrahydro pyridine-3-carboxylates are the mild conditions, high yields, short reaction times, simple experimental set-up and workup. Besides this, it is equally efficient and is compatible with different functional groups (electron releasing and electron attracting groups) and the approach proved to be of general applicability.

The plausible mechanism for the synthesis of bis-isoxazolyl-1,2,5,6-tetrahydro pyridine-3-carboxylates involves the formation of imine (**5**) and enamine (**6**) initially in the presence of IL. Then the enamine makes a nucleophilic attack on to imine, to form a Mannich type intermediate (**7**). This Mannich type intermediate (**7**) reacts with aldehyde to form (**8**) with the elimination of water molecule with the influence of IL. **8** undergoes spontaneous tautomerization and converts into (**9**). The tautomer (**9**) undergoes intramolecular Mannich type reaction to form intermediate (**10**), which tautomerises spontaneously to form bis-isoxazolyl-1,2,5,6-tetrahydro pyridine-3-carboxylate (**4**) (**Scheme 2**).

3. Conclusions

In conclusion, the present procedure that uses ammonium ionic liquid TEAA provides a fast and highly efficient methodology for the synthesis of bis-isoxazolyl-1,2, 5,6-tetrahydro pyridine-3-carboxylates. The significant improvements noticed in this reaction are high yields, reduction in reaction time and avoids the usage of hazardous organic solvents. These reactions were clean, and no by-products were detected under these conditions. To the best of our knowledge, this happens to be the first report for the construction of tetrahydropyridine ring by using isoxazole amine, aldehyde and EAA with TEAA ionic liquid, in which ionic liquid playing the dual role of solvent as well as catalyst. Moreover, this work clearly demonstrates that the potential of room temperature ionic liquid TEAA, to act as an efficient and recyclable catalyst, green solvent and shows much promise for future applications.

Table 3. ***Recycling of IL TEAA.***

Run	IL recycled (in mL)	Yield[a] (%)
1	5	97
2	5	97
3	4.8	95
4	4.6	94
5	4.2	93

[a]Isolated yields.

The authors are thankful to the Head, Department of Chemistry, Kakatiya University, Warangal for providing the facilities, the Director, Indian Institute of Chemical Technology, Hyderabad for recording ^{1}H NMR, and Mass Spectral data. B. Kishore is thankful to UGC, New Delhi, for the award of a fellowship (JRF).

4. Experimental Part

All the melting points were determined on a Fisher-Johns melting point apparatus and are uncorrected. Analytical TLC was performed on Merck precoated 60 F_{254} silica gel plates. Visualization was done by exposure to iodine vapor. IR spectra (KBr pellet) were recorded on a Perkin Elmer BX series FT-IR spectrometer. ^{1}H NMR spectra were recorded on a Varian Gemini 300 MHz spectrometer. ^{13}C NMR spectra were recorded on a Bruker 75 MHz spectrometer. Chemical shift values are given in ppm (δ) with tetramethyl silane as an internal standard. Mass spectral measurements were carried out by EI method on a Jeol JMC-300 spectrometer at 70 eV. Elemental analyses were performed on a Carlo Erba 106 and Perkin Elmer model 240 analyzers.

The IL TEAA was prepared according to the reported procedure [36-38].

4.1. Green Chemistry Approach of Fast and Highly Efficient One-Pot Synthesis of Bisisoxazolyl-1,2,5,6-Tetrahydro Pyridine-3-Carboxylates; Typical Procedure

4-Amino-3-methyl-5-styrylisoxazole **1** (1 mmol), benzaldehyde **2a** (1 mmol), and ethyl aceto acetate **3** (0.5 mmol) were taken in IL TEAA (5 mL), and then reaction mixture was stirred at room temperature for 10 min. After the completion of the reaction (monitored by TLC), the reaction mixture was extracted with diethyl ether. The combined layers was separated and dried over anhyd. Na_2SO_4, and evaporated under reduced pressure to afford the crude product **4a**. The ionic liquid left over in the reaction was washed with ethyl acetate and dried at 80˚C under vacuum and was reused for 5 consecutive runs. This procedure was followed for all the reactions listed in the **Table 4**.

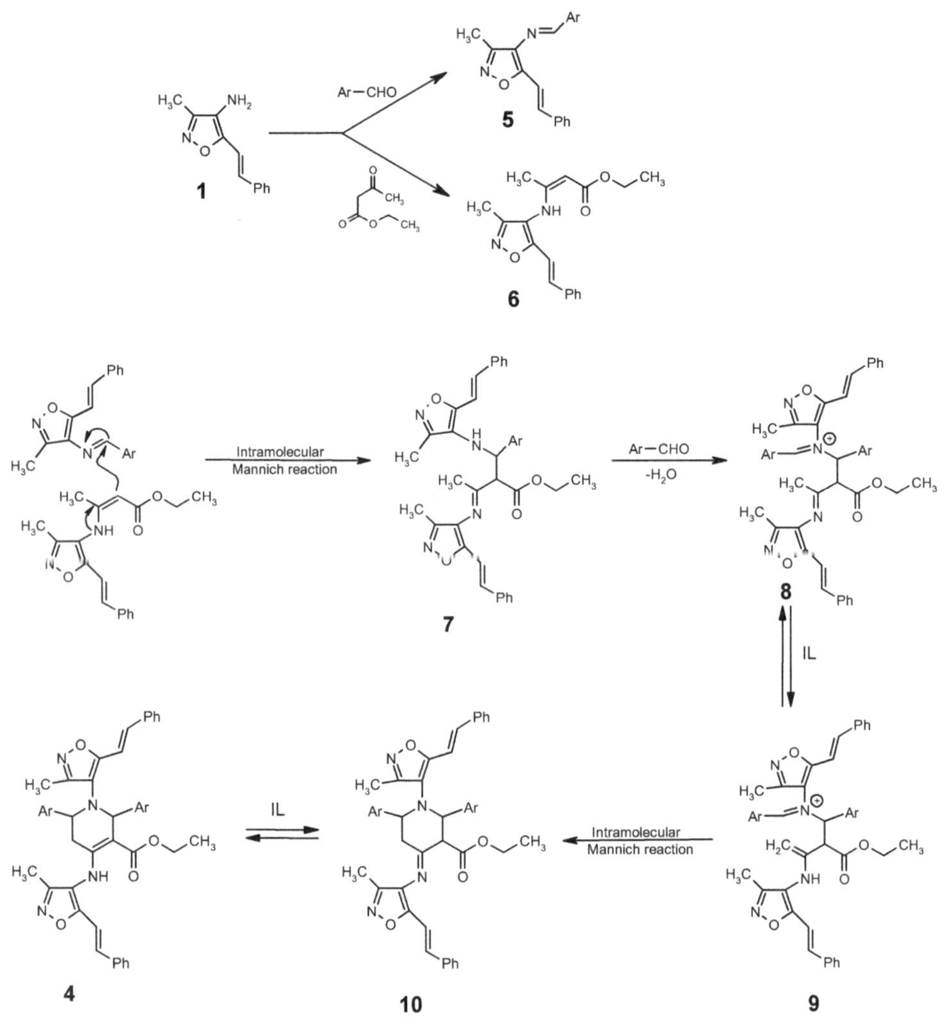

Scheme 2. Plausible mechanism for the formation of bis-isoxazolyl-1,2,5,6-tetrahydro pyridine-3-carboxylates.

Table 4. Synthesis of bis-isoxazolyl-1,2,5,6-tetrahydro pyridine-3-carboxylates by using TEAA IL.[a]

Entry	Isoxazole amine **1**	Aldehyde **2**	Product **4**	Time (min)	Yield[b] **4** (%)
1	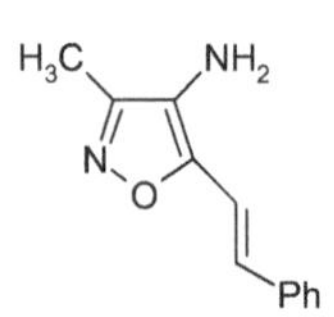 **1**	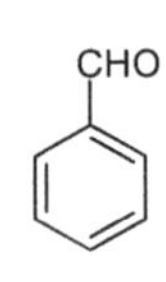 **2a**	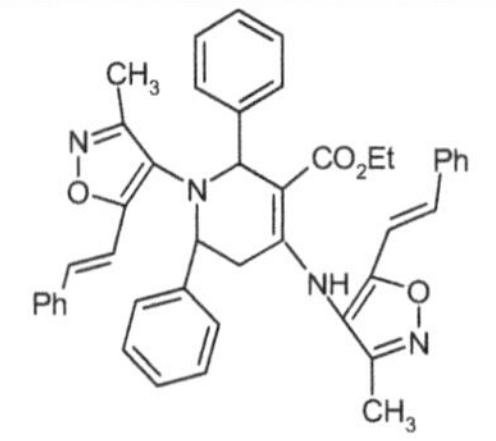 **4a**	10	**4a** (96)
2	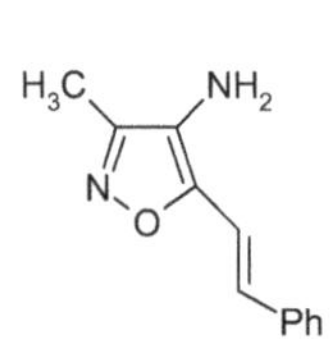 **1**	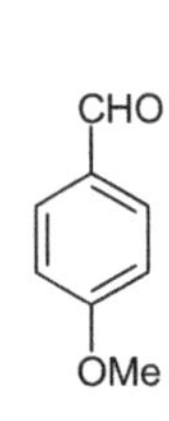 **2b**	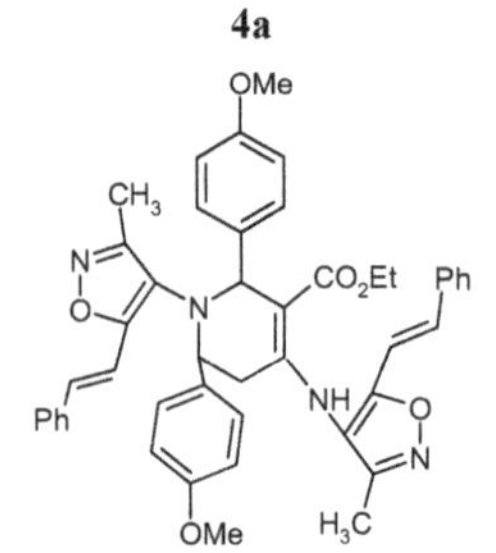 **4b**	8	**4b** (98)

Continued

3	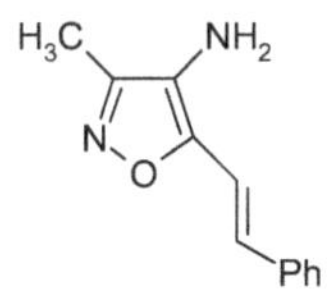	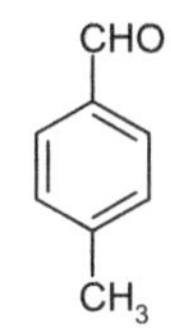	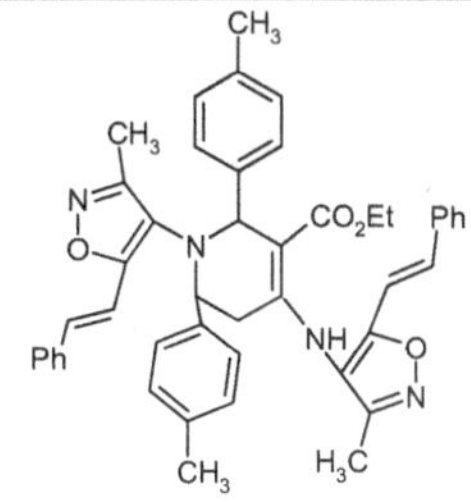	9	**4c** (95)
	1	**2c**	**4c**		
4	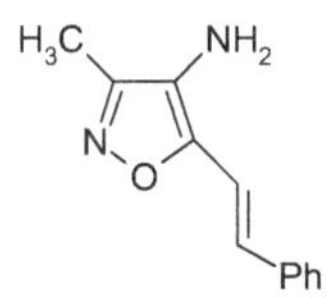	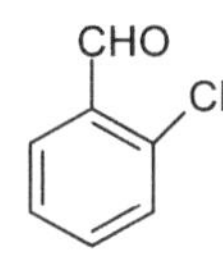	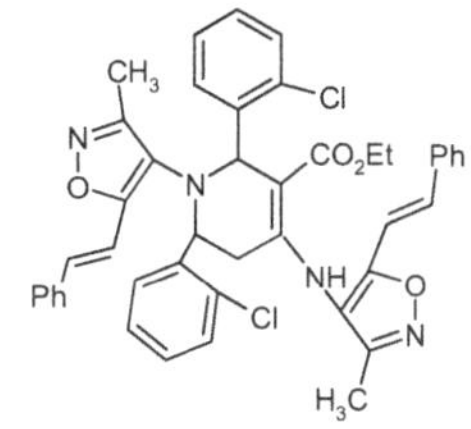	10	**4d** (96)
	1	**2d**	**4d**		
5	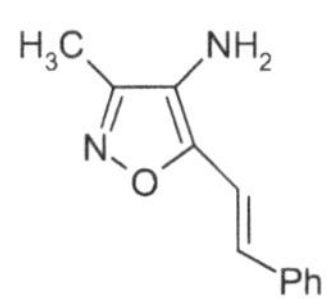	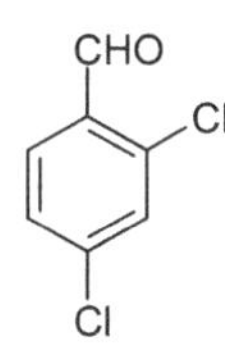	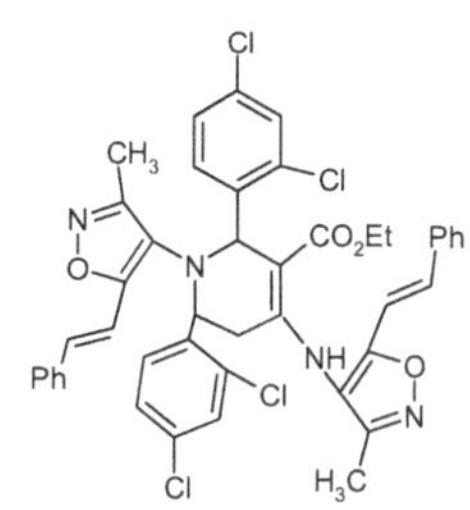	9	**4e** (94)
	1	**2e**	**4e**		
6	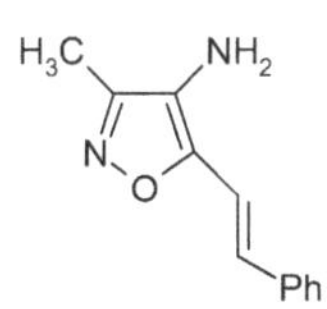	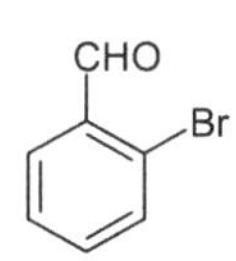	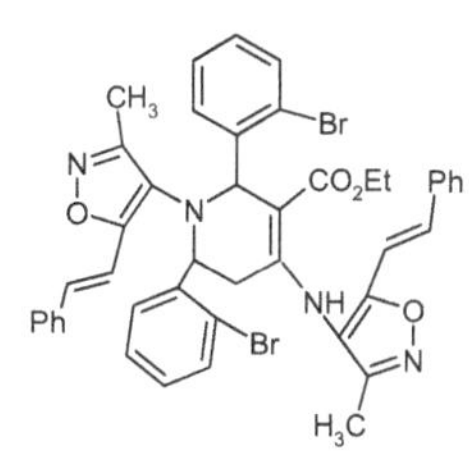	10	**4f** (94)
	1	**2f**	**4f**		
7	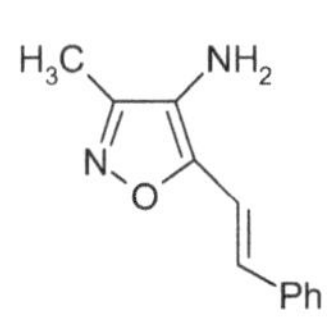	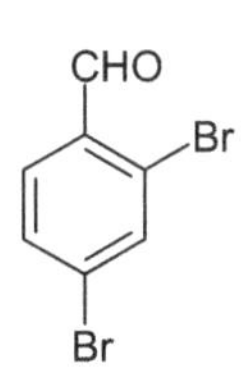	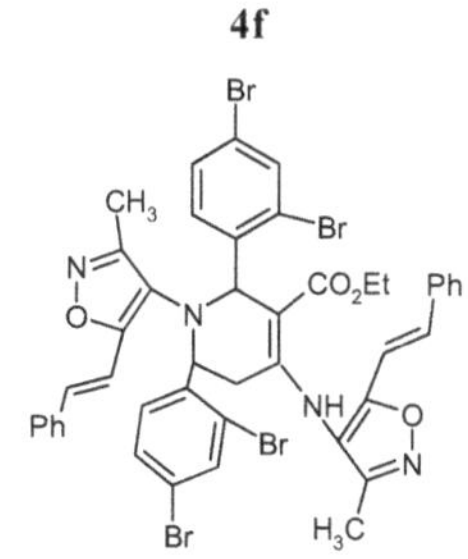	10	**4g** (97)
	1	**2g**	**4g**		
8	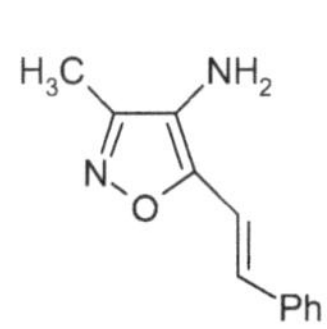	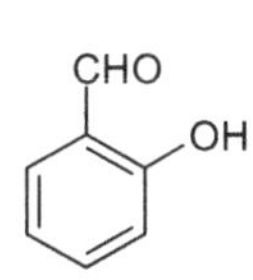	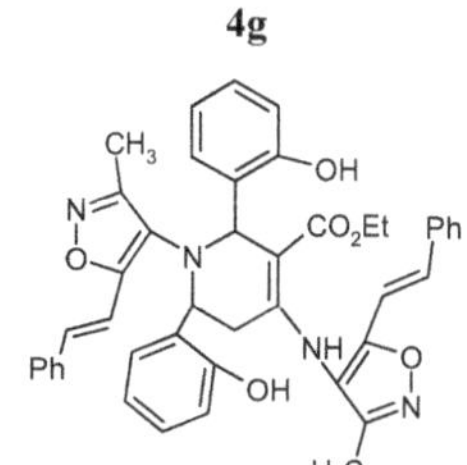	9	**4h** (96)
	1	**2h**	**4h**		

Continued

9	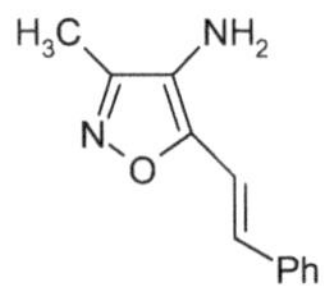 **1**	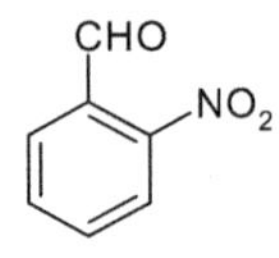 **2i**	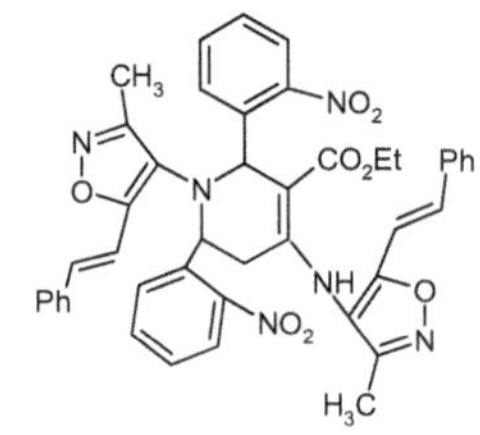 **4i**	10	**4i** (96)
10	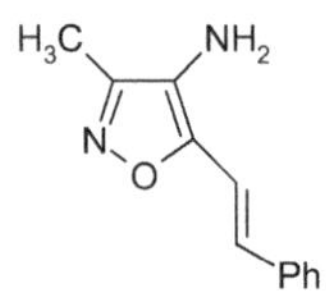 **1**	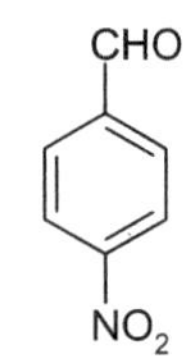 **2j**	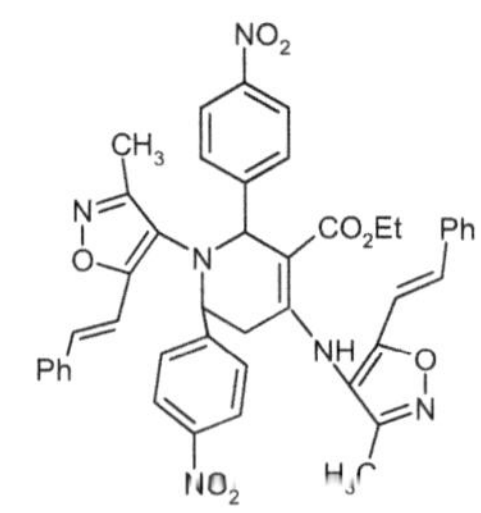 **4j**	8	**4j** (95)
11	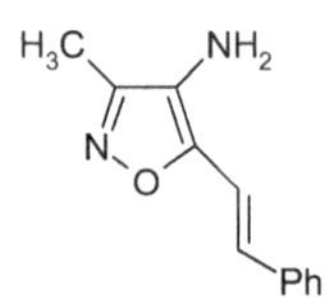 **1**	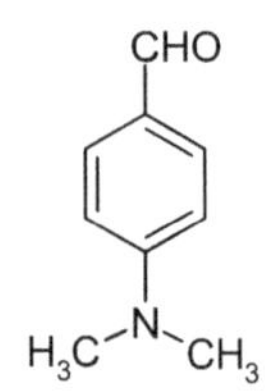 **2k**	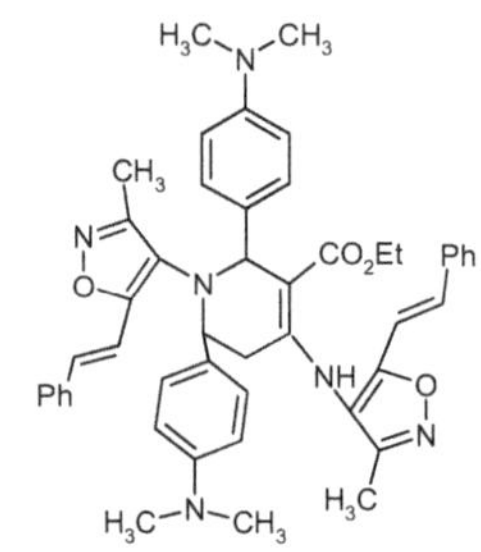 **4k**	10	**4k** (98)
12	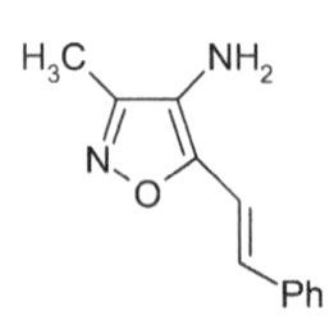 **1**	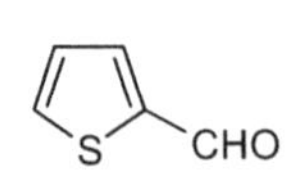 **2l**	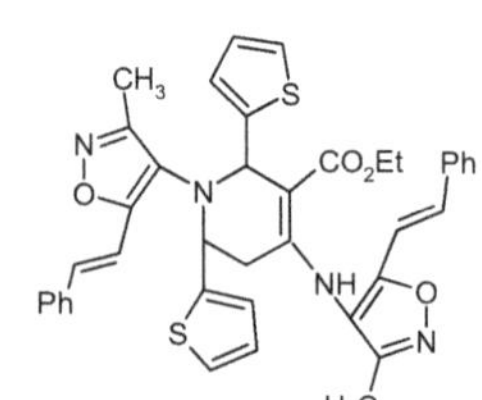 **4l**	8	**4l** (98)
13	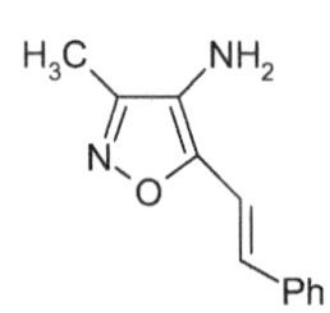 **1**	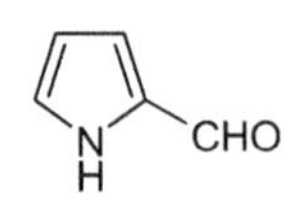 **2m**	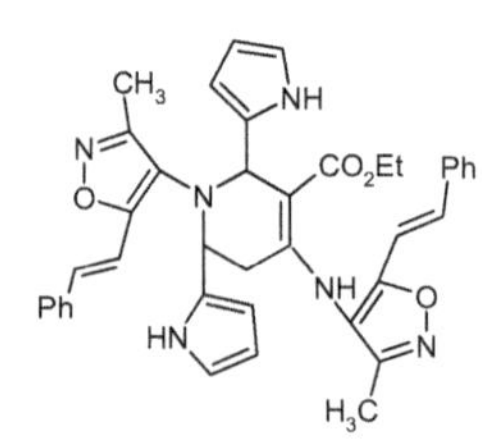 **4m**	8	**4m** (96)
14	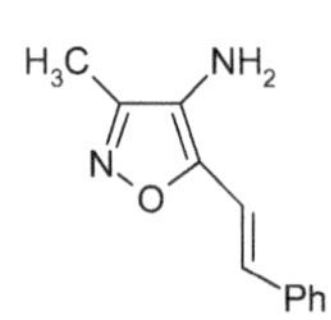 **1**	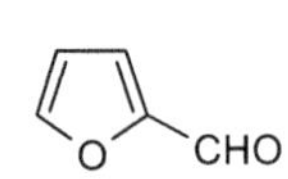 **2n**	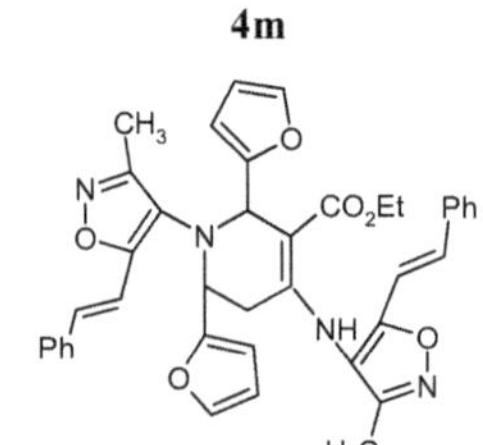 **4n**	9	**4n** (96)

[a]Reaction conditions: isoxazole amine **1** (2 mmol), aldehyde **2** (2 mmol), EAA **3** (1 mmol) and TEAA IL (5 mL), RT stirring. [b]Isolated yields.

4.2. Spectral Data of Compounds (4a-n)

Ethyl-4-(3-methyl-5-styrylisoxazol-4-ylamino)-1,2,5,6-tetrahydro-1-(3-methyl-5-styrylisoxazol-4-yl)-2,6-diphenyl pyridine-3-carboxylate (**4a**)

Brown solid; mp 221°C - 223°C. IR (KBr): 1665, 3428 cm^{-1}. ^{1}H NMR (300 MHz, $CDCl_3$): δ = 1.32 (t, 3 H, J = 7.2 Hz, OCH_2CH_3), 2.23 (s, 6 H, 2isoxazole-CH_3), 2.62 (dd, 1 H, CH, J = 15.2, 2.4 Hz), 2.76 (dd, 1 H, CH, J = 15.2, 5.7 Hz), 4.10 (q, 2 H, OCH_2CH_3), 5.23 (s, 1H, Ar-CH), 6.09 (t, 1 H, Ar-CH), 6.68 (d, 1 H, CH = CH, J = 12 Hz), 6.83 (d, 1 H, CH = CH, J = 12 Hz), 6.98 - 7.91 (m, 20 H, ArH), 8.93 (s, 1 H, NH, D_2O exchangeable). ^{13}C NMR (75MHz, DMSO-d_6): δ = 10.31, 12.36, 32.42, 48.46, 54.16, 63.45, 101.42, 104.26, 123.42, 126.13, 127.32, 128.06, 128.68, 129.21, 134.61, 136.52, 139.11, 141.46, 151.34, 158.56. MS (ESI): *m/z* = 689.36 $[M+H]^+$. Anal. Calcd for $C_{44}H_{40}N_4O_4$: C, 76.72; H, 5.85; N, 8.13. Found. C, 76.68; H, 5.81; N, 8.19.

Ethyl 4-(3-methyl-5-styrylisoxazol-4-ylamino)-1,2,5,6-tetrahdro-2,6-bis(4-methoxyphenyl)-1-(3-methyl-5-styrylisoxazol-4-yl)pyridine-3-carboxylate (**4b**)

Yellow solid; mp 215°C - 217°C. IR (KBr): 1661, 3442 cm^{-1}. ^{1}H NMR (300 MHz, $CDCl_3$): δ = 1.26 (t, 3 H, J = 7.2 Hz, OCH_2CH_3), 2.30 (s, 6 H, 2isoxazole-CH_3), 2.59 (dd, 1 H, CH, J = 15.2, 2.4Hz), 2.74 (dd, 1 H, CH, J = 15.2, 5.7 Hz), 3.62 (s, 6 H, 2 OCH_3), 4.08 (q, 2 H, OCH_2CH_3), 5.24 (s, 1 H, Ar-CH), 6.04 (t, 1 H, Ar-CH), 6.68 (d, 1 H, CH = CH, J = 12 Hz), 6.72 (d, 1 H, CH = CH, J = 12 Hz), 7.00 - 7.96 (m, 18 H, ArH), 9.00 (s, 1 H, NH, D_2O exchangeable). ^{13}C NMR (75 MHz, DMSO-d_6): δ = 10.26, 12.41, 32.40, 48.42, 54.28, 63.37, 63.81, 101.36, 104.55, 123.32, 126.41, 128.21, 128.57, 129.17, 134.48, 136.44, 139.23, 141.38, 151.40, 158.32, 159.13. MS (ESI): *m/z* = 749.37 $[M+H]^+$. Anal. Calcd for $C_{46}H_{44}N_4O_6$: C, 73.78; H, 5.92; N, 7.48. Found. C, 73.82; H, 5.87; N, 7.41.

Ethyl 4-(3-methyl-5-styrylisoxazol-4-ylamino)-1,2,5,6-tetrahydro-1-(3-methyl-5-styrylisoxazol-4-yl)-2,6-diptolylpyridine-3-carboxylate (**4c**)

Brown solid; mp 211°C - 213°C. IR (KBr): 1661, 3442 cm^{-1}. ^{1}H NMR (300 MHz, $CDCl_3$): δ = 1.09 (t, 3 H, J = 7.2 Hz, OCH_2CH_3), 2.25 (s, 6H, 2isoxazole-CH_3), 2.46 (s, 6 H, 2 Ar-CH_3), 2.50 (dd, 1 H, CH, J = 15.2, 2.4 Hz), 2.71 (dd, 1 H, CH, J = 15.2, 5.7 Hz), 4.11 (q, 2 H, OCH_2CH_3), 5.24 (s, 1 H, Ar-CH), 6.00 (t, 1 H, Ar-CH), 6.61 (d, 1 H, CH = CH, J = 12 Hz), 6.74 (d, 1 H, CH = CH, J = 12 Hz), 7.11 - 7.83 (m, 18 H, ArH), 8.90 (s, 1 H, NH, D_2O exchangeable). ^{13}C NMR (75 MHz, DMSO-d_6): δ = 10.21, 12.33, 24.23, 32.26, 48.39, 54.22, 63.40, 101.56, 104.31, 123.40, 126.23, 128.18, 128.52, 129.19, 134.68, 136.50, 138.37, 139.05, 141.42, 151.39, 158.44. MS (ESI): *m/z* = 717.17 $[M+H]^+$. Anal. Calcd for $C_{46}H_{44}N_4O_4$: C, 77.07; H, 6.19; N, 7.82. Found. C, 77.12; H, 6.16; N, 7.80.

Ethyl 4-(3-methyl-5-styrylisoxazol-4-ylamino)-2, 6-bis (2-chlorophenyl)-1,2,5,6-tetrahydro-1-(3-methyl-5-styrylisoxazol-4-yl)pyridine-3-carboxylate (**4d**)

Pale yellow solid; mp 235°C - 237°C. IR (KBr): 1668, 3440 cm^{-1}. ^{1}H NMR (300 MHz, $CDCl_3$): δ = 1.12 (t, 3 H, J = 7.2 Hz, OCH_2CH_3), 2.22 (s, 6 H, 2isoxazole-CH_3), 2.53 (dd, 1 H, CH, J = 15.2, 2.4 Hz), 2.69 (dd, 1 H, CH, J = 15.2, 5.7 Hz), 4.07 (q, 2 H, OCH_2CH_3), 5.20 (s, 1 H, Ar-CH), 5.98 (t, 1 H, Ar-CH), 6.64 (d, 1H, CH = CH, J = 12 Hz), 6.70 (d, 1 H, CH = CH, J = 12 Hz), 7.01 - 7.91 (m, 18 H, ArH), 8.83 (s, 1 H, NH, D_2O exchangeable). ^{13}C NMR (75MHz, DMSO-d_6): δ = 10.24, 12.33, 32.41, 48.46, 54.09, 63.31, 101.40, 104.19, 123.39, 126.24, 127.39, 128.61, 129.33, 134.21, 134.59, 136.47, 139.01, 141.39, 151.30, 158.62. MS (ESI): *m/z* = 757.29 $[M+H]^+$. Anal. Calcd for $C_{44}H_{38}Cl_2N_4O_4$: C, 69.75; H, 5.05; N, 7.39. Found. C, 69.81; H, 5.07; N, 7.33.

Ethyl 4-(3-methyl-5-styrylisoxazol-4-ylamino)-2,6-bis (2,4-dichlorophenyl)-1,2,5,6-tetrahydro-1-(3-methyl-5-styrylisoxazol-4-yl)pyridine-3-carboxylate (**4e**)

Pale yellow solid; mp 228°C - 230°C. IR (KBr): 1665, 3438 cm^{-1}. ^{1}H NMR (300 MHz, $CDCl_3$): δ = 1.10 (t, 3 H, J = 7.2 Hz, OCH_2CH_3), 2.20 (s, 6H, 2isoxazole-CH_3), 2.49 (dd, 1 H, CH, J = 15.2, 2.4 Hz), 2.60 (dd, 1 H, CH, J = 15.2, 5.7 Hz), 4.12 (q, 2 H, OCH_2CH_3), 5.11 (s, 1 H, Ar-CH), 5.86 (t, 1 H, Ar-CH), 6.60 (d, 1 H, CH = CH, J = 12 Hz), 6.73 (d, 1H, CH = CH, J = 12 Hz), 6.98 - 7.87 (m, 16 H, ArH), 9.10 (s, 1H, NH, D_2O exchangeable). ^{13}C NMR (75 MHz, DMSO-d_6): δ = 11.09, 12.41, 32.38, 48.67, 54.03, 63.29, 101.38, 104.20, 123.57, 126.26, 128.55, 129.36, 134.38, 134.56, 134.78, 136.42, 139.35, 141.51, 151.29, 158.58. MS (ESI): *m/z* = 825.40 $[M+H]^+$. Anal. Calcd for $C_{44}H_{36}Cl_4N_4O_4$: C, 63.93; H, 4.39; N, 6.78. Found. C, 63.97; H, 4.33; N, 6.74.

Ethyl 4-(3-methyl-5-styrylisoxazol-4-ylamino)-2,6-bis (2-bromophenyl)-1,2,5,6-tetrahydro-1-(3-methyl-5-styrylisoxazol-4-yl)pyridine-3-carboxylate (**4f**)

Pale brown solid; mp 224°C - 226°C. IR (KBr): 1670, 3441 cm^{-1}. ^{1}H NMR (300 MHz, $CDCl_3$): δ = 1.09 (t, 3 H, J = 7.2 Hz, OCH_2CH_3), 2.28 (s, 6 H, 2isoxazole-CH_3), 2.41 (dd, 1 H, CH, J = 15.2, 2.4 Hz), 2.63 (dd, 1 H, CH, J = 15.2, 5.7 Hz), 4.10 (q, 2 H, OCH_2CH_3), 5.25 (s, 1 H, Ar-CH), 5.78 (t, 1H, Ar-CH), 6.61 (d, 1 H, CH = CH, J = 12 Hz), 6.69 (d, 1 H, CH = CH, J = 12 Hz), 7.05 - 7.93 (m, 18 H, ArH), 9.08 (s, 1 H, NH, D_2O exchangeable). ^{13}C NMR (75 MHz, DMSO-d_6): δ = 10.16, 12.27, 32.53, 48.42, 54.17, 63.42, 101.33, 104.24, 123.48, 126.15, 127.46, 128.58, 129.45, 134.32, 134.68, 136.51, 139.12, 141.26, 151.44, 158.78. MS (ESI): *m/z* = 845.11 $[M+H]^+$. Anal. Calcd for $C_{44}H_{38}Br_2N_4O_4$: C, 62.42; H, 4.52; N, 6.62. Found. C, 62.38; H, 4.51; N, 6.65.

Ethyl 4-(3-methyl-5-styrylisoxazol-4-ylamino)-2,6-bis (2,4-dibromophenyl)-1,2,5,6-tetrahydro-1-(3-methyl-5-

styrylisoxazol-4-yl)pyridine-3-carboxylate (**4g**)

Brown solid; mp 241°C - 243°C. IR (KBr): 1668, 3442 cm^{-1}. ^{1}H NMR (300 MHz, $CDCl_3$): δ = 1.12 (t, 3 H, J = 7.2 Hz, OCH_2CH_3), 2.24 (s, 6 H, 2isoxazole-CH_3), 2.48 (dd, 1 H, CH, J = 15.2, 2.4 Hz), 2.59 (dd, 1 H, CH, J = 15.2, 5.7 Hz), 4.20 (q, 2 H, OCH_2CH_3), 5.19 (s, 1 H, Ar-CH), 5.64 (t, 1 H, Ar-CH), 6.57 (d, 1 H, CH = CH, J = 12 Hz), 6.70 (d, 1 H, CH = CH, J = 12 Hz), 7.00 - 7.87 (m, 16 H, ArH), 8.87 (s, 1 H, NH, D_2O exchangeable). ^{13}C NMR (75 MHz, DMSO-d_6): δ = 10.14, 12.27, 24.38, 32.39, 48.25, 54.20, 63.43, 101.51, 104.44, 123.57, 126.34, 128.11, 128.39, 129.10, 134.56, 136.41, 138.28, 139.22, 141.35, 151.46, 158.61. MS (ESI): *m/z* = 1001.08 $[M+H]^+$. Anal. Calcd for $C_{44}H_{36}Br_4N_4O_4$: C, 52.62; H, 3.61; N, 5.58. Found. C, 52.65; H, 3.68; N, 5.55.

Ethyl 4-(3-*methyl*-5-*styrylisoxazol*-4-*ylamino*)-1,2,5,6-*tetrahydro*-2,6-*bis*(2-*hydroxyphenyl*)-1-(3-*methyl*-5-*styrylisoxazol*-4-*yl*)*pyridine*-3-*carboxylate* (**4h**)

Brown solid; mp 203°C - 205°C. IR (KBr): 1657, 3445 cm^{-1}. ^{1}H NMR (300 MHz, $CDCl_3$): δ = 1.09 (t, 3 H, J = 7.2 Hz, OCH_2CH_3), 2.30 (s, 6 H, 2isoxazole-CH_3), 2.41 (dd, 1 H, CH, J = 15.2, 2.4 Hz), 2.60 (dd, 1 H, CH, J = 15.2, 5.7 Hz), 4.11 (q, 2 H, OCH_2CH_3), 5.18 (s, 1 H, Ar-CH), 5.52 (t, 1 H, Ar-CH), 6.62 (d, 1 H, CH = CH, J = 12 Hz), 6.71 (d, 1 H, CH = CH, J = 12 Hz), 7.04 - 7.93 (m, 18 H, ArH), 8.64 (s, 1 H, NH, D_2O exchangeable), 9.10 (s, 2 H, 2OH, D_2O exchangeable). ^{13}C NMR (75 MHz, DMSO-d_6): δ = 10.25, 12.31, 32.48, 48.36, 54.23, 63.56, 101.40, 104.38, 123.59, 126.11, 127.52, 128.47, 129.36, 134.41, 134.69, 136.72, 139.24, 141.31, 151.40, 158.55. MS (ESI): *m/z* = 721.37 $[M+H]^+$. Anal. Calcd for $C_{44}H_{40}N_4O_6$: C, 73.32; H, 5.59; N, 7.77. Found. C, 73.37; H, 5.54; N, 7.71.

Ethyl 4-(3-*methyl*-5-*styrylisoxazol*-4-*ylamino*)-1,2,5,6-*tetrahydro*-1-(3-*methyl*-5-*styrylisoxazol*-4-*yl*)-2,6-*bis*(2-*nitrophenyl*)*pyridine*-3-*carboxylate* (**4i**)

Brown solid; mp 230°C - 232°C. IR (KBr): 1667, 3448 cm^{-1}. ^{1}H NMR (300 MHz, $CDCl_3$): δ = 1.00 (t, 3 H, J = 7.2 Hz, OCH_2CH_3), 2.30 (s, 6 H, 2isoxazole-CH_3), 2.40 (dd, 1 H, CH, J = 15.2, 2.4 Hz), 2.59 (dd, 1 H, CH, J = 15.2, 5.7 Hz), 4.21 (q, 2 H, OCH_2CH_3), 5.19 (s, 1 H, Ar-CH), 5.63 (t, 1 H, Ar-CH), 6.58 (d, 1 H, CH = CH, J = 12 Hz), 6.61 (d, 1 H, CH = CH, J = 12 Hz), 7.00 - 7.85 (m, 18 H, ArH), 8.85 (s, 1 H, NH, D_2O exchangeable). ^{13}C NMR (75 MHz, DMSO-d_6): δ = 10.27, 12.46, 32.53, 48.55, 54.23, 63.38, 101.57, 104.36, 123.55, 126.24, 127.41, 128.55, 129.31, 134.56, 136.67, 139.26, 141.54, 148.53, 151.20, 158.51. MS (ESI): *m/z* = 781.26 $[M+H]^+$. Anal. Calcd for $C_{44}H_{40}N_6O_8$: C, 67.68; H, 5.16; N, 10.76. Found. C, 67.75; H, 5.19; N, 10.74.

Ethyl 4-(3-*methyl*-5-*styrylisoxazol*-4-*ylamino*)-1,2,5,6-*tetrahydro*-1-(3-*methyl*-5-*styrylisoxazol*-4-*yl*)-2,6-*bis*(4-*nitrophenyl*)*pyridine*-3-*carboxylate* (**4j**)

Pale brown solid; mp 237°C - 239°C. IR (KBr): 1655, 3438 cm^{-1}. ^{1}H NMR (300 MHz, $CDCl_3$): δ = 1.05 (t, 3 H, J = 7.2 Hz, OCH_2CH_3), 2.26 (s, 6 H, 2isoxazole-CH_3), 2.42 (dd, 1 H, CH, J = 15.2, 2.4 Hz), 2.51 (dd, 1 H, CH, J = 15.2, 5.7 Hz), 4.19 (q, 2 H, OCH_2CH_3), 5.25 (s, 1 H, Ar-CH), 5.59 (t, 1 H, Ar-CH), 6.60 (d, 1 H, CH = CH, J = 12 Hz), 6.73 (d, 1 H, CH = CH, J = 12 Hz), 6.97 - 7.85 (m, 18 H, ArH), 8.51 (s, 1 H, NH, D_2O exchangeable). ^{13}C NMR (75MHz, DMSO-d_6): δ = 10.22, 12.41, 32.58, 48.63, 54.25, 63.32, 101.48, 104.40, 123.57, 126.26, 128.11, 128.71, 129.32, 134.54, 136.62, 139.26, 141.56, 148.52, 151.49, 158.60. MS (ESI): *m/z* = 781.05 $[M+H]^+$. Anal. Calcd for $C_{44}H_{40}N_6O_8$: C, 67.68; H, 5.16; N, 10.76. Found. C, 67.70; H, 5.15; N, 10.77.

Ethyl 4-(3-*methyl*-5-*styrylisoxazol*-4-*ylamino*)-2,6-*bis* (4-(*dimethylamino*)*phenyl*)-1,2,5,6-*tetrahydro*-1-(3-*methyl*-5-*styrylisoxazol*-4-*yl*)*pyridine*-3-*carboxylate* (**4k**)

Orange solid; mp 219°C – 221°C. IR (KBr): 1660, 3441 cm^{-1}. ^{1}H NMR (300 MHz, $CDCl_3$): δ = 1.10 (t, 3 H, J = 7.2 Hz, OCH_2CH_3), 2.28 (s, 6 H, 2isoxazole-CH_3), 2.42 (dd, 1 H, CH, J = 15.2, 2.4 Hz), 2.53 (dd, 1 H, CH, J = 15.2, 5.7 Hz), 3.85 (s, 12 H, 2 $N(CH_3)_2$), 4.22 (q, 2 H, OCH_2CH_3), 5.27 (s, 1 H, Ar-CH), 6.00 (t, 1 H, Ar-CH), 6.63 (d, 1 H, CH = CH, J = 12Hz), 6.71 (d, 1 H, CH = CH, J = 12 Hz), 7.02 - 7.97 (m, 18 H, ArH), 8.50 (s, 1 H, NH, D_2O exchangeable). ^{13}C NMR (75 MHz, DMSO-d_6): δ = 10.23, 12.43, 32.37, 40.55, 48.51, 54.23, 63.38, 101.40, 104.32, 123.41, 126.26, 128.18, 128.54, 129.36, 134.58, 136.62, 139.21, 141.52, 149.71, 151.43, 158.67. MS (ESI): *m/z* = 775.11 $[M + H]^+$. Anal. Calcd for $C_{48}H_{50}N_6O_4$: C, 74.39; H, 6.50; N, 10.84. Found. C, 74.35; H, 6.51; N, 10.87.

Ethyl 4-(3-*methyl*-5-*styrylisoxazol*-4-*ylamino*)-1,2,5,6-*tetrahydro*-1-(3-*methyl*-5-*styrylisoxazol*-4-*yl*)-2,6-*di*(*thiophen*-2-*yl*)*pyridine*-3-*carboxylate* (**4l**)

Brown solid; mp 223°C - 225°C. IR (KBr): 1658, 3436 cm^{-1}. ^{1}H NMR (300 MHz, $CDCl_3$): δ = 1.08 (t, 3 H, J = 7.2 Hz, OCH_2CH_3), 2.20 (s, 6 H, 2isoxazole-CH_3), 2.45 (dd, 1 H, CH, J = 15.2, 2.4 Hz), 2.59 (dd, 1 H, CH, J = 15.2, 5.7 Hz), 4.08 (q, 2 H, OCH_2CH_3), 5.21 (s, 1 H, Ar-CH), 6.10 (t, 1 H, Ar-CH), 6.61 (d, 1 H, CH = CH, J = 12 Hz), 6.73 (d, 1 H, CH = CH, J = 12 Hz), 6.81 (dd, 2 H, 2thiophene-H), 7.23 (d, 2 H, 2thiophene-H), 7.31 (d, 2 H, 2thiophene-H), 7.52 - 7.97 (m, 10 H, ArH), 9.25 (s, 1 H, NH, D_2O exchangeable). ^{13}C NMR (75 MHz, DMSO-d_6): δ = 10.12, 12.22, 32.39, 48.40, 54.16, 63.55, 101.51, 104.23, 123.18, 123.50, 126.22, 126.86, 127.10, 129.31, 134.57, 136.43, 140.12, 141.61, 151.24, 158.50. MS (ESI): *m/z* = 701.33 $[M+H]^+$. Anal. Calcd for $C_{40}H_{36}N_4O_4S_2$: C, 68.55; H, 5.18; N, 7.99. Found. C, 68.57; H, 5.13; N, 7.96.

Ethyl 4-(3-*methyl*-5-*styrylisoxazol*-4-*ylamino*)-1,2,5,6-*tetrahydro*-1-(3-*methyl*-5-*styrylisoxazol*-4-*yl*)-2,6-*di*(1 *H*-*pyrrol*-2-*yl*)*pyridine*-3-*carboxylate* (**4m**)

Brown solid; mp 227°C - 229°C. IR (KBr): 1660, 3445

cm^{-1}. ^1H NMR (300 MHz, $CDCl_3$): δ = 1.16 (t, 3 H, J = 7.2 Hz, OCH_2CH_3), 2.30 (s, 6 H, 2isoxazole-CH_3), 2.48 (dd, 1 H, CH, J = 15.2, 2.4 Hz), 2.61 (dd, 1 H, CH, J = 15.2, 5.7 Hz), 4.11 (q, 2 H, OCH_2CH_3), 5.31 (s, 1 H, Ar-CH), 6.20 (t, 1 H, Ar-CH), 6.57 (d, 1 H, CH = CH, J = 12 Hz), 6.68 (d, 1 H, CH = CH, J = 12 Hz), 6.73 (dd, 2 H, 2pyrrole-H), 7.21 (d, 2 H, 2pyrrole-H), 7.30 (d, 2 H, 2pyrrole-H), 7.52 - 8.00 (m, 10H, ArH), 8.72 (s, 2 H, 2pyrrole-NH, D_2O exchangeable), 9.00 (s, 1 H, NH, D_2O exchangeable). ^{13}C NMR (75 MHz, DMSO-d_6): δ = 10.33, 12.30, 32.47, 48.50, 54.12, 63.42, 101.55, 104.32, 108.96, 109.11, 118.86, 123.57, 126.24, 129.32, 131.12, 134.58, 136.50, 141.40, 151.43, 158.66. MS (ESI): m/z = 667.10 $[M+H]^+$. Anal. Calcd for $C_{40}H_{38}N_6O_4$: C, 72.05; H, 5.74; N, 12.60. Found. C, 72.11; H, 5.71; N, 12.58.

Ethyl 4-(3-methyl-5-styrylisoxazol-4-ylamino)-2,6-di (furan-2-yl)-1,2,5,6-tetrahydro-1-(3-methyl-5-styryliso-xazol-4-yl)pyridine-3-carboxylate (**4n**)

Pale yellow solid; mp 209˚C - 211˚C. IR (KBr): 1655, 3432 cm^{-1}. ^1H NMR (300 MHz, $CDCl_3$): δ = 1.05 (t, 3 H, J = 7.2 Hz, OCH_2CH_3), 2.22 (s, 6 H, 2isoxazole-CH_3), 2.41 (dd, 1 H, CH, J = 15.2, 2.4 Hz), 2.59 (dd, 1 H, CH, J = 15.2, 5.7 Hz), 4.20 (q, 2 H, OCH_2CH_3), 5.26 (s, 1 H, Ar-CH), 6.18 (t, 1 H, Ar-CH), 6.60 (d, 1 H, CH = CH, J = 12 Hz), 6.74 (d, 1 H, CH = CH, J = 12 Hz), 6.69 (dd, 1 H, furan-H), 7.24 (d, 1 H, furan-H), 7.42 (d, 1 H, furan-H), 7.50 - 7.9 (m, 10 H, ArH), 8.92 (s, 1 H, NH, D_2O exchangeable). ^{13}C NMR (75 MHz, DMSO-d_6): δ = 10.26, 12.22, 32.34, 48.41, 54.10, 63.43, 101.41, 104.26, 106.77, 116.56, 123.39, 126.09, 129.22, 134.67, 136.55, 141.40, 142.10, 151.33, 152. 62, 158.52. MS (ESI): m/z = 669.46 $[M+H]^+$. Anal. Calcd for $C_{40}H_{36}N_4O_6$: C, 71.84; H, 5.43; N, 8.38. Found. C, 71.88; H, 5.41; N, 8.32.

REFERENCES

[1] T. Welton, "Room-Temperature Ionic Liquids: Solvents for Synthesis and Catalysis," *Chemical Reviews*, 1999, p. 2701.

[2] P. Wasserschid and W. Keim, "Ionic Liquids—New 'Solutions' for Transition Metal Catalysis," Angewandte Chemie International Edition, Vol. 39, No. 21, 2000, pp. 3772-3789.

[3] R. Sheldon, "Catalytic Reactions in Ionic Liquids," *Chemical Communications*, No. 23, 2001, pp. 2399-2407.

[4] R. T. Coutts and A. F. Casay, "Pyridine and Its Derivatives: A Supplement," Inter Science Publishers, New York, 1975.

[5] M. Ferles and J. Pliml, "Adavances in Heterocyclic Chemistry," Academic Press, New York, 1970.

[6] J. W. Langston, P. Ballard, J. W. Tetrud and I. Irwin, "Chronic Parkinsonism in Humans Due to a Product of Meperidine-Analog Synthesis," *Science*, Vol. 219, No. 4587, 1983, pp. 979-980.

[7] J. W. Langston, "MPTP Neurotoxicity: An Overview and Characterization of Phases of Toxicity," *Life Science*, Vol. 36, No. 3, 1985, pp. 201-206.

[8] C. C. Chiueh, R. S. Burns, S. P. Markey, D. M. Jacobowitz and I. J. Kopin, "A Primate Model of Parkinsonism: Selective Destruction of Dopaminergic Neurons in the Pars Compacta of the Substantia Nigra by N-Methyl-4-phenyl-1,2,3,6-tetrahydropyridine," *Life Science*, Vol. 36, 1985, pp. 213-217.

[9] J. W. Langston, L. S. Forno, C. S. Rebert and I. Irwin, "Selective Nigral Toxicity after Systemic Administration of 1-Methyl-4-phenyl-1,2,5,6-Tetrahydropyrine (MPTP) in the Squirrel Monkey," *British Research*, Vol. 292, 1984, pp. 390-394.

[10] J. E. Christie, A. Shering, J. Ferguson and A. I. M. Glen, "Physostigmine and Arecoline: Effects of Intravenous Infusions in Alzheimer Presenile Dementia," *The British Journal of Psychiatry*, Vol. 138, 1981, pp. 46-50.

[11] P. Sauerberg, J. J. Larsen, E. Falch and P. Krogsgaard-Larsen, "A Novel Class of Conformationally Restricted Heterocyclic Muscarinic Agonists," *Journal of Medicinal Chemistry*, Vol. 29, No. 6, 1986, pp. 1004-1009.

[12] P. Sauerberg, B. Fjalland, J. J. Larsen, T. Bach-Lauritsen, E. Falch and P. Krogsgaard-Larsen, "Pharmacological Profile of a Novel Class of Muscarinic Acetylcholine Receptor Agonists," *European Journal of Pharmacology*, Vol. 130, No. 1-2, 1986, pp. 125-131.

[13] P. Krogsgaard-Larsen, E. Falch, P. Sauerberg, S. B. Freedman, H. L. Lembùl and E. Meier, "In Subtypes of Muscarinic Receptors III," Elsevier, Amsterdam, 1988, pp. 69-74.

[14] P. Sauerberg, E. Falch, E. Meier, H. L. Lembùl and P. Krogsgaard-Larsen, "Heterocyclic Muscarinic Agonists. Synthesis and Biological Activity of Some Bicyclic Sulfonium Arecoline Bioisosteres," *Journal of Medicinal Chemistry*, Vol. 31, No. 7, 1988, pp. 1312-1316.

[15] T. Ogamino and S. Nishiyama, "A New Ring-Opening Access to Aeroplysinin-1, a Secondary Metabolite of Verongia Aerophoba," *Tetrahedron*, Vol. 59, 2003, pp. 9419-9423.

[16] G. Daidone, D. Raffa, B. Maggio, F. Plescia, V. M. C. Cutuli, N. G. Mangano and A. Caruso, "Synthesis and Antiproliferative Activity of Novel 3-(Indazol-3-yl)-quinazolin-4(3*H*)-one and 3-(Indazol-3-yl)-benzotriazin-4(3*H*)-one Derivatives," *Arch. Pharm. Med. Chem*, Vol. 50, 1999, p. 332.

[17] K. Tomita, Y. Takahi, R. Ishizuka, S. Kamamura, M. Nakagawa, M. Ando, T. Nakanishi, T. Nakamura and H. Udaira, Ann. Sankyo Res. Lab. Vol. 1, 1973, p. 25; Chem. Abstr. 1974, *80*, 120808.

[18] J. J. Talley, D. L. Brown, J. S. Carter, M. J. Graneto, C.

M. Koboldt, J. L. Masferrer, W. E. Perkins, R. S. Rogers, A. F. Shaffer, Y. Y. Zhang, B. S. Zweifel and K. Seibert, "4-[5-Methyl-3-phenylisoxazol-4-yl]-benzenesulfonamide, Valdecoxib: A Potent and Selective Inhibitor of COX-2," *Journal of Medicinal Chemistry*, Vol. 43, 2000, pp. 775-777.

[19] K. Haripara, S. Patel, A. Joshi and H. Parekh, "Synthesis and Biological Evaluation of Some Cyano Pyridines and Isoxazoles," *Indian J. Het. Chem*, Vol. 13, 2004, pp. 221-224.

[20] H. Uno, M. Kurokawa, Y. Masuda and H. Nishimura, "Studies on 3-Substituted 1,2-Benzisoxazole Derivatives. Syntheses of 3-(Sulfamoylmethyl)-1,2-Benzisoxazole Derivatives and Their Anticonvulsant Activities," *Journal of Medicinal Chemistry*, Vol. 22, No. 2, 1979, pp. 180-183.

[21] W.-T. Li, D.-R. Hwang, C.-P. Chen, C.-W. Shen, C.-L. Huang, T.-W. Chen, C.-H. Lin, Y.-L. Chang, Y.-Y. Chang, Y.-K. Lo, H.-Y. Tseng, C.-C. Lin, J.-S. Song, H.-C. Chen, S.-J. Chen, S.-H. Wu and C.-T. Chen, "Synthesis and Biological Evaluation of N-Heterocyclic Indolyl Glyoxylamides as Orally Active Anticancer Agents," *Journal of Medicinal Chemistry*, Vol. 46, No. 9, 2003, pp. 1706-1715.

[22] E. Rajanarendar, M. Nagi Reddy, K. Rama Murthy, K. Govardhan Reddy, S. Raju, M. Srinivas, B. Praveen and M. Srinivasa Rao, "Synthesis, Antimicrobial, and Mosquito Larvicidal Activity of 1-Aryl-4-Methyl-3,6-Bis-(5-Methylisoxazol-3-yl)-2-Thioxo-2,3,6,10b-Tetrahydro-1H-Pyrimido[5,4-c]quinolin-5-Ones," *Bioorganic & Medicinal Chemistry Letters*, Vol. 20, No. 20, 2010, pp. 6052-6055.

[23] E. Rajanarendar, M. Nagi Reddy, S. Rama Krishna, K. Rama Murthy, P. Surendar, R. N. Reddy and Y. N. Reddy, "Synthesis and *in Vitro* and *in Vivo* Anticancer Activity of Novel Phenylmethylene Bis-isoxazolo[4,5-b]azepines," *Bioorganic & Medicinal Chemistry Letters*, Vol. 22, No. 2012, pp. 149-153.

[24] E. Rajanarendar, S. Raju, M. Nagi Reddy, S. Rama Krishna, L. Hari Kiran, A. Ram Narasimha Reddy and Y Narasimha Reddy, "Multi-Component Synthesis and *in Vitro* and *in Vivo* Anticancer Activity of Novel Aryl- methylene Bis-isoxazolo[4,5-*b*]pyridine-*N*-Oxides," *European Journal of Medicinal Chemistry*, Vol. 50, 2012, pp. 274-279.

[25] E. Rajanarendar, M. Nagi Reddy, S. Rama Krishna, K. Govardhan Reddy, Y. N. Reddy and M. V. Rajam, "Design, Synthesis, *in Vitro* Antimicrobial and Anticancer Activity of Novel Methylenebis-Isoxazolo[4,5-b]azepines Derivatives," *European Journal of Medicinal Chemistry*, Vol. 50, 2012, pp. 344-349.

[26] E. Rajanarendar, M. Nagi Reddy, K. Govardhan Reddy and S. Rama Krishna, "l-Proline Catalyzed Efficient One-Pot Three-Component Aza-Diels–Alder Reactions on Nitrostyrylisoxazoles: A Facile Synthesis of New Isoxazolyl Tetrahydroquinolines and Isoxazolo[2,3-a]pyrimidines," *Tetrahedron Letters*, Vol. 53, No. 23, 2012, pp. 2909-2913.

[27] M. Misra, S. K. Pandey, V. P. Pandey, J. Pandey, R. Tripathi, R. P. Tripathi, "Organocatalyzed Highly Atom Economic One Pot Synthesis of Tetrahydropyridines as Antimalarials," *Bioorganic & Medicinal Chemistry*, Vol. 17, No. 2, 2009, pp. 625-633.

[28] A. T. Khan, T. Parvin and L. H. Choudhary, "Effects of Substituents in the β-Position of 1,3-Dicarbonyl Compounds in Bromodimethylsulfonium Bromide-Catalyzed Multicomponent Reactions: A Facile Access to Functionalized Piperidines," *The Journal of Organic Chemistry*, Vol. 73, No. 1, 2008, pp. 8398-8402.

[29] A. T. Khan, M. Lal and M. M. Khan, "Synthesis of Highly Functionalized Piperidines by One-Pot Multicomponent Reaction Using Tetrabutylammonium Tribromide (TBATB)," *Tetrahedron Letters*, Vol. 51, No. 33, 2010, pp. 4419-4424.

[30] A. T. Khan, M. M. Khan and K. R. B. Karthi, "Iodine Catalyzed One-Pot Five-Component Reactions for Direct Synthesis of Densely Functionalized Piperidines," *Tetrahedron*, Vol. 66, No. 39, 2010, pp. 7762-7772.

[31] P. A. Clarke, P. A. Zaytzev and A. C. Whitwood, "Pot, Atom and Step Economic (PASE) Synthesis of Highly Functionalized Piperidines: A Five-Component Condensation," *Tetrahedron Letters*, Vol. 48, No. 20, 2007, pp. 5209-5212.

[32] S. Mishra, R. Ghosh, "Efficient One-Pot Synthesis of Functionalized Piperidine Scaffolds via $ZrOCl_2 \cdot 8H_2O$ Catalyzed Tandem Reactions of Aromatic Aldehydes with Amines and Acetoacetic Esters," *Tetrahedron Letters*, Vol. 52, No. 22, 2011, pp. 2857-2861.

[33] W. Hong-Juan, M. Li-Ping and Z. Zhan-Hui, ACS Comb. Sci. Vol. 13, 2011, p. 181.

[34] R. Kumar, P. Chaudhary, S. Nimesh, A. K. Verma and R. Chandra, "An Efficient Synthesis of 1,5-Benzadiazepine Derivatives Catalyzed by Silver Nitrate," *Green Chemistry*, Vol. 8, No. 6, 2006, pp. 519-521.

[35] A. K. Verma, P. Attri, V. Chopra, R. K. Tiwari and R. Chandra, "Triethylammonium Acetate (*TEAA*): A Recyclable Inexpensive Ionic Liquid Promotes the Chemoselective Aza- and Thia-*Michael* Reactions," *Monatsh Chemistry*, Vol. 139, 2008, p. 1041

[36] C. Wang, L. Guo, H. Li, Y. Wang, J. Weng and L. Wu, "Preparation of Simple Ammonium Ionic Liquids and Their Application in the Cracking of Dialkoxypropanes," *Green Chemistry*, Vol. 8, No. 7, 2006, pp. 603-607.

[37] J. Weng, C. Wang, H. Li, Y. Wang, "Novel Quaternary Ammonium Ionic Liquids and Their Use as Dual Solvent-Catalysts in the Hydrolytic Reaction," *Green Chemistry*, Vol. 8, No. 1, 2006, pp. 96-99.

[38] H. Jiang, C. Wang, H. Li and Y. Wang, "Preparation of Dialkoxypropanes in Simple Ammonium Ionic Liquids," *Green Chemistry*, Vol. 8, No. 12, 2006, pp. 1076-1079.

Synthesis and Characterization of Novel Brønsted-Lewis Acidic Ionic Liquids

Hualiang An, Lijuan Kang, Wei Gao, Xinqiang Zhao*, Yanji Wang
Hebei Provincial Key Lab of Green Chemical Technology and High Efficient Energy Saving,
Hebei University of Technology, Tianjin, China

ABSTRACT

To meet the demands of some kinds of reactions catalyzed simultaneously by Brønsted acid and Lewis acid catalyst, two novel Brønsted-Lewis acidic ionic liquids, 1-carboxyethylene-3-(4-zinc acetate sulfobutyl) imidazolium chloride ($[CH_3COO\text{-}Zn\text{-}O_3S\text{-bim-}CH_2CH_2COOH]Cl$) and 1-(1,2-dicarboxy) ethylene-3-(4-zinc acetate sulfobutyl) imidazolium chloride ($[CH_3COO\text{-}Zn\text{-}O_3S\text{-bim-}C_4H_5O_4]Cl$) were synthesized, in which both Brønsted and Lewis acidic sites existed in the cation. The structures of the ionic liquids were determined by means of FT-IR, ^1H NMR and elemental analysis. The results of Py-IR analysis indicated that the two novel ionic liquids have both Brønsted and Lewis acid properties. The acid strength values (H_0) of the ionic liquids were measured utilizing the UV-visible spectroscopy combined with Hammett indicator method, and the acid amount of them was determined by acid-base titration.

Keywords: Brønsted-Lewis Acidic Ionic Liquids; Synthesis; Characterization

1. Introduction

Ionic liquids, due to their strong dissolving capacity, negligible vapor pressure, high thermal and chemical stability, have attracted much attention. The emergence of functionalized ionic liquids extends the application fields of ionic liquids. Acidic ionic liquids are a kind of functionalized ionic liquids, and recently have attracted more interests due to their advantages of both solid acid and liquid acid.

The traditional acidic ionic liquids include Brønsted acidic ionic liquids and Lewis acidic ionic liquids. Lewis acidic ionic liquids are usually prepared from metal halides and organic halides, and the most representative one of them is chloroaluminate ionic liquid. However, chloroaluminate ionic liquid is sensitive to moisture and unstable, so $FeCl_3$ or other metal halide gradually takes the place of $AlCl_3$ to prepare Lewis acidic ionic liquids which are not sensitive to water [1]. As compared with Lewis acidic ionic liquids, Brønsted acidic ionic liquids are synthesized later. Cole *et al.* [2] reported the synthesis of Brønsted acidic ionic liquids for the first time in 2002, and then the studies on Brønsted acidic ionic liquids developed rapidly. The traditional acidic ionic liquids mentioned above possess just single Lewis or Brønsted acid sites, which can not meet the demands of some reactions catalyzed simultaneously by Brønsted acid and Lewis acid catalyst, e.g. one-pot synthesis of methylene diphenyl dimethylcarbamate from aniline, dimethyl carbonate and formaldehyde [3]. Wang *et al.* [4] reported a dual acidic ionic liquid $[C_4SC_nim]Cl\text{-}AlCl_3$ first in 2005. The acidity of the dual acidic ionic liquid was determined by IR spectrography using pyridine as a probe, and the result showed that $[C_4SC_nim]Cl\text{-}AlCl_3$ was indeed a dual acidic ionic liquid. Liu *et al.* [5,6] synthesized a Brønsted-Lewis acidic ionic liquid $[HSO_3\text{-}(CH_2)_3\text{-}NEt_3]Cl\text{-}ZnCl_2$ and evaluated its catalytic performance in rosin dimerization and fatty acid methyl ester dimerization, respectively. They [7-9] also prepared another Brønsted-Lewis acidic ionic liquid $[HO_3S\text{-}(CH_2)_3\text{-mim}]Cl\text{-}ZnCl_2$ and used it to catalyze the two reactions mentioned above. Quan *et al.* [10], Liang *et al.* [11], and Jiang *et al.* [12] prepared Brønsted-Lewis acidic ionic liquids $[HO_3S\text{-}(CH_2)_4\text{-Py}][FeCl_4]$, $[HO_3S\text{-}(CH_2)_4\text{-IQ}][Cu_{1/2}SO_4]$ and $[HO_3S\text{-}(CH_2)_4\text{-Py}][Fe_{1/3}SO_4]$ respectively, and used them in the reactions of Michael addition. The Brønsted-Lewis acidic ionic liquids reported up till now are those in which the cation contains Brønsted acidity and the anion contains Lewis acidity. To the best of our knowledge, no

*Corresponding author.

publications about the synthesis of Brønsted-Lewis acidic ionic liquids with both Brønsted and Lewis acidic sites in the cation have been reported.

In this work, two novel Brønsted-Lewis acidic ionic liquids with both Brønsted and Lewis acidic sites in the cation were prepared first, and then their structures and acidity were characterized.

2. Results and Discussion

Two novel Brønsted-Lewis acidic ionic liquids, 1-carboxyethylene-3-(4-zinc acetate sulfobutyl) imidazolium chloride ([CH_3COO-Zn-O_3S-bim-CH_2CH_2COOH]Cl, BLILs-1) and 1-(1,2-dicarboxy) ethylene-3-(4-zinc acetate sulfobutyl) imidazolium chloride ([CH_3COO-Zn-O_3S-bim-$C_4H_5O_4$]Cl, BLILs-2), were synthesized and characterized in this work.

2.1. Synthesis Method

The synthesis approach of BLILs-1 consisted of four-step reactions (**Scheme 1**). 1-ethoxycarbonylethylene imidazole (Intermediate Product-1, IP-1) was obtained through an addition reaction of imidazole and ethyl acrylate first. Then IP-1 reacted with 1, 4-butane sultone to yield 1-ethoxycarbonylethylene-3-(4-sulfobutyl) imidazole (IP-2). The third step was the generation of 1-carboxye-thylene-3-(4-sulfobutyl) imidazolium chloride (IP-3). The Brønsted-Lewis acidic ionic liquid BLILs-1 was synthesized lastly by the reaction of IP-3 and zinc acetate, and the yield of the target product was 89.6%. In order to improve the acid strength and acid amount, diethyl maleate was used as a substitute for ethyl acrylate to introduce Bronsted acid site in the first step and other steps were similar to that for the synthesis of BLILs-1. Thus another Brønsted-Lewis acidic ionic liquid BLILs-2 was obtained (91.3% yield). The synthetic route was shown in **Scheme 2.**

2.2. Characterization Method

The structures of the intermediates (IP-2 and IP-3) were determined by means of FT-IR while the structures of intermediates (IP-4, IP-5 and IP-6) were determined by means of ^{1}H NMR analysis. The structures of BLILs-1 and BLILs-2 were characterized by IR, ^{1}H NMR and Elemental analysis, and the results indicated that the structures of the Brønsted-Lewis acidic ionic liquids synthesized in this work were corresponding to those of the target products.

2.3. Property of Ionic Liquids

The acidic type of the two ionic liquids was determined by FT-IR using pyridine as a probe (Py-IR). Pyridine can react separately with Brønsted and Lewis acids to produce the cation $[PyH]^+$ and the Py-Lewis complex, respectively. $[PyH]^+$ has an absorption peak near 1540 cm^{-1} in the FT-IR spectra, and the absorption peak of Py-

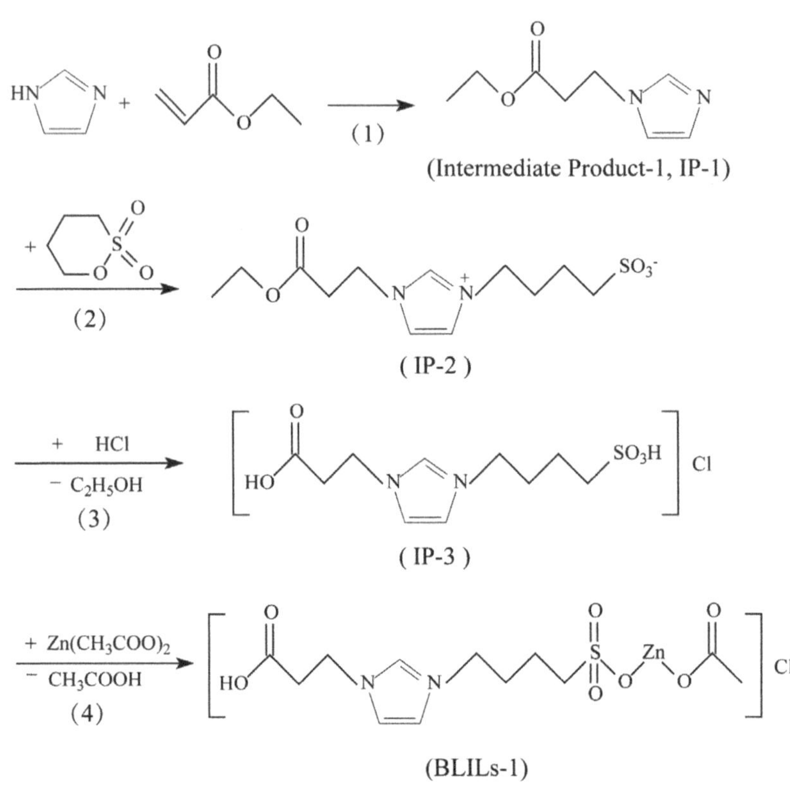

Scheme 1. Synthesis route of [CH_3COO-Zn-O_3S-bim-CH_2CH_2COOH]Cl.

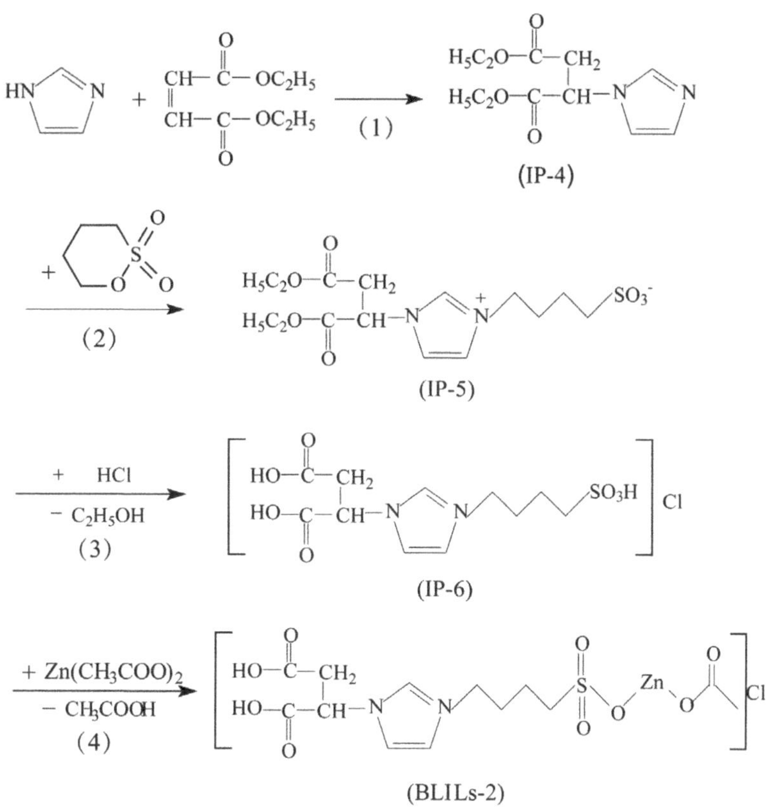

Scheme 2. Synthesis route of [CH_3COO-Zn-O_3S-bim-$C_4H_5O_4$]Cl.

Lewis complex is close to 1450 cm^{-1}. By observing these two peaks, the acidic type of a sample can be determined [6]. The characterization procedure in this work was as follows: the ionic liquids were heated and dissolved first, and then mixed with pyridine by a volume ratio of 5:2. The FT-IR spectra were finally obtained as shown in **Figure 1**. Compared with the spectra of pyridine, two new absorption peaks near 1540 and 1450 cm^{-1} appeared in the FT-IR spectra of Py + BLILs-1 and Py + BLILs-2, suggesting that both BLILs-1 and BLILs-2 show Brønsted and Lewis acidity.

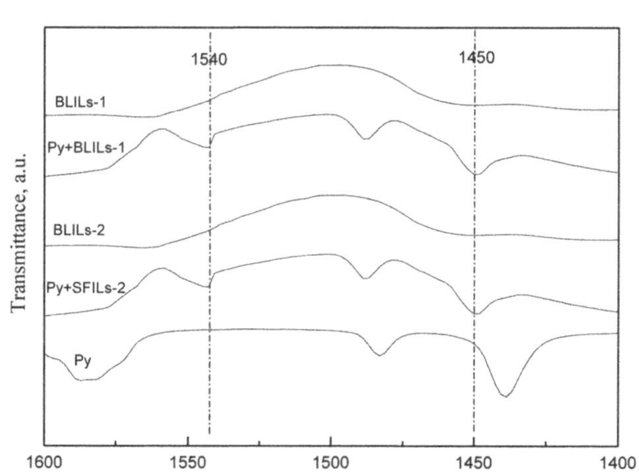

Figure 1. Py-IR spectra of samples.

The acid strength values (H_0) of the ionic liquids were determined by the UV-visible spectroscopy combined with Hammett indicator method at room temperature [13]. From **Table 1** it can be seen that H_0 values of the BLILs-1 and BLILs-2 were 5.6 and 4.2, respectively, indicating that the acid strength of BLILs-2 is higher than that of BLILs-1. The acid amount of the two Brønsted-Lewis acidic ionic liquids determined by acid-base titration was 2.0 mol NaOH/mol BLILs-1 and 3.0 mol NaOH/mol BLILs-2, respectively. The number of moles of -COOH in BLILs-2 (2.0 mol) were more than that in BLILs-1 (1.0 mol), which may be responsible for the higher acid strength and acid amount of BLILs-2.

The melting points of BLILs-1 and BLILs-2 were 80°C - 81°C and 85°C - 86°C, respectively, so the Brønsted-Lewis acidic ionic liquids with lower melting point

Table 1. The acid strength and acid amount of Brønsted-Lewis acidic ionic liquids[a].

BLILs	H_0	Acid amount (mol NaOH/mol BLILs)
BLILs-1	5.6[b]	2.0
BLILs-2	4.2[c]	3.0

[a] $H_0 = pK_a + \log([I]/[HI])$, acid amount was measured by acid-base titration; [b] Indicator: methyl red; [c] Indicator: dimethyl yellow.

are expected to be synthesized by changing the structures of anions or cations in the future, which are more beneficial for the applications of Brønsted-Lewis acidic ionic liquids in catalysis.

3. Conclusion

Two novel Brønsted-Lewis acidic ionic liquids, [CH_3-COO-Zn-O_3S-bim-CH_2CH_2COOH]Cl and [CH_3COO-Zn-O_3S-bim-$C_4H_5O_4$]Cl, were synthesized and characterized in this paper. Compared with [CH_3COO-Zn-O_3S-bim-CH_2CH_2COOH]Cl, [CH_3COO-Zn-O_3S-bim-$C_4H_5O_4$]Cl has a higher acid strength and acid amount. However, their melting points are rather high from the view of application as a solvent or a catalyst. A further attempt will be made to lower the melting point by adjusting the substitute group in cation or matching anion with cation.

4. Experimental

4.1. Materials

Imidazole was purchased from Tianjin Guangfu Fine Chemical Research Institute; ethyl acrylate was obtained from Tianjin Chemical Reagent Co., Ltd.; diethyl maleate was purchased from Tianjin Kemiou Chemical Reagent Co., Ltd.; 1,4-butane sultone was supplied by Wuhan Fengfan Chemical Co., Ltd.; $Zn(OAc)_2 \cdot 2H_2O$ was purchased from Tianjin Jiangtian Chemical Technology Co., Ltd. All the materials above were of analytical grade and used without further purification except that anhydrous zinc acetate was obtained by removal of crystal water from $Zn(OAc)_2 \cdot 2H_2O$ at 373 K for 2 h, and 1,4-butane sultone was purified by distillation before use.

4.2 Characterization Procedures

The Fourier transform infrared (FT-IR) spectra of the ionic liquids were recorded on a Bruker Vector 22 infrared spectroscopy (liquid film or KBr pellet) between 400 and 4000 cm^{-1} and collected at a resolution of 4 cm^{-1}. ^1H NMR spectra were recorded on a Bruker Avance 400 instrument in D_2O with TMS as the external standard. Elemental analysis of the ionic liquids was performed with a Vairo EL elemental analysis instrument. The acid strength of ionic liquids was determined by the UV-visible spectroscopy combined with Hammett indicator method, which was recorded on a Cary 300 UV-visible spectroscope at room temperature.

4.3. Preparation of Ionic Liquids

The synthesis approach of the Brønsted-Lewis acidic ionic liquids consists of four steps. Take BLILs-1 as an example.

4.3.1. 1-Ethoxycarbonylethylene Imidazole (IP-1)

Imidazole (0.5 mol), ethyl acrylate (0.55 mol) and solvent toluene were charged into a 250 mL three-necked flask equipped with a mechanical stirrer. Then the mixture was heated to 100°C for 24 h under vigorous stirring. After that, toluene and remaining ethyl acrylate were distilled off from the reaction mixture under vacuum at 100°C and a brown liquid IP-1 was obtained.

4.3.2. 1-Ethoxycarbonylethylene-3-(4-Sulfobutyl) Imidazole (IP-2)

IP-1 (0.2 mol) and 1,4-butane sultone (0.2 mol) were stirred at 50°C for 12 h, and a viscous liquid was obtained. The reaction mixture was washed by toluene and ethyl ether for three times, respectively. The viscosity got larger with washing operation, and then a gelatinous substance was obtained. The gelatinous substance was dried to constant weight in vacuum to afford IP-2 in the form of white solid at room temperature.

4.3.3. 1-Carboxyethylene-3-(4-Sulfobutyl) Imidazolium Chloride (IP-3)

IP-2 (0.1 mol) was dissolved in distilled water and then excessive HCl was added dropwise into the aqueous solution of IP-2 with vigorous stirring at room temperature. After that, the mixture was heated to 100°C and kept for 2 h. Water, remaining HCl and byproduct ethanol were removed by distillation under vacuum at 100°C, and the resultant viscous liquid was dried to constant weight under vacuum at 100°C to give a Brønsted acidic ionic liquid IP-3.

4.3.4. 1-Carboxyethylene-3-(4-Zinc Acetate Sulfobutyl) Imidazolium Chloride (BLILs-1)

IP-3 (0.1 mol) was dissolved in distilled water, and then anhydrous zinc acetate (0.1 mol) was added into the aqueous solution of IP-3. The mixture was stirred and heated to 50°C and kept for 4 h. Water was removed by vacuum distillation at 100°C, and the obtained viscous liquid was dried to constant weight under vacuum at 100°C to give the Brønsted-Lewis acidic ionic liquid BLILs-1 as a solid at room temperature.

BLILs-2 was prepared in a similar way and the difference was that diethyl maleate was used as a substitute for ethyl acrylate to introduce Bronsted acid site.

5. Characterization

5.1. 1-Carboxyethylene-3-(4-Zinc Acetate Sulfobutyl) Imidazolium Chloride (BLILs-1)

5.1.1. 1-Ethoxycarbonylethylene-3-(4-Sulfobutyl) Imidazole (IP-2)

FTIR(KBr): 1040, 1202, 1378, 1456, 1566, 1645, 1730, 2937, 2982, 3093, 3142 cm^{-1}.

5.1.2. 1-Carboxyethylene-3-(4-Sulfobutyl) Imidazolium Chloride (IP-3)

FTIR(KBr): 1038, 1162, 1414, 1453, 1566, 1725, 2853, 2924, 3104, 3145, 3406 cm^{-1}.

5.1.3. 1-Carboxyethylene-3-(4-Zinc Acetate Sulfobutyl) Imidazolium Chloride (BLILs-1)

FTIR(KBr): 1040, 1162, 1406, 1620, 1721, 2962, 3107, 3145, 3429 cm^{-1}. ^1H NMR (D_2O, δ, 400 MHz) δppm: 1.755 (2 H, m, $-CH_2-$), 1.975 (3 H, s, $-CH_3$), 2.032 (2 H, m, $-CH_2-$), 2.785[14] (2 H, t, $-CH_2$-COOH), 2.956[15] (2 H, t, $S-CH_2-$), 4.250[6] (2 H, t, $-CH_2$-N), 4.421[14] (2 H, t, $N-CH_2-$), 7.509[14] (1 H, d, C = CH-N), 7.515[14] (1 H, d, N-CH = C), 8.805[14] (1 H, s, N = CH-N). Anal.Calcd for BLILs-1: C 34.13%, H 4.58%, N 6.93%, S 7.34%, Cl 8.15%; Found: C 34.31%, H 4.81%, N 6.70%, S 7.05%, Cl 7.89%.

5.2. 1-(1,2-Dicarboxy) Ethylene-3-(4-Zinc Acetate Sulfobutyl) Imidazolium Chloride (BLILs-2)

5.2.1. 1-(1,2-Diethoxycarbonyl) Ethylene Imidazole (IP-4)

^1H NMR (D_2O, δ, 400 MHz) δppm: 1.092 (3 H, t, H_3C-), 1.191 (3 H, t, H_3C-), 3.176 (2 H, d, $-CH_2-$), 4.067 (2 H, q, $-CH_2-$), 4.140 (2 H, q, $-CH_2-$), 5.436 (1 H, t, -CH-), 6.958 (1 H, d, C = CH-N), 7.138 (1 H, d, N-CH = C), 7.700 (1 H, s, N = CH-N).

5.2.2. 1-(1,2-Diethoxycarbonyl) Ethylene-3-(4-Sulfobutyl) Imidazole (IP-5)

^1H NMR (D_2O, δ, 400 MHz) δppm: 1.184 (6 H, m, H_3C-), 1.688 (2 H, m, $-CH_2-$), 1.976 (2 H, m, $-CH_2-$), 2.830 (2 H, t, $-CH_2-$), 3.368 (2 H, d, $-CH_2-$), 4.129 (2 H, t, $-CH_2-$), 4.217 (4 H, m, $-CH_2-$), 5.701 (1 H, t, -CH-), 7.476 (1 H, d, C=CH-N), 7.561 (1 H, d, N-CH = C), 8.881 (1 H, s, N = CH-N).

5.2.3. 1-(1,2-dicarboxy) Ethylene-3-(4-Sulfobutyl) Imidazolium Chloride (IP-6)

^1H NMR (D_2O, δ, 400 MHz) δppm: 1.573 (2 H, m, $-CH_2-$), 1.819 (2 H, m, $-CH_2-$), 2.737 (2 H, t, $-CH_2-$), 3.187 (2 H, d, $-CH_2-$), 4.078(2 H, t, $-CH_2-$), 5.489 (1 H, t, -CH-), 7.400 (1 H, d, C = CH-N), 7.474 (1 H, d, N-CH = C), 8.851 (1 H, s, N = CH-N).

5.2.4. 1-(1,2-Dicarboxy) Ethylene-3-(4-Zinc Acetate Sulfobutyl) Imidazolium Chloride (BLILs-2)

FTIR (KBr): 728, 1038, 1160, 1399, 1570, 1630, 1730, 2968, 3147, 3416 cm^{-1}. ^1H NMR (D_2O, δ, 400 MHz) δppm: 1.660 (2 H, m, $-CH_2-$), 1.901 (2 H, m, $-CH_2-$), 1.929 (3 H, s, $-CH_3$), 2.897 (2 H, t, $-CH_2-$), 3.116 (2 H, d, $-CH_2-$), 4.138 (2 H, t, $-CH_2-$), 5.154 (1 H, t, -CH-), 7.412 (1 H, d, C = CH-N), 7.458 (1 H, d, N-CH = C), 8.856 (1 H, s, N = CH-N). Anal.Calcd for BLILs-2: C 32.56%, H 3.97%, S 6.68%, N 5.84%, Cl 7.31%; Found: C 30.81%, H 3.90%, S 6.76%, N 6.02%, Cl 7.47%.

6. Acknowledgements

The authors are grateful to the Special Program of National Basic Research Program of China (973 Program) (Grant 2010CB234602), National Natural Science Foundation of China (Grant 21076059), the Natural Science Foundation of Tianjin City (Grant 12JCYBJC12800), and the Key Basic Program of Applied Basic Research Plan of Hebei Province (Grant 12965642D).

REFERENCES

[1] L. J. Jia, Y. Y. Wang, H. Chen, Y. K. Shan and L. Y. Dai, "Alkylation of Benzene with 1-Hexene in Acidic Ionic Liquid Systems: $Et_3NHCl-FeCl_3$ and $Et_3NHCl-AlCl_3$ Ionic Liquids," *Reaction Kinetics and Catalysis Letters*, Vol. 86, No. 2, 2005, pp. 267-273.

[2] A. C. Cole, J. L. Jensen, I. Ntai, K. L. Tran, K. J. Weaver, D. C. Forbes and J. H. Davis Jr., "Novel Brønsted Acidic Ionic Liquids and Their Use as Dual Solvent-Catalysts," *Journal of the American Chemical Society*, Vol. 124, No. 21, 2002, pp. 5962-5963.

[3] L. M. Liu, F. Li, Y. J. Wang and X. Q. Zhao, "One Step Synthesis of Methylene Diphenyl Dicarbamate," *Chinese Journal of Catalysis*, Vol. 28, No. 8, 2007, pp. 667-669.

[4] X. H. Wang, G. H. Tao, Z. Y. Zhang and Y. Kou, "Synthesis and Characterization of Dual Acidic Ionic Liquids," *Chinese Chemical Letters*, Vol. 16, No. 12, 2005, pp. 1563-1565.

[5] S. W. Liu, C. X. Xie, S. T. Yu and F. S. Liu, "Dimerization of rosin using Brønsted-Lewis Acidic Ionic Liquid as Catalyst," *Catalysis Communication*, Vol. 9, No. 10, 2008, pp. 2030-2034.

[6] S. W. Liu, H. X. Zhou, S. T. Yu, C. X. Xie, F. S. Liu and Z. Q. Song, "Dimerization of Fatty Acid Methyl Ester Using Bronsted-Lewis Acidic Ionic Liquid as Catalyst," *Chemical Engineering Journal*, Vol. 174, No. 1, 2011, pp. 396-399.

[7] S. W. Liu, C. X. Xie, S. T. Yu, M. Xian and F. S. Liu, "A Bronsted-Lewis Acidic Ionic Liquid: Its Synthesis and Use as the Catalyst in Rosin Dimerization," *Chinese Journal of Catalysis*, Vol. 30, No. 5, 2009, pp. 401-406.

[8] S. T. Yu, S. W. Liu, F. S. Liu, C. X. Xie and K. H. Ji, "A Bronsted-Lewis Acidic Ionic Liquid and Its Application in the Reaction of Rosin Dimerization," CN Patent No. 101191035 B, 2010.

[9] S. W. Liu, L. Li, S. T. Yu, C. X. Xie, F. S. Liu and Z. Q. Song, "Polymerization of Fatty Acid Methyl Ester Using Acidic Ionic Liquid as Catalyst," *Chinese Journal of Ca-*

talysis, Vol. 31, No. 11-12, 2010, pp. 1433-1438.

[10] N. N. Quan, S. H. Bao and J. G. Yang, "Highly Efficient Procedure for the Conjugate Addition of Amines to Electron-Deficient Alkenes Catalyzed by Novel Ionic Liquid," *Chinese Science Bulletin*, Vol. 55, No. 23, 2010, pp. 2512-2516.

[11] X. Liang and C. Qi, "Synthesis of a Novel Ionic Liquid with both Lewis and Bronsted Acid Sites and Its Catalytic Activities," *Catalysis Communication*, Vol. 12, No. 9, 2011, pp. 808-812.

[12] X. Jiang, W. Ye, X. Song, W. Ma, X. Lao and R. Shen, "Novel Ionic Liquid with both Lewis and Bronsted Acid Sites for Michael Addition," *International Journal of Molecular Sciences*, Vol. 12, No. 11, 2011, pp. 7438-7444.

[13] Y. L. Geng, L. Y. Hu, X. Q. Zhao, H. L. An and Y. J. Wang, "Synthesis of 4, 4'-MDC in the Presence of Sulfonic Acid-Functionalized Ionic Liquids," *Chinese Journal of Chemical Engineering*, Vol. 17, No. 5, 2009, pp. 756-760.

[14] Z. F. Fei, D. B. Zhao, T. J. Geldbach, R. Scopelliti and P. J. Dyson, "Brønsted Acidic Ionic Liquids and Their Zwitterions: Synthesis, Characterization and pKa Determination," *Chemistry: A European Journal*, Vol. 10, No. 19, 2004, pp. 4886-4893.

[15] X. F. Liu, L. F. Xiao, H. Wu, Z. Li, J. Chen and C. G. Xia, "Novel Acidic Ionic Liquids Mediated Zinc Chloride: Highly Effective Catalysts for the Beckmann Rearrangement," *Catalysis Communications*, Vol. 10, No. 5, 2009, pp. 424-427.

Recent Applications of Electrocoagulation in Treatment of Water and Wastewater—A Review

Ville Kuokkanen[1*], Toivo Kuokkanen[1], Jaakko Rämö[2], Ulla Lassi[1,3]
[1]Department of Chemistry, University of Oulu, Oulu, Finland
[2]Thule Institute, University of Oulu, Oulu, Finland
[3]Kokkola University Consortium Chydenius, Kokkola, Finland

ABSTRACT

During the last two decades, and particularly during the last few years, the environmental sector has shown a largely growing interest in the treatment of different types of water and wastewater by electrocoagulation (EC). The aim of this work was to review studies, conducted mainly during 2008-2011, on the wide and versatile range of feasible EC applications employed in the purification of different types of water and wastewater. The EC applications discussed here were divided into 7 following categories: tannery, textile and colored wastewater; pulp and paper industry wastewater; oily wastewater; food industry wastewater; other types of industrial wastewater; surface water as well as model water and wastewater containing heavy metals, nutrients, cyanide and other elements and ions. In addition, this paper presents an overview of the optimum process conditions (treatment times, current densities, and initial pH) and removal efficiencies (mostly high) achieved for the EC applications discussed. In the vast majority of the studies discussed in this review, the aforementioned values were found to be in the range of 5 - 60 min (typically less than 30 min), 10 - 150 A/m^2 and near neutral pH, respectively. Both operating costs and electrical energy consumption values were found to vary greatly depending on the type of solution being treated, being between 0.0047 - 6.74 €/m^3 and 0.002 - 58.0 kWh/m^3, but in general they were rather low (typically around 0.1 - 1.0 €/m^3 and 0.4 - 4.0 kWh/m^3).

Keywords: Electrochemistry; Electrocoagulation (EC); Water Treatment; Wastewater Purification; Technoeconomic Evaluation

1. Introduction

Electrocoagulation (EC) is an emerging technology that combines the functions and advantages of conventional coagulation, flotation, and electrochemistry in water and wastewater treatment. Each of these fundamental technologies has been widely studied separately. However, a quantitative appreciation of the mechanism of interaction between these technologies employed in an electrocoagulation system is absent [1].

EC has been known for over a century. Aluminium/iron-based electrocoagulation was patented in the US already in 1909. EC was studied extensively in the latter half of the 20th century in both the US and the Soviet Union (former USSR), but at that time it was not found to be widely feasible for water treatment. This was mainly due to the then high electricity and investment costs [2].

Meanwhile, the demand for quality drinking water quality is increasing globally and environmental regulations regarding wastewater discharge are becoming increasingly stringent. Therefore, it has become necessary to develop more effective treatment methods for water purification and/or enhance the operation of current methods. This and eco-friendliness have led to increasing global interest in electrocoagulation as a research subject. Over the course of the last few decades, literature in the environmental sector has indeed shown a growing interest towards the treatment of different types of wastewater by EC. Particularly during the last few years, the amount of published literature on EC applications seems to have increased substantially.

Practical review papers on EC applications have been largely absent so far. To the best of our knowledge, only a few authors, e.g. [3,4] have addressed the subject recently, in addition to older reviews, (namely [1,2,5]), even though a significant number of studies on EC have

*Corresponding author.

been reported in the literature since then. Therefore, there is a need for an update on recent applications of EC. The aim of this work was to accomplish this, and based on the literature, to present an overview of practical optimum treatment times, current densities, electricity consumption, and operating costs in a wide and versatile range of feasible applications of EC in water and wastewater treatment, studied mainly during the years 2008-2011.

2. Background

2.1. Principles of Electrocoagulation

Electrolysis is a process in which oxidation and reduction reactions take place when electric current is applied to an electrolytic solution. Electrocoagulation is based on dissolution of the electrode material used as an anode. This so called "sacrificial anode" produces metal ions which act as coagulant agents in the aqueous solution *in situ* [1]. At its simplest, an electrocoagulation system consists of an anode and a cathode made of metal plates, both submerged in the aqueous solution being treated [3]. The electrodes are usually made of aluminum, iron, or stainless steel (SS), because these metals are cheap, readily available, proven effective, and non-toxic. Thus they have been adopted as the main electrode materials used in EC systems [6,7]. The configurations of EC systems vary. An EC system may contain either one or multiple anode-cathode pairs and may be connected in either a monopolar or a bipolar mode [3]. During EC, the following main reactions take place at the electrodes. Anodic reactions [2]:

$$Al(s) \rightarrow Al^{3+} + 3e^- \quad E^0 = +1.66\ V \tag{1}$$

$$Fe(s) \rightarrow Fe^{2+} + 2e^- \quad E^0 = +0.44\ V \tag{2}$$

$$2H_2O(l) \rightarrow O_2(g) + 4\,H^+ + 4e^- \quad E^0 = -1.23V \tag{3}$$

Ferrous iron may be oxidized to Fe^{3+} by atmospheric oxygen or anode oxidation, and may be considered as [8]:

$$Fe^{2+} \rightarrow Fe^{3+} + e^- \quad E^0 = -0.77\ V \tag{4}$$

$$2Fe^{2+} + \frac{1}{2}O_2(g) + H_2O(l) \rightarrow 2Fe^{3+} + 2\,OH^- \quad E_0 = -0.37\ V \tag{5}$$

Cathodic reactions [2]:

$$2H_2O + 2e^- \rightarrow H_2(g) + 2OH^- \quad E^0 = -0.83\ V \tag{6}$$

Additionally, when chloride is present and the anode potential is sufficiently high, the following reactions may take place in the EC cell [9]:

$$2Cl^- \rightarrow Cl_2 + 2e^- \quad E^0 = -1.36\ V \tag{7}$$

$$Cl_2 + H_2O \rightarrow HClO + H^+Cl^- \quad E^0 = -0.93\ V \tag{8}$$

$$HClO \rightarrow H^+ + OCl^- \tag{9}$$

The formation of active chlorine species (Cl_2, HClO, OCl^-) enhances the performance of the EC reactor through oxidation reactions. The dissolution of the anode metal is based on Faraday's law:

$$m_{metal} = \frac{I \times t_s \times M}{z \times F} \tag{10}$$

where I is the applied current (A), t_s is the treatment time (s), M is the molar mass of the electrode material (M_{Al} = 26.982 g/mol, M_{Fe} = 55.845 g/mol), z is the valency of ions of the electrode material ($z_{Al} = 3$, $z_{Fe} = 2$), and F is Faraday's constant (96485 C/mol).

It has been found that the theoretical amount of anodic dissolution is exceeded in real EC applications. This phenomenon is referred to as superfaradaic efficiency, and the experimental values of anode metal dissolution have varied between 105% and 190% of the theoretically expected value [10-15]. This phenomenon is thought to be attributed to pitting corrosion, especially in the presence of chlorine ions [2].

Electrochemically generated metal cations will react spontaneously, forming various monomeric species such as $Al(OH)^{2+}$, $Al(OH)^{2+}$, $Al_2(OH)_2^{4+}$, and $Al(OH)^{4-}$ and polymeric species such as $Al_6(OH)_{15}^{3+}$, $Al_7(OH)_{17}^{4+}$, $Al_8(OH)_{20}^{4+}$, $Al_{13}O_4(OH)_{24}^{4+}$, and $Al_{13}(OH)_{34}^{5+}$, which finally transform into $Al(OH)_3$ according to complex precipitation kinetics [16]. Ferric ions generated electrochemically may form monomeric ions, ferric hydroxo complexes with OH^- ions, and polymeric species. These species/ions are: $FeOH^{2+}$, $Fe(OH)_2^+$, $Fe_2(OH)_2^{4+}$, $Fe(OH)_4^-$, $Fe(H_2O)_5OH^{2+}$, $Fe(H_2O)_4(OH)_2^+$, $Fe(H_2O)_8(OH)_2^{4+}$, and $Fe_2(H_2O)_6(OH)_4^{2+}$, which further react to form $Fe(OH)_3$ [17-19]. The formation of these complexes depends strongly on the pH of the solution. Above pH 9, $Al(OH)^{4-}$ and $Fe(OH)^{4-}$ are the dominant species [20].

Aluminum and iron hydrolysis products then destabilize pollutants present in the solution, allowing agglomeration and further separation from the solution by settling or flotation. Destabilization is achieved mainly by means of two distinct mechanisms, *i.e.* 1) charge neutralization of negatively charged colloids by cationic hydrolysis products; and 2) "sweep flocculation", where impurities are trapped and removed in the amorphous hydroxide precipitate produced. Several factors such as pH and coagulant dosage have an impact on the relative importance of charge neutralization and sweep flocculation. Microbubbles (H_2 and O_2) released at the electrode surfaces bring about electroflotation by adhering to agglomerates and carrying them to the water surface [21].

The most important factors influencing the efficiency of the EC process are the electrode materials used, applied current density, treatment time, and solution chemistry, including initial pH and the chemical composition of the aqueous solution being removed. The solution temperature, type of salt used to raise conductivity, presence of chloride, electrode gap, passivation of the anode, and water flow rate also have an impact on the removal efficiency and economic durability of a given EC application.

The advantages of EC over conventional coagulation (CC) include economic aspects (relatively low investment, maintenance, energy, and treatment costs), significantly lower volume of sludge produced, better sludge quality (lower water content, much larger and more stable flocs with better settlability), similar or slightly better efficiency, avoidance of chemical additions, ease of automation, simple equipment and compact size of EC systems (allowing decentralized treatment), greater functional pH range and pH neutralization effect, and the presence of electroflotation (EF) [5].

2.2. Evaluation Principles

The removal efficiencies (R%) presented in Chapters 3.1 - 3.7 have been calculated with the Equation (11):

$$R\% = \frac{c_0 - c_1}{c_0} \times 100 \quad (11)$$

where c_0 and c_1 are pollutant concentrations before and after EC treatment, respectively. Hydraulic retention times (HRT, min) were calculated with Equation (12):

$$HRT = V/Q \quad (12)$$

where Q is the flow rate (l/min). Current densities (i, A/m^2) can be calculated with the Equation (13):

$$i = I/A_{eff} \quad (13)$$

where A_{eff} is the effective, submerged area of the anode (m^2). When no current density values were given by the authors, either the current or voltage value in optimum conditions is presented in **Tables 1-7**. The electrical energy consumption (EEC, kWh/m^3) presented in Chapters 3.1 - 3.7 has been calculated with the Equation (14):

$$EEC = \frac{U \times I \times t}{60 \times V} \quad (14)$$

where U is the applied voltage (V), t is the treatment time (min) and V is the volume of the treated water (dm^3). Operating costs (OC, €/m^3) have been calculated with the Equation (15):

$$OC = a \times EEC + b \times EMC \quad (15)$$

where a and b are the current market prices of electricity (€/kWh) and electrode materials (€/kg), respectively, and EMC (kg/m^3) is electrode material consumption. The market prices used in calculating economic value have varied slightly from one paper to another, depending on the country and year of publication (prices have been on the rise over time), but they have been within a very similar range. Derived from Equations (14) and (15), EEC and OC per kg of specific pollutant/parameter (kWh/kg_x and €/kg_x) can also be readily calculated based on the removal percentage and initial/final concentrations of the pollutant. These values have been presented also by some of the authors. The pH values presented in **Tables 1-7** represent the range in which the EC application performed the best (the highest value is marked in brackets), even though the process would have performed nearly as well with pH values outside of this range. If the natural pH of the aqueous solution was inside this range, it has been noted separately.

In a handful of papers, the author did not present economic values and current densities in optimum conditions, but they could be approximated by using the values (total submerged anode surface area, applied current and voltage or current density, volume of the wastewater treated, treatment time, initial concentration, removal efficiency, etc.) given in the article. It should be strongly underlined that these rather simple calculations were done only when the authors of the corresponding papers had not presented the numbers themselves but had clearly stated the values needed for the calculations, or when the values could be easily deducted from the publication in question. The approximation calculations were based on Equations (11) and (13)-(15). Current market prices were estimated to be approximately 0.10 - 0.11 €/kWh (in Finland in July 2011, including electrical energy, distribution of electricity, and taxes) and 1.6 - 1.7 €/kg for aluminum and 0.33 - 0.37 €/kg for iron. These values are also similar to those used in the reviewed papers. All currencies (usually United States dollar) given in the papers for OC values have been converted to euros (in **Tables 1-7**) using up-to-date exchange rates.

Additionally, in some articles all the optimum values were not clearly stated or no specific values were given. In such cases, the missing values have been estimated, if possible, and if the results were reasonable and in line with the text. Estimations were based on the figures, tables, and text presented. Also, in a few papers additional optimum parameters were taken into account, meaning e.g. that when a major drop in treatment time, current density, or EEC-value was found to correspond with only a slight reduction in removal efficiency, the lesser removal efficiency (and corresponding other) values were also considered optimal in economic terms. Whenever any of these actions have been performed, it has

Table 1. Recent applications of EC in the treatment of tannery, textile and colored wastewater.

Water & wastewater types used	Genuine (G)/ Synthetic (S) water	Anode/ Cathode material	Reactor type	Volume treated [ml]	Optimum electrode gap [mm]	Optimum current density, treatment time and initial pH [A/m^2], [min], []			Initial pollutant levels [mg/l]	Optimum removal efficiency [%]	Optimum EEC [kWh/m^3]	Optimum EEC [kWh/kg_X]	Optimum OC [€/m^3]	Optimum OC [€/kg_X]	Research group & Publication year
Tannery wastewater containing organic and inorganic pollutants	G	Fe	Batch	3000	50	22.4	20	(7 - 9)[c]	COD: 4100 - 6700 BOD: 630 - 975 Cr: 11.5 - 14.3 TSS: 600 - 955 TKN: 144 - 170 TDS: 13300 - 19700 O & G: 638 - 780 Color: 3800 - 6330 [Pt-Co]	COD: 95 BOD: 96 Cr: 100 TSS: 96 TKN: 62 TDS: 50 O & G: 99 Color: ~98[g]	0.13	n.d.	0.25	n.d.	Kongjao *et al.* 2008 [11]
Tannery wastewater	G	Fe	Batch	400	60	333	5/30[h]	7.4[d]	COD: 3700 Sulfide: 440 Cr_{total}: 22 SS: 2690	COD: 46[g]/56[h] Sulfide: 25/97[h] Cr_{total}: n.d./97 SS: n.d./70	~(3.13/ 15.63 - 16.49)[e]	1.8/~(9.0 - 9.5[h]) kWh/kg_{COD}	n.d.	n.d.	Apaydin *et al.* 2009 [26]
Industrial textile wastewater	G	Al	Batch	2200	20	80	70	7[c]	COD: 1260 Turbidity: 1310 [NTU] TS: 1750	COD: 70 Turbidity: 90 TS: 50	n.d.	n.d.	n.d.	n.d.	Zodi *et al.* 2010 [28]
Wastewater containing dyes from textile industry (Direct Red 81, azo dye)	S	Al	Batch	500	15	18.75	60	5 - 9 [6]	Dye: 25 - 200 (optimum 50)	Dye: 98	n.d.	n.d.	n.d.	n.d.	Aoudj *et al.* 2010 [29]
Dye-containing wastewater (Direct Red 23 & Reactive Blue 140)	S × 2	Fe	Continuous + hydrogen gas collecting	4420	8	30/40	5/5	n.d.	Dye: 100/100	COD: 93/n.d. Color: 99/89 TS: 89/n.d.	0.69/1.42	n.d.	>0.12	n.d.	Phalakornkule *et al.* 2010 [30]
Blue reactive (Reactive Blue 140), red disperse (Disperse Red 1), mixed dyes and a real textile wastewater	S × 3 + G	Al Fe[b] Al-Fe	Batch	1800	8	30 - 40	5	(7 - 9.6)[c]	COD: 278 - 736 Color: n.d. TSS: 85 - 354 TDS: 1715 - 6106 TS: 1800 - 6460	[COD: ~90/ (55 - 79) Color: ~100/ (79 - 97) TSS: n.d./ (55 - 96) TDS: n.d./ (21 - 23) TS: n.d./ (26 - 28)][g]	~(0.3 - 1.0)[g]/ (0.42 - 1.62)[g]	n.d.	n.d.	n.d.	Phalakornkule *et al.* 2010 [31]
Crystal violet solution	S	Al/SSFe /SS[b]	Batch	500	11	28	5	5.8	COD: n.d. Dye: 50 - 200 (optimum 200)	COD: ~100[g] Dye: ~96[g]	0.4	n.d.	n.d.	n.d.	Durango-Usuga *et al.* 2010 [32]
Acid green dye 50	S	Al	Batch	n.d.	10	16.7	21	6.9 - 11 [9]	COD: n.d. Dye: 100 - 300 (optimum 100)	COD: 87 Color: 96[g]	~0.48[e]	3.82 kWh/kg_{COD} ~4.8 kWh/kg_{dye}[g]	n.d.	n.d.	El-Ashtoukhy & Amin 2010 [33]
Acid brown 14 (azo dye)	S	Al Al/Fe[b] Fe	Batch Batch (pilot scale)	500 9000	10 30	6.329 2.58	18 198	4 - 6.4 [6.4][c]	COD: n.d. Dye: 30 - 100 (optimum 50)	COD: 87 64 Color: 91 80	~(0.053 - 0.037)[e]	~(1.21 - 1.15)[e] kWh/kg_{COD} ~(1.16 - 0.92)[e] kWh/kg_{dye}	n.d.	n.d.	Parsa *et al.* 2011 [34]
Acid red 14 (azo dye)	S	Fe/Steel	Batch	250	10	100	4	4 - 10 [7]	Dye: 50	Color: 91	n.d.	n.d.	n.d.	n.d.	Aleboyeh *et al.* 2008 [35]

Continued

Orange II (azo) dye solution	S	Al	Continuous	3000	6	120	8.6[a] (0.35 l/min)	6.5	Dye: 10 - 50 (optimum 10)	Color: 95	n.d.	n.d.	n.d.	n.d.	Mollah *et al.* 2010 [36]
Reactive black 5 (azo dye)	S	Al Fe[b]	Batch	500	25	45.75	5	5 - 9 [5]	Dye: 40 - 200 (optimum 100)	Color: 99	~0.50[e]	4.96 kWh/kg_{dye}	n.d.	n.d.	Şengil & Özacar 2009 [12]
Indigo carmine (Acid Blue 74)	S	Mild steel/SS	Batch	1000	3	10.91/ 54.57[h]	180/ 15[h]	5 - 9 [8]	Dye: 25 - 100 (optimum 40)	Color: 99/91[h]	~0.68/0.72[e]	17/18[h] kWh/kg_{color}	n.d.	n.d.	Secula *et al.* 2011 [37]
Red dye solution mixture of 2-naphtoic acid & 2-naphtol	S	Al	Continuous	n.d.	10	312.5	5[a] (0.62 l/min)	6 - 9 [6]	COD: 2500 Dye: 25 - 200 (optimum 80 - 100)	COD: >80/>80[h] Color: 95/~91[h]	~0.30[e]/n.d.	3.2/1.6[h] kWh/kg_{dye} ~0.15[e]/n.d. kg_{COD}	n.d.	n.d.	Merzouk *et al.* 2009 [38]

[a] = HRT (hydraulic retention time) in EC systems with continuous mode of operation [min]; [b] = Observed as the best electrode configuration of those tested; [c] = The natural, unmodified pH value of the water or wastewater (found optimal); [d] = The natural, unmodified pH value of the water or wastewater (the effect of pH not researched); [e] = Approximation calculation based on values given in the article at issue; [g] = Optimum value estimated from the data in the article (precise value not given); [h] = Additional "optimum value" estimated from the data in the article; n.d. = Not determined.

been marked accordingly in **Tables 1-7**.

3. Overview of Different Types of Water and Wastewater Recently Treated by Electrocoagulation

Chapters 3.1 - 3.7 present a summary of recent applications of EC with different types of water and wastewater divided into categories by topic. Removal efficiencies, economic values and essential operational parameters in optimum process conditions are presented in **Tables 1-7** along with other specifications (*i.e.* electrode materials, genuineness/artificiality and initial pollutant levels of the water, reactor type, volume of water treated, electrode gap) of the research in question. If multiple electrode materials were tested, the optimum values presented are for the electrode configuration found best (if any), which is noted in the corresponding columns in **Tables 1-7**.

3.1. Tannery, Textile and Colored Wastewater

The global tannery industry represents an important economic sector in many countries. The quantity of effluent generated is about 30 l for every kilogram of hide or skin processed and it contains high concentrations of organic pollutants and Cr (III), which could be oxidized to highly toxic and carcinogenic Cr (VI) [22,23].

Dye-containing wastewaters are a major environmental concern because of their unaesthetic nature and their ability to hinder the penetration of light into water, which is detrimental to living organisms in bodies of water [24,25]. Azo dyes are one of the most widely used synthetic dyes. They can be toxic and mutagenic to aquatic life and are molecularly stable, rendering them resistant to biological and even chemical degradation [24]. **Table 1** presents a summary of recent applications of EC in the treatment of tannery, textile and colored wastewater.

A study has been conducted on the treatment of wastewater from a tannery plant using the EC technique. A bench-scale system with iron electrodes was employed. The wastewater had high initial pollution parameter levels (see **Table 1**). After optimization, the EC treatment was found very effective and produced clear water (see **Figure 1**). The natural pH of the wastewater (7.0 - 8.7) was found to be within the optimum range of values. The EEC and OC values were found to be low, 0.13 kWh/m^3 and 0.25 $€/m^3$, respectively. Parallel monopolar connections were found to be more suitable for the treatment process than monopolar serial and dipolar parallel connection modes [11]. Another investigation compared EC and EF (Electro-Fenton process - the addition of H_2O_2 to an EC process to bring about oxidation reactions) in treating genuine, highly polluted tannery wastewater [26]. It is worth mentioning that treatment of tannery wastewater by conventional biological methods is often inadequate for complete purification, especially of ammo-

Figure 1. Tannery wastewater (a) before and (b) after treatment by electrocoagulation. Adapted from [11].

Table 2. Recent applications of EC in the treatment of paper industry wastewater.

Water & wastewater types used	Genuine (G)/ synthetic (S) water	Anode/ cathode material	Reactor type	Volume treated [ml]	Optimum electrode gap [mm]	Optimum current density, treatment time and initial pH [A/m²], [min], []			Initial pollutant levels [mg/l]	Optimum removal efficiency [%]	Optimum EEC [kWh/m³]	Optimum EEC [kWh/kgX]	Optimum OC [€/m³]	Optimum OC [€/kgX]	Research group & publication year
Black liquor wastewater (from pulp and paper industry)	G	Al[b] Fe	Batch	300	5	140	50	5 - 7 [7]	COD: 7960 Polyphenols: 3220 Color: n.d. TSS: 1160	COD: 98 Polyphenols: 92 Color: 99 TSS: n.d.	n.d.	n.d.	n.d.	n.d.	Zaied & Bellakhal 2009 [45]
Bleached kraft mill effluent (pre-treated by sedimentation & aeration)	G	Al[b] Fe	Batch	250	20	48	2/7.5[h]	7.6[d]	COD: 426 BOD: 26 Lignin: 13514 Phenol: 0.535	COD: ~46[g]/75[h] BOD: ~34[g]/70[h] Lignin: ~79[g]/80[h] Phenol: ~96[g]/98[h]	~(0.12/ 0.46)[e]	~120/~350 kWh/kg$_{COD}$[g] ~3.2/~9.5 kWh/kg$_{lignin}$[g]	n.d.	n.d.	Uğurlu *et al.* 2008 [46]
Paper mill wastewater (gravitationally & biologically pre-treated)	G	Al[b, g] Fe	Batch	2500	20	150[g]	90[g]	7.7[d]	COD: 285 DOC: 75 Turbidity: 35 [NTU] Lignin: n.d. As: 3.8	COD: 68 DOC: 46 Turbidity: 100 Lignin: 30 (Fe: 75) As: 92	n.d.	n.d.	n.d.	n.d.	Zodi *et al.* 2011 [47]
Paper mill wastewater	G	Al Al/Fe Fe/Al Fe	Batch	1500	10	700	30	(5 - 7)[c]	COD: 1700 Color: n.d. Phenol: 34 BOD: 850 TOC: 910 TSS: 1060 TS: 9801	COD: 86[g] (Fe) Color: 92[g] (Al) Phenol: 96[g] (Fe) BOD: n.d. TOC: n.d. TSS: n.d. TS: n.d.	n.d.	n.d.	n.d.	n.d.	Katal & Pahlavanzadeh 2011 [48]
Tissue paper wastewater	G	Al	Batch	300 4500 (scale-up)	10 10	10.8 - 16.2 10.8	30 60	(6.7 - 7)[d]	Turbidity: 80 - 120 [NTU] COD: 2645 BOD: 1688 TSS: 205 TDS: 1200 - 1600	Turbidity: 92 - 97 92 COD: n.d. 50 BOD: n.d. 60 TSS: n.d. n.d. TDS: n.d. n.d.	0.62 - 1.17 0.68	n.d.	n.d.	n.d.	Terrazas *et al.* 2010 [10]

[b] = Observed as the best electrode configuration of those tested; [c] = The natural, unmodified pH value of the water or wastewater (found optimal); [d] = The natural, unmodified pH value of the water or wastewater (the effect of pH not researched); [e] = Approximation calculation based on values given in the article at issue; [g] = Optimum value estimated from the data in the article (precise value not given); [h] = Additional "optimum value" estimated from the data in the article; n.d. = Not determined.

nia and tannins (low biodegradability due to polyphenolic compounds) [27]. Biological treatment of wastewater containing resistant and toxic compounds requires long processing times and large treatment areas and generates high amounts of low-density sludge.

The experiments were repeated three times and the experimental error was found to be around 3%. The results presented are for the EC process only. It was found that the EF process was 10% more efficient in removing pollutants while its energy consumption was 20% lower. However, the cost of adding hydrogen peroxide was not considered. It was concluded that both processes showed fast and efficient purification of tannery industry wastewater. Note that the latter number presented in **Table 1** is a visual approximation of **Figure 1** in the article, because the precise value was not given in the text [26].

Genuine dark-grey colored, turbid (initially 1310 NTU, nephelometric turbidity unit) textile wastewater was treated by a recirculated batch EC process using Al electrodes. The natural pH of the wastewater was 7, which was also the optimum value, making addition of chemicals unnecessary. Optimization of removal efficiency with response surface methodology (RSM) corresponding to the Box-Behnken experimental design was successfully performed. Statistical testing of the model obtained was conducted using Fisher's statistical test for analysis of variance (ANOVA). Percentages of COD (chemical oxygen demand), TS (total solids) and turbidity removed were taken as the system responses, while current density, initial pH, and treatment time were the input parameters [28].

A batch EC system with Al electrodes was proposed

for decolorizing synthetic azo-dye-containing industrial wastewater. Direct Red 81-dye concentration was measured using ultraviolet-visible spectrophotometry (maximum absorbance at wavelength I_{max} = 522 nm). It should be noted that this procedure was also used to estimate dye concentrations in other works presented in **Table 1** when synthetic dye solutions were used, employing relevant maximum intensities. In this study, however, four supporting electrolyte types were compared; of these, NaCl showed the best performance. This was suggested to be due to Cl^- anions destroying the passivation layer formed on the aluminum electrode, leading to a greater rate of anodic dissolution. A decolorization rate of 98% was reached in optimum conditions [29].

Two different types of synthetic dye effluents were prepared and treated with an EC apparatus working in a continuous upflow mode. This EC system also applied a hydrogen gas collecting system. The dyes used to prepare the wastewater were azo-based Direct Red 23 and Reactive Blue 140. In **Table 1**, the first number presents the wastewater containing Direct Red 23; the latter is the wastewater with Reactive Blue 140. Experimental and theoretical maximum hydrogen yields were compared, denoting 89% - 94% efficiency in the gas collecting system. The results showed that the energy yield of harvested hydrogen (converted to electricity for the EC process with an assumed efficiency of 50%) could reduce the EEC-value of the EC process by 13% and 8.5% for Direct Red 23 and Reactive Blue 140 solutions, respectively. It was also stated that the high-quality hydrogen collected could also have been saved for use as a reactant in industrial processes. Decolorization rates of 99% and 89% for a 5-min EC-run applying a current density of 30 A/m^2 and 40 A/m^2 were found to be the optimum conditions for the Direct Red 23 and Reactive Blue 140 solutions, respectively. EEC-values were found to be low and the OC of the EC system was calculated as being less than 0.12 €/m^3 [30].

A typical textile effluent may have fluctuating properties because it contains various types of dye molecules. Therefore, a study was set up to investigate the decolorization of two different dyes (Reactive Blue 140 and azo-based Disperse Red 1) and a mixed dye made of them. Real textile wastewater was also treated with the same batch-EC system. All three synthetic wastewaters had results very close to each other (both EEC-values and removal efficiencies, which showed complete decolorization), therefore their values are given as their approximate averages in **Table 1**, followed by the results for real wastewater. Three different electrode configurations were tested, of which iron electrodes proved to be superior. The same optimum parameter values that were used for the synthetic dyed wastewaters prior to it were also used for the treatment of real wastewater, thus these values might have not been optimal for it. Of the five EC tests run in these conditions, one seemed to have failed (it was not in line with the others) and was therefore ruled out [31].

A batch EC system was employed to treat a synthetic crystal violet (CV) solution. Na_2SO_4 was used as the support electrolyte in this work, which claimed that NaCl was a controversial choice because of its possibility to form organic chlorine by-products. In this study, a two-level full factorial experimental design (2^k) was employed to evaluate decolorization of the CV wastewater by EC. The levels of four variables (initial pH, CV concentration, supporting electrolyte concentration, current density) were studied. Reduced empirical models for both Al and Fe anodes were proposed for CV removal with EC. The correlation percentages were 96% and 83% for Al and Fe anodes, respectively. Iron was found superior to aluminum in this application, and residual amounts of less than 1 mg/l of iron were detected after an optimal 5-min EC run, while color and COD were fully removed [32].

A comparison of EC and EO (electro-oxidation) in treating Acid Green dye 50-based synthetic wastewater was done. Both processes were carried out in novel cathodic H_2-gas-stirred batch reactors. This was proposed to reduce the capital and operating costs of the reactor by making mechanical stirring unnecessary, provided that the EC cell is designed properly. EC was found more efficient: in optimum conditions, practically complete decolorization and a COD reduction of 87% versus 68% were accompanied by clearly lower energy consumption than what was achieved with EO. However, it was concluded that further studies on different types of dyes are needed to confirm this [33].

EC was employed to remove Acid Brown 14 from an aqueous solution by bench-(500 ml) and pilot-scale (9000 ml) batch processes. This type of dye was chosen because of its high level of usage in the textile industry. Aluminum was found superior to iron as the anode material for this application. Rather low values of current density (6.329/2.58 A/m^2) were found optimal for the batch- and pilot-scale systems, leading to very low EEC-values (0.053/0.037 kWh/m^3) while achieving substantial removal efficiencies for COD and color. The EC process was concluded to be promising in treating azo-dye-containing wastewater [34].

Optimization of Acid Red 14 azo dye removal by a batch-EC (Fe/steel electrodes) process with RSM was performed. Evaluation was based on the simple and combined effects of three main independent parameters: current density, treatment time, and initial pH of the wastewater. The study showed clearly that RSM was suitable for optimizing the EC process operating conditions and maximizing dye removal (91.27% in 4 min, whereas 93.93% in 4.47 min was predicted by the model, see **Figure 2**). A high coefficient of determination (R^2 = 0.928) ensured satisfactory adjustment of the model derived from the experimental data [35].

A continuous EC system with a 450-ml electrolytic cell bearing aluminum electrodes was utilized in treating synthetic wastewater polluted with azo-based Orange II dye. In this experiment, it was found clearly evident that when operating in optimum conditions (near-neutral initial pH, 350 ml/min flow rate, 4 g/l added NaCl), the color of the dye solution had almost completely vanished. The EC apparatus was summarized to be simple to design and operate and is an inexpensive tool for treatment of dye-containing textile wastewater [36].

Treatment of other artificially colored wastewater by batch EC was investigated, with iron used as the electrode material due to its clear superiority over aluminum here. The dye used in the experiments was azo-based Reactive Black 5, which was chosen because of its extensive annual consumption rate. Complete decolorization of 100 mg/l initial dye concentration was achieved rapidly in 5 min, by adding of 3 g/l NaCl and applying a current density of 45.75 A/m^2. Optimal initial pH was found to cover a broad range of pH values [12].

An Indigo Carmine (Acid Blue 74)-based aqueous solution was treated with a batch EC system employing mild steel/SS electrodes. The gap between the vertically positioned electrodes was only 3 mm. Generation of green iron(II) hydroxide into the solution changed its color from dark blue to dark green and further to yellow brown. At the end of the EC process, stirring was halted and sedimentation and flotation occurred, and a removal efficiency of 96% was achieved in optimal conditions. The observed color change during the EC process is depicted in **Figure 3**. A wide range of initial pH values were found to be suitable for this application. It was also shown that controlling pH (preventing it from rising by acid addition) during the EC process was detrimental and limited the development of flocs, at least in this case. NaCl was proposed as a better choice than Na_2SO_4 as a supporting electrolyte [37].

A red dye solution of 2-naphtoic acid and 2-naphtol was treated with Al-EC in a continuous mode. The reactor consisted of two compartments, the first being the actual electrolytic cell and the second a settling compartment filled by overflow from the first one. The initial COD value of the solution was 2500 mg/l when the total concentration of the dye mixture was 100 mg_{dye}/l. Optimal results were achieved with a 14-min residence time (of which only 5 min in the EC compartment), resulting in 95% dye removal efficiency and a 3.2 kWh/m^3 EEC-value. The EEC-value could still be halved by increasing the conductivity of the water by adding NaCl, but this led to a slight reduction in removal efficiency. The high performance of EC in a continuous mode in removing this type of dye from wastewater was proven in this pa-

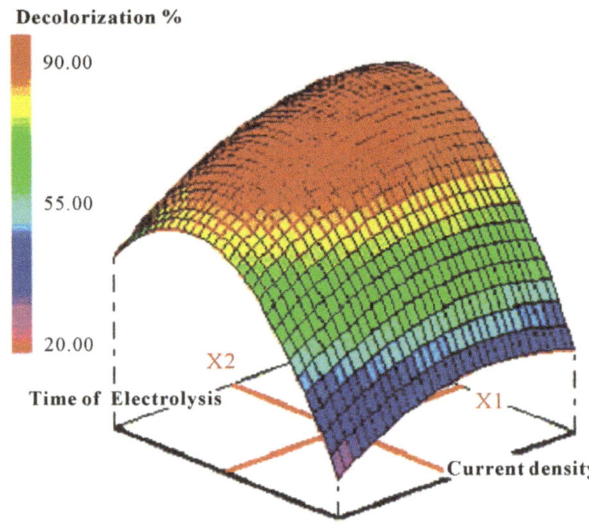

Figure 2. Three-dimensional contour plot obtained from the experimental data of color removal efficiency vs. current density (X1) and time of electrolysis (X2). Adapted from [35].

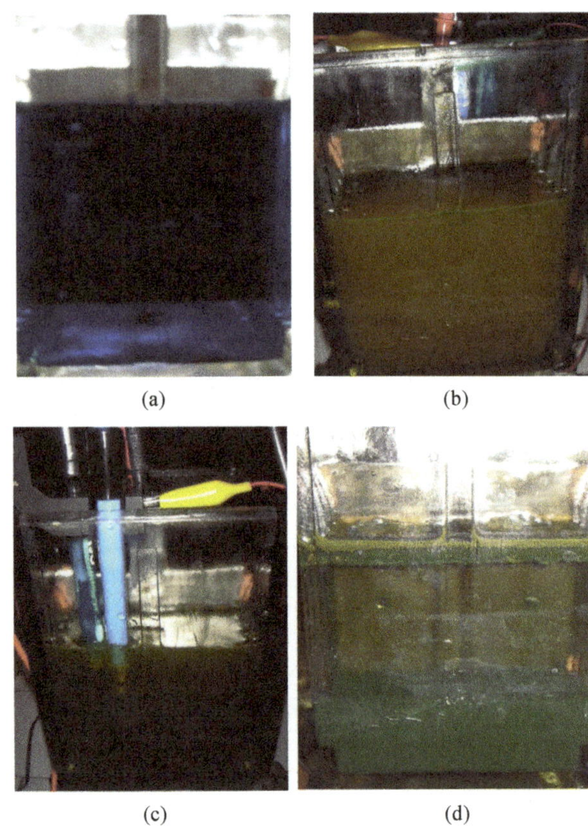

Figure 3. Evolution of the electrocoagulation process: (a) initial state; (b) 20 min, pH = 8.2; (c) 40 min, pH = 9; (d) 180 min of electrocoagulation and 5 min after stirring was turned off (initial concentration = 50 mg/l, $pH_{initial}$ = 7.1, current density = 10.91 A/m^2. Adapted from [37].

per [38].

To summarize, EC treatment of tannery, textile, and colored waters has been under extensive development in recent years, with promising results. Wide research has been done in this field in addition to the studies presented in **Table 1**; e.g. those published in [8,25,39-43].

3.2. Pulp and Paper Industry Wastewater

The pulp and paper industry is one of the major water-intensive chemical process industries, contributing significantly to environmental pollution, e.g. in the form of black liquor. Blackish color, high amounts of organic load, suspended solids (mainly fibers), COD, and BOD (biological oxygen demand) are characteristic of effluents from pulp and paper industry. Arsenic may also be present. The strong color of the wastewater derives mainly from polymerization between lignin-degraded products and tannins. The drawbacks associated with conventional treatment techniques have made it necessary to develop more effective methods for treating this type of wastewater. **Table 2** presents a summary of recent applications of EC in the treatment of pulp and paper industry wastewater [44].

Genuine black liquor wastewater with a high concentration of pollutants and a high pH value of 12 was treated with a batch EC system. The main characteristics of the wastewater before treatment are presented in **Table 2**. Repeatability tests were performed under the same experimental conditions. A relative standard deviation (R.S.D) value of less than 3% was achieved, proving good repeatability of the EC process. Aluminum and iron were very close to each other in efficiency with different pollutants being removed. However, aluminum was slightly better overall and was chosen as the optimum. Also, Fe electrodes caused the water to turn green at first, and then to yellow and turbid due to Fe(II) and Fe(III) species generated, whereas the resulting effluent treated with Al was found very clear and stable [45].

The paper mill effluents from a modern bleached kraft mill were utilized in an EC study. The lignin concentration of the pretreated, brown effluent was extremely high, 13514 mg/l. All experiments were repeated twice, and the experimental error was approximately 4%. Aluminum was found superior to iron as the electrode material. The effect of initial pH was not studied in this work (the near-neutral original value was used). The high EEC-value of COD compared with that of lignin is related to its significantly lower initial concentration in the wastewater. The results of this study were found to suggest that EC is an effective alternative in paper mill effluent treatment [46].

Pre-treated wastewater from another paper mill with color, pH, and COD similar to those in the previous study was treated by EC in a batch mode. The wastewater was constantly circulated in the EC system with a peristaltic pump. A settling test was also conducted in a separate 46-cm-high glass column at the end of the test. All tests were done in triplicate. No clearly superior single electrode configuration could be found here, considering all aspects. Aluminium had somewhat better removal efficiency (except for lignin) than Fe, but the flocs it produced were clearly weaker in quality (SVI index values of 0.081 - 0.091 l/g and 0.207 - 0.310 l/g for Fe and Al, respectively) and more difficult to handle. The research group concluded that they will therefore use Fe in their consecutive studies. However, Al was chosen as the optimum here based on raw efficiency numbers. EC was proposed as a very effective tool for treating wastewater of the paper industry [47].

Highly polluted paper mill wastewater was treated by EC. No clearly superior single electrode configuration could be found of the four that were tested. Using Fe caused color reduction to be only 62%, and when using Al, COD and phenol reductions were 77% and 91%, respectively. Hybrid electrodes were the most constant in every parameter investigated, with results between those of Al and Fe. It should be pointed out that the optimum current density suggested (700 A/m^2) was the highest one presented in **Table 2** and significantly higher than in other studies. It was found that the water could be purified optimally directly without pH adjustment. After duplicating all tests, the experimental error was found to be below 4% [48].

Wastewater from the tissue paper industry was treated by EC with an aim to obtain water quality acceptable for reuse. Real wastewater with high pollution levels (see **Table 2**) was employed. A scale-up system of the Al-batch EC process was also tried, with similar efficiencies and EEC-values. The results showed that a separation gap of 10 mm produces a faster build-up of sludge between electrodes. However, it yields more efficient removal of turbidity and lower energy consumption than larger gaps. EC was concluded to have proven to be an efficient method for removing turbidity from this type of wastewater, producing water of quality (8 NTU) suitable for reuse in the paper bleaching stage [10].

In addition to the studies presented in **Table 2**, at least [44,49,50] have also recently contributed to EC research in this particular sector. On the whole, treating paper industry wastewater by EC seems to be a feasible alternative and a subject of interest.

3.3. Oily Wastewater

Oily wastewaters with greatly varying compositions and very high pollutant levels are generated by various sources, such as petroleum refineries, discharge of bilge and ballast water, workshops, petrol stations, rolling mills, restaurants, edible oil and soap factories, as well as other general industrial sources. Oil-in-water can be found as free-floating oil, as an unstable oil/water emulsion, and also as a highly stable oil/water emulsion, which are all difficult to treat [7,51].

Table 3 presents a summary of recent applications of EC in the treatment of oily wastewater. In addition to the studies presented in **Table 3**, also [14,52-61] have recently studied oily wastewater purification by EC with promising results, which shows that there is a high interest in EC research in the field of oily wastewater purification.

A synthetic industrial oil-in-water emulsion was prepared and treated with EC. The original oil concentration of the emulsion was 5%, which corresponded to extremely high COD and turbidity values of 62300 mg/l and 29700 NTU, respectively. Nevertheless, very high RSM-optimized removal efficiencies of 90% for COD and 99% for turbidity were achieved in less than 22 min with ANOVA showing a high variance coefficient (R^2 = 0.998), ensuring satisfactory adjustment of the model with the experimental data [62].

Purification of oily wastewater resulting from washing the bilges of boats using EC working in a batch mode was studied. Iron was found to best Al as the electrode material. EC treatment with both materials was tested using monopolar (MP) and bipolar (BP) electrode configurations. The EEC- and OC-values for a MP configuration were found to be significantly lower than those of a BP configuration, whereas COD reductions were found rather similar, thus making Fe-MP the most feasible solution. The initial green color of the wastewater disappeared after EC treatment and the effluent became more transparent. A very low OC value was achieved alongside high removal efficiency. Note that in **Table 3**, C_{10}-C_{50} stands for C_{10}-C_{50}-hydrocarbons (indicators of raw oil-based hydrocarbons) [63].

Sunflower oil refinery wastewater with a natural pH value as low as 1.4 and COD as high as 15000 mg/l was treated in a batch EC reactor with aluminum electrodes. Na_2SO_4 and PAC (poly aluminum chloride) were added to the water to increase its conductivity and enhance coagulation. The addition of 0.5 mg/l PAC raised the removal efficiency of COD from 94.5% to 98.9% in optimum conditions (see **Table 3**). A significant initial pH adjustment to 5 - 7 was found to be required for optimal functionality. The treated effluent was very clear and its quality exceeded the local direct discharge standard and therefore EC was found very efficient in treating this type of wastewater [64].

An industrial waste emulsion containing fluorescent penetrating oil used in aeronautics was treated with EC using Al electrodes. The parameters shown in **Table 3** are for the industrial-scale EC system only, as in this study a successful two-step scale-up (from a batch mode laboratory system to continuous pilot- and industrial-scale systems) of the EC process with very similar performance parameters was conducted. The values given for the aforementioned EC process in **Table 3** were ones achieved after additional sand and carbon filters, which contributed approximately 5 - 10 additional percentage points to removal efficiency. In this study, the industrial-scale system used utilized an innovative partial re-circulation of sludge supernatant (see **Figure 4** for a presentation of the EC process pilot prototype), which was found to enhance coagulation (increasing removal efficiency) without increasing the EEC-value and to diminish the sludge generation rate. This phenomenon was proposed to be due to introduction of a basal quantity of

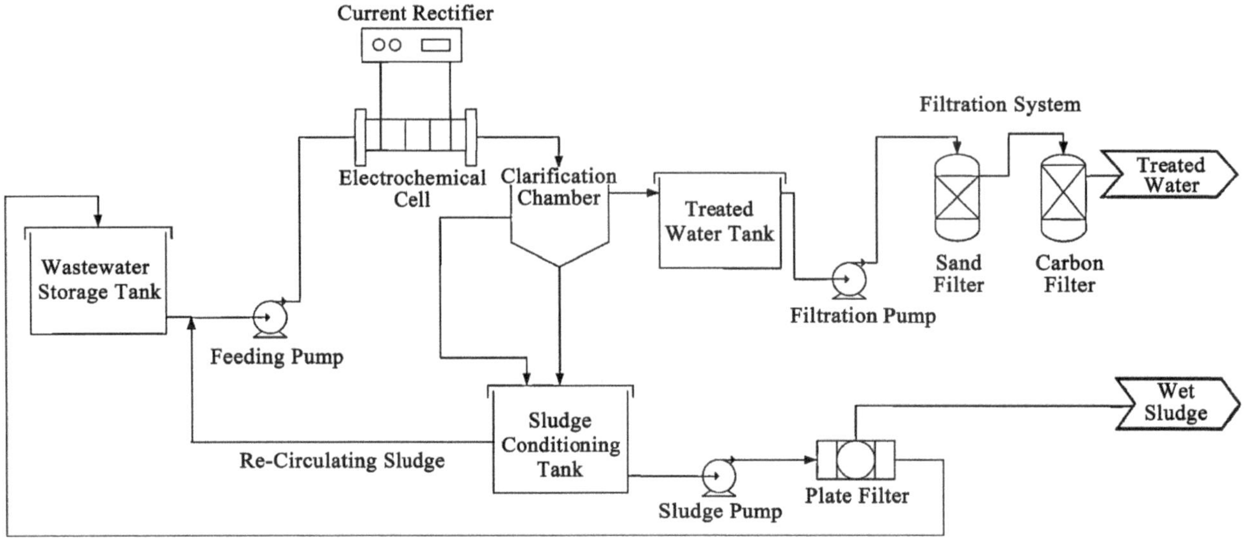

Figure 4. Scheme of the EC pilot prototype set up (see above). Adapted from [65].

Table 3. Recent applications of EC in the treatment of oily wastewater.

Water and wastewater types used	Genuine (G)/ Synthetic (S) water	Anode/ cathode material	Reactor type	Volume treated [ml]	Optimum electrode gap [mm]	Optimum current density, treatment time and initial pH [A/m^2], [min], []			Initial pollutant levels [mg/l]	Optimum removal efficiency [%]	Optimum EEC [kWh/m^3]	Optimum EEC [kWh/kg$_X$]	Optimum OC [€/m^3]	Optimum OC [€/kg$_X$]	Research group
Industrial oil-in-water emulsion	S	Al/SS	Batch	400	10	250	22	7	COD: 62300 Turbidity: 29 700 [NTU]	COD: 90 Turbidity: 99	3	n.d.	n.d.	n.d.	Tir & Moulai-Mostefa 2008 [62]
Oily bilgewater	G	Al Fe[b]	Batch	1700	15	1.5 [A] = ~34[e]	60	4	COD: 3350 - 3450 TOC - 468 BOD$_5$: 118 - 216 Turbidity: 2210 [NTU] C_{10}-C_{50}: 422 - 460 O & G: 800 TSS: 501 - 585 TS: 2160 - 2400	COD: 78 TOC: n.d. BOD$_5$: 90 - 96 Turbidity: 98 - 99 C_{10}-C_{50}: 99 - 100 O & G: 95 - 96 TSS: 99 - 100 TS: 32 - 43	2.07 - 2.25	n.d.	0.032 - 0.033	n.d.	Asselin *et al.* 2008 [63]
Vegetable oil refinery wastewater	G	Al	Batch Batch + PAC 0,5 mg/l	300	8	350	90	5 - 7 [7]	COD: 15000	COD: 95 99	n.d.	96 kWh/kg$_{COD}$ 42 kWh/kg$_{COD}$	~(0.39 - 1.2)[e]	0.026 - 0.08 €/kg$_{COD}$	Un *et al.* 2009 [64]
Industrial wastewater containing a fluorescent penetrant oil (emulsion)	G	Al	Continuous + sand- and AC-filtration	23000	5	110	5[a] (3.2 l/min)	(6.5 - 7.5)[d]	COD: 1000 - 2500 Color: 350 [Pt-Co] Turbidity: 600 [NTU]	COD: 95 Color: 99 Turbidity: 99	~(0.5 - 0.7)[g]	n.d.	0.18[g]	n.d.	Meas *et al.* 2010 [65]
Olive mill wastewater	G (diluted to 20 %)	Al	Batch	100	28	250	15	(4 - 6)[c]	COD: 20000 Polyphenols: 260 Color: n.d.	COD: 84 Polyphenols: 87 Color: 92	~5.26[e]	2.63 kWh/kg$_{COD}$	~0.54[e]	0.27 €/kg$_{COD}$	Hanafi *et al.* 2010 [66]
Waste metal cutting fluids	G	Al Fe[b]	Batch	800	10	60	25	6 - 7 [7[c]]	COD: 17312 TOC: 3155 Turbidity: 15350 [NTU]	COD: 92 TOC: 82 Turbidity: 100	n.d.	n.d.	0.348	0.0016 €/kg$_{COD}$ 0.0101 €/kg$_{TOC}$	Kobya *et al.* 2008 [67]
Contaminated groundwater (petroleum hydrocarbons)	G	(Al, Fe & SS in different combinations) SS/Fe[b]	Batch + aeration Continuous + aeration	200	30	180	(20/40 20/40[a] [0.01/ 0.005 l/min])[g]	(6 - 9)[c] [7]	TPH: 64 TDS: 1178	(TPH: ~85/~91 ~81/ ~91)[g] TDS: n.d.	n.d.	n.d.	n.d.	n.d.	Moussavi *et al.* 2011 [68]
Palm oil-based biodiesel wastewater	G	Al/ Graphite	Batch	1000	15	20 [V] = ~135 - 140[e]	25	6	COD: 30980 O & G: 6020 SS: 340 Methanol: 10667	COD: 55 O & G: 97 SS: 98 Methanol: n.d.	5.57	n.d.	n.d.	n.d.	Chavalparit & Ongwandee 2009 [70]
Rose oil processing wastewater	G	Fe	Batch	400	65	80	20	6.4 - 7.1 [64]	COD: 9500 BOD: 4950 Turbidity: 750 [NTU] TS: 7690	COD: 80 BOD: n.d. Turbidity: 81 TS: n.d.	6.25	0.825	n.d.	n.d.	Avsar *et al.* 2007 [71]

[a] = HRT (hydraulic retention time) in EC systems with continuous mode of operation [min]; [b] = Observed as the best electrode configuration of those tested; [c] = The natural, unmodified pH value of the water or wastewater (found optimal); [d] = The natural, unmodified pH value of the water or wastewater (the effect of pH not researched); [e] = Approximation calculation based on values given in the article at issue; [g] = Optimum value estimated from the data in the article (precise value not given); n.d. = Not determined.

coagulant working as a coagulant initiator to the wastewater. The removal efficiency was high while the OC value was found to be very low. Cost estimates for the process indicated an investment-return time for the EC system of only about 17 weeks (at a rate of 8 m^3 wastewater generated weekly at the plant) compared with the plant's current policy of simply sending it out for disposal [65].

Genuine olive mill wastewater (OMW), diluted with water to a fifth of the original concentration, was treated with EC an system using aluminum electrodes. Even after diluting the OMW, initial pollution concentrations were extremely high, as can be seen in **Table 3**. All experiments were tripled, achieving accuracy better than

4%. The naturally occurring pH of OMW and a 2-g/l NaCl addition were found appropriate for achieving effective treatment, in which a significantly low EEC-value of 2.63 kWh/kg_{COD} and very high removal efficiency (see **Table 3**) could be reached. The final pH of the OMW was nearly neutral. It was also found that the EC treatment reduced the toxicity of OMW for *Bacillus cereus* by 70%. The growth of bacteria was nearly similar to a standard medium in the treated OMW, whereas in untreated OMW, bacterial growth was impossible. Consequently, EC was considered a viable pre-treatment step prior to a biological process for treating of OMW [66].

Metal cutting fluids are widely used for cooling and lubrication in metal industries. Batch EC treatment was performed on an extremely highly polluted (see **Table 3**) white-colored waste metal cutting fluid (WMCF) obtained from one such company's production of automotive engines, transmissions, and stamping plants. Iron was found to be clearly more economical than aluminum in treating the WMCF, even though removal efficiencies were very similar and high for both electrode materials. Operating costs were found to be low and the natural pH of WMCF was found to be optimal, thus addition of chemicals to alter the initial pH was unnecessary. It was concluded that despite achieving high COD reductions, further treatment is nonetheless needed because of the residual COD values, which still exceeded the local discharge standards [67].

The efficacy of EC for treating petroleum- contaminated groundwater (from a site close to a local petroleum refinery) was evaluated and quantified as total petroleum hydrocarbons (TPH) removed. Al, Fe, and SS in different combinations were tested as electrode materials, of which a SS/Fe-combination proved to be the most suitable. The EC process was studied with systems working in both batch and continuous modes. The natural neutral pH of the groundwater was found to be optimal and increasing HRT was found to improve TPH removal in the continuous EC systems. Note that the removal efficiencies shown in **Table 3** were achieved with aeration in the EC cell, which added approximately 22 percentage points to the removal efficiencies (see **Figure 5**). This was proposed to be due to aeration-transferred oxygen accelerating the oxidation of Fe^{2+} (dissolved from the anode) in solution to Fe^{3+}, which in turn resulted in formation of greater $Fe(OH)_3$ flocs, thus improving TPH removal by adsorption. EC was summarized to be a promising technique in eliminating TPH from groundwater [68].

Biodiesel production generates large amounts of alkaline and highly oily wastewater with low nitrogen and phosphorus content, rendering it unable to be purified by biological treatment [69]. Such wastewater (with frying oil waste and crude palm oil used as biodiesel feedstock)

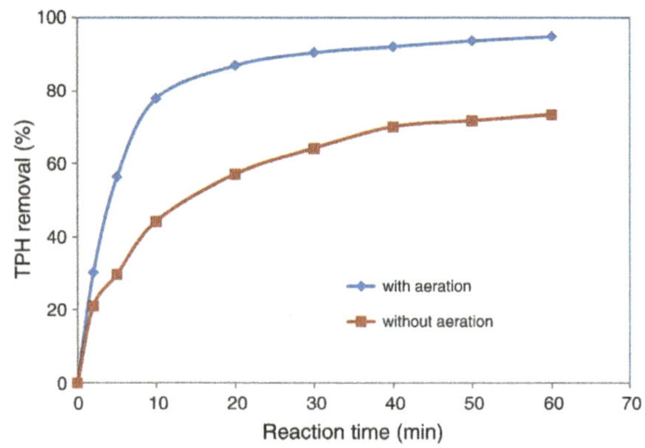

Figure 5. Effect of aeration on the removal of TPH in a batch-EC process (using Fe electrodes) as a function of reaction time. Adapted from [68].

was treated with a batch EC system using aluminum as the anode and (rather rarely) graphite as the cathode material. Box-Behnken design-based RSM optimization was applied to evaluate the best process conditions. The RSM results matched experimental values very well; the predicted optimum values for initial pH, voltage, and reaction time were 6.06, 18.2 V, and 23.54 min, respectively, with removal efficiencies all within 0.7 percentage points of those obtained experimentally. Ultimately, EC was found suitable as a primary treatment of biodiesel wastewater, which, however, still requires further biological treatment [70].

Industrial wastewater generated in rose oil processing was purified by EC. In a previous study by the same research group, Fe was proven better as the electrode material than Al for this type of wastewater. This was due to clearly lower residual metal concentrations, which were less than 0.2 mg/l with Fe. In this paper, CC, EF, and two different Fenton processes were compared. It was concluded that the EC process was most suitable for this type of wastewater, while CC performed the worst. Initial pollutant concentrations in the wastewater were very high (see **Table 3**). Note that even though 6.4 was found to be the optimum initial pH value, an acidic natural pH value of 4 was used to achieve the efficiencies presented in **Table 3**, because the difference in removal efficiencies was negligible [71].

3.4. Food Industry Wastewater

Wastewaters from agro-industries come from a myriad of sources and their compositions vary greatly. On the whole they are, however, characterized by high COD and BOD due to their high level of organic content [72]. **Table 4** presents a summary of recent applications of EC in the treatment of food industry wastewater.

The dairy industry is associated with generation of huge amounts of wastewater (approximately 0.2 l to 10 l of effluent per liter of processed milk) that is high in nu-

Table 4. Recent applications of EC in the treatment of food industry wastewater.

Water and wastewater types used	Genuine (G)/ Synthetic (S) water	Anode/ cathode material	Reactor type	Volume treated [ml]	Optimum electrode gap [mm]	Optimum current density, treatment time and initial pH [A/m^2], [min], []			Initial pollutant levels [mg/l]	Optimum removal efficiency [%]	Optimum EEC [kWh/m^3]	Optimum EEC [kWh/kg$_X$]	Optimum OC [€/m^3]	Optimum OC [€/kg$_X$]	Research group
Dairy wastewater	S	Fe	Batch	1500	10	270	50	7[c]	COD: 3900 Turbidity: 1744 [NTU] TS: 3090 TN: 113	COD: 70 Turbidity: 100 TS: 48 TN: 93	~(0.83 - 30.0)[e]	~2.76 kWh/kg$_{COD}$	0.051 - 1.80	n.d.	Kushwaha *et al.* 2010 [77]
Dairy wastewater	G	Fe	Batch	650	25	60	1	6 - 8 [7[c]]	COD: 18300 O & G: 4570 TSS: 10200	COD: 98 O & G: 99 TSS: n.d.	~0.055[e]	0.003 kWh/kg$_{COD}$	n.d.	n.d.	Şengil & Özacar 2006 [73]
Potato chips manufacturing wastewater	G	Al[b] Fe	Batch	250	11	20 - 300	5 - 40	4 - 6	COD: 2200 - 2800 Turbidity: 260 - 610 [NTU] BOD: 1650 - 2150	COD: 60 Turbidity: 98 BOD: n.d.	4.0	n.d.	0.36 - 4.10	n.d.	Kobya *et al.* 2006 [78]
Almond industry wastewater	G	Al/Fe[b] Fe/Al	Batch Continuous (pre-industrial scale-up)	700 ~54000[e]	10 15	50 45[g]	15 8.2[a, g] (6.67 l/min)	4 - 8 [5.7[c]]	COD: 5300 TOC: 1400 BOD$_5$: 1000 Color: 18000 [Pt-Co] Turbidity: 3200 [FTU] N$_{total}$: 240 P$_{total}$: 3 TSS: 3400	COD: 81/81 TOC: 74/79 BOD$_5$: 80/67 Color: 100/98 Turbidity: 99/98 N$_{total}$: 75/85 P$_{total}$: 100/99 TSS: 100/99	n.d./2.8[g]	n.d.	n.d.	n.d.	Valero *et al.* 2011 [79]
Pasta and cookie processing wastewater	G	Al	Batch Batch + H_2O_2 (= EF)	1500	n.d.	0.0182	60	3 - 8 [4]	COD: 7500 BOD$_5$: 3445 Color: 35 [Pt-Co] Turbidity: 1153 [NTU] TS: 2905 Fecal coliforms, MPN: 11000	COD: 80/90 BOD$_5$: n.d./96 Color: n.d./57 Turbidity: n.d./97 TS: n.d./95 Fecal coliforms: n.d./100	n.d.	n.d.	n.d.	n.d.	Roa-Morales *et al.* 2007 [80]
Poultry slaughterhouse wastewater	G	Al Fe[b]	Batch + polymer (LPM 9511, 10 mg/l)	1700	15	0.3 [A] = ~7[e]	60	(6.1 - 6.5)[c]	COD: 2700 - 3100 BOD: n.d. Turbidity: n.d. O & G: 720 - 950 TSS: n.d. TS: 1440 - 2380	COD: 80 - 84 BOD: 84 - 88 Turbidity: 86 - 94 O & G: 98 - 100 TSS: 85 - 93 TS: 58 - 70	4.07 - 4.31	n.d.	0.49 - 0.51	n.d.	Asselin *et al.* 2008 [81]
Poultry manure wastewater (UASB pretreated)	S	Al[b] Fe	Batch	400	62	150	20	4 - 6 [5]	COD: 4120 Color: 3390 [HU]	COD: 90 Color: 92	n.d.	2.6 kWh/kg$_{COD}$	n.d.	n.d.	Yetilmezsoy *et al.* 2009 [15]
Egg processing wastewater	S + G	Al Fe SS[b]	Batch Batch + coagulant (200 mg/l bentonite)	1000	70	15 [V]	30/ (35, 26)	(4.5 - 5)[c]	COD: 8637 - 8983/ 4068 - 4132 TSS: 1651 - 1953/930 - 1086 Turbidity: 933 - 1267/ 1340 - 2060 [FTU]	COD: 97/(92, 95) TSS: 97/(97, 97) Turbidity: 99/(99, 99)	n.d.	n.d.	n.d.	n.d.	Xu *et al.* 2002 [82]
Baker's yeast wastewater	G	Al Fe	Batch	800	2	70	50	5.5 - 6.5 [6.5]/ 6 - 8 [7]	COD: 2485 TOC: 1061 Turbidity: 2075 [NTU] TSS: 503	COD: 71/69 TOC: 53/52 Turbidity: 90/56 TSS: n.d.	n.d.	n.d.	1.08/0.36	0.58/0.19 €/kg$_{COD}$	Kobya & Delipinar 2008 [83]
Tea factory wastewaters	G	Steel	Batch	400	5	24 [V]	n.d.	6[c]	COD: 293/607 BOD$_5$: 42/193 Color: 2004/9210 [Pt/Co]	COD: 91/97 BOD$_5$: 84/42 Color: 100/100	2.27/2.76	n.d.	n.d.	n.d.	Maghanga *et al.* 2009 [84]

[a] = HRT (hydraulic retention time) in EC systems with continuous mode of operation [min]; [b] = Observed as the best electrode configuration of those tested; [c] = The natural, unmodified pH value of the water or wastewater (found optimal); [e] = Approximation calculation based on values given in the article at issue; [g] = Optimum value estimated from the data in the article (precise value not given); n.d. = Not determined.

trients and has highly varied pH [73,74]. In addition to those presented in **Table 4**, EC-treatment of dairy wastewater has been recently investigated by at least [75,76]. In a study on simulated dairy wastewater (SDW), a batch Fe-EC system was used coupled with a four-factor, five-level, full-factorial central composite design (CCD) based on RSM [77]. In optimum conditions, removal efficiency was high while economical values were low. Initial pollution levels of the SDW were high, with COD, TS, and turbidity being 3900 mg/l, 3090 mg/l and 1744 NTU, respectively. The authors suggested that, based on the results of TGA/DTA (thermal gravimetric analysis/ differential thermal analysis), the electrogenerated SDW-sludge could be dried and used as fuel in boilers/incinerators or in fuel briquette production.

Real dairy effluent was treated by a Fe-EC system. Initial pollutant concentrations were very high; 18300 mg/l COD, 4570 mg/l O & G (oil and grease), and 10200 mg/l TSS (total suspended solids). Optimum treatment occurred very rapidly, as merely 1 min was found sufficient, while the EEC-value was extremely low (0.003 kWh/kg_{COD}) and removal efficiencies were very high (98% and 99% of COD and O & G, respectively). The equilibrium data obtained by the authors were found to fit very well into the Freundlich adsorption isotherm model ($R^2 = 0.99$) [73].

Potato chip manufacturing wastewater (PCW) was treated by batch-EC. Aluminum was found clearly superior to iron as the electrode material in this application. The removal efficiencies of COD and turbidity (having high initial values of 2200 - 2800 mg/l and 260 - 610 NTU, respectively) were high; however no clear single set of optimum process conditions was proposed. The natural pH of the PCW was 6.2 - 6.5, which was virtually optimal; therefore no pH adjustment was necessary. The results from kinetic studies showed that the kinetic data fit the second-order kinetic model very well ($R^2 > 0.96$) [78].

The results of batch and further continuous (removal efficiencies presented in **Table 4** are in the form batch/ continuous) pre-industrial scale-up experiments showed that EC is an effective technology for treating wastewater from the almond industry, as the initially brown and murky wastewater turned colorless and clear as a result of EC treatment. Initial pollutant values were very high: e.g. 5300 - 6300 ppm COD and 3200 - 4000 FTU (formazin turbidity unit) turbidity. The results obtained for Al/Fe and Fe/Al electrodes were very similar, however the first was chosen over the latter mainly for practical reasons. The treated water was found to satisfy wastewater discharge legislation and the electrogenerated sludge was discovered to be neutral and non-toxic. This work also proved that EC can also be transferred (see **Table 4**, very similar results) to an automated industrial scale. No pH adjustment was required because the raw effluent had a pH value of 5.7, which was found optimal [79].

Wastewater from pasta and cookie processing was treated by an aerated batch -EC reactor after which the process was also investigated with added H_2O_2 (EF). The removal efficiencies presented in **Table 4** are in the form EC/EF. The addition of H_2O_2 was found to raise the removal efficiency of COD from 80% to 90% with otherwise similar process conditions (values for other parameters with EC only were not given). The raw wastewater had very high initial pollutant values (see **Table 4**). A very interesting effect of total wastewater disinfection by EC was observed. The optimum initial pH range of EC was found to be very large in this application, being 3 - 8 [80].

Genuine red-orange-colored poultry slaughterhouse wastewater containing high amounts of organic matter (e.g. proteins, blood, fat) was treated by batch-EC using Al and Fe in monopolar (MP) and bipolar (BP) electrode configurations. The Fe-BP electrode configuration was found to be the most suitable here. Applying EC made the effluent more transparent and thus EC was found efficient for decolorization and clarification of poultry wastewater. After the optimum process conditions had been found, reproducibility tests were performed by repeating the EC test in triplicate. During these tests, 10 mg/l of a cationic polymer was added to the treated effluents. It was thus verified that the EC process is highly effective, economical, and repeatable (see **Table 4**) for treating this type of wastewater [81].

The performance of a batch-EC system was investigated on UASB (up-flow anaerobic sludge bed)-pretreated artificially generated poultry manure wastewater. High removal efficiencies of 90% COD and 92% color were achieved with 20-min EC-treatment. In this study, a toxicity test was also conducted using small fish, *Lebistes reticulatus*, which showed that the EC-treated water did not cause any of the fish to die or behave abnormally during the 48-h test (a local environmental requirement) or even after extended 120-h exposure to the water. Aluminum was found to outperform Fe as the electrode material, mainly because of the greatly lower color reduction efficiency of iron, while for COD removal, aluminum was found only slightly more effective. The initial COD concentration of the water was 4120 mg/l and it was dark brown in color (3390 Hazen units). The EEC-value was given only at the natural initial pH of 8.82, at which the results obtained were, however, very similar to those of pH 5. The results showed that almost 99.5% of the Al released precipitated in the form of EC sludge and the supernatant had an Al content of about 2.3 mg/l [15].

Simulated (SWS) and genuine egg processing industry wastewaters (EPW) with very high pollutant levels (see **Table 4**) were treated with batch-EC with multiple electrode compositions, of which SS was considered most suitable. The effect of adding 200 mg/l of bentonite coagulant was also studied using the real wastewater. The results presented in **Table 4** are in the form SWS/(EPW, EPW + coagulant). The results obtained for both wastewaters showed very similar high removal efficiencies in a short treatment time. Additionally, EC yielded valuable by-products bearing high digestible protein and fat values. Addition of the coagulant further enhanced the EC process slightly. An economic analysis of EC indicated that this treatment is economically feasible and capital investments in equipment for a large-scale commercial egg processing plant could be recovered in 14 months [82].

Dark brown baker's yeast wastewater (BYW) with high pollution levels (see **Table 4**) was treated by batch-EC. The results are presented in the form Al/Fe, because neither of the electrode materials used was proven to be universally superior to the other. Al was found to achieve slightly better removal efficiencies but at significantly higher operating costs. It was suggested that albeit the EC process could be adapted effectively for treatment of BYW, the effluent still contained a large amount of COD, which needed to be further treated by a secondary process [83].

Four wastewater samples (SP_1-SP_4) were taken from different points in constructed wetlands following the outlet of a tea factory, and their treatment by batch-EC with steel electrodes was tested. The results presented in **Table 4** are those of waters SP_1/SP_4. The results of waters SP_2 and SP_3 were found rather similar. The waters were highly colored before the EC-treatment (2004/9210 mg/l Pt-Co), however 100% color removal in all the samples, SP_1-SP_4, was recorded. No NaCl was added to the wastewater samples even though their conductivities were low (134 - 317 µS/cm). Thus, it can be assumed that the energy consumption values could have been lower than the ones obtained. Diluting the waters prior to EC was found to be detrimental in terms of the EEC-value [84].

3.5. Other Types of Industrial Wastewater

Table 5 presents a summary of recent applications of EC in the treatment of other types of industrial wastewater.

Treatment of highly complex and highly polluted industrial wastewater (a mixture of wastewaters from 144 different factories received at a wastewater treatment plant) by batch-EC was studied. Three different electrode combinations were studied. Using aluminum and iron anodes simultaneously (corresponding cathodes were made of the same material as the anode) was found to outperform (better removal efficiencies, less sludge produced) the use of either metal alone as the anode material, combining the advantages of both. Complete disinfection of the wastewater was achieved. A short EC treatment time of 30 min resulted in almost similar (removal efficiencies only a few percentage points less) performance as a 60-min treatment time, for which the results are given in **Table 5** [85].

Electroplating, metal finishing, and mining are industrial process in which large volumes of hazardous wastes containing heavy metals and free and metal cyanides are generated. Generally, cyanide removal from wastewater is carried out by chlorination, requiring high operating costs [86]. EC-treatment studies of separate cadmium and nickel electroplating rinse wastewaters (results in this order in **Table 5**) also containing high amounts of cyanide were carried out using iron electrodes in a batch mode. The raw pH values of both the wastewaters were found optimal. In optimum conditions (see **Table 5**), complete metal and cyanide removal was observed for both wastewaters, with OC-values of 1.05 €/m^3 (cadmium-laden wastewater) and 2.45 €/m^3 (nickel-laden wastewater) [87].

Removal of Ni, Cu, and Cr from very heavily polluted industrial galvanic wastewater was carried out by batch-EC. The raw pH of the wastewater was 1.5, which is strongly acidic, and adjustment to 5 was found necessary. The wastewater had very high conductivity (41 mS/cm) and its metal content was extremely high; around 2 g/l Ni, 2.5 g/l Cu, and 0.7 g/l Cr (70% present as Cr(VI)). The optimum electrode configuration of the EC system consisted of two separate anode-cathode-pairs used simultaneously instead of a single one made of Al or Fe. This novel EC process was found very efficient (see **Table 5**) in removing metals from galvanic wastewater. It was concluded that EC could be a good alternative or an after-treatment (the varying composition of such wastewaters may limit the feasibility of EC as a primary method) to conventional methods in this application [88].

Another study of Ni, Cu, and Cr removal from real industrial metal plating wastewater by batch-EC was conducted. The natural initial pH of the wastewater was 3 and it was chosen as "optimal" (optimal values presented in **Table 5** given for this pH value) to avoid major addition of chemicals for pH adjustment, even though higher initial pH values (7 - 9) resulted in higher metal removal efficiencies, which were also obtained significantly faster. A Fe/Al electrode configuration was found most suitable of the four combinations of Al and Fe tested, however Fe/Fe was found to be nearly as efficient. In the end, it was concluded that EC in optimum process conditions could effectively reduce metal ions in the metal plating wastewater to a very low level, yielding 100% removal efficiencies for Ni, Cu, and Cr in 20 min [89].

Complexing agent and heavy metal (Zn, Ni) removal from a genuine, highly polluted metal plating effluent

Table 5. Recent applications of EC in the treatment of other types of industrial wastewater.

Water and wastewater types used	Genuine (G)/ Synthetic (S) water	Anode/ cathode material	Reactor type	Volume treated [ml]	Optimum electrode gap [mm]	Optimum current density, treatment time and initial pH [A/m^2], [min], []			Initial pollutant levels [mg/l]	Optimum removal efficiency [%]	Optimum EEC [kWh/m^3]	Optimum EEC [kWh/kg_X]	Optimum OC [€/m^3]	Optimum OC [€/kg_X]	Research group
Highly complex industrial wastewater (a mixture of wastewaters of 144 different factories)	G	Al Fe (Al + Fe)[b]	Batch	4000[f]	20	45.45	60	6 - 8 [8[c]]	COD: 2000 - 2500 BOD_5: 900 - 1050 Turbidity: 1400 - 1800[NTU] Color: 2500 - 4750 [Pt-Co] Fecal coliforms, MPN: 111000 TS: 5360	COD: 69 BOD_5: 71 Turbidity: 80 Color: 83 Fecal coliforms: >99 TS: n.d.	n.d.	n.d.	n.d.	n.d.	Linares-Hernández *et al.* 2009 [85]
Electroplating rinse wastewaters containing cadmium, nickel and cyanide	G × 2	Fe	Batch	650	10	30/60	30/80	(7.6 - 10.6)[c]/ (8 - 10)[c] [10]	Cd: 102/- Ni: -/175 Cyanide: 120/261 COD: 180/220 TSS: 175/185	Cd: 99/- Ni: -/99 Cyanide: 100/100 COD: n.d./n.d. TSS: n.d./n.d.	6.13/11.94	n.d.	1.05/2.45	n.d.	Kobya *et al.* 2010 [87]
Galvanic wastewater	G	Al Fe (Al +Fe)[b]	Batch	1200	20	(1.0 [A] + 0.05 [A])[g]	180	5	Ni: 2000 Cu: 2500 Cr: 700	(Ni: 95 Cu: 100 Cr: 95)[g]	4.3[g]	n.d.	n.d.	n.d.	Heidmann & Calmano 2010 [88]
Metal plating wastewater	G	Al/Al Al/Fe Fe/Fe Fe/Al[b]	Batch	650	10	100	20	7 - 9 [9]	Ni: 394 Cu: 45 Cr: 44.5	Ni: 100 Cu: 100 Cr: 100	10.07	n.d.	n.d.	n.d.	Akbal & Camcı 2011 [89]
Complexed wastewater from metal plating industry	G	SS	Batch	1800[f]	3	90	180	6 - 8 [6[c]]	TOC: 170 - 173 Zn: 217 - 236 Ni: 248 - 282	TOC: 66 Zn: 100 Ni: 100	n.d.	160 kWh/kg_{TOC} 70 kWh/kg_{Ni}	n.d.	n.d.	Kabdaşlı *et al.* 2009 [90]
Automotive assembly plant rinse waters from zinc phosphate coating	G	Al[b,g] Fe	Batch Continuous	850 3500	11 20	60 60	25 (0.4/0.1 l/min)[g]	3 - 6 [5]	COD: 150 TOC: 20 Turbidity: 80 [NTU] Zn^{2+}: 40 Phosphate: 120 O & G: 10 TSS: 240	(COD: 77 TOC: 65 Turbidity: 98 Zn^{2+}: 97 Phosphate: 99 O & G: 100 TSS-97)[g]	~4.8[e] n.d.	n.d. n.d.	n.d. 2.98/6.74	n.d. n.d.	Kobya *et al.* 2010 [91]
Chemical mechanical polishing (CMP) wastewater from semiconductor fabrication	G	Fe/Al	Batch	500	20	20 [V]	20	(8 - 9)[d]	COD: 400 - 600 Turbidity: 200 - 300 [NTU] TOC: 3 - 5 TSS: 4000 - 5000	COD: 85 Turbidity: n.d. TOC: n.d. TSS: n.d.	0.64	n.d.	n.d.	n.d.	Wang & Chou 2009 [93]
Carwash wastewater	G	Fe + BDD	Batch (EC) + Batch (EO)	300	15	20	6	6 - 9 [6.4[c]]	COD: 572 Surfactants: 95.5 BOD: 178	COD: 75/97 Surfactants: 100/100 BOD: n.d.	0.14/12.0	n.d.	n.d.	n.d.	Panizza & Cerisola 2010 [94]
Alcohol distillery wastewater	G	Fe	Batch	1500	20	44.65	120	8	COD: 15600 BOD: 7200 TS: 34100 TDS: 2290	COD: 51 Color: 95 TS: n.d. TDS: n.d.	n.d.	n.d.	n.d.	n.d.	Kumar *et al.* 2009 [95]
Fermentation industry molasses process water (biologically pre-treated)	G	Al/SS[b] Fe/SS	Batch	6250	6	137	57	7.5[d]	COD: 4500 Color: n.d. TDS: 1600	COD: 70 Color: 95 TDS: n.d.	n.d.	n.d.	n.d.	n.d.	Ryan *et al.* 2008 [96]
Coking wastewater	G	Fe/Ti	Continuous	2000	10	300	80[a] (0.025 l/min)	8 - 11 [8]	Color: n.d. COD: 91 - 111 BOD_5: 18 - 28 NH_3-N: 4.8 - 5.8	Color: 91 COD: n.d. BOD_5: n.d. NH_3-N: n.d.	n.d.	n.d.	n.d.	n.d.	Zhang *et al.* 2011 [97]
Antimony mine flotation wastewater	G	Al	Batch	500	10	166.67	60	2 - 6 [2]	Sb_{total}: 10.4 - 28.6 As_{total}: 0.010 - 0.025 SBX: 0.376 - 0.434 Ca^{2+}: 360 - 389 Mg^{2+}: 25	Sb_{total}: 97 - 98 As_{total}: 100 SBX: 71 - 77 Ca^{2+}: n.d. Mg^{2+}: n.d.	n.d.	n.d.	n.d.	n.d.	Zhu *et al.* 2011 [98]

Continued

Laundry wastewater	G	Al	Batch	1500	15	1.32 [A]	45	(6 - 8)[c]	COD: 4155 Turbidity: 245 [NTU] P: 27.6 Detergent: 463 Color: 1430 [n.d.] SS: 987 Pb: 4.35 Zn: 3.2	COD: 93 Turbidity: 96, P: 97 Detergent: 94 Color: 90 SS: n.d. Pb: n.d. Zn: n.d.	~19.8[e]	n.d.	n.d.	n.d.	Janpoor *et al.* 2011 [99]
Paint manufacturing wastewater	G	Al[b] Fe	Batch	800	10	35	15	4 - 8 [7][c]	COD: 19700 TOC: n.d. SS: 1100 Pb: 1.44 Fe_{tot}: 4.82	COD: 94 TOC: 89 SS: n.d. Pb: n.d. Fe_{tot}: n.d.	n.d.	n.d.	0.129	n.d.	Akyol 2012 [101]
Marble processing wastewaters	G	Al[b] Fe	Batch	250	5	15	0.5[h]/2	(8 - 9)[c] [9]	Turbidity: 2640 [FTU] SS: 5178 TDS: 0.21 O & G: 20	Turbidity: 99[h]/100 SS: 99[h]/100 TDS: n.d./n.d. O & G: n.d./n.d.	~0.0568[e]/~0.227[e]	~0.0091[e]/0.0363 kWh/kg_{SS}	~0.0155[e]/0.062	~0.0025[e]/0.01	Solak *et al.* 2009 [102]
Industrial aqueous solution containing polyvinyl alcohol (PVA)	S	Al/Al Al/Fe Fe/Al[b] Fe/Fe	Batch	500	20	10 [V]	120	6.5[d]	PVA: 100	PVA: 77	~1.57[e]	15.68 kWh/kg_{PVA}[g]	n.d.	n.d.	Chou *et al.* 2010 [103]
Industrial aqueous solution containing salisylic acid (SA)	S	Al	Batch	500	20	12	60[g]	n.d.	SA: 100	SA: 87[g]	n.d.	0.00037 kWh/kg_{SA}[g]	n.d.	n.d.	Chou *et al.* 2011 [104]
Water distribution system water	G	Fe	Batch	1300	20	12 [V]	60	10	Calcium hardness: 138 [mg/l $CaCO_3$] Total hardness: 300 [mg/l $CaCO_3$] Turbidity: 3 [NTU]	Calcium hardness: 98 Total hardness: 97 Turbidity: n.d.	n.d.	n.d.	n.d.	n.d.	Malakootian *et al.* 2010 [105]
Municipal wastewater	G	SS	Batch	1300[f]	30	0.8 [A] = ~182[e]	5	7[d]	$BOD_{particulate}$: 51 - 84 Turbidity: 49 - 53 [NTU] TSS: 126 - 160	$BOD_{particulate}$: 99 Turbidity: 93 TSS: 95	n.d.	n.d.	n.d.	n.d.	Bukhari 2008 [106]
Underground water used as feed water of a reverse osmosis (RO) desalination plant	G	SS + sand filter (2 L) & Birm iron filter	Batch	5000[f]	n.d.	1[h]/1.75 [A] = ~(10 - 16/18 - 28)[e]	6	7[d]	Turbidity: 150 [NTU] TSS: 300 TDS: 1800	Turbidity: 92[h]/98 TSS: 94[h]/99 TDS: n.d./n.d.	n.d.	n.d.	n.d.	n.d.	Sadeddin *et al.* 2011 [108]
Mineral treatment wastewater containing ultrafine quartz suspensions	S	Al/SS	Batch	250	21	23.9[g]	10	9	Turbidity: n.d.	Turbidity: 90[g]	1.87[g]	n.d.	n.d.	n.d.	Kılıç *et al.* 2009 [109]
Bentonite suspensions (representing clay-polluted waters)	S	Fe	Batch	100[d]	30	4	40	(3/12)	Turbidity: 18 - 24 [NTU]	Turbidity: 85/80	n.d.	n.d.	n.d.	n.d.	Ghernaout *et al.* 2008 [110]

[a] = HRT (hydraulic retention time) in EC systems with continuous mode of operation [min]; [b] = Observed as the best electrode configuration of those tested; [c] = The natural, unmodified pH value of the water or wastewater (found optimal); [d] = The natural, unmodified pH value of the water or wastewater (the effect of pH not researched); [e] = Approximation calculation based on values given in the article at issue; [f] = Reactor volume (sample volume not mentioned); [g] = Optimum value estimated from the data in the article (precise value not given); [h] = Additional "optimum value" estimated from the data in the article; n.d. = Not determined.

with SS-EC in a batch mode was studied. The wastewater consisted of carrier, brightener, and metal chlorides strongly bound to the organic complexing agent (which comprised 90 % of TOC, total organic carbon). In optimum conditions (see **Table 5**), complete Zn and Ni removal was achieved and 66% of TOC was also removed.

The natural pH of the wastewater was found to be optimal. NaCl additions were also found unnecessary due to the high electrolyte (1.5 - 1.7 g/l chloride) concentration in the wastewater. EC was concluded to be a promising treatment method for complexed metal removal from wastewater originating in the metal plating industry [90].

Zinc phosphate rinse water (ZPO) from an automotive assembly plant was treated by EC. Use of Al and Fe in batch and continuous modes was studied. The removal efficiencies of batch and continuous modes were very similar, thus the values presented in **Table 5** are for the batch -Al process only. The removal efficiencies of Al and Fe were also very close to each other and no superior electrode material was suggested in the paper. However, Al was chosen as the optimum here because of its slightly better removal efficiencies and economic values. Al was also found to perform reasonably well in a significantly wider pH range than Fe. Furthermore, because the optimum initial pH for Fe would have been around 3, acid addition would have been necessary (the initial pH of ZPO was 3.8), whereas Al performed optimally with unmodified pH. In optimum process conditions, EC treatment was able to achieve high removal efficiencies with the pollutant parameters studied (see **Table 5**) [91].

Treatment of genuine chemical mechanical polishing (CMP) wastewater generated in the semiconductor fabrication industry by a batch -EC system was studied and found applicable. In another study on EC -treatment of CMP wastewater by the same research group, it was observed that Fe/Al is the most suitable electrode configuration (out of four different configurations tested) for such wastewater [92]. The CMP wastewater was highly alkaline and turbid (200 - 300 NTU), having a milky appearance while its mean particle size was as minuscule as 85 - 95 nm. Under optimum conditions, the EC process was found to remove 85% of COD in 20 min with a low EEC -value of 0.64 kWh/m^3. The very fast COD removal by the EC process was considered a great advantage of EC. The kinetic data obtained matched the pseudo first-order kinetic model well ($R^2 = 0.97$) [93].

Real carwash wastewater was treated by a combined EC/EO batch process. After Fe-EC, a 90-min EO-step at 100 A/m^2 current density was conducted using a boron-doped diamond (BBD) anode. The removal efficiencies and EEC-values presented in **Table 5** are in the form EC/EC + EO total; other values are given for the EC-step only. After the EC-step in optimum conditions, 75% of COD was removed rapidly with low energy consumption (0.14 kWh/m^3). Total surfactant removal was also noted after the EC -step. No pH alterations were found necessary. After the additional EO-step, 97% of COD was removed with a total EEC-value of 12.0 kWh/m^3. In an earlier study by the same authors, only EO was used to treat the same wastewater, achieving complete COD removal, but at a cost of enormous energy consumption of 375 kWh/m^3. Thus, adding EC prior to EO had lowered the EEC -value significantly while achieving similar removal efficiency [94].

High-strength (see **Table 5**) dark-black-colored bio-digester effluent (BDE) from an alcohol distillery was treated by a batch -EC system using iron electrodes, employing RSM to optimize the process. The created model had a high R^2 value of 0.8547 and in optimum process conditions, 51% of COD and 95% of color were removed. It was also proposed that the EC-sludge of BDE could be used in making blended fuel briquettes along with other organic fuels, as its heating value was found to be 5.3 MJ/kg [95].

Variable and highly colored and polluted molasses process water (MPW) from the discharge outlet of an anaerobic/aerobic effluent treatment facility attached to a large industrial fermentation plant was treated by a batch EC system. In this study, CC using $FeCl_3$ and aluminum sulfate was compared with EC, which was tested with Al/SS (found superior) and Fe/SS electrode combinations. Both CC processes lowered the pH of MPW to strongly acidic values of 2.4 and 3.8, respectively (the natural pH of MPW was 7.5). However, EC raised the pH to mildly alkaline 8.6 - 8.8, so it was concluded by the authors that this makes EC a significantly better option. Also, at a major industrial plant with a 1000 m^3/d output, 6 t of coagulant chemical would be needed, whereas only 300 kg of electrode material would produce similar results. Removal efficiencies were found to be in the same range with CC and EC. The EC process was not optimized properly at all, as the purpose of the work was solely to compare CC and EC. However, the results were still promising (see **Table 5**) and the reactor used was noticeably larger than in most other studies presented in **Tables 1-7** [96].

A study on decolorization of coking wastewater containing inorganic pollutants and organic contaminants by a continuous EC process using Fe/Ti electrodes was conducted. The results showed great potential in EC-based decolorization (91% color removal efficiency) of coking wastewater with only a slight initial pH alteration from the natural 7 to 8 needed. Adding NaCl to the water showed a substantial increase in removal efficiency, possibly due to electrogeneration of Cl_2, a strong oxidant [97].

An EC technique with aluminum electrodes was used in a batch mode to remove toxic and carcinogenic antimony from antimony mine flotation wastewater. The antimony concentration of the water being treated was 10 - 30 mg/l. The water also contained As (10 - 25 μg/l), SBX (sodium butyl xanthate, 380 - 430 μg/l), and substantial amounts of cations; thus no NaCl was added to increase conductivity. The EC process performed almost as well in the initial pH range of 6 - 10 as in the optimal range of 2 - 6, indicating a wide scale of applicability and the redundancy of pH adjustment (the pH value of raw water was near 7). In optimum conditions (60 min of electrolysis at 166.67 A/m^2 current density), complete As removal and nearly complete antimony removal efficiencies were achieved along with 71% - 77% removal efficiency for SBX, indicating that EC is a promising technology for removing antimony from industrial wastewater [98].

Laundry-based wastewater accounts for approximately 10 % of municipal sewer discharges; thus the efficiency of a batch electrochemical system using aluminum electrodes in treating real laundry wastewater (see **Table 5** for composition) was investigated [99]. All experiments were repeated twice and the experimental error was below 3%; average data are reported. The unaltered, near-neutral initial pH value was found optimal. Removal efficiencies in optimal conditions were high, being 90% - 97%. Therefore, it was concluded that when compared with other treatment processes, EC is more effective in treating laundry wastewater. In another study (not presented in **Table 5**) on EC treatment of (artificial) laundry wastewater, 62% COD removal efficiency was reached [100]. In this study, the application of ultrasound was studied and found to clearly enhance the EC process.

The treatability of paint manufacturing wastewater (PMW) by EC in a batch mode was investigated and found economic and feasible. The performance of Al electrodes was found to be better than that of Fe electrodes in terms of removal efficiency and OC. Initial pollutant concentrations in PMW were very high (COD 19700 mg/l, BOD 2800 mg/l, SS 1100 mg/l), however in optimal process conditions (fast 15-min treatment at a low current density of 35 A/m^2), very high removals of 94% for COD and 89% for TOC were achieved. Absorbance decreased substantially as a result of the EC treatment, indicating a significant change in the color of the water. No pH alteration of the PMW was found necessary [101].

Removal of suspended solids and turbidity from marble processing wastewater by EC was studied using a batch laboratory-scale (250 ml solution) reactor. Both Al and Fe were tested as electrode materials. When iron was used as the electrode material, removal efficiencies were found to be only slightly lower than those of Al, but the OC values were significantly higher. Therefore, Al (monopolar parallel connection) was chosen as the better option. The initial concentrations of the wastewater were very high (turbidity 2640 FTU, TSS 5178 mg/l). EC treatment neutralized the wastewater, slightly lowering its pH value towards 7 from the initial optimal value of 9 (naturally 8.23). Complete removal of TSS and turbidity was achieved rapidly within 2 min and even only 0.5-min EC treatment was able to provide 99% removal efficiencies for TSS and turbidity. Therefore, the EEC and OC-values were very low (see **Table 5**) and the EC process was concluded to be highly effective in this application [102].

Polyvinyl alcohol (PVA) is a well-known water-solu ble polymer that is hazardous and barely biodegradable. PVA is found in wastewaters of a wide range of industries; thus the feasibility of batch-EC in removing PVA from a synthetic (100 mg/l) aqueous solution was investigated. Of the four different electrode combinations tested, Fe/Al was found clearly the most efficient. The effect of altering the initial pH of the PVA solution was not studied. The experimental results showed that the kinetics of PVA removal by EC could be described with a pseudo-second-order model (R^2 = 0.99). In optimum conditions, 77% of the initial PVA was removed [103].

Salicylic acid (SA) is widely used in the pharmaceutical and cosmetic industries and it potentially has adverse health effects in animals and humans. EC -removal of SA (100 mg/l) from a synthetic industrial aqueous solution using aluminum electrodes in a batch mode was investigated and suggested to be promising. The effect of the initial pH of the SA solution was not studied and no base pH value was mentioned. Solution temperature was found to slightly affect removal efficiencies (up to about 9 percentage points in otherwise similar process conditions); 298 K was found optimal. In optimum conditions (applying a low current density of 12 A/m^2), 87% SA removal efficiency was documented, also providing an extremely low EEC -value. According to the kinetic data obtained, a pseudo-second-order kinetic model described SA removal best (R^2 = 0.98) [104].

The performance of Fe-EC in a batch mode in removing hardness from drinking water was evaluated. The water distribution system water used in this study had a pH value of 8.35 and total and calcium hardness values of 300 mg/l $CaCO_3$ and 138 mg/l $CaCO_3$, respectively. In optimum conditions, 98% of the former and 97% of the latter were removed, thus it was shown that ions responsible for water hardness could be removed by EC [105].

Raw municipal wastewater was electrocoagulated in a batch mode using SS-electrodes [106]. The EC-treatment was found effective and rapid, as it took only 5 min to achieve 99%, 93%, and 95% removal efficiencies for $BOD_{particulate}$, turbidity, and TSS, respectively. EC test runs were conducted using only raw wastewater with a pH of 7. In another study (not presented in **Table 5**) on actual municipal wastewater, it was concluded that by using very low currents (10 A/m^2), EC can reduce phosphorus and pollution associated with colloids, helping to diminish the organic load of the effluent [107].

Underground water (containing colloidal particles which cause membrane fouling in reverse osmosis, RO) used as the feed water of a RO desalination plant was treated with a batch EC-system using electrodes made of SS. A sand filter (2 l) also containing a Birm (a solid similar to active carbon, used for iron removal) filter was added to the process line after the EC unit to remove coagulated matter. The removal efficiencies given in **Table 5** were achieved after the whole process. Before EC, the water had turbidity and TSS values of 150 NTU and 300 mg/l, respectively, and its pH was 7.0. Experiments were conducted using this initial pH only. For both parameters, very high removal efficiencies were achieved rapidly (6 min) and with low current values. Further RO studies conducted using EC-pretreated water proved that all fouling indicators such as flow, pressure drop, and silt density index (SDI) showed less fouling when EC was added prior to EO [108].

Mineral treatment processes produce wastewater containing suspended and stable colloidal particles which degrade recirculation of water in processing plants. Such synthetic aqueous solutions containing quartz were treated by batch-EC using Al/SS electrodes. The median particle size of the quartz-in-water (320 mg/l quartz, initial pH 4) was 11.61 μm. A comparison between EC and CC (using aluminum sulfate in jar tests) was made, achieving similar removal efficiencies (around 90%) when similar amounts of aluminum were added to the water. The optimum pH range of CC was found to be 6 - 9, which was wider than that of EC. However, CC was found to acidify the water, whereas EC treatment shifted the initial suspension pH towards neutral. A 10-min treatment was sufficient for both methods and the kinetics of EC could be modeled with a second-order rate equation. No clearly superior treatment method for the wastewater in question could be proposed within the scope of the study, as no economic values were presented [109].

In another study, EC-treatment of synthetic wastewater was carried out in a batch electrochemical cell equipped with iron electrodes. Bentonite suspensions (~20 NTU turbidity) represented colloid-polluted wastes, as clays behave like hydrophobic colloids in water. Turbidity removals of 80% - 85% could be obtained with very low currents (4 A/m^2, 40-min EC-run). The effect of initial pH was explored with values of 3, 7, and 12, and only the neutral initial pH gave poor results. This was explained as being due to different destabilization mechanisms being prevalent in medias of different pH values. In acidic media, charge neutralization was considered to be the main removal mechanism, whereas sweep flocculation would be dominant in an alkaline solution of this type. It was concluded that the EC process can be applied to treatment of wastes polluted with colloids [110].

3.6. Surface Water and Other Natural Water

Table 6 presents a summary of recent applications of EC in the treatment of surface water and other natural water.

Treating simulated surface water containing algae (one of the most dominant cyanobacteria, *Microcystis aeruginosa*) by batch -EC was studied. The initial cell density used in the experiments was maintained at 1.2×10^9 - 1.4×10^9 cells/l. Aluminum was found to be an excellent electrode material for this application compared with iron (no coloration of water and substantially greater removal efficiency). Interestingly, it was found that algae removal was accelerated dramatically with increased water temperature. Ultimately, complete algae removal was achieved with low values of current density and EEC: 10 A/m^2 and 0.4 kWh/m^3, respectively. Thus, the results were proposed to indicate the effectiveness of EC in algae removal, from both the technical and economic points of view [111].

Laboratory experiments were carried out to investigate the effectiveness of disinfection by EC in a batch mode using artificial wastewater containing *Escherichia coli.* Real north-Algerian dam water from Ghrib (known for having high hardness) and Keddara (high algae content) dams were also used. The EC parameters presented for the three waters in **Table 6** are in the same order as mentioned above. Aluminum electrodes were found slightly more efficient than ordinary steel (Fe) and stainless steel electrodes. Electrochemical disinfection was proven effective, because the treatment times were rather low and total disinfection and algae removal were achieved [112].

In another study (not presented in **Table 6**), Al -EC in a batch mode was found to be a suitable process for decreasing hardness and removing bacteria, algae, and bacterial nutriments from two different raw surface waters. The water samples originated from a river and a pond. Complete disinfection was achieved [113].

Reduction of humic acids (HA) from 1000 mg/l synthetic solutions by a batch Al-EC system was studied [114]. The effect of applying electromagnetic (EM) treatment prior to EC was also investigated in both batch and continuous modes, of which the latter was found more suitable. EM is an attractively simple approach in which the water being treated flows through a magnetic field, and it consequently slightly changes some of its physicochemical properties. Both EM and EC processes were found to perform best at neutral pH. The 10-min EM-pretreatment was found to slightly further increase the removal efficiency of HA by EC from 96% to 100% [114].

Removal of NOM (natural organic matter) from surface water (inlet flow of a Finnish paper mill) by a batch EC process using aluminum electrodes has been studied using RSM and ANOVA. In modeled optimum conditions, the applied current density, treatment time, and EEC-value were low, with simultaneous high removal efficiency. A slight lowering of initial pH (naturally 5.8)

Table 6. Recent applications of EC in the treatment of surface water and other natural water.

Water and wastewater types used	Genuine (G)/ Synthetic (S) water	Anode/ cathode material	Reactor type	Volume treated [ml]	Optimum electrode gap [mm]	Optimum current density, treatment time and initial pH [A/m^2], [min], []			Initial pollutant levels [mg/l]	Optimum removal efficiency [%]	Optimum EEC [kWh/m^3]	Optimum EEC [kWh/kg$_X$]	Optimum OC [€/m^3]	Optimum OC [€/kg$_X$]	Research group
Water containing algae (cyanobacteria)	S	Al[b] Fe	Batch	1000	10	10	45	4 - 7	Cyanobacteria: 1.2×10^9 - 1.4×10^9 [cells/l]	Cyanobacteria: 100	0.4	n.d.	n.d.	n.d.	Gao *et al.* 2010 [111]
Dam waters	S + G × 2	Al[b] Steel SS	Batch	500	50	(1[g]/0.8/0.25) [A] = ~(202/162/51)[e]	10[g]/35/30	7 - 9.5	*E. coli*: n.d./n.d./n.d.	*E. coli*: 98[g]/100/100	~23.6[e]/16.8/~4.7[e]	n.d.	n.d.	n.d.	Ghernaout *et al.* 2008 [112]
Water containing humic acids (HA)	S	Al	Batch Continuous (EM) + Batch (EC)	500	40	33.3	30 10[a] (0.01 l/min) + 30	7 7 + 7	HA: 1000	HA: 96 100	n.d.	n.d.	n.d.	n.d.	Ghernaout *et al.* 2009 [114]
Surface water (river) containing a high concentration of NOM (paper mill inlet flow)	G	Al	Batch	500[i]	10[i]	4.8	12[i]	4.3	DOC: 18.35 UV 254 nm: 0.64 [absorbance] Turbidity: 0.51 [NTU]	DOC: 80 UV_{254}: 91 Turbidity: n.d.	0.4	n.d.	n.d.	n.d.	Vepsäläinen *et al.* 2009 [115]
Micro-polluted raw water	G	Al[b] Fe	Batch	1000	10	50	20	5 - 7.5 [7.5[c]]	TOC: 8.5 - 16.2 Oil: 0.8 - 1.5 NH_3-N: 0.75 - 1.26	TOC: 70 Oil: 86 NH_3-N: 75	n.d.	n.d.	n.d.	n.d.	Li *et al.* 2008 [116]
Marine water containing microalgae (for biodiesel production)	S × 2	SS	Batch	300	48	10 [V]	15	4 - 9	Microalgal: 600/300	Microalgal: 98/99	n.d.	4.44/9.16 kWh/$kg_{Microalgal}$	n.d.	n.d.	Uduman *et al.* 2011 [117]
Freshwater and marine water containing microalgae (for biodiesel production)	S × 2	Al/(IrO_2/TiO_2)[b] Fe/(IrO_2/TiO_2)	Batch	1000	44	(6/15)[g]	(30/50)[g]	4 - 6 [4]	Microalgal: 300 - 600/300 - 600	Microalgal: (80/95)[g]	n.d.	~0.3/~(1.5 - 2.0) kWh/$kg_{Microalgal}$	n.d.	n.d.	Vandamme *et al.* 2011 [118]
Pesticide-contaminated (metribuzin, MB) groundwater	S	Fe/SS	Batch Batch + UV	1300[f]	n.d.	18	80	5 - 6 [6[c]] 6 - 7 [6[c]]	MB: 50 - 300 (optimum 200)	MB: 89 95	n.d.	n.d.	n.d.	n.d.	Yahiaoui *et al.* 2011 [119]
Geothermal waters containing boron	G	Al	Batch	1500	5	15[h]/30[h]/60	30	8	B: 24	B: 73[h]/84[h]/96	0.73[h]/~2.3[h]/12.8	n.d.	n.d.	n.d.	Yilmaz *et al.* 2008 [120]
Riverwater containing mercury(II)	S × 2	Al Fe[b]	Batch	100	30	125	15	3 - 7 [7]	Hg^{2+}: 4 Hg^{2+}: 4 COD: 378	Hg^{2+}: 100 Hg^{2+}: 100 COD: 90	n.d.	n.d.	n.d.	n.d.	Nanseu-Njiki *et al.* 2009 [121]

[a] = HRT (hydraulic retention time) in EC systems with continuous mode of operation [min]; [b] = Observed as the best electrode configuration of those tested; [c] = The natural, unmodified pH value of the water or wastewater (found optimal); [e] = Approximation calculation based on values given in the article at issue; [f] = Reactor volume (sample volume not mentioned); [g] = Optimum value estimated from the data in the article (precise value not given); [h] = Additional "optimum value" estimated from the data in the article; [i] = Queried from the author; n.d. = Not determined.

was found beneficial. The above values are presented in **Table 6** for water at room temperature (23˚C). The water samples taken from the river were at 3˚C and EC -runs were also performed with water of this temperature. The removal efficiency was then only four percentage points lower, thus it was concluded that EC is a feasible treatment process for removal of NOM also during the cold water period in the Nordic countries [115].

Batch-EC was used and found feasible for treating micro-polluted surface water in laboratory-scale experiments. Aluminum was selected as the electrode material, because although iron produced nearly similar removal efficiencies, it also colored the water (to greenish at first and then to brown). Initial pollutant concentrations were very low (see **Table 6**), but so were also the optimal current density and treatment times (50 A/m^2 and 20 min, respectively). Also, no pH adjustment was found necessary [116].

Chlorococcum sp. and *Tetraselmis* sp. (results in this order in **Table 6**) were cultivated to produce artificial marine water to be separated by batch-EC from microalgae for use in biodiesel production. Both species were quite different but had approximate cell sizes of 10 µm. Altering the initial pH value between 4 - 9 had no clear impact on removal efficiencies; EC was found applicable over the whole scale tested. Both waters were naturally within this pH range: the first water had a pH value of 9.1 and the latter, 8.3. High recovery efficiencies were obtained, up to 98% and 99% for *Chlorococcum* sp. and *Tetraselmis* sp., respectively. Microalgae flotation due to hydrogen bubble attachment was documented using a hi-speed camera; this is presented in **Figure 6** [117].

In another similar study, using EC to harvest synthetic marine water (containing *Phaeodactylum tricornutum*) and freshwater (containing *Chlorella vulgaris*) for biodiesel production was evaluated (results in this order in **Table 6**). Using Al as anodes was found clearly more efficient than using Fe. Both electrode configurations had cathodes made of IrO_2/TiO_2, which is fairly uncommon. The aluminum content of the harvested microalgal biomass was less than 1%, while the aluminum concentration of the process water was below 2 mg/L for *C. Vulgaris* and below 0.5 mg/l for *P. tricornutum*. Rather rarely, the effect of stirring the water was tested, within a range of 0 - 200 rpm. It was found that increasing the stirring speed significantly increased the performance of the EC process up to a value of 150 rpm, enhancing contact rates between coagulants and microalgal cells. However, further increasing the stirring rate was found to decrease the performance of the EC process to near the level of 0 rpm. This was proposed to be due to the break-up of flocs because of the high shear forces applied. Under optimal conditions, the EEC-values were around 0.3 kWh/kg_{algae} harvested for *P. tricornutum* and approximately 1.5 - 2.0 kWh/kg_{algae} for *C. vulgaris*, while the respective microalgal recovery rates were 80% and 95%. In specific triplication tests, the process was found to be repeatable. Compared to centrifugation, EC was thus suggested as substantially more energy-efficient. Finally, it was concluded that EC is a promising technology for harvesting marine microalgae, but tests with large-scale pilot EC reactors need to be done to confirm this [118].

A batch EC system using a Fe/SS electrode configuration was used to treat model pesticide-contaminated (metribuzin, MB) groundwater. Metribuzin is considered a general-use pesticide which belongs to the group of triazinone herbicides, and it is highly water-soluble (1.05 g/l). The initial MB values used in the study (50 - 300 mg/l) were similar to those measured in discharges from MB manufacturers. The performance of the EC process was compared with that of a batch combined EC + UV (ultraviolet) process, and the latter was found slightly more effective (MB removal efficiencies of 89% and 95%, respectively). Using a batch EF + UV process to treat the water was also investigated, but it resulted in lower removal efficiencies than the EC + UV process, and thus the results are not shown in **Table 6**. When the water was treated with UV alone for 80 min (optimal value found for EC + UV), approximately 12% removal efficiencies were achieved (see **Figure 6** in the original paper). The natural initial pH value (6) of the water was found optimal for both the EC and EC + UV processes, while low pH values would have been needed for optimal functioning of the EF + UV process. In the end, it was proposed that the process studied may be employed successfully to remove pesticides from water [119].

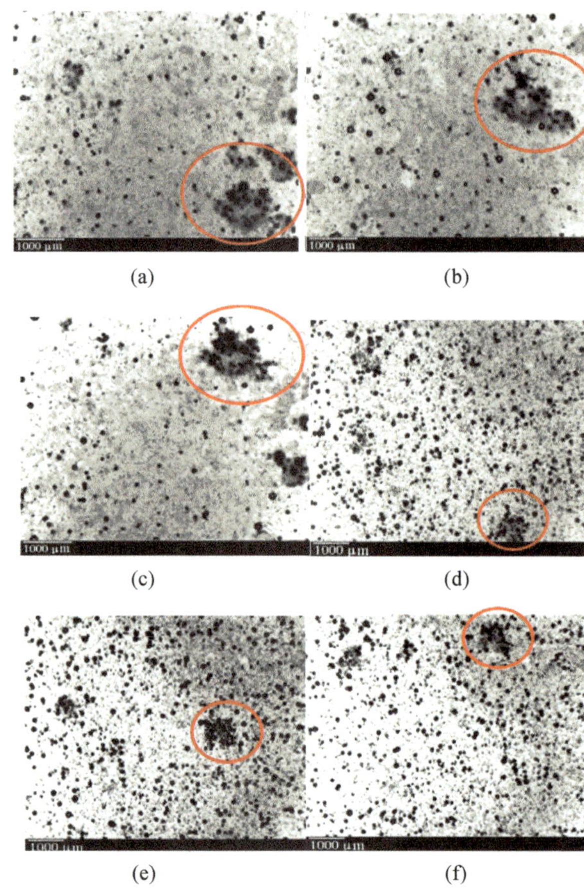

Figure 6. Hydrogen bubble-floc interaction at different times: For *Chlorococcum* sp.: (a) 0 s; (b) 0.4 s; and (c) 0.8 s after the current was turned off. For *Tetraselmis* sp.; (d) 0.4 s; (e) 0.7 s; and (f) 0.8 s after the current was turned off. (Main bubble-microalgae floc aggregates are circled in each figure). Adapted from [117].

Although boron is a vital micronutrient for plant and crop growth, high boron concentrations in irrigation water are known to be detrimental to them. Therefore, treatment of genuine geothermal water containing 24 mg/l boron was studied with an EC system working in a batch mode and using aluminum electrodes. An initial pH value of 8 was found optimal, which was near to the natural value of 6.5. It was found that decreasing the current density value from 60 A/m^2 (proposed as optimal) to 30 A/m^2 or 15 A/m^2 corresponds to a tremendous decrease in the EEC-value from 12.8 kWh/m^3 to 2.3 kWh/m^3 and 0.73 kWh/m^3, respectively, while boron removal efficiencies decreased from 96% to 84% and 73 %, respectively. It was concluded that after the EC process, the effluent water could be used for irrigation [120].

Studies on mercury (II) removal from water were conducted using a laboratory-scale (100 ml) batch-EC system. A synthetic mercury-containing (4 mg/l) solution was first used to optimize the process. Iron was found to perform better than aluminum; however, both electrodes achieved very high removal efficiencies over a wide pH scale. Mercury was then added to a river water sample achieved very high removal efficiencies over a wide pH scale. Mercury was then added to a river water sample and the solution was treated in the previously determined optimum process conditions (see **Table 6**). Complete mercury removal was achieved also from the semi-synthetic mercury-contaminated river water, with 90% COD removal, as well. It was concluded that EC can be effective in the treatment of water polluted by mercury(II) ions [121].

3.7. Model Water and Wastewater Containing Heavy Metals, Nutrients, Cyanide and Other Elements and Ions

EC studies presented in this category of waters can be simplified to have been conducted using small-scale laboratory Al, Fe, or SS batch reactors and synthetic model water or wastewater with only one particular pollutant removed (see **Table 7**). Therefore, not all of the studies [13,122-137] are discussed in the following text (although the optimum conditions found are presented in **Table 7**), but only those which differ notably from this in some way. **Table 7** presents a summary of recent applications of EC in the treatment of waters containing heavy metals, cyanide, and other elements and ions.

Arsenic removal batch-EC experiments using iron were conducted in a laboratory with synthetic solutions, providing complete As(V) removal with a short treatment time. The experiments were further expanded to field tests in which 50-l batches of real arsenic- and phosphate-containing groundwater were treated. In the field tests, a filtration step was also added after the EC, raising the removal efficiency of total arsenic from 97% to 100%. The economic values were found to be very low and the naturally neutral initial pH of the water was found optimal [122].

Iron-containing (25 mg/l) synthetic solutions were treated by batch-EC. Magnesium was used as the anode material (iron as the cathode), which is rare, but the EC system performed well and in optimum conditions very high removal efficiencies with very low current density values were achieved (see **Table 7**) [123].

Magnesium was used as the anode material (with a SS cathode) in yet another study. The synthetic water being treated contained boron and the batch EC system performed well in removing it. A scale-up batch EC system with an 8.5-l cell volume was also built and tested. It was found to produce exactly similar removal efficiencies with similar (optimum) process conditions (86% of initial boron removed at an initial pH of 7, using a low current density of 20 A/m^2). This was concluded to show the robustness of the EC process [125].

Removal of cadmium from aqueous solutions was studied with Al-EC in a batch mode. In addition to using a regular DC power source, application of an alternating pulse current (APC) to prevent passivation of the electrodes was studied. EC in the APC mode was found to perform slightly better than in the DC mode, with near-complete cadmium removal and a significantly shorter treatment time required. Also, a pilot-scale batch-EC system with a 2000-l cell volume was built and tested in this study. The results were consistent with the results obtained from the laboratory scale, showing that the process was technologically feasible and scalable. Adding high concentrations of co-existing ions (carbonate, phosphate, silicate, or arsenate) to the solution was found to decrease cadmium removal efficiency significantly due to competition for adsorption binding sites [126].

Removal of chromium from synthetic solutions with concentrations of 50, 100, 500, and 1000 mg/l (values given in this order in **Table 7**) were treated with a batch-EC system using aluminium/iron electrodes. Use of APC was also studied in addition to the DC mode. During these experiments the polarity of the anode and cathode was thus changed every 4 min. It was found that the APC mode was more efficient than the DC mode here, also, with operating times 3%, 6%, 15%, and 25% shorter when treating initial Cr(VI) concentrations of 50, 100, 500, and 1000 mg/l, respectively. This makes the APC mode more cost-effective. Also, the turbidity values of the treated water were 1 NTU and 20 NTU with APC and DC, respectively. NaCl, KCl, PAC, and $NaNO_3$ were tested as a supporting electrolyte; NaCl and KCl were found most suitable in every aspect. In optimum conditions, complete removal of chromium was achieved [127].

Solutions containing radioactive strontium were treated with a batch-EC system. Initial concentrations of

Table 7. Recent applications of EC in the treatment of waters containing heavy metals, cyanide, and other elements and ions.

Water and wastewater types used	Genuine (G)/ Synthetic (S) water	Anode/ cathode material	Reactor type	Volume treated [ml]	Optimum electrode gap [mm]	Optimum current density, treatment time and initial pH [A/m²], [min], []			Initial pollutant levels [mg/l]	Optimum removal efficiency [%]	Optimum EEC [kWh/m³]	Optimum EEC [kWh/kg$_X$]	Optimum OC [€/m³]	Optimum OC [€/kg$_X$]	Research group
Contaminated groundwater (As)	S	Al Fe[b]	Batch	650	13	2.5	2.5	5.5 - 7.5 [6.5]	As: 0.15	As: 99	0.014	n.d.	0.0047	n.d.	Kobya *et al.* 2011 [13]
Contaminated drinking water (As)	S G	Fe	Batch Batch (scale-up/ field test)	1000 50000	20 5	0.022 [A] = 3.9[e] 2 [A]	(5/15)[g] 180	5 - 7 [7[c]]	As(V): 0.1 - 1.0 As_{total}: 0.45 - 0.67 Phosphate: 0.18 - 0.75	As(V): (~97/100)[g] As_{total}: 100 Phosphate: 100	0.5 0.72 - 0.78	n.d.	n.d. 0.077	n.d.	Wan *et al.* 2011 [122]
Contaminated drinking water (Fe)	S	Mg/Fe	Batch	900	5	2[h]/6	35	6	Fe: 5 - 25 (optimum 25)	Fe: 92[h]/98	n.d.	n.d.	n.d.	n.d.	Vasudevan *et al.* 2009 [123]
Contaminated drinking water (Fe)	S	Al	Batch	1000	5	12.5	40	8	Fe: 2 - 15 (optimum 10)	Fe: 99	~9.4[g]	n.d.	0.21	n.d.	Ghosh *et al.* 2010 [124]
Drinking water containing boron	S	Mg/SS	Batch	900	5	20	30	7	B: 3 - 7 (optimum 5)	B: 86	n.d.	n.d.	n.d.	n.d.	Vasudevan *et al.* 2010 [125]
Water containing cadmium	S	Al-alloy	Batch	1000[f]	5	10[h]/20	30 (AC) 45 (DC)	6 - 8 [7]	Cd: 10 - 50 (optimum 20)	Cd: 94[h]/98 (AC) 92[h]/96 (DC)	0.227[h]/0.454 (AC) 0.881[h]/1.002 (DC)	n.d.	n.d.	n.d.	Vasudevan *et al.* 2011 [126]
Water containing Cr(VI)	S	Fe/Al - Al/Fe (APC)	Batch	700[f]	15	56/153/ 153/222	20/25/ 55/110	3 - 5 [5[c]]	Cr(VI): 50/100/ 500/1000	Cr(VI): 98/98/ 99/100	4.0/16.3/ 20.2/58.0	n.d.	n.d.	n.d.	Keshmirizadeh *et al.* 2011 [127]
Water containing copper	S	Al	Batch	560	15	(9.1/ 9.0)[g] [V]	(10.4/ 10.2)[g]	6 - 8	Cu: 2.5 - 32.5 (optimum 14.2/15.0)[g]	Cu: (81/90)[g]	(4.07/ 6.32)[e,g]	n.d.	n.d.	n.d.	Bhatti *et al.* 2011 [128]
Water containing fluoride	S	Al	Batch	2000	30	111	25	7	Fluoride: 25 - 125 (optimum 25)	Fluoride: 95	n.d.	n.d.	0.90 - 1.05	n.d.	Behbahani *et al.* 2011 [129]
Aqueous solution containing indium	S	Fe/Al	Batch	500	20	64	50/60[h]/ 108[h]	2.3[d]	In: 20 - 80 (optimum 20/40[h]/80[h])	In: 90/90[h]/90[h]	~0.002[e]/ n.d./n.d.	0.085/n.d. /n.d.	n.d.	n.d.	Chou & Huang 2009 [130]
Wastewater containing manganese	S	Al	Batch + Aerator	250	10	62.5	(30/ 60)[g]	7	Mn^{2+}: 25 - 400 (optimum 100)	Mn^{2+}: (78/94)[g]	n.d.	n.d.	n.d.	n.d.	Shafaei *et al.* 2010 [131]
Water containing phosphate	S	Al[b] Fe	Batch	5000	9	(1/10/ 30/50)[g]	~(380/ 150/ 80/25)[g]	n.d.	Phosphate: 27	Phosphate: (100/100/ 100/100)[g]	0.06/~0.38[g]/ ~0.58[g]/ 0.73	n.d.	n.d.	n.d.	Lacasa *et al.* 2011 [132]
Phosphate-contaminated water	S	Al/SS Al-alloy /SS[b] Fe/SS	Batch	900	5	20	30	6 - 9 [7]	Phosphate: 100	Phosphate: 99	n.d.	n.d.	n.d.	n.d.	Vasudevan *et al.* 2009 [133]
Water containing strontium	S	Al SS[b]	Batch	200	60	80	50	5 - 7 [7]	Sr(II): 10 - 100 (optimum 10)	Sr(II): 93	n.d.	n.d.	~2[g]	n.d.	Murthy & Parmar 2011 [134]
Cyanide-laden wastewater	S	Al Al-Fe Fe Fe-Al[b]	Batch Continuous	250 250	30 30	150	(20/30)[g] 140[a] (0.00671 /min)	9.5[d]	Cyanide: 300	Cyanide: (98/100)[g] 100	n.d.	n.d.	n.d.	n.d.	Moussavi *et al.* 2011 [135]
Contaminated soil (Pb, Zn, Cd) washing solution containing EDTA	S	Al/SS	Batch	500	10	160[g]	30[g]	7.1[d]	Pb: 1220 - 1620 Zn: 230 - 290 Cd: 8 - 10 EDTA: 16500 - 19700	Pb: 95 Zn: 68 Cd: 66 EDTA: ~50	~8[e]	n.d.	n.d.	n.d.	Lestan & Pociecha 2010 [136]

Continued

Aqueous solution containing tannic acid (TA)	S	Fe	Batch	200	20	100	75[e]	7 - 9 [9]	TA: 50	TA: 99	n.d.	2.4 kWh/kg_{COD}	n.d.	n.d.	Mansouri *et al.* 2011 [137]

[a] = HRT (hydraulic retention time) in EC systems with continuous mode of operation [min]; [b] = Observed as the best electrode configuration of those tested; [c] = The natural, unmodified pH value of the water or wastewater (found optimal); [d] = The natural, unmodified pH value of the water or wastewater (the effect of pH not researched); [e] = Approximation calculation based on values given in the paper; [f] = Reactor volume (sample volume not mentioned); [g] = Optimum value estimated from the data in the article (precise value not given); [h] = Additional "optimum value" estimated from the data in the article; n.d. = Not determined.

up to 100 mg/l were investigated. Even though according to **Figure 5** in the article and as mentioned in the text, a neutral initial pH value was found to be the most efficient, nonetheless pH 5 was chosen as the optimum by the authors. However, the difference between these values was negligible. Raising the temperature above 30°C was found to decrease efficiency. Stainless steel was found to be a clearly better electrode material than Al for this application, with optimal removal efficiency of 93%. In this study, removal efficiency was found to greatly improve when the distance between the electrodes was increased [134].

Synthetic wastewater containing 300 mg/l cyanide was treated by EC operating in both batch and continuous modes. Out of the four electrode combinations tested, Fe/Al was found the best-performing, with Fe/Fe nearly as effective. When Al was used as the anode material, substantially lower removal efficiencies for cyanide were achieved. Both the batch and continuous EC experiments were conducted using additional aeration in the EC reactor. Aeration was found to improve the efficiency of cyanide removal in the 30-min batch EC test by approximately 6 percentage points, raising it from 94% to 100%. Cyanide could be removed completely from the wastewater with EC using both operating modes, although the batch mode was found more efficient (see **Table 7**). Therefore, EC was concluded to be a promising technique for treating cyanide-laden wastewater [135].

4. Discussion

Research on various applications of EC has been conducted extensively around the world recently. A large number of these studies have been conducted in the Middle Eastern countries and India. In these studies EC has been found to be a feasible, economical and ecological alternative in the treatment of various types of water and wastewater with promising results. The interest in EC seems to be on the rise. Apart from a handful of articles, this paper discusses EC literature published in 2008-2011.

A few of the EC systems studied worked in a hybrid treatment mode (EM, EF, addition of polymer, UV-light, aeration, filtration). RSM has been successfully applied by several authors to optimize the EC process, providing high levels of significance and very low percentages of experimental error (related to theoretical models) in the papers discussed in this study. Therefore, RSM could be applied with EC to find case-specific optimum operating conditions. Duplication or triplication of all test runs was performed in a number of papers. They confirmed that the EC process is repeatable with low (a few percent) experimental error.

Innovations such as collecting the hydrogen gas produced during EC and utilizing it for the EC process's own energy demand (leading to a 13% decrease in the energy costs of the EC process), solar-powered EC systems, partially recirculating EC sludge supernatant, applying sonar or magnetic field treatment prior to EC (inducing changes in the structure of the studied aqueous solutions, which in turn leads to enhancement of the EC process by a few percentage points), and applying APC have been studied, with promising results.

Most of the authors have conducted their EC studies using small (250 - 2000 ml) laboratory-scale reactors with magnetic stirring and virtually all of the EC systems operated in a batch mode. Systems working in a continuous mode and larger scale-up systems have also been investigated, showing promising treatment results and the scalability of the process. Such EC systems should be applied more in future studies.

Slightly over half of the EC studies scrutinized in this paper were carried out using genuine water or wastewater. In a few studies both genuine and synthetic wastewaters modeling a certain similar type of wastewater were used. The wastewater from paper the industry used by the authors was all genuine. Nearly all of the wastewaters from the oil and food industries were also genuine. Synthetic wastewaters have been used much more extensively in the EC studies of tannery, textile, and colored wastewaters, as only a third of such studies discussed here were done using real wastewater. The other types of industrial wastewater studied were mostly genuine.

In nearly all of the studies discussed here, the electrode materials used were made of aluminum, iron, or SS in different combinations, with only a few exceptions. As it can be seen from **Tables 1-7**, the superiority of different electrode materials seems to vary between different types of aqueous solutions being treated and must therefore always be researched case-specifically. In some of the studies, aluminum and iron performed so similarly that a clear choice of the superior material could not be suggested. It must also be taken into consideration that treatment costs and efficiencies are not always the sole factors when choosing between different electrode materials to be used, because other technical aspects (e.g. floc properties, coloration of water by iron, etc.) may also affect the decision-making.

The distance between the electrodes fluctuated between 2 - 70 mm; however, most setups used by the authors considered here had as electrode gap of 5 - 20 mm. The effect of this parameter on the feasibility of the EC treatment was rarely studied; e.g. [10,29,124,134]. This also applies to the effect of temperature and stirring, although all of these have been shown to have a varied effect on the removal efficiencies of EC [31,48,115,117, 120,134,138].

Even though the EC process seems to function well over a wide range of pH values in most studies, generally a relatively narrow pH range (depending on the electrode materials used) where the process performed optimally could be found. This pH range was mostly found to be close to neutral pH values, as observed in **Tables 1-7**.

The treatment costs and electricity consumption of the EC process in optimal process conditions were not calculated and presented by all the authors, but the aforementioned values were mostly somewhat low when they were presented (typically around 0.1 - 1.0 €/m^3 and 0.4 - 4.0 kWh/m^3, respectively, see **Tables 1-7**). On the whole, the values for optimal treatment costs and electricity consumption varied greatly between different studies and different types of aqueous solutions, the aforementioned values fluctuating between 0.0047 - 6.74 €/m^3 and 0.002 - 58.0 kWh/m^3, respectively. However, both values were close to the lower end of the scale presented above.

In most of the papers reviewed in chapter 3.1 the EEC -values were given, most of them being inside or close to a range of 0.1 - 0.8 kWh/m^3. Operating costs were calculated in only two of the papers, however, they were found to be similar in degree: 0.12 and 0.24 €/m^3, which can be considered very low. Oily wastewaters had economic numbers close to each other. As for OC, 0.2 - 0.4 €/m^3 seemed to be an average value, and correspondingly, 2 - 6 kWh/m^3 was the average EEC -value. The results of a very recent investigation of EC treatment of bio oil-in-water and synthetic oil-in-water emulsions support these findings [139].

Food industry wastewater treated with EC had OC -values in a range of 0.051 - 4.1 €/m^3 (given in less than half of the papers). The papers presented in chapter 3.5 had economic numbers presented in about half of the papers and their fluctuation was considerable (between 0.062 - 6.74 €/m^3 and 0.14 - 19.8 kWh/m^3). This can be considered to be mainly due to the significant variance of water types, compositions, and their pollutant concentrations. Nearly all of the EC treatment results for waters presented in chapter 3.7 had very low OC-values when they were presented; the same is true for EEC-values (see **Table 7**). This could be mainly due to the fact the waters in this category were synthetic, modeling the removal of only one pollutant at a time and at mostly relatively low concentrations. The publications reviewed in chapters 3.2 and 3.6 had no OC-values given. Paper industry wastewater EEC-values were also absent, whereas in surface water category they fluctuated substantially (between 0.4 - 23.6 kWh/m^3).

The observed optimal current densities varied greatly, but in most studies they were found to be in the range of 10 - 150 A/m^2. When the various waters were divided into categories (chapters 3.1 - 3.7), it seems the optimal current density values of oily wastewaters were the highest of all the categories, on average (approx. 160 A/m^2). Waters from other industries also had optimal current density values often higher than average. The corresponding value for waters and wastewaters presented in chapters 3.6 and 3.7 were the lowest, on average: approx. 45 - 50 and 60 - 65 A/m^2, respectively. Both water categories had most of their optimal current values inside the range of 5 - 60 A/m^2. For wastewaters presented in chapter 3.5, optimal current density values were largely in the range of 100 - 150 A/m^2. These observations could be related to the high levels of pollutants in oily and other industrial wastewaters and, on the contrary, to surface waters and modeled waters and wastewaters being only mildly polluted in comparison, with relatively low concentrations of pollutants.

One of the advantages of the EC process is its fast treatment capability, and in a vast majority of the studies discussed in this paper, optimal treatment times were found to be in the range of 5 - 60 min (these figures do not take required sedimentation times into account, which are generally rather short). More than half of the authors found an optimum treatment time of 30 min or less. In some cases a two-to three-hour EC treatment time was needed, however some wastewater could be purified rapidly in a few minutes. As with other functional parameters, treatment duration seems to be strongly dependent on the type of water being purified and its concentration. The applied current density also has a significant impact on treatment duration.

5. Conclusions

The range of feasible EC applications is expanding. In a vast majority of the studies discussed in this paper, optimal treatment times were found to be in the range of 5 - 60 min (not taking into account required sedimentation times). More than half of the authors found an optimum treatment time of 30 min or less. Observed optimal current densities varied greatly, but in most studies they were found to be in the range of 10 - 150 A/m^2. Even though the EC process seems to function well over a wide range of pH values in most studies, generally a relatively narrow pH range could be found where the process performed optimally. This pH range was mostly found to be close to the neutral pH value.

The superiority of different electrode materials seems to vary between different types of aqueous solutions being treated, and must therefore always be studied case-specifically. Both OC and EEC-values were found to fluctuate greatly between different water types being treated, between 0.0047 - 6.74 €/m^3 and 0.002 - 58.0 kWh/m^3, but they were generally rather low (typically around 0.1 - 1.0 €/m^3 and 0.4 - 4.0 kWh/m^3, respectively). To conclude, EC has great potential in purification of various types of water and wastewater and seems to be a feasible and economical alternative in this field, although more research is needed, especially using larger-scale and/or continuous systems and focusing on the fundamentals of the EC process.

6. Acknowledgements

This study was supported by Maa—ja vesitekniikan tuki ry. and Academy of Finland (AquAlSi-project). Expression of gratitude is presented to Authorized Translator Keith Kosola for revising the language.

REFERENCES

[1] P. K. Holt, G. W. Barton and C. A. Mitchell, "The Future for Electrocoagulation as a Localised Water Treatment Technology," *Chemosphere*, Vol. 59, No. 3, 2005, pp. 355-367.

[2] G. Chen, "Electrochemical Technologies in Wastewater Treatment," *Separation and Purification Technology*, Vol. 38, No. 1, 2004, pp. 11-41.

[3] M. M. Emamjomeh and M. Sivakumar, "Review of Pollutants Removed by Electrocoagulation and Electrocoagulation/Flotation Processes," *Journal of Environmental Management*, Vol. 90, No. 5, 2009, pp. 1663-1679.

[4] E. Butler, Y.-T. Hung, R. Y.-L. Yeh and M. S. Al Ahmad, "Electrocoagulation in Wastewater Treatment," *Water*, Vol. 3, No. 2, 2011, pp. 495-525.

[5] M. Y. A. Mollah, R. Schennach, J. R. Parga and D. L. Cocke, "Electrocoagulation (EC)—Science and Applications," *Journal of Hazardous Materials*, Vol. 84, No. 1, 2001, pp. 29-41.

[6] P. R. Kumar, S. Chaudhari, K. C. Khilar and S. P. Mahajan, "Removal of Arsenic from Water by Electrocoa- gulation," *Chemosphere*, Vol. 55, No. 9, 2004, pp. 1245-1252.

[7] X. Chen, G. Chen and P. L. Yue, "Separation of Pollutants from Restaurant Wastewater by Electrocoagu- lation," *Separation and Purification Technology*, Vol. 19, No. 1-2, 2000, pp. 65-76.

[8] S. Zodi, O. Potier, F. Lapicque and J. Leclerc, "Treatment of the Textile Wastewaters by Electrocoagulation: Effect of Operating Parameters on the Sludge Settling Characteristics," *Separation and Purification Technology*, Vol. 69, No. 1, 2009, pp. 29-36.

[9] S. Gao, M. Du, J. Tian, J. Yang, J. Yang, F. Ma and J. Nan, "Effects of Chloride Ions on Electro-Coagulation-Flotation Process with Aluminum Electrodes for Algae Removal," *Journal of Hazardous Materials*, Vol. 182, No. 1-3, 2010, pp. 827-834.

[10] E. Terrazas, A. Vázquez, R. Briones, I. Lázaro and I. Rodríguez, "EC Treatment for Reuse of Tissue Paper Wastewater: Aspects That Affect Energy Consumption," *Journal of Hazardous Materials*, Vol. 181, No. 1-3, 2010, pp. 809-816.

[11] S. Kongjao, S. Damronglerd and M. Hunsom, "Simultaneous Removal of Organic and Inorganic Pollutants in Tannery Wastewater Using Electrocoagulation Technique," *Korean Journal of Chemical Engineering*, Vol. 25, No. 4, 2008, pp. 703-709.

[12] İ. A. Şengil and M. Özacar, "The Decolorization of C.I. Reactive Black 5 in Aqueous Solution by Electrocoa- gulation Using Sacrificial Iron Electrodes," *Journal of Hazardous Materials*, Vol. 161, No. 2-3, 2009, pp. 1369-1376.

[13] M. Kobya, F. Ulu, U. Gebologlu, E. Demirbas and M. S. Oncel, "Treatment of Potable Water Containing Low Concentration of Arsenic with Electrocoagulation: Different Connection Modes and Fe-Al Electrodes," *Separation and Purification Technology*, Vol. 77, No. 3, 2011, pp. 283-293.

[14] G. Mouedhen, M. Feki, M. D. P. Wery and H. F. Ayedi, "Behavior of Aluminum Electrodes in Electrocoagulation Process," *Journal of Hazardous Materials*, Vol. 150, No. 1, 2008, pp. 124-135.

[15] K. Yetilmezsoy, F. Ilhan, Z. Sapci-Zengin, S. Sakar and M. T. Gonullu, "Decolorization and COD Reduction of UASB Pretreated Poultry Manure Wastewater by Electrocoagulation Process: A Post-Treatment Study," *Journal of Hazardous Materials*, Vol. 162, No. 1, 2009, pp. 120-132.

[16] M. Rebhun and M. Lurie, "Control of Organic Matter by Coagulation and Floc Separation," *Water Science and Technology*, Vol. 27, No. 11, 1993, pp. 1-20.

[17] M. Kobya, O. T. Can and M. Bayramoglu, "Treatment of Textile Wastewaters by Electrocoagulation Using Iron and Aluminum Electrodes," *Journal of Hazardous Materials*, Vol. 100, No. 1-3, 2003, pp. 163-178.

[18] L. D. Benefield, J. F. Judkins and B. L. Weand, "Process Chemistry for Water and Wastewater Treatment," Prentice-Hall, Englewood Cliffs, 1982.

[19] A. J. Rubin, "Aqueous-Environmental Chemistry of Me-

tals," Ann Arbor Science Publishers, Ann Arbor, 1974.

[20] F. Akbal and S. Camcı, "Comparison of Electrocoagulation and Chemical Coagulation for Heavy Metal Removal," *Chemical Engineering & Technology*, Vol. 33, No. 10, 2010, pp. 1655-1664.

[21] J. Duan and J. Gregory, "Coagulation by Hydrolysing Metal Salts," *Advances in Colloid and Interface Science*, Vol. 100-102, 2003, pp. 475-502.

[22] A. Kumar, "Tannery Effluent," In: M. Doble and A. Kumar, Eds., *Biotreatment of Industrial Effluents*, Butterworth-Heinemann, Burlington, 2005, pp. 133-143.

[23] R. M. Makdisi, "Tannery Wastes Definition, Risk Assessment and Cleanup Options, Berkeley, California," *Journal of Hazardous Materials*, Vol. 29, No. 1, 1991, pp. 79-96.

[24] N. M. Mahmoodi, M. Arami, N. Y. Limaee and N. S. Tabrizi, "Decolorization and Aromatic Ring Degradation Kinetics of Direct Red 80 by UV Oxidation in the Presence of Hydrogen Peroxide Utilizing TiO_2 as a Photocatalyst," *Chemical Engineering Journal*, Vol. 112, No. 1-3, 2005, pp. 191-196.

[25] Y. Yalçın Şevki, "Optimization of Bomaplex Red CR-L Dye Removal from Aqueous Solution by Electrocoagulation Using Aluminum Electrodes," *Journal of Hazardous Materials*, Vol. 153, No. 1-2, 2008, pp. 194-200.

[26] O. Apaydin, U. Kurt and M. T. Gönüllü, "An Investigation on the Treatment of Tannery Wastewater by Electrocoagulation," *Global Nest Journal*, Vol. 11, No. 4, 2009, pp. 546-555.

[27] L. Szpyrkowicz, S. N. Kaul and R. N. Neti, "Tannery Wastewater Treatment by Electro-Oxidation Coupled with a Biological Process," *Journal of Applied Electrochemistry*, Vol. 35, No. 4, 2005, pp. 381-390.

[28] S. Zodi, O. Potier, F. Lapicque and J. Leclerc, "Treatment of the Industrial Wastewaters by Electrocoagulation: Optimization of Coupled Electrochemical and Sedimen- tation Processes," *Desalination*, Vol. 261, No. 1-2, 2010, pp. 186-190.

[29] S. Aoudj, A. Khelifa, N. Drouiche, M. Hecini and H. Hamitouche, "Electrocoagulation Process Applied to Wastewater Containing Dyes from Textile Industry," *Chemical Engineering and Processing: Process Intensification*, Vol. 49, No. 11, 2010, pp. 1176-1182.

[30] C. Phalakornkule, P. Sukkasem and C. Mutchimsattha, "Hydrogen Recovery from the Electrocoagulation Treatment of Dye-Containing Wastewater," *International Journal of Hydrogen Energy*, Vol. 35, No. 20, 2010, pp. 10934-10943.

[31] C. Phalakornkule, S. Polgumhang, W. Tongdaung, B. Karakat and T. Nuyut, "Electrocoagulation of Blue Reactive, Red Disperse and Mixed Dyes, and Application in Treating Textile Effluent," *Journal of Environmental Management*, Vol. 91, No. 4, 2010, pp. 918-926.

[32] P. Durango-Usuga, F. Guzmán-Duque, R. Mosteo, M. V. Vazquez, G. Peñuela and R. A. Torres-Palma, "Experimental Design Approach Applied to the Elimination of Crystal Violet in Water by Electrocoagulation with Fe or Al Electrodes," *Journal of Hazardous Materials*, Vol. 179, No. 1-3, 2010, pp. 120-126.

[33] E. Z. El-Ashtoukhy and N. K. Amin, "Removal of Acid Green Dye 50 from Wastewater by Anodic Oxidation and Electrocoagulation—A Comparative Study," *Journal of Hazardous Materials*, Vol. 179, No. 1-3, 2010, pp. 113-119.

[34] J. B. Parsa, H. R. Vahidian, A. R. Soleymani and M. Abbasi, "Removal of Acid Brown 14 in Aqueous Media by Electrocoagulation: Optimization Parameters and Minimizing of Energy Consumption," *Desalination*, Vol. 278, No. 1-3, 2011, pp. 295-302.

[35] A. Aleboyeh, N. Daneshvar and M. B. Kasiri, "Optimization of C.I. Acid Red 14 Azo Dye Removal by Electrocoagulation Batch Process with Response Surface Methodology," *Chemical Engineering and Processing: Process Intensification*, Vol. 47, No. 5, 2008, pp. 827-832.

[36] M. Y. A. Mollah, J. A. G. Gomes, K. K. Das and D. L. Cocke, "Electrochemical Treatment of Orange II Dye Solution—Use of Aluminum Sacrificial Electrodes and Floc Characterization," *Journal of Hazardous Materials*, Vol. 174, No. 1-3, 2010, pp. 851-858.

[37] M. S. Secula, I. Creţescu and S. Petrescu, "An Experimental Study of Indigo Carmine Removal from Aqueous Solution by Electrocoagulation," *Desalination*, Vol. 277, No. 1-3, 2011, pp. 227-235.

[38] B. Merzouk, B. Gourich, A. Sekki, K. Madani, C. Vial and M. Barkaoui, "Studies on the Decolorization of Textile Dye Wastewater by Continuous Electrocoagulation Process," *Chemical Engineering Journal*, Vol. 149, No. 1-3, 2009, pp. 207-214.

[39] F. Zidane, P. Drogui, B. Lekhlif, J. Bensaid, J. F. Blais, S. Belcadi and K. El Kacemi, "Decolourization of Dye-Containing Effluent Using Mineral Coagulants Produced by Electrocoagulation," *Journal of Hazardous Materials*, Vol. 155, No. 1-2, 2008, pp. 153-163.

[40] A. R. Khataee, V. Vatanpour and A. R. Amani Ghadim, "Decolorization of C.I. Acid Blue 9 Solution by UV/Nano-TiO_2, Fenton, Fenton-Like, Electro-Fenton and Electrocoagulation Processes: A Comparative Study," *Journal of Hazardous Materials*, Vol. 161, No. 2-3, 2009, pp. 1225-1233.

[41] A. H. Essadki, M. Bennajah, B. Gourich, C. Vial, M. Azzi and H. Delmas, "Electrocoagulation/Electroflotation in an External-Loop Airlift Reactor—Application to the Decolorization of Textile Dye Wastewater: A Case Study," *Chemical Engineering and Processing: Process Intensification*, Vol. 47, No. 8, 2008, pp. 1211-1223.

[42] B. Merzouk, M. Yakoubi, I. Zongo, J.-P. Leclerc, G. Pa-

ternotte, S. Pontvianne and F. Lapicque, "Effect of Modification of Textile Wastewater Composition on Electrocoagulation Efficiency," *Desalination*, Vol. 275, No. 1-3, 2011, pp. 181-186.

[43] A. Benhadji, M. Taleb Ahmed and R. Maachi, "Electrocoagulation and Effect of Cathode Materials on the Removal of Pollutants from Tannery Wastewater of Rouïba," *Desalination*, Vol. 277, No. 1-3, 2011, pp. 128-134.

[44] K. S. P. Kalyani, N. Balasubramanian and C. Srinivasakannan, "Decolorization and COD Reduction of Paper Industrial Effluent Using Electro-Coagulation," *Chemical Engineering Journal*, Vol. 151, No. 1-3, 2009, pp. 97-104.

[45] M. Zaied and N. Bellakhal, "Electrocoagulation Treatment of Black Liquor from Paper Industry," *Journal of Hazardous Materials*, Vol. 163, No. 2-3, 2009, pp. 995-1000.

[46] M. Uğurlu, A. Gürses, Ç. Doğar and M. Yalçın, "The Removal of Lignin and Phenol from Paper Mill Effluents by Electrocoagulation," *Journal of Hazardous Materials*, Vol. 87, No. 3, 2008, pp. 420-428.

[47] S. Zodi, J. Louvet, C. Michon, O. Potier, M. Pons, F. Lapicque and J. Leclerc, "Electrocoagulation as a Tertiary Treatment for Paper Mill Wastewater: Removal of Non-Biodegradable Organic Pollution and Arsenic," *Sepa- ration and Purification Technology*, Vol. 81 , No. 1, 2011, pp. 62-68.

[48] R. Katal and H. Pahlavanzadeh, "Influence of Different Combinations of Aluminum and Iron Electrode on Electrocoagulation Efficiency: Application to the Treatment of Paper Mill Wastewater," *Desalination*, Vol. 265, No. 1-3, 2011, pp. 199-205.

[49] M. Vepsäläinen, H. Kivisaari, M. Pulliainen, A. Oikari and M. Sillanpää, "Removal of Toxic Pollutants from Pulp Mill Effluents by Electrocoagulation," *Separation and Purification Technology*, Vol. 81, No. 2, 2011, pp. 141-150.

[50] S. Bellebia, S. Kacha, A. Z. Bouyakoub and Z. Derriche, "Experimental Investigation of Chemical Oxygen Demand and Turbidity Removal from Cardboard Paper Mill Effluents Using Combined Electrocoagulation and Adsorption Processes," *Environmental Progress & Sustainable Energy*, Vol. 31, No. 3, 2011, pp. 361-370.

[51] M. Hanafy and H. I. Nabih, "Treatment of Oily Wastewater Using Dissolved Air Flotation Technique," *Energy Sources, Part A: Recovery, Utilization and Environmental Effects*, Vol. 29, No. 2, 2007, pp. 143-159.

[52] E. GilPavas, K. Molina-Tirado and M. Á. Gómez-García, "Treatment of Automotive Industry Oily Wastewater by Electrocoagulation: Statistical Optimization of the Operational Parameters," *Water Science and Technology*, Vol. 60, No. 10, 2009, pp. 2581-2588.

[53] M. Saeedi and A. Khalvati-Fahlyani, "Treatment of Oily Wastewater of a Gas Refinery by Electrocoagulation Using Aluminum Electrodes," *Water Environment Research*, Vol. 83, No. 3, 2011, pp. 256-264.

[54] P. Drogui, M. Asselin, S. K. Brar, H. Benmoussa and J.-F. Blais, "Electrochemical Removal of Organics and Oil from Sawmill and Ship Effluents," *Canadian Journal of Civil Engineering*, Vol. 36, No.3, 2009, pp. 529-539.

[55] P. Cañizares, F. Martínez, C. Jiménez, C. Sáez and M. A. Rodrigo, "Coagulation and Electrocoagulation of Oil-in-Water Emulsions," *Journal of Hazardous Materials*, Vol. 151, No. 1, 2008, pp. 44-51.

[56] P. Cañizares, C. Jiménez, F. Martínez, M. A. Rodrigo and C. Sáez, "The pH as a Key Parameter in the Choice between Coagulation and Electrocoagulation for the Treatment of Wastewaters," *Journal of Hazardous Materials*, Vol. 163, No. 1, 2009, pp. 158-164.

[57] C. J. Izquierdo, P. Cañizares, M. A. Rodrigo, J. P. Leclerc, G. Valentin and F. Lapicque, "Effect of the Nature of the Supporting Electrolyte on the Treatment of Soluble Oils by Electrocoagulation," *Desalination*, Vol. 255, No. 1-3, 2010, pp. 15-20.

[58] K. Bensadok, S. Benammar, F. Lapicque and G. Nezzal, "Electrocoagulation of Cutting Oil Emulsions Using Aluminium Plate Electrodes," *Journal of Hazardous Materials*, Vol. 152, No. 1, 2008, pp. 423-430.

[59] M. H. El-Naas, S. Al-Zuhair, A. Al-Lobaney and S. Makhlouf, "Assessment of Electrocoagulation for the Treatment of Petroleum Refinery Wastewater," *Journal of Hazardous Materials*, Vol. 91, No. 1, 2009, pp. 180-185.

[60] M. B. Agustin, W. P. Sengpracha and W. Phutdhawong, "Electrocoagulation of Palm Oil Mill Effluent," *International Journal of Environmental Research and Public Health*, Vol. 5 , No. 3, 2008, pp. 177-180.

[61] T. J. Ahmad and D. Zawawi, "Palm Oil Mill Effluent (POME) Treatment by Using an Electrocoagulation Method," *Research Journal of Chemistry and Environment*, Vol. 15, No. 2, 2011, pp. 606-609.

[62] M. Tir and N. Moulai-Mostefa, "Optimization of Oil Removal from Oily Wastewater by Electrocoagulation Using Response Surface Method," *Journal of Hazardous Materials*, Vol. 158, No. 1, 2008, pp. 107-115.

[63] M. Asselin, P. Drogui, S. K. Brar, H. Benmoussa and J. Blais, "Organics Removal in Oily Bilgewater by Electrocoagulation Process," *Journal of Hazardous Materials*, Vol. 151, No. 2-3, 2008, pp. 446-455.

[64] U. Tezcan Un, A. S. Koparal and U. Bakir Ogutveren, "Electrocoagulation of Vegetable Oil Refinery Wastewater Using Aluminum Electrodes," *Journal of Environmental Management*, Vol. 90 , No. 1, 2009, pp. 428-433.

[65] Y. Meas, J. A. Ramirez, M. A. Villalon and T. W. Chap-

man, "Industrial Wastewaters Treated by Electrocoagulation," *Electrochimica Acta*, Vol. 55, No. 27, 2010, pp. 8165-8171.

[66] F. Hanafi, O. Assobhei and M. Mountadar, "Detoxification and Discoloration of Moroccan Olive Mill Wastewater by Electrocoagulation," *Journal of Hazardous Materials*, Vol. 174, No. 1-3, 2010, pp. 807-812.

[67] M. Kobya, C. Ciftci, M. Bayramoglu and M. T. Sensoy, "Study on the Treatment of Waste Metal Cutting Fluids Using Electrocoagulation," *Separation and Purification Technology*, Vol. 60, No. 3, 2008, pp. 285-291.

[68] G. Moussavi, R. Khosravi and M. Farzadkia, "Removal of Petroleum Hydrocarbons from Contaminated Groundwater Using an Electrocoagulation Process: Batch and Continuous Experiments," *Desalination*, Vol. 278, No. 1-3, 2011, pp. 288-294.

[69] K. Suehara, Y. Kawamoto, E. Fujii, J. Kohda, Y. Nakano and T. Yano, "Biological Treatment of Wastewater Discharged from Biodiesel Fuel Production Plant with Alkali-Catalyzed Transesterification," *Journal of Bioscience and Bioengineering*, Vol. 100, No. 4, 2005, pp. 437-442.

[70] O. Chavalparit and M. Ongwandee, "Optimizing Electrocoagulation Process for the Treatment of Biodiesel Wastewater Using Response Surface Methodology," *Journal of Environmental Sciences*, Vol. 21, No. 11, 2009, pp. 1491-1496.

[71] Y. Avsar, U. Kurt and T. Gonullu, "Comparison of Classical Chemical and Electrochemical Processes for Treating Rose Processing Wastewater," *Journal of Hazardous Materials*, Vol. 148, No. 1-2, 2007, pp. 340-345.

[72] B. Demirel, O. Yenigun and T. T. Onay, "Anaerobic Treatment of Dairy Wastewaters: A Review," *Process Biochemistry*, Vol. 40, No. 8, 2005, pp. 2583-2595.

[73] İ. A. Şengil and M. Özacar, "Treatment of Dairy Wastewaters by Electrocoagulation Using Mild Steel Electrodes," *Journal of Hazardous Materials*, Vol. 137, No. 2, 2006, pp. 1197-1205.

[74] B. Balannec, M. Vourch, M. Rabiller-Baudry and B. Chaufer, "Comparative Study of Different Nanofiltration and Reverse Osmosis Membranes for Dairy Effluent Treatment by Dead-End Filtration," *Separation and Purification Technology*, Vol. 42, No. 2, 2005, pp. 195-200.

[75] K. Bensadok, N. El Hanafi and F. Lapicque, "Electrochemical Treatment of Dairy Effluent Using Combined Al and Ti/Pt Electrodes System," *Desalination*, Vol. 280, No. 1-3, 2011, pp. 244-251.

[76] S. Tchamango, C. P. Nanseu-Njiki, E. Ngameni, D. Hadjiev and A. Darchen, "Treatment of Dairy Effluents by Electrocoagulation Using Aluminium Electrodes," *Science of the Total Environment*, Vol. 408, No. 8, 2010, pp. 947-952.

[77] J. P. Kushwaha, V. C. Srivastava and I. D. Mall, "Organics Removal from Dairy Wastewater by Electrochemical Treatment and Residue Disposal," *Separation and Purification Technology*, Vol. 76, No. 2, 2010, pp. 198-205.

[78] M. Kobya, H. Hiz, E. Senturk, C. Aydiner and E. Demirbas, "Treatment of Potato Chips Manufacturing Wastewater by Electrocoagulation," *Desalination*, Vol. 190, No. 1-3, 2006, pp. 201-211.

[79] D. Valero, J. M. Ortiz, V. García, E. Expósito, V. Montiel and A. Aldaz, "Electrocoagulation of Wastewater from Almond Industry," *Chemosphere*, Vol. 84, No. 9, 2011, pp. 1290-1295.

[80] G. Roa-Morales, E. Campos-Medina, J. Aguilera-Cotero, B. Bilyeu and C. Barrera-Díaz, "Aluminum Electrocoagulation with Peroxide Applied to Wastewater from Pasta and Cookie Processing," *Separation and Purification Technology*, Vol. 54, No. 1, 2007, pp. 124-129.

[81] M. Asselin, P. Drogui, H. Benmoussa and J. Blais, "Effectiveness of Electrocoagulation Process in Removing Organic Compounds from Slaughterhouse Wastewater Using Monopolar and Bipolar Electrolytic Cells," *Chemosphere*, Vol. 72, No. 11, 2008, pp. 1727-1733.

[82] L. J. Xu, B. W. Sheldon, D. K. Larick and R. E. Carawan, "Recovery and Utilization of Useful By-Products from Egg Processing Wastewater by Electrocoagulation," *Poultry Science*, Vol. 81, No. 6, 2002, pp. 785-792.

[83] M. Kobya and S. Delipinar, "Treatment of the Baker's Yeast Wastewater by Electrocoagulation," *Journal of Hazardous Materials*, Vol. 154, No. 1-3, 2008, pp. 1133-1140.

[84] J. K. Maghanga, F. K. Segor, L. Etiégni and J. Lusweti, "Electrocoagulation Method for Colour Removal in Tea Effluent: A Case Study of Chemomi Tea Factory in Rift Valley, Kenya," *Bulletin of the Chemical Society of Ethiopia*, Vol. 23, No. 3, 2009, pp. 371-381.

[85] I. Linares-Hernández, C. Barrera-Díaz, G. Roa-Morales, B. Bilyeu and F. Ureña-Núñez, "Influence of the Anodic Material on Electrocoagulation Performance," *Chemical Engineering Journal*, Vol. 148, No. 1, 2009, pp. 97-105.

[86] M. R. V. Lanza and R. Bertazzoli, "Cyanide Oxidation from Wastewater in a Flow Electrochemical Reactor," *Industrial and Engineering Chemistry Research*, Vol. 41, No. 1, 2002, pp. 22-26.

[87] M. Kobya, E. Demirbas, N. U. Parlak and S. Yigit, "Treatment of Cadmium and Nickel Electroplating Rinse Water by Electrocoagulation," *Environmental Technology*, Vol. 31, No. 13, 2010, pp. 1471-1481.

[88] I. Heidmann and W. Calmano, "Removal of Ni, Cu and Cr from a Galvanic Wastewater in an Electrocoagulation System with Fe- and Al-Electrodes," *Separation and Purification Technology*, Vol. 71, No. 3, 2010, pp. 308- 314.

[89] F. Akbal and S. Camcı, "Copper, Chromium and Nickel Removal from Metal Plating Wastewater by Electrocoa-

gulation," *Desalination*, Vol. 269, No. 1-3, 2011, pp. 214-222.

[90] I. Kabdaşlı, T. Arslan, T. Ölmez-Hancı, I. Arslan-Alaton and O. Tünay, "Complexing Agent and Heavy Metal Removals from Metal Plating Effluent by Electrocoa- gulation with Stainless Steel Electrodes," *Journal of Hazardous Materials*, Vol. 165, No. 1-3, 2009, pp. 838-845.

[91] M. Kobya, E. Demirbas, A. Dedeli and M. T. Sensoy, "Treatment of Rinse Water from Zinc Phosphate Coating by Batch and Continuous Electrocoagulation Processes," *Journal of Hazardous Materials*, Vol. 173, No. 1-3, 2010, pp. 326-334.

[92] C. Wang, W. Chou, L. Chen and S. Chang, "Silica Particles Settling Characteristics and Removal Performances of Oxide Chemical Mechanical Polishing Wastewater Treated by Electrocoagulation Technology," *Journal of Hazardous Materials*, Vol. 161, No. 1, 2009, pp. 344-350.

[93] C.-T. Wang and W.-L. Chou, "Performance of COD Removal from Oxide Chemical Mechanical Polishing Wastewater Using Iron Electrocoagulation," *Journal of Environmental Science and Health—Part A Toxic/Hazardous Substances and Environmental Engineering*, Vol. 44, No. 12, 2009, pp. 1289-1297.

[94] M. Panizza and G. CerisolaApplicability of Electrochemical Methods to Carwash Wastewaters for Reuse. Part 2: Electrocoagulation and Anodic Oxidation Integrated Process," *Journal of Electroanalytical Chemistry*, Vol. 638, No. 2, 2010, pp. 236-240.

[95] M. Kumar, F. I. A. Ponselvan, J. R. Malviya, V. C. Srivastava and I. D. Mall, "Treatment of Bio-Digester Effluent by Electrocoagulation Using Iron Electrodes," *Journal of Hazardous Materials*, Vol. 165, No. 1-3, 2009, pp. 345-352.

[96] D. Ryan, A. Gadd, J. Kavanagh, M. Zhou and G. Barton, "A Comparison of Coagulant Dosing Options for the Remediation of Molasses Process Water," *Separation and Purification Technology*, Vol. 58, No. 3, 2008, pp. 347-352.

[97] L. Zhang, G. Xue, N. Zhang, S. Liu, A. Duan and L. Wang, "Decolorization Study of Coking Wastewater by Continuous Electrocoagulation Process," *ICMREE2011—Proedings* 2011 *International Conference on Materials for Renewable Energy and Environment*, Shanghai, May 2011, pp. 860-863.

[98] J. Zhu, F. Wu, X. Pan, J. Guo and D. Wen, "Removal of Antimony from Antimony Mine Flotation Wastewater by Electrocoagulation with Aluminum Electrodes," *Journal of Environmental Sciences*, Vol. 23, No. 7, 2011, pp. 1066-1071.

[99] F. Janpoor, A. Torabian and V. Khatibikamal, "Treatment of Laundry Waste-Water by Electrocoagulation," *Journal of Chemical Technology and Biotechnology*, Vol. 86, No. 8, 2011, pp. 1113-1120.

[100] C. Wang, W. Chou and Y. Kuo, "Removal of COD from Laundry Wastewater by Electrocoagulation/Electroflotation," *Journal of Hazardous Materials*, Vol. 164, No. 1, 2009, pp. 81-86.

[101] A. Akyol, "Treatment of Paint Manufacturing Wastewater by Electrocoagulation," *Desalination*, Vol. 285, 2012, pp. 91-99.

[102] M. Solak, M. Kılıç, Y. Hüseyin and A. Şencan, "Removal of Suspended Solids and Turbidity from Marble Pro- cessing Wastewaters by Electrocoagulation: Comparison of Electrode Materials and Electrode Connection Systems," *Journal of Hazardous Materials*, Vol. 172, No. 1, 2009, pp. 345-352.

[103] W. Chou, C. Wang and K. Huang, "Investigation of Process Parameters for the Removal of Polyvinyl Alcohol from Aqueous Solution by Iron Electrocoagulation," *Desalination*, Vol. 251, No. 1-3, 2010, pp. 12-19.

[104] W. Chou, C. Wang, K. Huang and T. Liu, "Electrochemical Removal of Salicylic Acid from Aqueous Solutions Using Aluminum Electrodes," *Desalination*, Vol. 271, No. 1-3, 2011, pp. 55-61.

[105] M. Malakootian, H. J. Mansoorian and M. Moosazadeh, "Performance Evaluation of Electrocoagulation Process Using Iron-Rod Electrodes for Removing Hardness from Drinking Water," *Desalination*, Vol. 255, No. 1-3, 2010, pp. 67-71.

[106] A. A. Bukhari, "Investigation of the Electro-Coagulation Treatment Process for the Removal of Total Suspended Solids and Turbidity from Municipal Wastewater," *Bioresource Technology*, Vol. 99, No. 5, 2008, pp. 914-921.

[107] M. A. Rodrigo, P. Cañizares, C. Buitrón and C. Sáez, "Electrochemical Technologies for the Regeneration of Urban Wastewaters," *Electrochimica Acta*, Vol. 55, No. 27 2010, pp. 8160-8164.

[108] K. Sadeddin, A. Naser and A. Firas, "Removal of Turbidity and Suspended Solids by Electro-Coagulation to Improve Feed Water Quality of Reverse Osmosis Plant," *Desalination*, Vol. 268, No. 1-3, 2011, pp. 204-207.

[109] M. G. Kılıç, Ç. Hoşten and Ş. Demirci, "A Parametric Comparative Study of Electrocoagulation and Coagulation Using Ultrafine Quartz Suspensions," *Journal of Hazardous Materials*, Vol. 171, No. 1-3, 2009, pp. 247-252.

[110] D. Ghernaout, B. Ghernaout and A. Boucherit, "Effect of pH on Electrocoagulation of Bentonite Suspensions in Batch Using Iron Electrodes," *Journal of Dispersion Science and Technology*, Vol. 29, No. 9, 2008, pp. 1272-1275.

[111] S. Gao, J. Yang, J. Tian, F. Ma, G. Tu and M. Du, "Electro-Coagulation-Flotation Process for Algae Removal," *Journal of Hazardous Materials*, Vol. 177, No. 1-3, 2010, pp. 336-343.

[112] D. Ghernaout, A. Badis, A. Kellil and B. Ghernaout, "Application of Electrocoagulation in *Escherichia coli* Culture and Two Surface Waters," *Desalination*, Vol. 219, No. 1-3, 2008, pp. 118-125.

[113] C. Ricordel, A. Darchen and D. Hadjiev, "Electrocoagulation-Electroflotation as a Surface Water Treatment for Industrial Uses," *Separation and Purification Technology*, Vol. 74, No. 3, 2010, pp. 342-347.

[114] D. Ghernaout, B. Ghernaout, A. Saiba, A. Boucherit and A. Kellil, "Removal of Humic Acids by Continuous Electromagnetic Treatment Followed by Electrocoagulation in Batch Using Aluminium Electrodes," *Desalination*, Vol. 239, No. 1-3, 2009, pp. 295-308.

[115] M. Vepsäläinen, M. Ghiasvand, J. Selin, J. Pienimaa, E. Repo, M. Pulliainen and M. Sillanpää, "Investigations of the Effects of Temperature and Initial Sample pH on Natural Organic Matter (NOM) Removal with Electrocoagulation Using Response Surface Method (RSM)," *Separation and Purification Technology*, Vol. 69, No. 3, 2009, pp. 255-261.

[116] X. Li, Q. Feng, Q. Meng and Y. Ceng, "Electrocoa- gulation for the Drinking Water Treatment of Polluted Surface Water Supplies," *2nd International Conference on Bioinformatics and Biomedical Engineering, iCBBE* 2008, Shanghai, May 2008, pp. 3091-3094.

[117] N. Uduman, V. Bourniquel and M. K. Danquah and A. F. A. Hoadley, "A Parametric Study of Electrocoagulation as a Recovery Process of Marine Microalgae for Biodiesel Production," *Chemical Engineering Journal*, Vol. 174, No. 1, 2011, pp. 249-257.

[118] D. Vandamme, S. C. V. Pontes, K. Goiris, I. Foubert, L. J. J. Pinoy and K. Muylaert, "Evaluation of Electro-Coagulation-Flocculation for Harvesting Marine and Freshwater Microalgae," *Biotechnology and Bioengineering*, Vol. 108, No. 10, 2011, pp. 2320-2329.

[119] O. Yahiaoui, L. Aizel, H. Lounici, N. Drouiche, M. F. A. Goosen, A. Pauss and N. Mameri, "Evaluating Removal of Metribuzin Pesticide from Contaminated Groundwater Using an Electrochemical Reactor Combined with Ultraviolet Oxidation," *Desalination*, Vol. 270, No. 1-3, 2011, pp. 84-89.

[120] A. E. Yilmaz, R. Boncukcuoğlu, M. M. Kocakerim, M. T. Yilmaz and C. Paluluoğlu, "Boron Removal from Geothermal Waters by Electrocoagulation," *Journal of Hazardous Materials*, Vol. 153, No. 1-2, 2008, pp. 146-151.

[121] C. P. Nanseu-Njiki, S. R. Tchamango, P. C. Ngom, A. Darchen and E. Ngameni, "Mercury(II) Removal from Water by Electrocoagulation Using Aluminium and Iron Electrodes," *Journal of Hazardous Materials*, Vol. 168, No. 2-3, 2009, pp. 1430-1436.

[122] W. Wan, T. J. Pepping, T. Banerji, S. Chaudhari and D. E. Giammar, "Effects of Water Chemistry on Arsenic Removal from Drinking Water by Electrocoagulation," *Water Research*, Vol. 45, No. 1, 2011, pp. 384-392.

[123] S. Vasudevan, J. Lakshmi and G. Sozhan, "Studies on the Removal of Iron from Drinking Water by Electrocoa- gulation—A Clean Process," *Clean—Soil, Air, Water*, Vol. 37, No. 1, 2009, pp. 45-51.

[124] D. Ghosh, C. R. Medhi and M. K. Purkait, "Treatment of Drinking Water Containing Iron Using Electrocoagulation," *International Journal of Environmental Engineering*, Vol. 2, No. 1-3, 2010, pp. 212-227.

[125] S. Vasudevan, S. M. Sheela, J. Lakshmi and G. Sozhan, "Optimization of the Process Parameters for the Removal of Boron from Drinking Water by Electrocoagulation—A Clean Technology," *Journal of Chemical Technology and Biotechnology*, Vol. 85, No. 7, 2010, pp. 926-933.

[126] S. Vasudevan, J. Lakshmi and G. Sozhan, "Effects of Alternating and Direct Current in Electrocoagulation Process on the Removal of Cadmium from Water," *Journal of Hazardous Materials*, Vol. 192, No. 1, 2011, pp. 26-34.

[127] E. Keshmirizadeh, S. Yousefi and M. K. Rofouei, "An Investigation on the New Operational Parameter Effective in Cr(VI) Removal Efficiency: A Study on Electrocoagulation by Alternating Pulse Current," *Journal of Hazardous Materials*, Vol. 190, No. 1-3, 2011, pp. 119-124.

[128] M. S. Bhatti, D. Kapoor, R. K. Kalia, A. S. Reddy and A. K. Thukral, "RSM and ANN Modeling for Electrocoagulation of Copper from Simulated Wastewater: Multi Objective Optimization Using Genetic Algorithm Approach," *Desalination*, Vol. 274, No. 1-3, 2011, pp. 74-80.

[129] M. Behbahani, M. R. A. Moghaddam and M. Arami, "Techno-Economical Evaluation of Fluoride Removal by Electrocoagulation Process: Optimization through Response Surface Methodology," *Desalination*, Vol. 271, No. 1-3, 2011, pp. 209-218.

[130] W. Chou and Y. Huang, "Electrochemical Removal of Indium Ions from Aqueous Solution Using Iron Electrodes," *Journal of Hazardous Materials*, Vol. 172, No. 1, 2009, pp. 46-53.

[131] A. Shafaei, M. Rezayee, M. Arami and M. Nikazar, "Removal of Mn^{2+} Ions from Synthetic Wastewater by Electrocoagulation Process," *Desalination*, Vol. 260, No. 1-3, 2010, pp. 23-28.

[132] E. Lacasa, P. Cañizares, C. Sáez, F. J. Fernández and M. A. Rodrigo, "Electrochemical Phosphates Removal Using Iron and Aluminium Electrodes," *Chemical Engineering Journal*, Vol. 172, No. 1, 2011, pp. 137-143.

[133] S. Vasudevan, J. Lakshmi, J. Jayaraj and G. Sozhan, "Remediation of Phosphate-Contaminated Water by Electrocoagulation with Aluminium, Aluminium Alloy and Mild Steel Anodes," *Journal of Hazardous Materials*, Vol. 164, No. 2-3, 2009, pp. 1480-1486.

[134] Z. V. P. Murthy and S. Parmar, "Removal of Strontium by Electrocoagulation Using Stainless Steel and Aluminum Electrodes," *Desalination*, Vol. 282, 2011, pp. 63-67.

[135] G. Moussavi, F. Majidi and M. Farzadkia, "The Influence of Operational Parameters on Elimination of Cyanide from Wastewater Using the Electrocoagulation Process,"

Desalination, Vol. 280, No. 1-3, 2011, pp. 127-133.

[136] M. Pociecha and D. Lestan, "Using Electrocoagulation for Metal and Chelant Separation from Washing Solution after EDTA Leaching of Pb, Zn and Cd Contaminated Soil," *Journal of Hazardous Materials*, Vol. 174, No. 1-3, 2010, pp. 670-678.

[137] K. Mansouri, K. Elsaid, A. Bedoui, N. Bensalah and A. Abdel-Wahab, "Application of Electrochemically Dissolved Iron in the Removal of Tannic Acid from Water," *Chemical Engineering Journal*, Vol. 172, No. 2-3, 2011, pp. 970-976.

[138] N. Mameri, A. R. Yeddou, H. Lounici, D. Belhocine, H. Grib and B. Bariou, "Defluoridation of Septentrional Sahara Water of North Africa by Electrocoagulation Process Using Bipolar Aluminium Electrodes," *Water Research*, Vol. 32, No. 5, 1998, pp. 1604-1612.

[139] M. Karhu, V. Kuokkanen, T. Kuokkanen and J. Rämö, "Bench Scale Electrocoagulation Studies of Bio Oil-in-Water and Synthetic Oil-in-Water Emulsions," *Separation and Purification Technology*, Vol. 96, 2012, pp. 296-305.

Nomenclature

ANOVA	analysis of variance
APC	alternating pulse current
BOD	biological oxygen demand
CC	conventional coagulation
COD	chemical oxygen demand
DTA	differential thermal analysis
EC	electrocoagulation
EEC	electrical energy consumption
EF	electro-Fenton process
EM	electromagnetic treatment
EMC	electrode material consumption
EO	electro-oxidation
FTU	formazin turbidity unit
HRT	hydraulic retention time
NOM	natural organic matter
NTU	nephelometric turbidity unit
OC	operating costs
O & G	oil and grease
PAC	poly aluminium chloride
RO	reverse osmosis
RSM	response surface methodology
TGA	thermal gravimetric analysis
TOC	total organic carbon
TS	total solids
TSS	total suspended solids
UV	ultraviolet

Ionic Liquid-Based Ultrasound-Assisted Emulsification Microextraction Coupled with HPLC for Simultaneous Determination of Glucocorticoids and Sex Hormones in Cosmetics

Xiaoji Cao*, Lingxiao Shen, Xuemin Ye, Feifei Zhang, Jiaoyu Chen, Weimin Mo*
Research Center of Analysis and Measurement, Zhejiang University of Technology, Hangzhou, China

ABSTRACT

An effective and environmentally friendly method based on ionic liquid-based ultrasound-assisted emulsification microextraction (IL-USAEME) and high performance liquid chromatography (HPLC) has been developed for the determination of nine hormones including three glucocorticoids, one androgen and five progestogens in cosmetics. Several factors that affect the extraction efficiency, such as the kinds and volume of ionic liquid, salt concentration, ultrasonic power and time, and centrifugation time were investigated and optimized. Under the optimum extraction condition, the recoveries of analytes ranged from 85.97% to 108.84% except prednisolone (62.30%). The intraday and interday precision was below 2.51% and 3.30%, respectively.

Keywords: Ionic Liquid; Ultrasound-Assisted Emulsification Microextraction; Hormones; Cosmetics; High Performance Liquid Chromatography

1. Introduction

Cosmetics have become more and more important in our daily life for skin whitening, anti-wrinkle, removing freckle, moisture, and so on [1]. In order to achieve these effects remarkably, illicit chemicals such as hormones are added to the cosmetics. And the potential accumulation hazard of these additives is terrible [2]. Although the Hygienic Standard for Cosmetic of China and European Union Directive 76/768/EEC has banned the addition of hormones to commercial cosmetics, cosmetics containing hormones are still in the market.

At present, the high performance liquid chromatography and its combination with mass spectrometry are the most common methods for determination of trace hormones in the complex matrices. For these analytic techniques, in order to decrease the interference of matrix and concentrate the targeted compounds, a sample pretreatment is indispensable, which is often laborious and consumes a large amount of organic solvents [3-5].

Ionic liquids (ILs) composed of organic cations and various types of anions have been successfully applied to various areas of analytical chemistry as environmental benign solvents [6-10]. Ionic liquid-based ultrasound- assisted emulsification microextraction (IL-USAEME) is an efficient, simple, rapid and cheap extraction technique, in which applying ultrasound radiation to facilitate the emulsification of a microvolume of ionic liquid extractant in an aqueous sample, and subsequent separation of the two immiscible liquid phases by centrifugation [11, 12]. This method has been used for the extraction of organic compounds in environmental samples. However, to the best of our knowledge, IL-USAEME has not been applied for the extraction of hormones from cosmetics.

In this paper, an ionic liquid-based ultrasound-assisted emulsification microextraction (IL-USAEME) combined with HPLC method for simultaneous determination of glucocorticoids and sex hormones in cosmetics will be investigated.

2. Materials and Methods

2.1. Reagents and Chemicals

The pharmaceutical standards (purity, 97.5% - 99.5%) of androstendione, hydroxyprogesterone acetate, megestrol

*Corresponding author.

acetate and melengestrol acetate were purchased from Dr. Ehrenstorfer GmbH (Augsburg Germany). Northindrone was purchased from Toronto Research Chemicals Inc. (North York, Canada). Prednisolone and Beclometasone Dipropionate were purchased from the National Institutes for Food and Drug Control (NIFDC) (Beijing, China). Dexamethasone was purchased from Zhejiang Xianju Pharmaceutical Co., Ltd. (Zhejiang, China) and 17 a-hydroxyprogesterone caproate was purchased from Tokyo Chemical Industry Co., Ltd. (Tokyo, Japan). Chromatographic grade methanol and acetonitrile were supplied by Merck (Darmstadt, Germany). Deionized water was purified by a Milli-Q water purification system (18 MΩ, Millipore, Bedford, MA, USA). [C_6MIM] [PF_6] and [C_7MIM][PF_6] were gained from Chengjie Co. Ltd. (Shanghai, China) and used as received. The cosmetic products Kose were purchased from local supermarket and stored at room temperature.

2.2. Ionic Liquid-Based Ultrasound-Assisted Emulsification Microextraction

Ionic liquid-based ultrasound-assisted emulsification microextraction was performed in KQ-100 DA and KQ-500 DA ultrasonic water baths (Kunshan, Jiangsu, China). A 0.2 g cosmetic sample, 125 μl [C_7MIM][PF_6], 62.5 μl saturated NaCl aqueous solution and 187.5 μl water were introduced to a 1.5 ml eppendorf centrifuge tube. And the mixed solution was ultrasonically extracted at ultrasonic power of 100 W for 15 min. The emulsification solution obtained after ultrasonication was centrifuged at 10000 rpm for 12 min by the Centrifuge 5804 (Eppendorf AG, Hamburg, Germany). After centrifugation, the IL extraction solution (~100 μl) which was deposited at the bottom of the centrifuge tube was collected using a 200 μl pipette and then diluted by methanol to 1ml. Finally, the solution was filtered through 0.22 μm nylon filter membrance.

2.3. HPLC Analyses

The HPLC analysis was performed using a Waters 2695 HPLC system (Waters, Milford, MA, USA) equipped with a quaternary solvent delivery system, an auto-sampler with a 100 μl sample loop, a PDA detector, a column oven, and a data station running the Empower data software. A CAPCELL PAK C18 column (150 mm × 4.6 mm, 3 μm, Chuo-ku, Tokyo, Japan) was used for separation and was maintained at 25˚C. The mobile phase was composed of water (eluent A) and acetonitrile (eluent B), and the elution was carried out in gradient mode at a flow rate of 0.7 ml/min. The gradient program was as follows: 0 min - 2 min, 30% B; 2 min - 6 min, 30% - 70% B; 6 min - 15 min, 70% B; 15 min - 16 min, 70% - 100% B; 16 min - 25 min, 100% B. Finally, the initial condition was re-introduced over 0.01 min, and the equilibration time was 5 min. The injection volume was 10 μl. The detection wavelength was 242 nm for all the analytes except megestrol acetate and melengestrol acetate (288 nm).

3. Results and Discussion

3.1. Optimization of Extraction Conditions

To acquire the high recoveries and enrichment efficiency of the studied hormones (shown as **Figure 1**), the effects of several factors, such as kinds of ionic liquid, volume of [C_7MIM][PF_6], salt concentration, ultrasonic power and time, and centrifugation time, were examined and optimized.

3.1.1. Screening of the IL-Based Extraction Solvent

In the case of USAEME, the suitable extraction solvent should have density greater than that of water and also a low solubility in water [14]. And the structures of ILs have a significant effect on its physicochemical proper-

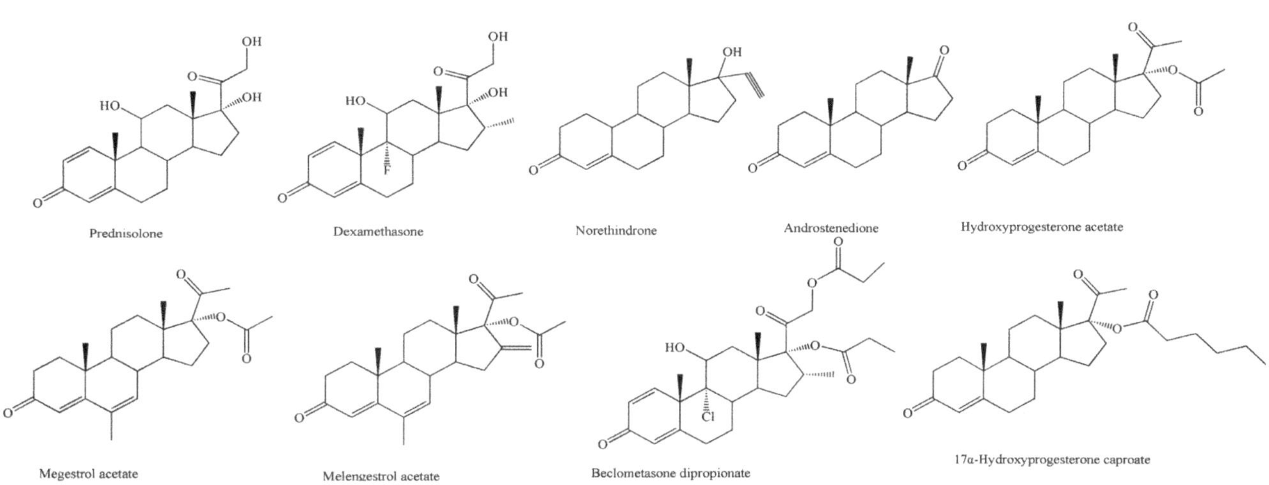

Figure 1. The chemical structures of the studied hormones.

ties, which might greatly affect the extraction efficiency of targeted compounds [15]. In this study, two types of IL, including [C_6MIM][PF_6] and [C_7MIM][PF_6] which are hydrophobic were examined. As shown in **Figure 2**, the [C_7MIM][PF_6] extraction phase produced the better extraction performance for all hormones than that of the [C_6MIM][PF_6], attributed to the stronger hydrophobic interactions between the longer alkyl chain of the [C_7MIM][PF_6] ionic liquid and the hydrophobic hormones. Thus, [C_7MIM][PF_6] was selected for all of the subsequent experiments.

3.1.2. Effect of the Volume of [C_7MIM][PF_6]

To study the influence of the volume of [C_7MIM][PF_6] on the extraction efficiency, experiments were carried out by using different volumes of [C_7MIM][PF_6] (from 50 to 150 μl) as the extraction solvent. As shown in **Figure 3**, the extraction efficiency increased when the volume of [C_7MIM][PF_6] increased from 50 to 125 μl, and decreased obviously above 125 μl. The reason could be that the appropriate volume of [C_7MIM][PF_6] could be dispersed sufficiently into sample solution under ultrasound, but with the volume of [C_7MIM][PF_6] increased, the hormones could not be transferred into the IL phase sufficiently for its high viscosity and density [12]. Therefore, 125 μl of [C_7MIM][PF_6] was selected for the further study.

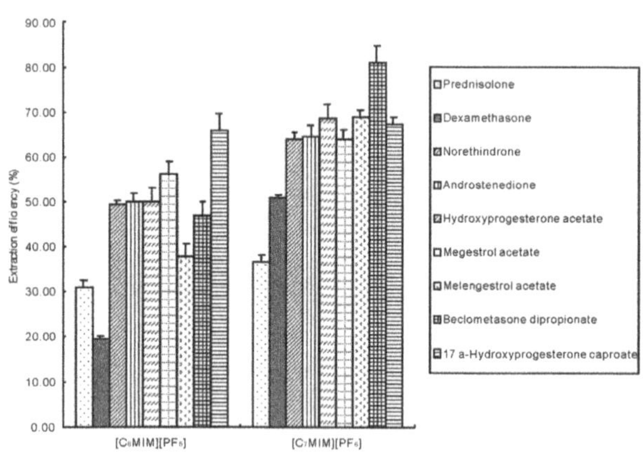

Figure 2. Effect of extraction solvent. ILs volume: 100 μl; extraction time: 15 min; ultrasonic power: 100 W; salt addition: 0 mg/ml. centrifugation time: 10 min (10,000 rpm); spiked concentration: 10 μg/ml.

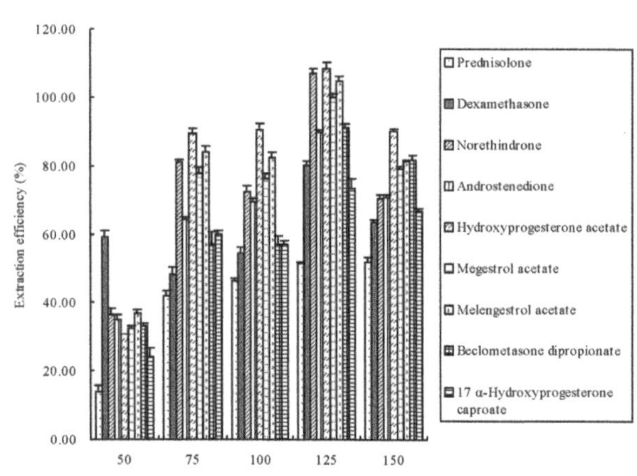

Figure 3. Effect of the volume of [C_7MIM][PF_6]. extraction time: 15 min; ultrasonic power: 100 W; salt addition: 0 mg/ml. centrifugation time: 10 min (10,000 rpm); spiked concentration:10 μg/ml.

3.1.3. Effect of Salt Addition

The influence of salt addition on the performance of IL-USAEME was investigated by adding different amount of NaCl (0 to 9%, w/v) to solution. The results demonstrated that the salt addition had neglectful effect on the extraction efficiency of analytes. On the other hand, the adequate salt addition is helpful to the formation of IL extraction phase. Consequently, the 9% NaCl was used in subsequent experiments.

3.1.4. Effect of Ultrasonic Power and Time

In this work, the emulsification of the ionic liquid into sample solution was facilitated by ultrasound radiation. The ultrasonic power can affect the mass transfer rate and the contact area between the sample solution and the ionic liquid extractant. When the ultrasonic power was tested at two different levels: 100 W, 500 W, it was obvious that the suitable ultrasonic power should be 100 W.

Additionally, the effect of ultrasonic time on extraction efficiency was evaluated in the range of 5 min - 25 min. The results shown in **Figure 4** indicated that the extraction efficiency achieved the maximum at the ultrasonic time of 15 min. The main reason may be that adequate ultrasonic time is needed to ensure the equilibrium of extraction. However, prolonging the ultrasonic time induced the descent of extraction efficiency because of the dissolution of IL in sample solution. Consequently, the ultrasonic time of 15 min was enough.

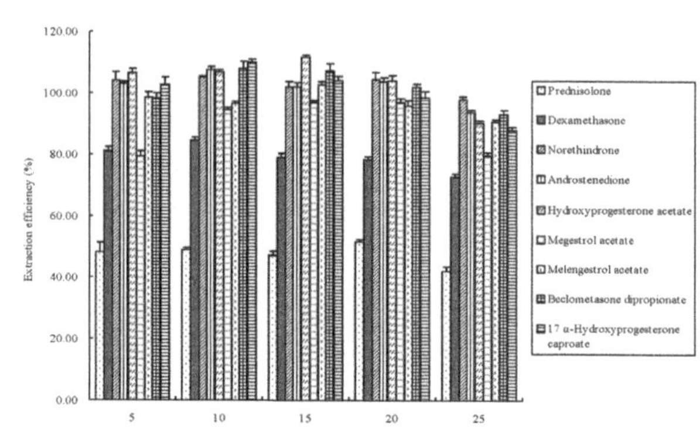

Figure 4. Effect of extraction time. [C_7MIM][PF_6] volume: 125 μl; salt addition: 9% (m/v); ultrasonic power: 100 W; centrifugation time: 10 min (10,000 rpm); spiked concentration: 10 μg/ml.

3.1.5. Effect of Centrifugation Time

The centrifugation plays an important role on the sedimentation of IL phase. The effect of centrifugation time ranging from 9 to 21 min was tested. As observed in **Figure 5**, the extraction efficiency of the most analytes increased with the increase of the centrifugation time from 9 to 12 min. When the centrifugation time was longer than 12 min, the extraction efficiency of dexamethasone, hydroxyprogesterone acetate and beclometasone dipropionate decreased and no significant change was observed for others. Therefore, 12 min was chosen as the centrifugation time.

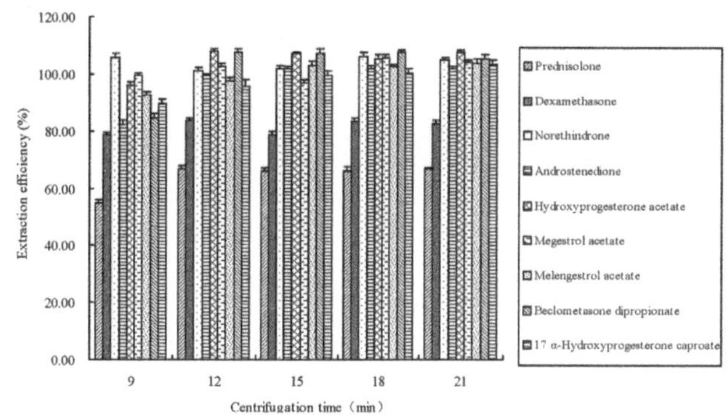

Figure 5. Effect of centrifugation time. [C_7MIM][PF_6] volume: 125 µl; salt addition: 9% (m/v); ultrasonic power: 100 W; extraction time: 15 min; spiked concentration: 10 µg/ml.

3.2. Evaluation of the method

The chromatogram of a spiked sample obtained by IL-USAEME under the optimum conditions was shown in **Figure 6**. Under the optimum conditions, linear range, limits of detection (LODs), recovery and precision were determined and shown in **Tables 1** and **2**. As can be seen, good linearities were observed for all analytes with the coefficient of correlation (r^2) ranging from 0.9980 to 0.9994. The LODs were calculated at a signal-to-noise (S/N) ratio of 3. The mean recoveries were 62.30% - 108.84% and the RSDs were 0.26% - 4.69%. The intraday and interday precisions varied from 0.10% to 2.51 % and from 0.33% to 3.30 %, respectively. The results demonstrated that the method is potential for the detection of hormones fortified in cosmetics.

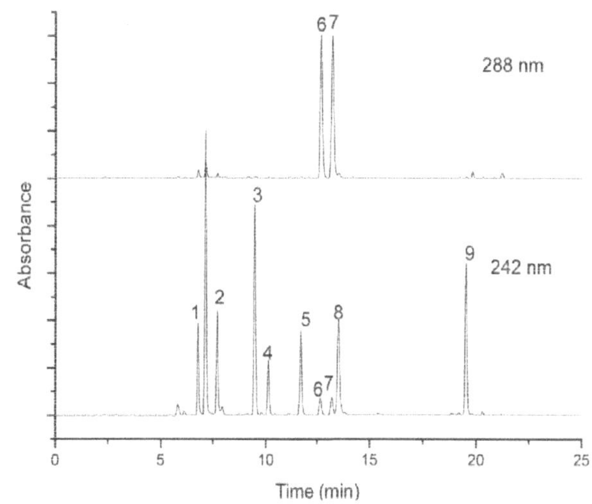

Figure 6. Chromatogram of spiked sample obtained at wavelength 242 and 288 nm. (1) prednisolone, (2) dexamethasone, (3) norethindrone, (4) androstenedione, (5) hydroxyprogesterone acetate, (6) megestrol acetate, (7) melengestrol acetate, (8) beclometasone dipropionate, (9) 17 a-hydroxyprogesterone caproate.

4. Conclusion

In this study, a simple, cheap and green IL-USAEME combined with HPLC method was developed and successfully applied to simultaneously determine three glucocorticoids, one androgen and five progestogens in cosmetics, in which using ultrasound radiation to facilitate

Table 1. Method calibration data and recoveries of analytes.

Analyte	Linear range (µg/ml)	Correlation coefficient (r^2)	LOD (µg/g)	Mean recovery (%) (n = 3)	
				1 (µg/ml)	10 (µg/ml)
Prednisolone	0.50 - 100.00	0.9994	0.15	62.30 ± 1.63	66.45 ± 1.23
Dexamethasone	0.60 - 120.00	0.9991	0.40	98.34 ± 0.26	95.11 ± 0.76
Norethindrone	0.20 - 100.00	0.9990	0.10	99.86 ± 0.24	92.12 ± 3.60
Androstenedione	0.10 - 50.00	0.9992	0.20	101.75 ± 1.69	88.11 ± 2.87
Hydroxyprogesterone acetate	0.60 - 120.00	0.9980	0.40	98.07 ± 3.50	97.16 ± 3.38
Megestrol acetate	0.10 - 50.00	0.9991	0.05	105.09 ± 2.82	85.97 ± 2.02
Melengestrol acetate	0.10 - 50.00	0.9991	0.10	100.21 ± 2.61	92.49 ± 1.73
Beclometasone dipropionate	1.00 - 200.00	0.9991	0.85	108.84 ± 3.42	106.86 ± 4.69
17α-Hydroxyprogesterone caproate	0.30 - 150.00	0.9992	0.25	95.26 ± 3.91	98.19 ± 4.38

Table 2. Intraday and interday precision.

Analyte	Intraday[a] (RSD%)			Interday[b] (RSD%)		
	1 (μg/ml)	5 (μg/ml)	25 (μg/ml)	1 (μg/ml)	5 (μg/ml)	25 (μg/ml)
Prednisolone	1.91	0.45	0.57	1.54	0.94	0.68
Dexamethasone	1.82	0.88	0.61	2.01	0.40	0.70
Norethindrone	1.74	0.99	0.38	0.68	0.36	0.57
Androstenedione	1.33	0.89	0.45	0.87	0.96	0.45
Hydroxyprogesterone acetate	0.26	0.47	0.08	1.52	0.58	0.44
Megestrol acetate	0.85	0.47	0.10	1.65	0.73	0.50
Melengestrol acetate	1.05	0.09	0.64	1.97	0.87	0.42
Beclometasone dipropionate	0.60	0.52	0.34	3.30	0.69	0.57
17α-Hydroxyprogesterone caproate	2.51	1.01	0.88	0.90	0.33	0.90

[a]The intraday precision was gained by analyzing a sample three times in one day. [b]The interday precision was received by analyzing a sample once a day over three consecutive day.

the emulsification of a microvolume of ionic liquid instead of organic solvent into the sample solution. The experimental conditions for the extraction of these analytes have been investigated. Under the optimized conditions, satisfactory extraction efficiency of the analytes was obtained.

5. Acknowledgements

We gratefully acknowledge the financial support from socially useful projects under the Zhejiang Province Science and Technology Department (2010C33074) and the project supported by the Zhejiang University of Technology Foundation (20070167).

REFERENCES

[1] Y. Liu, X. L. Zhang, Y. Z. Ouyang, Z. HU, L. Ma, *et al.*, "Trace Detection of Hormones and Sulfonamides in Viscous Cosmetic Products by Neutral Desorption Extractive Electrospray Ionization Tandemmass Spectrometry," *Journal of Mass Spectrometry*, Vol. 46, No. 8, 2011, pp. 794-803.

[2] M. Plotan, C. Frizzell, V. Robinson, C. T. Elliott and L. Connolly, "Endocrine Disruptor Activity in Bottled Mineral and Flavored Water," *Food Chemistry*, Vol. 136, No. 3-4, 2013, pp. 1590-1596.

[3] C. Kai, C. T. Elliott, D. H. Phillips, M.-L. Scippo, M. Muller, *et al.*, "Treatment of Estrogens and Androgens in Dairy Wastewater by a Constructed Wetland System," *Water Research*, Vol. 46, No. 7, 2012, pp. 2333-2343.

[4] C. S. Tamvakopoulos, J. M. Neugebauer, M. Donnelly and P. R. Griffin, "Analysis of Betamethasone in Rat Plasma Using Automated Solid-Phase Extraction Coupled with Liquid Chromatography-Tandem Mass Spectrometry. Determination of Plasma Concentrations in Rat Following Oral and Intravenous Administration," *Journal of Chromatography B*, Vol. 776, No. 2, 2002, pp. 161-168.

[5] Z. G. Zhang, B. Shi, G. F. Wang, L. Y. Lin and Y. Zhou, "Determination of Medroxyprogesterone Acetate in Aquatic Product by Liquid Chromatography with Tandem Mass Spectrometric Detection," *Journal of Chinese Mass Spectrometry Society*, Vol. 27, No. 1, 2006, pp. 36-39.

[6] X. J. Cao, X. M. Ye, Y. B. Lu, Y. Yu and W. M. Mo, "Ionic Liquid-Based Ultrasonic-Assisted Extraction of Piperine from White Pepper," *Analytica Chimica Acta*, Vol. 640, No. 1-2, 2009, pp. 47-51.

[7] X. J. Cao, J. F. Qiao, L. P. Wang, X. M. Ye, L. B. Zheng, *et al.*, "Screening of Glycoside Isomers in P. Scrophulariiflora Using Ionic Liquid-Based Ultrasonic-Assisted Extraction and Ultra-performance Liquid Chromatography/electrospray Ionization Quadrupole Time-of-Flight Tandem Mass Spectrometry," *Rapid Communications in Mass Spectrometry*, Vol. 26, No. 7, 2012, pp. 740-748.

[8] R. Zhang, N. Li, C. L. Wang, Y. P. Bai, R. B. Ren, *et al.*, "Ionic Liquid Foam Floatation Coupled with Solid Phase Extraction for Separation and Determination of Hormones by High-Performance Liquid Chromatography," *Analytica Chimica Acta*, Vol. 704, No. 1-2, 2011, pp. 98-109.

[9] M. Q. Kang, S. Sun, N. Li, D. H. Zhang, M. Y. Chen, *et al.*, "Extraction and Determination of Hormones in Cosmetics by Homogeneous Ionic Liquid Microextraction High-Performance Liquid Chromatography," *Journal of Separation Science*, Vol. 35, No. 16, 2012, pp. 2032-2039.

[10] L. Fischer, T. Falta, G. Koellensperger, A. Stojanovic, D. Kogelnig, *et al.*, "Ionic Liquids for Extraction of Metals and Metal Containing Compounds from Communal and

Industrial Waste Water," *Water Research*, Vol. 45, No. 15, 2011, pp. 4601-4614.

[11] D. D. Ge and H. K. Lee, "A New 1-Hexyl-3-Methylimidazolium tris (Pentafluoroethyl)Trifluorophosphate Ionic Liquid Based Ultrasound-Assisted Emulsification Microextraction for the Determination of Organic Ultraviolet Filters in Environmental Water Samples," *Journal of Chromatography A*, Vol. 1251, 2012, pp. 27-32.

[12] P. Liang, F. Wang, Q. Wan, "Ionic Liquid-Based Ultrasound-Assisted Emulsification Microextraction Coupled with High Performance Liquid Chromatography for the Determination of Four Fungicides in Environmental Water Samples," *Talanta*, Vol. 105, 2013, pp. 57-62.

[13] N. M. Najafi, H. Tavakoli, Y. Abdollahzadeh and R. Alizadeh, "Comparison of Ultrasound-Assisted Emulsification and Dispersive Liquid-Liquid Microextraction Methods for the Speciation of Inorganic Selenium in Environmental Water Samples Using Low Density Extraction Solvents," *Analytica Chimica Acta*, Vol. 714, No. 7, 2012, pp. 82-88.

[14] V. Herrera-Herrera, M. Asensio-Ramos, J. Hernández-Borges and M. Á. Rodríguez–Delgado, "Dispersive Liquid Liquid Microextraction for Determination of Organic Analytes," *Trends in Analytical Chemistry*, Vol. 29, No. 7, 2010, pp. 728-751.

[15] C. F. Poole and S. K. Poole, "Extraction of Organic Compounds with Room Temperature Ionic Liquids," *Journal of Chromatography A*, Vol. 1217, No. 16, 2010, pp. 2268-2286.

Oligoalkylnaphthenic (C_6-C_{12}) Oils Obtained in the Presence of Ti-Containing Ionic-Liquid Catalysts

Akif Hamid Azizov, Reyhan Valy Aliyeva, Khayala Haydar Asgerova*
The Y.H. Mamedaliyev Institute of Petrochemical Processes of Azerbaijan National Academy of Sciences, Baku, Azerbaijan

ABSTRACT

In this present paper, the result of C_6-C_{12} α-olefins oligomerizaton in the presence of recyclable chloroaluminate type ionic-liquid catalysts and their combination with Ti-containing complexes is presented. By different methods of analysis the composition and structure of oligoalkylnaphtenic products obtained in the presence of ionic-liquid catalysts and Ti-containing modifiers have been studied.

Keywords: Ionic Liquid; Oligomerization; PAO

1. Introduction

The continued use of large quantities of organic solvents as liquid media for chemical reactions, extractions and formulations is a major problem in nowadays chemical industry. Here E-factor, as it was defined by Roger Sheldon, is the environmental factor of a process, e.g. the ratio (by weight) of the by-products to the desired product(s). As can be seen, the oil refining and bulk chemicals industries are the least polluting industries, while thee fine chemicals and pharmaceutical companies, using inefficient and dirty processes, albeit on a much smaller scale, are polluting the environment severely. In addition to the problematic volatility of these solvents, they often present additional risk due to toxicity and/or flammability [1]. Thus, the discovery of surrogates to traditional organic solvents which offer reduced environmental impact and safer processes is of great importance.

J. D. Holbrey and K. R. Seddon consider four main alternate strategies for this: solvent-free synthesis, the use of water, the use of supercritical fluids as solvents and the use of ionic liquids as solvents.

The option of the use of ionic liquid as solvents seems to be one of the most promising methods in green chemistry. The unique peculiarities of low temperature ionic liquids such as non-volatility, non-flammability, low melting points, high thermal stability in a wide temperature range, relatively low viscosity, inexpensivity and diversity to manufacture and so on, have attracted interests of investigators from industrial and academic research centers from all around the world. They are highly solvating liquids, composed of bulky cations and anions, which dissolve many organic and inorganic substances. In the last few years there appeared a lot of papers and patents, devoted to preparation methods, solvent properties and application of ionic liquids as the solvents in different processes. Some of them have been overviewed in [1-10].

Low temperature ionic liquids are widely investigated as replacement for environmentally hazardous volatile organic solvents in the processes of organic synthesis and petrochemistry.

They have been used primarily as replacements for conventional media in green chemical processes or as solvent catalysts. The first use of ionic liquids in homogeneous catalysis was reported in 1990 for dimerisation and polymerization reactions [11,12]. Since then, a rapidly growing number of transition metal catalysed reactions have been described in various ionic liquids.

Ionic liquids are often called "green solvents", and indeed the link between ionic liquids and green chemistry is mainly related to the characteristic of low volatility. The review presents C_6-C_{12} α-olefins oligomerizaton in the presence of recyclable chloroaluminate type ionic-liquid catalysts and their combination with Ti-containing complexes. Many of methods of analysis are used to study the composition and structure of oligoalkylnaphtenic products.

*Corresponding author.

2. Experimental

All operations on the catalyst synthesis and oligomerization reactions were performed in an inert gas (nitrogen, argon) atmosphere using anhydrous solvents (benzene, toluene) and freshly distilled C_6-C_{12} α-olefins.

α-olefins—before the use was dried over calcinated Al_2O_3 or $CaCl_2$ and purified by distillation in atmospheric pressure.

Waterless $AlCl_3$ was stored in an inert atmosphere in the hermetic vessels.

Diethylamine hidrocloride (DEAHC) used for complex ammonium salts preparation were purified by distillation at atmospheric pressure or by recrystallization.

Chloroaluminate ionic—liquids were prepared by interacting $AlCl_3$ with the corresponding complex ammonium salt taken at molar ratios 1.5 - 2.1.

Synthesis of the ionic-liquid catalytic systems and also oligomerization of α-olefine in their presence were carried out in the thermostarted dried up glass reactor in the atmosphere of dry nitrogen or argon.

With that end in view the reactor was preliminary carefully vacuumized at warming up in a nitrogen current before preparation of the complex catalyst. Then after cooling to a room temperature the corresponding amine were entered into it.

The needed quantity of $AlCl_3$ was added to the obtained complexes in a nitrogen current at intensive mixing. Catalyst preparation was accomplished at 30°C - 35°C during 10 - 30 minutes.

The ionic liquids obtained represented themselves viscous liquids at room temperature with a yellowish shade.

After termination of chloroaluminate ionic-liquid catalysts (ILC) synthesis α-olefin was introduced into the reactor at mixing through a drop funnel.

$TiCl_4$ was applied in the form of a mobile colourless liquid smoking strongly on air. Molecular weight (M) 189.71, density (D^{25}) 1.726, fusion temperature (T_m)-23°C, boiling temperature (T_b) 135.8°C (760 mmHg).

2,6-di-(2',6'-diisopropyl-1-aminomethyl)-4-(1-methylc yclo-pentile)-phenol (M = 554, $T_{boiling\ point}$ = 205 - 207°C, $T_{melting\ point}$ = 91°C) has been prepared using the known method by interaction of para-(1-methylcyclopenthyl)-phenol

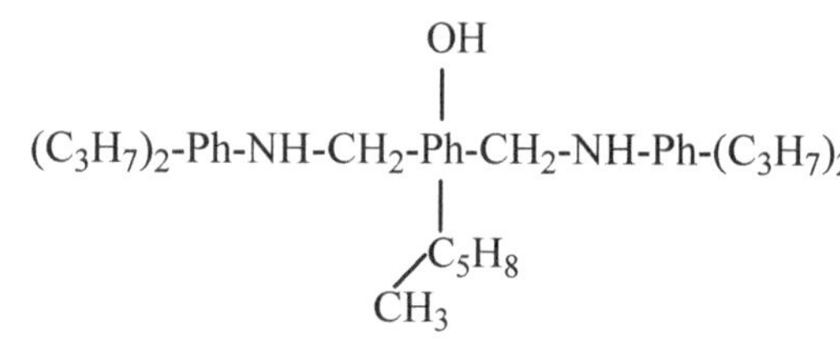

with formaldehyde and 2,6-diisopropylaniline according to Mannich reaction.

The titanium-phenolate complex (TF) was synthesized by the simple one-step method of interacting 2,6-di-(2' 6'-diisopropyl-1-aminomethyl)-4-(1-methylcyclopentile)-phenol with $TiCl_4$:

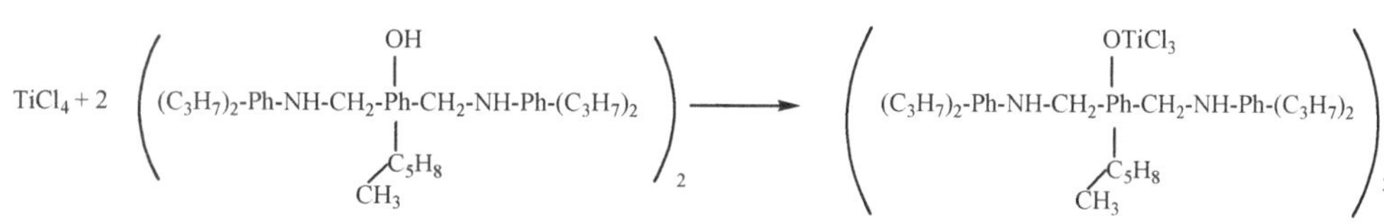

HCl released from the reaction was bound to the amine functionality of the phenolic compound in situ and remained as a part of the formed Ti-containing complex, which, after removal of the solvent, was represented as the brown colored powder, supposedly, sof zwitterionic type complex of the following formula:

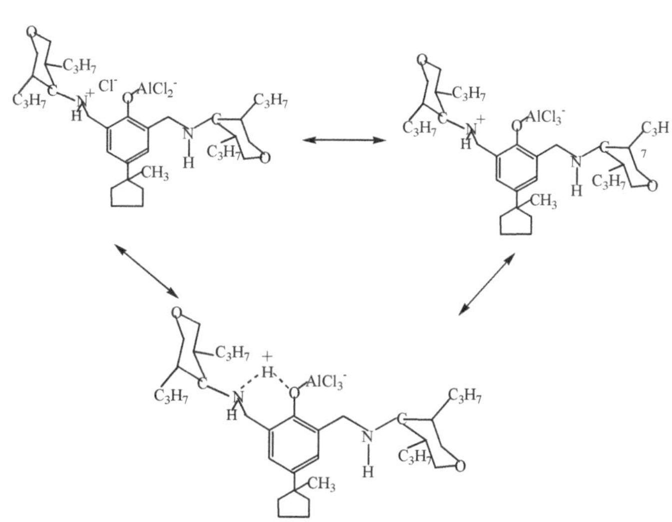

The following ionic-liquid catalysts were prepared by this method:

$$\left[(C_2H_5)_2NH_2\right]^+ \cdot \left[Al_2Cl_7\right]^- \qquad \text{(ILC-I)}$$

$$\left[(C_2H_5)_2NH_2\right]^+ \cdot \left[Al_2Cl_7\right]^- + TF \qquad \text{(ILC-II)}$$

For the preparation of TF was dissolved in ILC-I at the weight ratio TF:ILC-I = 0.001:1.

The experiments on oligomerization were conducted at the temperatures 60°C over 3 h with intensive agitating of reaction mass by a electromagnetic stirrer. After the termination of oligomerization the reaction products were separated from the catalyst by washing off it with 10% solution of NaOH and then by the distilled water till the neutral reaction or by decantation. After drying the unreacted monomer was distilled off from the obtained oil in vacuum and synthetic oil fraction having b.p. 250 - 350°C or >350°C was isolated.

In case of application in the quality of ionic-liquid catalyst of the catalytic system, prepared on aluminum chloride basis, the oligomerization process was conducted at the fusion temperature of the specified ILC. In this case, upon termination of reaction the received reaction mixture was divided into two layers at cooling: the bottom layer represented ILC layer, the top layer – the oligomerizate layer. After separation of ILC from oligomerizate by decantation it was stored in an inert gas atmosphere for further reusing it repeatedly.

Oligomerization of C_6-C_{12} α-olefins was carried out with application of the synthesized ionic-liquid catalytic systems in quantities, in which $AlCl_3$ content was within 2% mass, counted on the olefin. The received results are presented in **Table 1**.

Oligomerization of C_6-C_{12} α-olefins was conducted in the presence of ILCS-I and ILCS-II. A way of preparation of ILCS, a course of carrying out of the oligomerization processes and the analysis methods.

The results of researches on oligomerization of α-olefin in the presence of ILCS-I and ILCS-II at 60°C, identical concentration of catalyst (0.2% $AlCl_3$ masses, counted on olefin) are resulted in **Table 1**.

Apparently, the total amount of ILCS-I and ILCS-II in all cases of oligomerization makes 6% of mass., and a yield of oligomer amounts 85 - 103.92 gr oligomer/gr Al × hour, 77.1 - 90.3 × 10^3 gr oliqomer/gr Ti × hour, correspondingly. After the end of reaction, oligomerizate easily separates from catalyzate because of diphasic character of the process. ILCS is kept in the form of uncolored bed in an inert medium. A yield of products, within 3 h. amounts 68% - 85%.

3. Methods of Analysis

Oligomerization products were analyzed by exclusion chromatography, DSC, IR- and PMR-spectroscopy.

Molecular-weight distribution (MWD) of the obtained products was studied by size exclusion chromatography method using high performance "Kovo" (Czech Republic) liquid chromatograph with a refractive index detector. Two 3.3 mm - 150 mm columns packed with the "Separon-SGX" stationary phase with a particle size of 7 mm and a porosity of 100 A° were used. Dimethylformamide was used as an eluent (flow rate 0.3 ml/min, temperature 20 - 25°C). A calibration plot of log M versus VR in the range M = 2 – 100 - 102 was obtained using polyethylene glycol standards and transformed to the common dependence of fraction (%) of chains from their molecular weights (**Figure 1**). Calculations of the average molecular weights and MWD characteristics were made from the data of size exclusion chromatographic analysis in accordance with the procedure described in [14]. Average molecular weights (Mw and Mn) were calculated using the following formulas:

$$M_w = \sum M_i \omega_i \,, \quad M_n = 1/\sum \omega_i / M_i$$

where M_i is the molecular weight, corresponding to the ith slice of the chromatogram; ω_i is the area fraction of the ith slice.

Table 1. Oligomerization of α-olefin depending on composition of ILCS [Conditions: aminophenol: $TiCl_4$ = 1:1 mol., medium: argon].

№	Monomer	ILC	[Al], mol/l,	[Ti], mol/l·10^{-4}	Ti/Al, mol·10^{-4}	Yield*		
						qr of oliqomer /qr Al × hour	qr of oliqomer/qr Ti·hour × 10^3	mass. %
1	C_6**	I	0.19	—	—	85	—	68
2	C_7	I	0.2	—	—	87	—	70
3		II	0.2	1.3	6.8	93	79,7	75
4	C_8	I	0.2	—	—	88	—	71
5		II	0.2	1.3	6.8	91	78.5	73.5
6	C_{10}	I	0.22	—	—	89	—	72
7		II	0.22	1.5	6.8	90	77,1	80
8	C_{12}	I	0.22	—	—	100.25	—	82
9		II	0.22	1.4	6.8	103.92	90.3	85

*Yield after 3 h. **Oligomerizatssion conditions: t = 25°C; time = 45 min; amount of ILC = 8 wt% of 1-hexene [13].

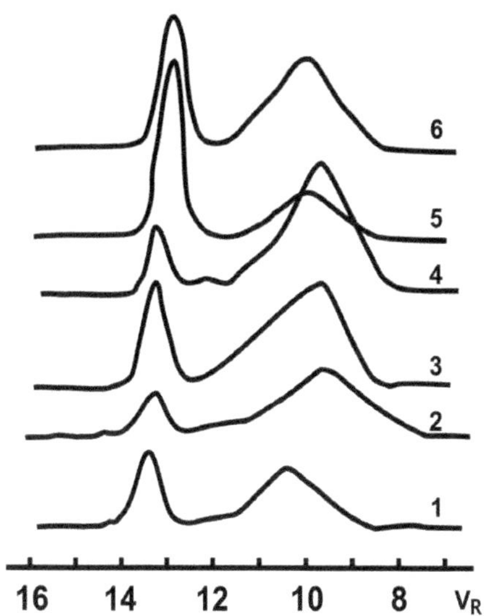

Figure 1. Exclusion chromatograms of oliqoalkylnaphtenic (C_6-C_{12}) fraction obtained in the presence of ILCS-I and ILCS-II.

DSC analysis was carried out on a Thermoelectron Q-20 Differential Scanning Calorimeter, USA (Joint Grant of NASA and CRDF no. SIP-03), at a heating rate of 10˚C/min in the air or nitrogen atmosphere.

IR-spectroscopy analysis of the samples was performed on a Spectrum BX, FT-IR system "Perkin-Elmer" (USA) in the range 650 - 4400 sm^{-1}.

PMR spectra were recorded on a Brunker pulsing Fourier spectrometer (Germany) operating at the frequency of 300 MHz and at room temperature. Deuterated acetone was used as a solvent. The relative amount of protons in various structural fragments was calculated by the integration of the corresponding resonance absorption bands.

4. Result and Discussion

4.1. Molecular Weight Characteristics of the Oligomers

The results of exclusion chromatographic analysis of C_6-C_{12} α-olefins oligomers (Fr > 350˚C) are presented on **Figure 1**. As is seen from **Table 2**, the oligomerization products of dodecene—1 received in the presence of ILCS-I and ILCS-II have the MM indexes within the limits—M_w = 1560 - 6890, M_n = 1400 - 5300, where MMD coefficient changes within M_w/M_n = 1.11 - 1.19. Anyway in each system the received MMD indexes change in the narrow limits.

4.2. Structure

The received oligomers have also been investigated by IR-spectroscopy (**Figure 2**). In the IR spectrums of initial monomer—α-olefin the absorption band was observed at 721 sm^{-1}, responsible for the pendular fluctuations of the methylene groups $(-CH_2-)_n$ with n > 4. There are also the bands of deformation at 1377.1 sm^{-1} and valence vibrations at 2853.5 sm^{-1}, characteristic for methyl groups ($-CH_3$). Deformation absorption bands at 1465.64 sm^{-1} and valence vibrations bands at 2922.7 sm^{-1} correspond also to the methylene groups ($-CH_2-$). Except the listed absorption bands, there are in the spectrum the deformation bands at 1300.30 sm^{-1} and the valence vibration bands at 2955.60 sm^{-1}, characteristic for -CH groups. Besides, there are the absorption bands at 908.3 sm^{-1}, 992.05 sm^{-1} corresponding to deformation fluctuations of the end vinyl groups. And also the bands s with maxima at 1821.1 sm^{-1} are displayed, responsible for an overtone

Table 2. Molecular mass indexes of oligolefin fractions received in the presence of ILCS.

№	Monomer	ILCS	M_w	M_n	M_w/M_n
1	C_6^*	I	690	640	1.08
2	C_7	I	1560	1400	1.11
3		II	3770	2385	1.16
4	C_8	I	3000	2500	1.2
5		II	3900	3700	1.05
6	C_{10}	I	4450	3740	1.19
7		II	5180	4540	1.14
8	C_{12}	I	4820	4310	1.12
9		II	6890	5300	1.3

*Oligomerizatssion conditions: t = 25˚C; time = 45 min; amount of ILC = 8 wt% of 1-hexene.

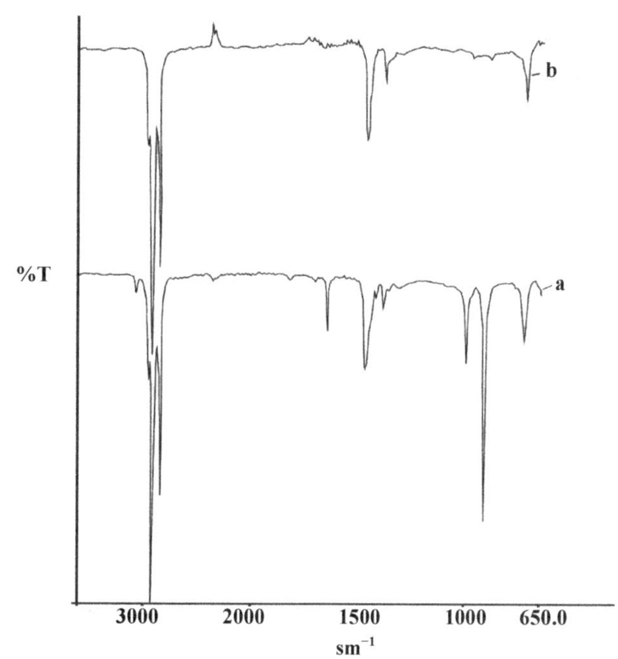

Figure 2. IR spectrum of dodecene-1 (a) and of oliqododecene fraction (b), produced in the presence of ILCS-I.

of nonplanar deformation fluctuations of the end vinyl groups. Absorption band with a maximum at 1641.6 sm^{-1} is characteristic for the valence vibrations of C=C bonds of nonsaturated hydrocarbons, and at 3076.4 sm^{-1} for the valence vibrations of -CH = group in an alkene molecule.

Analysis of the IR spectrums of C_6-C_{12} oliqoalkyl fraction (OAF) received in the presence of ILCS has shown that intensity of absorption bands in them with maxima at 908.3 sm^{-1}, 992. sm^{-1}, 1641.6 sm^{-1}, 1821.4 sm^{-1}, characteristic for nonsaturated hydrocarbons is much less than in the initial monomer. There are in the spectrums the absorption bands at 980.3 sm^{-1}, 1020.2 sm^{-1}, characterizing the oscillating motions of the carbon atoms in the naphthenic rings, and at 900.2 sm^{-1}, responsible for hydrogen atom of the isolated naphthenic ring [15,16].

For α-olefins and oliqoalkyl fraction, received in the presence of ILCS were calculated the optical densities of the bands characterizing oscillating motions of paraffinic, naphthenic, nonsaturated fragments. The obtained data are presented in **Table 3**.

The results of definition of structure oligoalkylfraction synthesized in the presence of ILCS (**Tables 4, 5** and **Figure 3**) by NMR-spectroscopy method have also confirmed the data of the IR-spectral analysis. The PMR-SPECTRA of oligoalkylfraction samples, received in the presence of ILCS-I and ILCS-II, are also characterized by the presence of the resonance absorption bands of the protons of methyl (CH_3-) (0.89 - 0.98 ppm) and methylene $(-CH-)_n$ (1.28 ppm) groups, and also resonance absorption bonds in the field of 1.50 - 1.80 ppm, corresponding to chemical shifts of the protons in the naphthenic frames.

After strong amplification (16 times) the signals in the field of 4.8 - 5.5 ppm, corresponding to nonsaturated structures, very weak absorption bands that specified the insignificant presence of olefinic linkages have been found. At significant intensifying of resonance absorption by accumulation in the PMR spectra of oligodecene

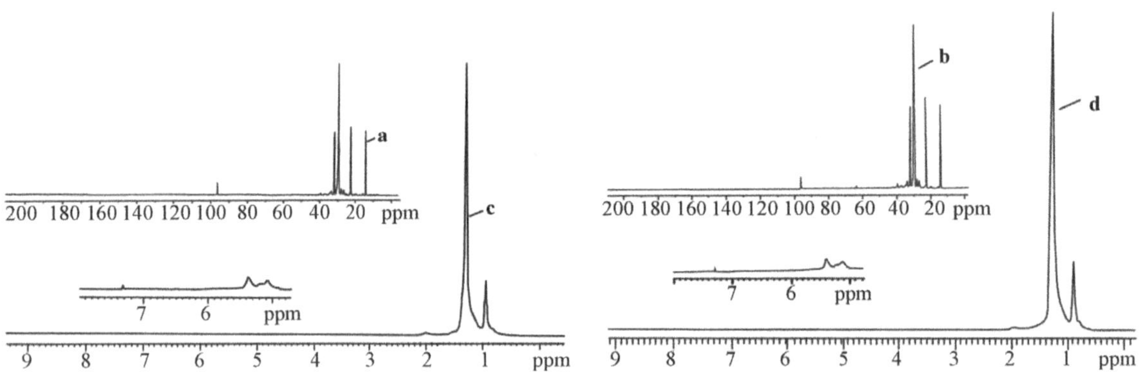

Figure 3. ^{13}**C NMR (a, b) and PMR (c, d) spectra of oligododecene fraction produced in the presence: ILCS-I (a, c) and ILCS-II (b, d).**

Table 3. Optical densities calculated from IR spectra.

Sample	ILCS	D_{920}	D_{1000}	D_{1650}	D_{900}	D_{980}	D_{720}	D_{1380}
C_{10}	-	1.9138	1.1357	0.5916	–	–	0.4367	0.4225
	II	0.0512	–	0.0649	0.0939	0.0792	0.6532	0.8159
C_{12}	-	1.9243	1.1416	0.6232	–	–	0.5441	0.5093
	I	0.0435	–	–	0.0550	0.0669	0.5819	0.7202
	II	0.1684	0.0792	0.0550	0.0918	0.0918	0.5819	0.6232

Table 4. Structural parameters determined from PMR spectral data of dodecene-1in the presence of ILCS oligomerization products.

ILCS	Distribution of hydrogen atoms in structural fragments, %				
	-CH = CH_2 (5.0 - 5.3 ppm) -CH = CH- (5.3 - 5.5 ppm)	CH_2 at double bond (1.90 - 2.15 ppm)	CH at 5 - 7 naphtenic ring (1.45 - 1.80 ppm)	β CH_2, in the chain (1.05 - 1.45 ppm)	Terminal CH_3 group (0.50 - 1.05 ppm)
I	0.25	2.03	3.19	75.48	17.93
II	0.39	2.65	5.35	72.41	16.18

Table 5. Assignment of the signals in the ^{13}C NMR spectra of dodecene-1 oligomerization products.

ILCS- I	C_1	C_2	C_3	C_4	C_5	C_6	C_7	C_8	C_9	C_{10}	C_{11}	C_{12}
Chain	CH_3- 14.265	-CH- **29.834**	**-CH_2**	**-CH_2- 29.499**	**-CH_2 32.045**	**-CH_2-**	**-CH_2-**	**-CH_2- 32.045**	**-CH_2- 29.499**	**CH_2-**	**-CH_2- 22.793**	**CH_3 24.265**
5-membered naphthenic ring 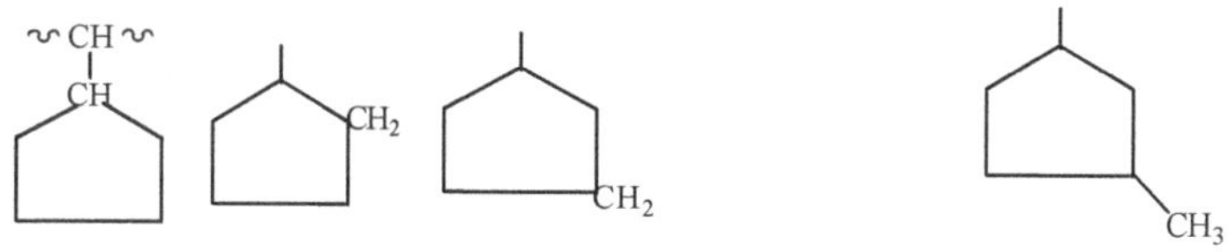	28.277	--	26.826			19.879						
6-membered naphthenic ring 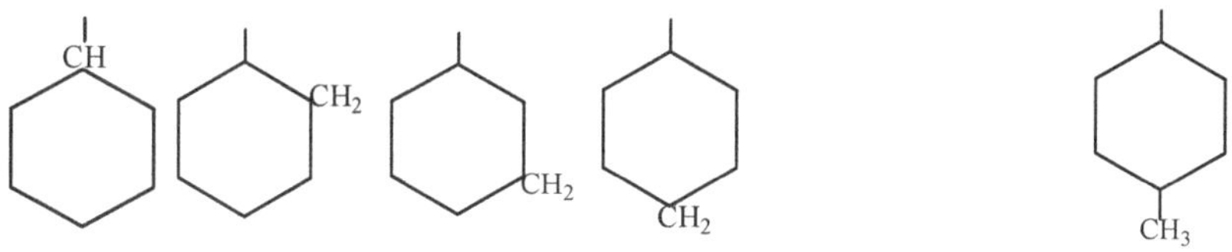	28.277	39.557	26.826	26.826			22.793					
7-membered naphthenic 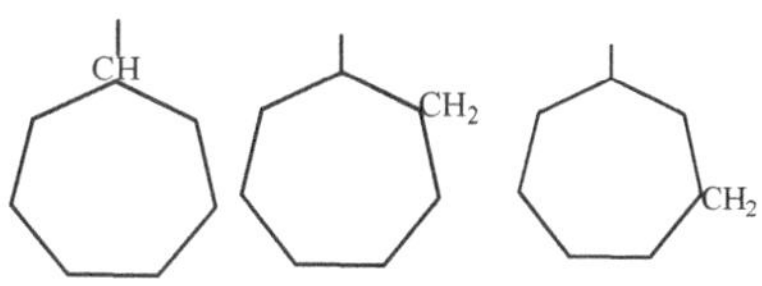	34.899	37.205	26.826									

ILCS- II	C_1	C_2	C_3	C_4	C_5	C_6	C_7	C_8	C_9	C_{10}	C_{11}	C_{12}
Chain	CH_3- 14.229	-CH- **29.845**	**-CH_2**	**-CH_2- 29.509**	**-CH_2- 32.053**	**-CH_2-**	**-CH_2-**	**-CH_2- 32.053**	**-CH_2- 29.509**	**-CH_2-**	**-CH_2- 22.793**	**-CH_3 14.229**
5-membered naphthenic ring 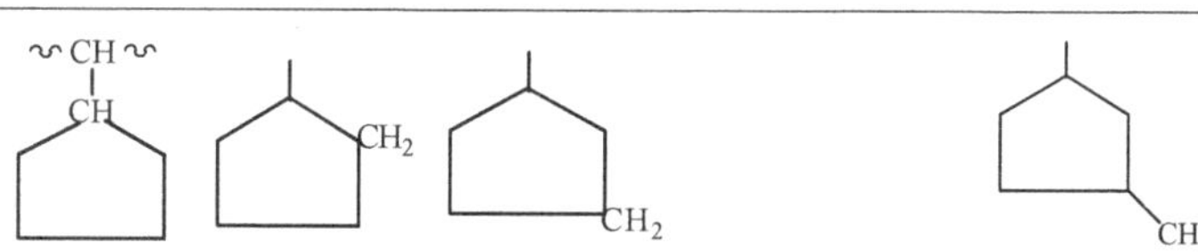	28.272	--	26.820			19.838						
6- naphthenic ring 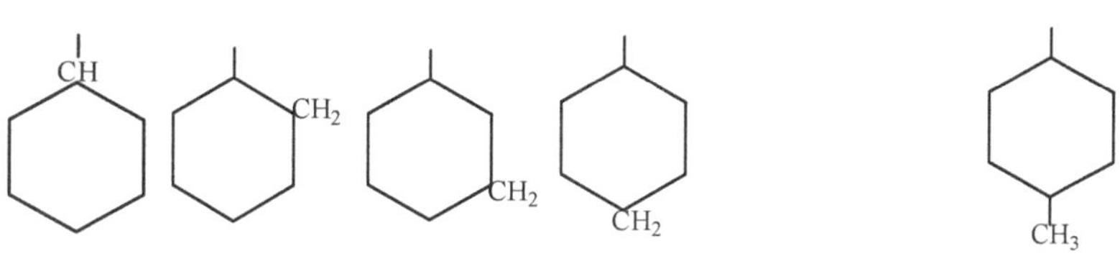	28.272	39.547	26.820	26.820			22.793					
7-membered naphthenic 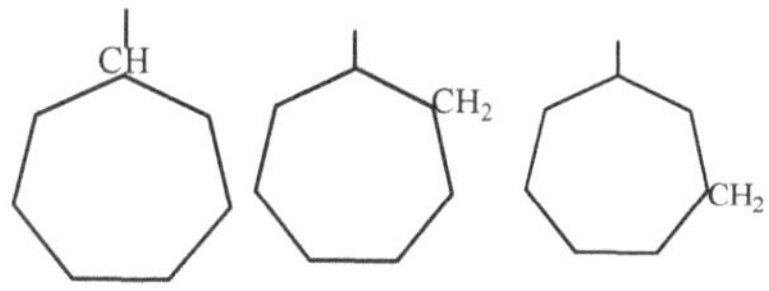	34.856	37.222	26.820									

weak absorption bands of protons of the CH = CH_2 methylene groups (at 5.0 - 5.5 ppm), and of vinylene CH = CH groups at 5.7 - 5.7 ppm) are observed also [17].

Research of the structural parameters of C_6-C_{12} α-olefins oligomerization products, specify passing of the process with formation of the oligomers enriched with the fragments of (iso) paraffinic and naphthenic nature which concentration changes within 96.3 - 97.3 and 2.4% - 3.0% mass depending on the composition of catalytic system. Accordingly, concentration of the fragments containing multiple bonds make only 0.3% - 1.0% mass., which is much more low than in initial olefine (**Table 5**).

This fact once again obliquely testifies to passing of ring formation in the last part of a growing chain. With growth in the length of an alkyl radical in the olefin molecule the probability of bigger naphthenic ring formation increases.

The results of IR-, PMR and NMR spectral analyses have shown identity of frames of all investigated oligoalkylnaphtenic samples, synthesized in the presence of chloroaluminate ILCS. Though, the weak absorption bands observed in both spectra, characterizing multiple bonds, testify to the presence in the synthesized products of the trace amounts of nonsaturated fragments.

4.3. Some Physical and Chemical Properties Oligoalkylnaphtenic Oils

Some physical-chemical indexes of oligoalkylnaphtenic product, produced in the presence of ILCS-I and ILCS-II, have also been estimated. The results are presented in **Table 6**. The values of viscosity index (VI) have been calculated in accordance with GOST 25371-97. Physical - chemical properties of the synthesized oligoalkylnaphtenic product were defined according to the procedures [18,19].

From the data of **Table 6** it is evident that the samples of the synthesized oligoalkylnaphtenic oils have low density values, high viscosity and viscosity index (≥108), low freezing point (−33˚C - −45˚C), high ignition temperature (≥172˚C), good antioxidative stability, detergency and low vapor ability, do not contain in their composition sulphurous compounds and metal-complexes.

Thus the results of conducted researches show that upon oligomerization of C_6-C_{12} *α*-olefins with participation of the chloroaluminate type ionic liquid catalysts may be synthesized the oligomeric molecules having predominantly hybridized paraffinic-naphthenic structures, free of unsaturated fragments. This may favor to use them as a component for preparation of synthetic and semisynthetic base oil compositions without use of hydrogenation step or providing it in a mild condition.

With that end in view hydrogenation of oligododecenenaphtenic oil was conducted using industrial Ni/Cr/Pd catalyst at P = 30 атм., T = 120˚C, time = 4 h. It has been found that even in that comparatively hard conditions the VI indexes of the tested samples do not change practically after hydrogenation (VI = 159 before, and 158 after hydrogenation). But some destruction of the high molecular weight molecules takes place during hydrogenation.

In the DSC curves the shift in the position of high molecule peak at V_R = 10 - 12 to V_R = 14 takes place. After hydrogenation the share of the oligomer molecules with V_R = 14 decreases. The destruction process leads to some changes in the thermal indexes of oligomerization product oligododecenenaphtenic defined by DSC method (Termoelectron Corporation Q-20, USA). After hydrogenation of oligododecenenaphtenic the *heat flow value* (Wt/g) *decreases a* little (**Figure 4**). The hydrogenated product characterizes by less induction period and more complicated course of its oxidation (**Figure 5**), which confirms occurrence of destruction process taken place in the hydrogenation conditions.

5. Conclusions

C_6-C_{12} *α*-olefins oligomerization in the presence of chloroaluminate ionic liquids and their combination with Ti-containing complexes used as modifiers have been studied. The structure, molecular-weight characteristics and low-temperature properties of produced oligomers have been examined by means of IR, PMR spectroscopy, size exclusion chromatography and the DSC.

It was shown that in all cases of using ILCs, the obtained oligomers contain predominantly hybrid hydro-

Table 6. Physical - chemical properties of the synthesized oligoalkylnaphtenic oils.

Monomer	ILCS	Viscosity index (VI)	Temperature, ˚C		Kinematic viscosity, mm²/c, at:		Density at 20˚C, kg/m³	Colour on calorimeter, SNA, unit SNA
			Flash point	Freezing point	V_{100}	V_{40}		
C_6	I	108	220	−35	12.20	103.23	838.2	1.5
C_7	I	116	224	−33	80.45	10.83	835.0	1.5
	II	120	218	−33	90.0	12.0	838.0	1.5
C_{10}	I	128	172	−28	218.09	26.23	870.0	1.5
	II	124	174	−30	285.94	30.50	875.0	<2
C_{12}	I	159	212	−38	145	19	870.0	1.5
	II	162	210	−45	137.7	19.6	873.0	1.5

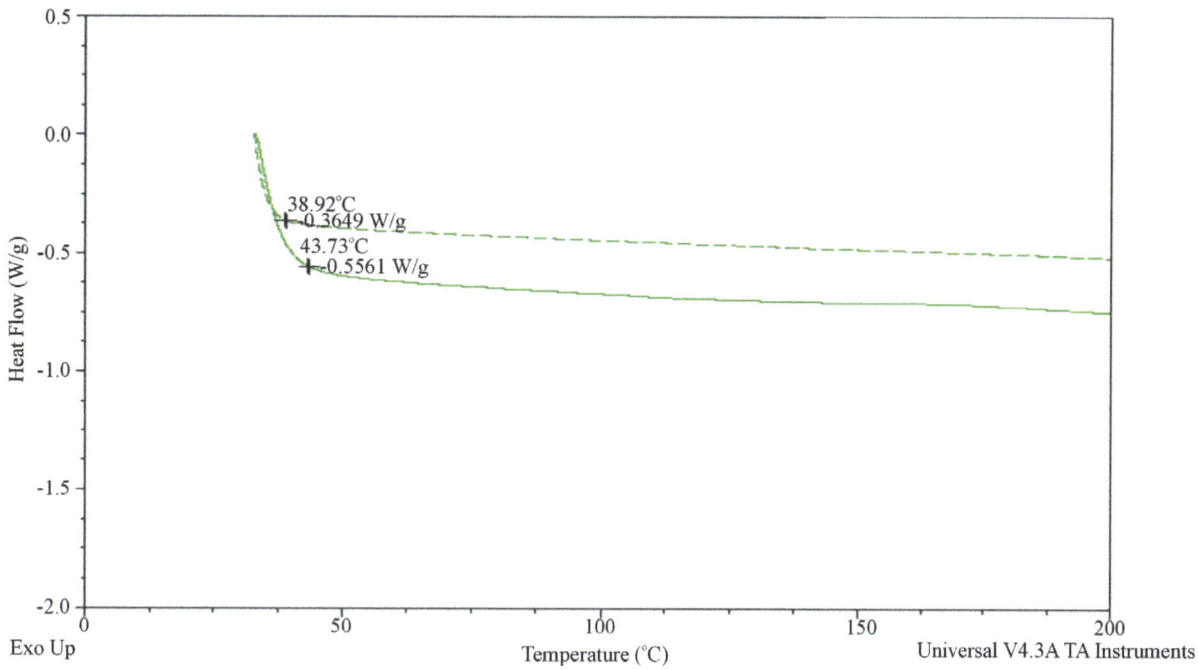

Figure 4. DSC thermo-stability curves (in nitrogen atmosphere) of oligododecenenaphtenic ODF before (---) and after (–) hydrogenation.

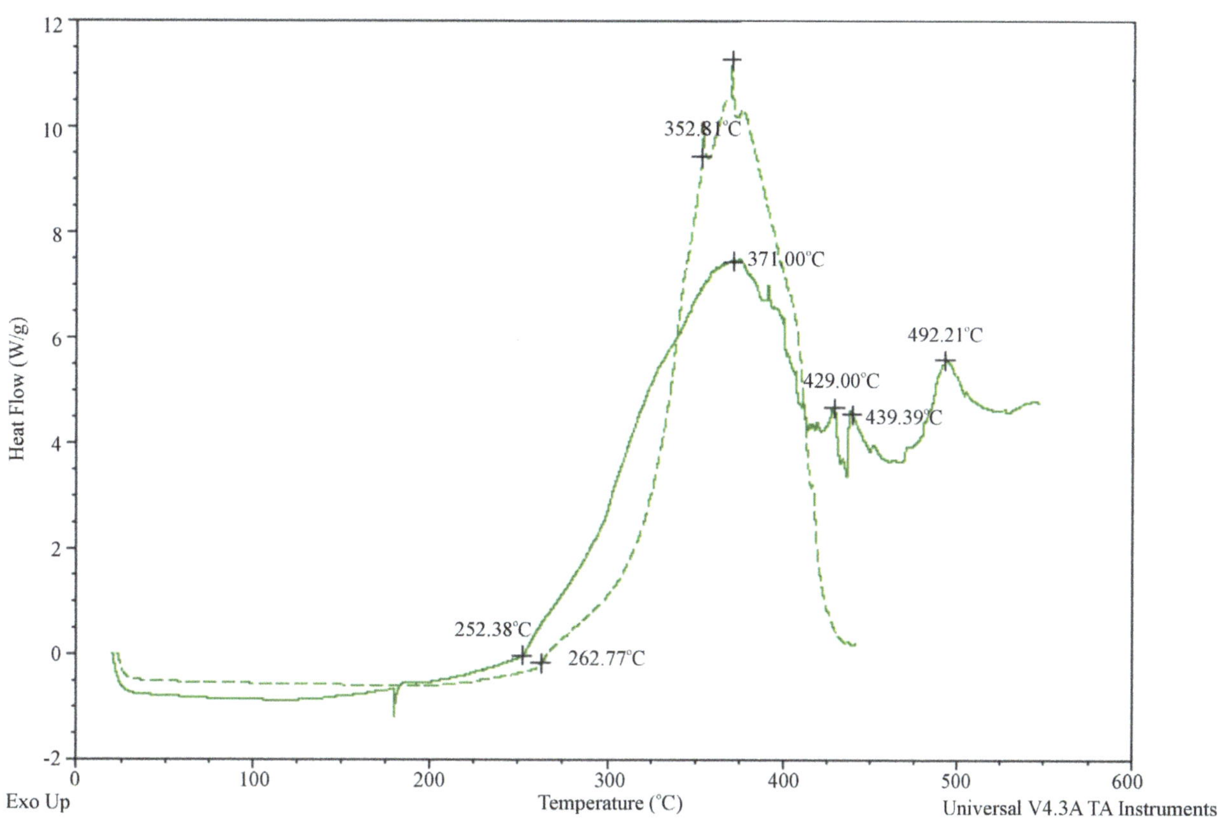

Figure 5. DSC thermo-oxidation (in oxygen atmosphere) curves of non-hydrogenated(---) and hydrogenated (–) oligododecenenaphtenic.

carbons of paraffinic-naphthenic structure with long alkyl radical branching termed as oligoalkylnaphthenic hydrocarbons.

The synthesized oligoalkylnaphtenic oils are characterized by low density values, high viscosity and viscosity index (≥108), low freezing point (−33°C - −45°C), high ignition temperature (≥172°C), good antioxidative stability, detergency and low volatility. They may be used as the high viscosity base oils with good thermo-oxidative stability or as a component for preparation of the synthetic and semisynthetic base oil compositions without use of additional hydrogenation step or by conducting it in a mild condition.

REFERENCES

[1] J. D. Holbrey and K. R Seddon, *Clean Product and Processes*, Vol. 1, 1999, pp. 223-236.

[2] A. H. Azizov and Z. H. Asadov, "Production and Usage of α-Olefins and Advantages in the Metal-Complex Catalyzed Selective Oligomerization of Ethylene," *Processes of Petrochemistry and Oil Refining*, Vol. 1, 2003, pp. 44-67.

[3] T. Welton, "Room-Temperature Ionic Liquids. Solvents for Synthesis and Catalysis," *Chemical Reviews*, Vol. 99, No. 8, 1999, pp. 2071-2083.

[4] A. H. Azizov, "Preparation Methods and Solvent Properties of Low Temperature Ionic Liquids," *Process of Petrochemistry and Oil Refening*," Vol. 1, No. 8, 2002, pp. 90-115.

[5] A. H. Azizov, "Low Temperature Ionic Liquids in Organic and Petrochemical Synthesis," *Processes of Petrochemistry and Oil Refining*, Vol. 2, No. 9, 2002, pp. 6-48.

[6] R. V. Aliyeva, "Ionic Liquids in Catalysis Petrochemical Processes," *Processes of Petrochemistry and Oil Refining*, Vol. 1, No. 16, 2004, pp. 26-46.

[7] R. V. Aliyeva, "Ionic Liquids during Separation and Extraction," *Processes of Petrochemistry and Oil Refining*, Vol. 2, No. 25, 2006, pp. 30-39.

[8] A. A. Xanmetov, Kh. H. Asgerova, R. V. Aliyeva and A. H. Azizov, "Ionic Liquids to Oligomerization and Polymerization Processes of Olefins," *Processes of Petrochemistry and Oil Refining*, 11, 42, 2010, pp. 309-328 (in Russian).

[9] P. Wasserscheid, "*Green Industrial Applications of Ionic Liquids*," Kluwer Academic Publishers, Boston, 2003, pp. 29-47.

[10] K. R. Seddon, "Ionic Liquids for Clean Technology," *Journal of Chemical Technology and Biotechnology*, Vol. 68, No. 4, 1997, pp. 351-356.

[11] H. Olivier, "Recent Developments in the Use of Non-Aqueous Ionic Liquids for Two-Phase Catalysis," *Journal of Molecular Catalysis A*, 146, 1999, pp. 285-289.

[12] H. Olivier, J. A. Chodorge and P. Travers, "Olefin Oligomerization with Homogeneous Catalysis," *Petroleum Technology Quarter* (*PTQ*), *Autumn*, 1999, p. 141.

[13] A. H. Azizov, *et al.*, "Selective Synthesis and the Mechanism of Formation of the Oligoalkylnaphthennic Oils by Oligocyclization of 1-Hexene in the Presence of Ionic-Liquid Catalysts," *Applied Catalysis A*: *General*, Vol. 375, 2010, pp. 70-77.

[14] N. R. Bektashi, A. M. Mustafayev, Qusyynov, *et al.*, "Chromatographic Investigation of Regularities of Synthesis of Halogen-Containing Unsaturated Polyesters," *Journal Prikladnoy Chimii*, Vol. 84, No. 7, 2011, pp. 1211-1217 (in Russian).

[15] K. Nakanisi, "Infrakrasnie Sprctri I Stroenie Organiçeskixch Soedineniy, M.," 1965, p. 210 (in Russian).

[16] L. Bellami, "The Infra-Red Spectra of Complex Molecules," Publishing House of Foreign Literature, Moscow City, 1963, p. 25 (in Russian).

[17] E. Breitmayer, "Carbon-13 NMR Spectrocopy," 3rd Edition, VCH, Weinheim, pp. 183-196.

[18] "Nefteprodukti Metodi Ispitaniy Part 1, İzdatelstvo Standartovç Moskva," 1977 (in Russian).

[19] B. V. Belyanin and V. N. Erikh, "Tekhniçeskiy Analiz Nefteproduktov i Qaza," Leninqrad, Chimiya, 1979, pp 111-112 (in Russian).

22

Interactions in Aqueous Solution of a Zwitterionic Surfactant with a Water Treatment Protein from *Moringa oleifera* Seeds Studied by Surface Tension and Ultrasonic Velocity Measurements*

Habauka M. Kwaambwa[1#], Fiona M. Nermark[2]
[1]School of Health and Applied Sciences, Polytechnic of Namibia, Windhoek, Namibia
[2]Department of Chemistry, University of Botswana, Gaborone, Botswana

ABSTRACT

The interaction of a zwitterionic surfactant with the water treatment protein extracted from *Moringa oleifera* has been investigated by surface tension and ultrasonic velocity measurements. The critical micelle concentration (CMC) of zwitterionic surfactant was determined to be 2.4 ± 0.3 mM by both techniques and the partial specific volume $v = 0.78 \pm 0.06$ cm^3/g for the protein was found. There seems to be a mild interaction between the protein and the surfactant as shown by surface tension measurements. The ultrasonic velocity was found to decrease in the vicinity of the critical micelle concentration which may be due to micelle aggregates formation and the protein caused a shift of the surfactant's CMC to a higher concentration.

Keywords: Critical Micelle Concentration; Density; Surface Tension; Ultrasonic Velocity; Zwitterionic Surfactant

1. Introduction

It is well-known that solution environment such as pH, salt and surfactant can significantly alter the behaviour and overall performance of biomolecules. In our earlier research of the protein from *Moringa oleifera* (MO) seeds, we have used a number of techniques to study their effects on the properties of the protein in order to understand the nature and mechanism of this protein in water treatment [1-7]. Most of these studies have used anionic surfactants particularly sodium dodecyl sulphate (SDS) although some of them used also cationic and non-anionic surfactants but not with zwitterionic surfactant. Generally, zwitterionic surfactants have been much less studied than other classes of surfactants. Their behavior is often described in relation to ionic and nonionic surfactants but the electrically neutral nature of zwitterionic surfactants clearly distinguishes them from ionic surfactants. Studies of the MO protein-surfactant interactions have been performed in our labs aiming to understand how the surfactant binding may affect the protein structure and function and hence its efficient use for water treatment.

Zwitterionic surfactants, which contain both positively and negatively charged head groups, are interesting molecules because of their many unique properties. In general, they are mild to the skin and eyes, exhibit low toxicity, display excellent water solubility, broad isoelectric ranges, high foam stability, and resistance to hard water and to degradation by oxidizing agents [8]. Also, changes in temperature, pH, and added electrolyte have been found to have minimal effect on zwitterionic surfactants. In order to obtain better insight into their interactions with aqueous MO protein systems, we report here the results of an investigation of the interactions of a zwitterionic surfactant with a water treatment protein extracted from MO seeds, studied by surface tension and ultrasonic velocity techniques.

*Competing interests: the authors declare that no competing interests exist.
#Corresponding author.

2. Materials and Methods

2.1. Materials

A zwitterionic surfactant *N*-dodecyl-*N*,*N*-dimethyl-3-ammonio-1-propanesulfonate (DDAPS) (also named *n*-dodecyl sulfobetaine) was supplied by Sigma Aldrich (≥98% assay, critical micelle concentration of 2 - 4 mM in the range 20 - 25°C, Mol. Wt. = 335.55 g·mol^{-1}) and was used without any further purification. The structure formula of DDAPS is shown in **Figure 1**. Sodium chloride (99% purity) was supplied by Rochelle Chemicals. Distilled water was used for all experiments.

2.2. Extraction and Purification of Protein

The extraction and purification of protein powder was done using the method of Ndabigengesere and Narasiah [9], and the experimental details are as described previously by Kwaambwa and Maikokera [1]. To minimize the number of components, no buffer was used. This is with the view that the protein investigated is used for water treatment and buffer conditions would not reflect the environment is it intended. Furthermore, previous studies have shown that the protein is insensitive up to as high as pH 9 [2,4,7] and the isoelectric has been reported to be between pH 10 and 11 [9], suggesting that it is very stable and to a large extent considered self-buffering.

2.3. Surface Tension Measurements

Surface tension of a series of differing concentrations of surfactant (DDAPS) in the absence and presence of the freeze dried protein, were measured with a Krüss K9 tensiometer using the Du Nouy platinum ring detachment method which was calibrated using distilled water. Readings were taken in triplicates. The platinum ring was cleaned before each measurement by heating on a Bunsen burner flame to remove any residual deposits.

A series of different concentrations of surfactant were prepared from the stock solution by dilution using distilled water and varying protein concentrations in distilled water as required. The surface tension measurements were measured at room temperature immediately after sample preparation. Surfactant solutions in the presence of the protein were used to study the protein-surfactant interactions.

The surface tension data was analysed using the simplest form of the Gibbs equation written as

$$-d\gamma = RT\sum_i \Gamma_i d\ln a_i \quad (1)$$

where γ is the surface tension, R is the universal gas constant, and T is the temperature in kelvins (K), whereas Γ_i and a_i are the surface excesses and activities of all the independent components. For measurements on a non-

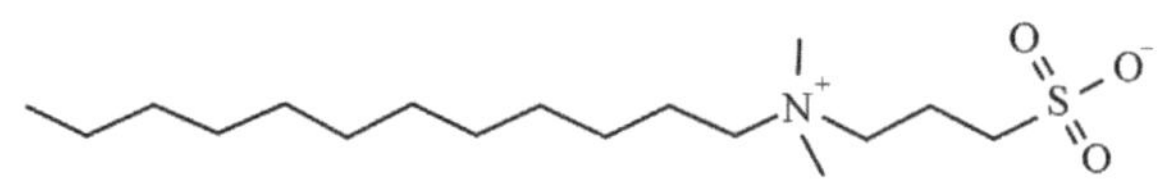

Figure 1. Schematic structure of N-Dodecyl-N,N-dimethyl-3-ammonio-1-propane sulphonate, showing the location of positive and negative charges.

ionic surfactant solution, this generally reduces to

$$-d\gamma = RT\Gamma \ln a \quad (2)$$

There are some considerations concerning the definition of Γ but these are generally unimportant in the context of surfactant systems. For nonionic surfactants, the usual practice is to use the approximation of concentrations (c) rather than activities except at or above the CMC. The slope of the plot of γ against ln c gives the corresponding surface excess Γ at that point using Equation (2).

2.4. Density Measurements

For a macromolecule such as a polyelectrolyte, a detailed understanding of the solution requires information on a variety of physical properties such as partial specific volume. The partial specific volume of a protein (or other macromolecule) is needed for a number of physicochemical techniques such as analytical ultracentrifugation, sedimentation equilibrium, and small-angle scattering. Specific volumes may be obtained from density measurements. However, partial specific volumes are not always accurately determined because experimental precision may be limited by the preparation of solutions of different concentrations. In this study, density data of the solutions were needed in the ultrasonic velocity measurements.

The measurements were performed at a constant temperature of 298.15 K using an Anton Parr DMA 4500 vibrating-tube precision density meter. The measurements were taken in triplicates to get an average value. The density meter was calibrated using distilled water as its density is accurately known from literature.

2.5. Ultrasonic Velocity Measurements

All measurements were performed at 298.15 K. The speed of sound was measured with uncertainty of ±0.3% using a multi frequency ultrasonic interferometer (Mittal Enterprises, New Delhi, India), operating at 1 MHz, which was calibrated with water. Since the speed of sound of our samples spans an interval ranging approximately from 1500 to 1525 ms^{-1}, the apparatus was calibrated with pure water for which there was reliable speed of sound values in the literature. The temperature stability was maintained within ±0.2 K by circulating thermostated water around the cell with a circulating pump.

Ultrasonic waves of known frequency are produced by

a quartz crystal fixed at the bottom of the ultrasonic cell. These waves are reflected by a movable metallic plate kept parallel to the quartz crystal and standing waves are formed in the medium. This acoustic resonance gives rise to an electrical reaction on the generator driving the quartz crystal.

If the distance between crystal and plate is now increased or decreased and the variation is exactly one half wavelength, $\lambda/2$, or multiple of it, anode current becomes maximum or minimum. In order to minimize the uncertainty of the measurement, their number n is counted. Each maximum is recorded with the highest swing of the needle on the micrometer scale. The total distance, d, moved by the reflector is given by

$$d = n\lambda/2 \quad (3)$$

where λ is the wavelength. A linear plot of d versus n gives a slope of $\lambda/2$. The frequency of the cell crystal v being accurately known (1 MHz), the speed of sound, u, is calculated by using the relation:

$$u = \lambda v \quad (4)$$

3. Results and Discussion

3.1. Surface Tension Measurements

Figure 2 shows the plot of γ as a function of ln[DDAPS] with and without the protein based on Gibbs equation. This method is widely used to determine the critical micelle concentration (CMC) on a wide variety of surfactants [10]. The CMC of DDAPS was found to be 2.6 ± 0.3 mM, which fall within the range given by the chemical supplier (*i.e.* 2 - 4 mM) and also reported in literature [8,11]. The shapes of the plots around the CMC show that the surfactant is impure. A pure surfactant will have a net break at the CMC but the plots do not show a net break but rather it reduces and then increases steadily, that is, it has a minimum which results from the impurity.

The effect of the protein extracted from protein on the

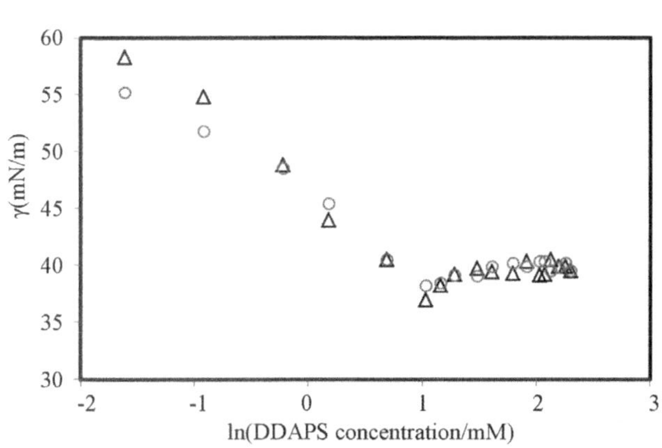

Figure 2. Variation of the surface tension a zwitterionic surfactant, DDAPS, in the presence (circles) and absence (triangles) of 0.02% MO protein.

zwitterionic surfactant, DDAPS, surface tension was also studied and is shown in the same figure. The results show that the protein interacts mildly with the surfactant in the concentration range studied. In other words, the protein does not interact significantly with the zwitterionic surfactant to affect its surface activity. In fact, the behaviour observed for the protein/zwitterion system is similar to what Kwaambwa and Maikokera observed for the protein and nonionic surfactant triton X-100 (TX-100) [3]. This is not surprising because zwitterionic surfactants, although have ionic groups, are overall neutral and hence behave like nonionic surfactants.

Although an analysis of the surface tension curves can, in principle, provide quantitative information on surfactant-protein interactions, practically speaking, complete interpretation of the data becomes more difficult when the protein itself is appreciably surface active as it is the case for the protein from MO seeds. The surface tension profiles were obtained both for the surfactant systems and in the presence of protein, showing the characteristic break points in mixed protein/surfactant systems. For ease of interpretation of the analysis, the experiments were done in distilled water rather than in a buffer.

In addition to the typical CMC observed for DDAPS solutions, the profiles of the protein/surfactant mixtures show inflection points commonly known as the critical micelle concentration in the presence of protein (CMC*). The values are determined in a similar way as CMC, *i.e.* the surfactant concentration at which the surface tension reaches a minimum value. The CMC* values for different protein concentrations are normally higher than in pure water, which can be explained by considering the amount of surfactant that is bound to the protein. In our study, surface tension at CMC*, *i.e.* when the protein is saturated by surfactant and free micelles start to form, reached value close to that obtained for the surfactant solution alone as shown in **Figure 2**. It should be noted that this value is not influenced by the protein concentration as it is envisaged that protein molecules are totally displaced from the interface when the surfactant concentration reaches the CMC*.

Addition of NaCl seems to lower the surface tension but not the CMC as shown in **Figure 3**. The addition of salt reduced the surface tension more in the low DDAPS concentration region of the plot implying that more protein molecules were adsorbed onto the air/water interface to cause a reduction in surface tension (*i.e.* the salt makes the protein more surface active). The protein is known to be cationic (implying that the positive charges are more than the negative charges or the positive charges of the protein are more exposed) and so addition of salt results in neutralisation of the positive ions and hence making the protein more surface active.

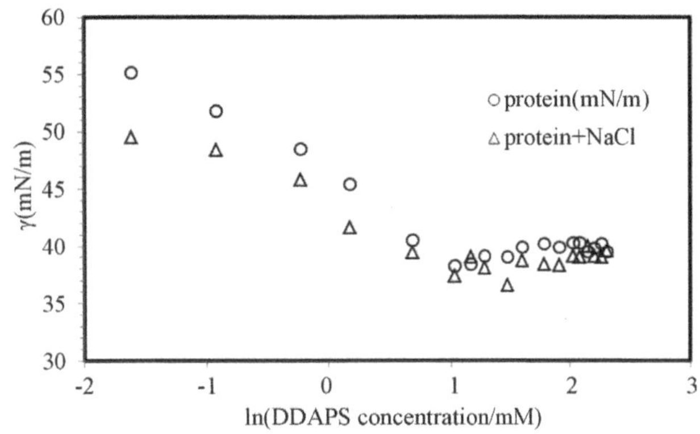

Figure 3. Variation of the surface tension of a zwitterionic surfactant, DDAPS, in 0.02% protein (circles) and in a mixture of 0.02% protein and 0.01 M NaCl (triangles).

3.2. Density Measurements

Figure 4 shows the linear variation of density of DDAPS as its function of concentration in the presence and absence of the protein. The partial specific volume of DDAPS was estimated from density measurements. Density values, ρ, increase with the concentration of the polyelectrolyte. If the partial specific volume v is independent of the concentration, the following equation is obeyed

$$\rho = \rho_o + (1 - v\rho_o)c \qquad (5)$$

where ρ_o is the water density and c is the polymer concentration in g/cm^3.

The dependence of the density on the concentration is linear with statistical correlation parameter, R^2, greater than 0.99 and Equation (5) was applied to the data to obtain v from the slope. The data yielded a partial specific volume $v = 0.78 \pm 0.06$ cm^3/g for the protein. Experimental values for the partial specific volume of native proteins in aqueous solution cluster around 0.74 cm^3/g [12]. The value in water obtained for the protein agrees well with the literature although the method used for preparing solutions of different concentrations had the limitation of not knowing precisely the protein concentration.

The protein increases the density of the DDAPS proportionally. There is no break in the plot at the CMC (determined to 2.6 mM) hence showing that the micellisation does not affect the density. The slopes gave the same value of partial specific volume of $v = 1.0027 \pm 0.0002$ cm^3/g since the linear plots of the density versus DDAPS concentration in the presence and absence of protein are parallel in the concentration range studied (see **Figure 4**).

3.3. Ultrasonic Velocity Measurements

Figure 5 shows the plots of the ultrasonic velocities of DDAPS surfactant with and without protein. The plot of

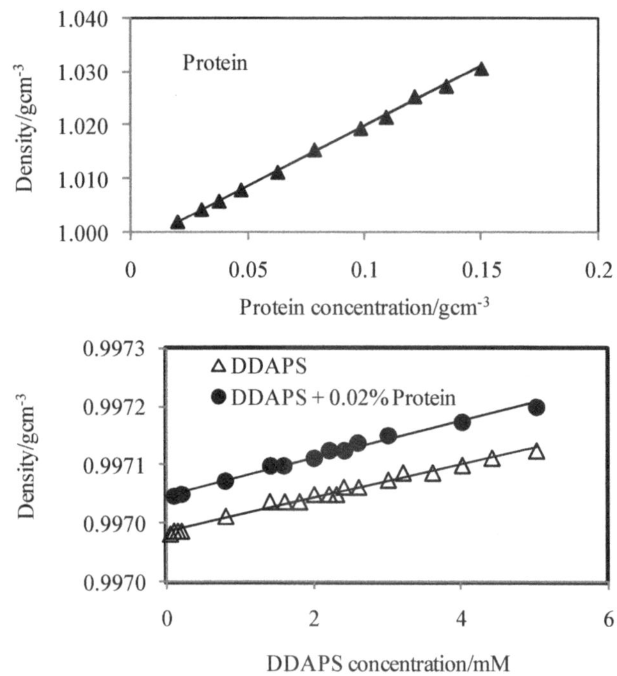

Figure 4. Variation of density of DDAPS as a function of concentration in the presence and absence protein.

surfactant with protein deviates from that without the protein but the ultrasonic velocity initially increases with increase in surfactant concentration and reaches a maximum before beginning to go down to a minimum value. After the minimum, the ultrasonic velocity begins to increase again. The minima occurred at DDAPS concentrations of 2.3 mM and 3.0 mM in the absence and presence of MO protein, respectively. The value for minimum for DDAPS in the absence of protein corresponds to CMC value determined from the surface tension measurements. In fact, the minimum for the plot of the surfactant with protein of 3.0 mM falls within the range for DDAPS reported by the supplier. The MO protein causes an increase in the CMC of the surfactant which could be due to the interference of micelle aggregates formed at the CMC.

In this study, density measurements showed to increase with increase in concentration, as shown in **Figure 4**, and speed of sound is said to increase with increase in density. This trend changes towards the CMC showing a decrease in the speed of sound and then an increase after the CMC. The break in the ultrasonic velocity as a function of surfactant concentration plot shows that the CMC can be determined using this technique. This technique was counterchecked using sodium dodecyl sulphate (SDS) and the results are shown in **Figure 6**. The ultrasonic velocity behaviour is the same as that of DDAPS and the inflection point minimum occurs at 8.3 mM, which corresponds with the reported literature CMC of SDS of 8.25 mM at 25°C [13].

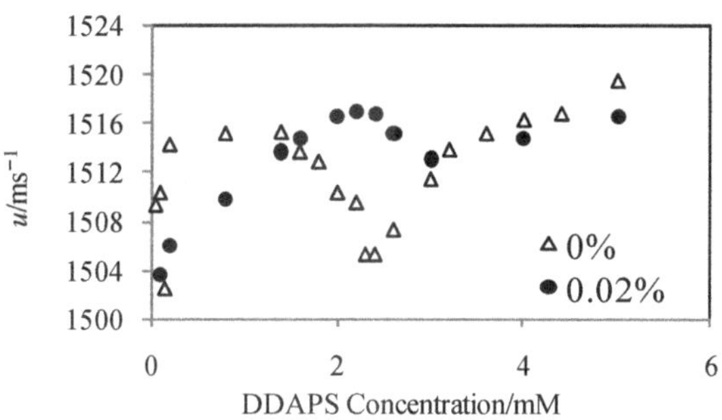

Figure 5. Ultrasonic velocity of the surfactant DDAPS in the presence (0.02%) and absence (0%) of the protein.

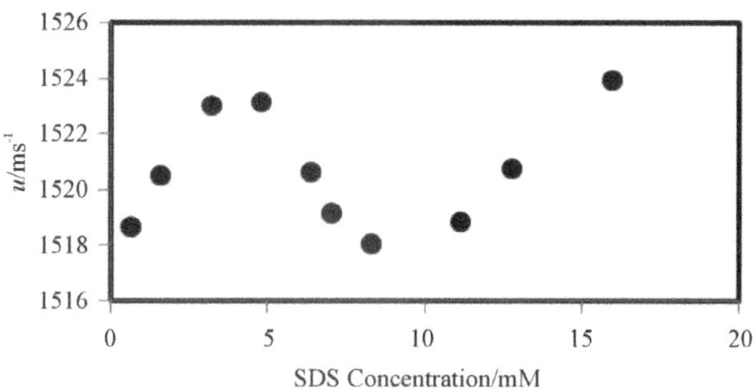

Figure 6. Ultrasonic velocity of anionic surfactant SDS as a function of concentration.

4. Conclusion

The zwitterionic surfactant DDAPS interacts mildly with cationic protein extracted from MO seeds as evidenced from both surface tension and ultrasonic velocity measurements. The ultrasonic velocity technique seems to be more sensitive to micellisation. The interaction behaviour of the protein with zwitterionic surfactant is similar to that observed with nonionic surfactants. The protein increases the CMC of the surfactant when investigated by ultrasonic velocity. There is a reasonable agreement in the CMC values determined by the two techniques and average result was 2.4 ± 0.3 mM. Formation of micelles causes reduction of ultrasonic velocity through surfactant solutions whereas the protein increases proportionally the density of the DDAPS solutions with a linear relationship with surfactant concentration. The protein extracted from MO seeds slightly increases the CMC of the surfactant. There is minimum effect of pH on the interaction of the protein with surfactant.

5. Acknowledgements

The authors are grateful to the Polytechnic of Namibia IRPC and University of Botswana ORD for supporting this study.

REFERENCES

[1] R. Maikokera and H. M. Kwaambwa, "Interfacial Properties and Fluorescence of a Coagulating Protein Extracted from *Moringa oleifera* Seeds and Its Interaction with Sodium Dodecyl Sulphate," *Colloids & Surfaces B: Biointerfaces Journal*, Vol. 55, No. 2, 2007, pp. 173-178.

[2] H. M. Kwaambwa and R. Maikokera, "A Fluorescence Spectroscopic Study of a Coagulating Protein Extracted from *Moringa oleifera* Seeds," *Colloids & Surfaces B: Biointerfaces Journal*, Vol. 60, No. 2, 2007, pp. 213-220.

[3] H. M. Kwaambwa and R. Maikokera, "Air-Water Interface Interaction of Anionic, Cationic, and Non-Ionic Surfactants with a Coagulant Protein Extracted from *Moringa oleifera* Seeds Studied Using Surface Tension Probe," *Water SA*, Vol. 33, No. 4, pp. 583-588.

[4] H. M. Kwaambwa and R. Maikokera, "Infrared and Circular Dichroism Spectroscopic Characterisation of Secondary Structure Components of a Water Treatment Coagulant Protein Extracted from *Moringa oleifera* Seeds," *Colloids & Surfaces B: Biointerfaces Journal*, Vol. 64, No. 1, 2008, pp. 118-125.

[5] R. Maikokera and H. M. Kwaambwa, "Use of Viscosity to Probe The Interaction of Anionic Surfactants with a Coagulant Protein from *Moringa oleifera* Seeds," *Research Letters in Physical Chemistry*, Vol. 2009, 2009, Article ID: 927329, 5 Pages.

[6] H. M. Kwaambwa, M. Hellsing and A. R. Rennie, "Adsorption of a Water Treatment Protein from *Moringa oleifera* Seeds to a Silicon Oxide Surface Studied by Neutron Reflection," *Langmuir*, Vol. 26, No. 6, 2010, pp. 3902-3910.

[7] H. M. Kwaambwa and A. R. Rennie, "Interactions of Surfactants with a Water Treatment Protein from *Moringa oleifera* Seeds in Solution Studied by Zeta Potential And Light Scattering Measurements," *Biopolymers*, Vol. 97, No. 4, 2012, pp. 209-218.

[8] D. G. Marangoni and A. A. McLachlan, "Interactions between Zwitterionic and Conventional Anionic and Cationic Surfactants," *Journal of Colloid and Interface Science*, Vol. 295, No. 1, 2006, pp. 243-248.

[9] A. Ndabigengesere, K. S. Narasiah and B. G. Talbot, "Active Agents and Mechanism of Coagulation of Turbid Waters Using *Moringa oleifera*," *Water Research*, Vol. 29, No. 2, 1995, pp. 703-710.

[10] P. X. Li, Z. X. Li, H.-H. Shen, R. K. Thomas, J. Penfold and J. R. Lu, "Application of the Gibbs Equation to the Adsorption of Nonionic Surfactants and Polymers at the Air-Water Interface: Comparison with Surface Excesses Determined Directly Using Neutron Reflectivity," *Langmuir*, Vol. 29, No. 30, 2013, pp .9324-9334.

[11] A. Amin-Alami., N. Kamenka and S. Partyka, "Interfacial Properties of Zwitterionic Surfactants: N-Dodecylbetaine," *Thermochimica Act*, Vol. 122, No. 1, 1987, pp. 171-179.

[12] L. R. Murphy, N. Matubayasi, V. A. Payne and R. M. Levy, "Protein Hydration and Unfolding-Insights from Experimental Partial Specific Volumes and Unfolded Protein Models," *Folding & Design*, Vol. 3, No. 2, 1998, pp. 105-118.

[13] E. A. El-Hefian and A. H. Yahaya, "Investigation on Some Properties of SDS Solutions," *Australian Journal of Basic and Applied Sciences*, Vol. 5, No. 7, 2011, pp. 1221-1227.

Characterization of Some Jordanian Crude and Exhausted Olive Pomace Samples

Khalid M. Tawarah*, Rajaa A. Rababah
Department of Chemistry, Yarmouk University, Irbid, Jordan

ABSTRACT

In an attempt to provide a set of specifications for the use and trade of the olive pomace in Jordan, several samples of the crude and laboratory-prepared exhausted (oil-free) olive pomace have been subjected to a thorough thermochemical characterization. Such characterization included determination of fat content by using n-hexane and Soxhlet extractor, ultimate and proximate analyses obtained by using an elemental analyzer and a thermogravimetric procedure, respectively, gross and net calorific values obtained by using adiabatic oxygen bomb calorimetry, mineral content (ash), and analysis of the pyrolysis thermograms in terms of specified temperatures and residual masses associated with such temperatures as obtained under an inert atmosphere of nitrogen gas at a flow rate of 100 ml per min and a heating rate of 20°C per min from room temperature up to 600°C. The properties of both the crude and the exhausted olive pomace were compared. The gross calorific values and the results of the ultimate analyses for the two pomace types were found to correlate very well as indicated by the use of a literature correlation formula usually used for estimating the gross calorific value of a fuel when its ultimate analysis is known. Other literature correlation formulas used for estimating the gross calorific value from the proximate analysis data were also used to check the adequacy of our procedure for getting the proximate analysis from the thermogravimetric pyrolysis thermograms.

Keywords: Olive Pomace; Heat of Combustion; Thermogravimetric Analysis; Ultimate Analysis; Proximate Analysis; Ash; Pyrolysis

1. Introduction

Jordan is one of the countries of the Mediterranean basin that are considered the home land of the olive tree. The olive oil plays an important role in the nutritional needs of the people of these countries. The olive oil and the pickled olives are nearly part of everyday meal in Jordan. There are about 11.5 million olive trees in Jordan [1]. The average annual olive fruit production in Jordan in the period 2000-2008 is about 170 thousand tons [2]. An average of about 83% of this amount goes for olive oil production and the rest is mainly used for table olive [2]. There are about 130 olive mills in Jordan; most of them are three-phase centrifugal mills [2]. The native name of the solid by-product of the olive mill is *Jift*; however *olive pomace* and *olive cake* are commonly used in the literature to denote this solid by-product. The major components of the olive pomace are the crushed pits, pulp, and skin of the olive fruit while minor components include minerals and some unrecovered olive oil.

The available statistics [2] covering the 2000-2008 olive harvest seasons in Jordan indicate that the annual average production of sun-dried olive pomace is about 35 thousand tons. Nowadays this amount is expected to be substantially higher than this figure because of the growing interest in planting olive trees in Jordan.

Until very recently, only a limited amount of olive pomace was used for space heating in little towns and villages throughout the country and the rest was disposed randomly. Consequently, the improper disposal of olive pomace was a serious environmental issue in Jordan because of its negative effects on soil and ground water. Nowadays, the olive pomace, with a calorific value of about 23 kJ/g [3,4], is becoming a valuable renewable energy source and an affordable substitute for liquid fuels usually used in domestic space heating. This shift from traditional liquid fuels (kerosene and diesel) to the

*Corresponding author.

renewable solid olive pomace fuel is mainly due to the high prices of liquid fuels. About ten years ago, the prices of liquid fuels were nearly 25% of the current prices (current prices are about $1.0 per liter as of November, 2012) and at that time few people were interested in using olive pomace as a source of energy. Likewise, the prices of olive pomace increased by a factor of 4 over the same period of time because of the heavy demand on the use of olive pomace for space heating especially in central and northern parts of Jordan. In the winter of 2012, the price of olive pomace reached 120 JD per ton (about $170). In addition to using olive pomace for space heating, there are other reported potential uses of olive pomace such as fertilizer compost [5], animal fodder [6], source for the manufacture of activated carbon [7], source of bio-pesticides [8], co-firing with coal in power stations [9], source of olive pomace oil for soap industry [10], and biosorbent for handling liquid waste [11,12].

The common types of olive pomace are the crude olive pomace which retains a certain amount of residual oil after olive pressing and the exhausted olive pomace which might have insignificant amount of residual oil. The exhausted olive pomace is usually obtained from crude olive pomace by a solvent extraction process that removes most of its residual oil content. The crude olive pomace is the most common type in Jordan. The main reason for converting the crude olive pomace into the exhausted form is to separate its residual olive oil where it can be used in soap manufacturing [10] and other value-added applications. The residual olive oil in the crude olive pomace is usually extracted by subjecting the olive pomace to mechanical pressing (mechanical method) or to extraction by a low-boiling, non-polar solvent, such as hexane (extraction method) [13]. Solvent extraction is basically a process of diffusion of a solvent into the oil-bearing cells of the raw material resulting in a solution of the oil in solvent [13].

The relevant literature information concerning olive pomace in Jordan has focused on its combustion for energy production [14], combustion performance of diesel fuel and olive pomace slurry [15], an energy source and catalyst for oil shale production of energy [16], physical durability and stability of olive pomace briquettes [17], direct combustion of olive pomace using fluidized bed combustor [18], thermochemical characterization of Jordanian olive pomace [3], and description of the burning behaviour of olive pomace briquettes [4].

The lack of an official set of specifications concerning the use and trade of olive pomace in Jordan as a solid biomass fuel has motivated us to conduct the present study. The major components of the specification of a solid fuel are the listing of the percentages of carbon, hydrogen, nitrogen, sulfur, ash, moisture, as well as the gross calorific value (GCV) and the net calorific value (NCV). In addition to these essential aspects of the present study, we have included data on the pyrolysis of olive pomace as obtained from analysis of the thermogravimetric (TG) curves recorded under inert atmosphere and covering a temperature range from room temperature up to 600°C.

2. Description of Samples and Experimental Procedures

Six olive pomace samples with initial total wet mass of about 25 Kg that belong to the olive harvest seasons of 2009, 2010, and 2011 were considered in the present study. A brief description of the nature of these samples is shown in **Table 1**. The olive pomace samples (RR1 and RR2) that belong to the 2011 olive harvest season were collected from the operation lines of two different olive mills and were rich in moisture by the time when they were brought to the laboratory. They were air-dried in their loose form. These two samples showed a significant mass loss over a period of about 2 months. The three fire briquettes that belong to the 2009 harvest season (sample KT120) were family donation purchased by the donating family in the summer of 2010. The moisture content of this sample was about 1% by the time we received it in March, 2011. The other three samples (KT121, KT122, and KT123) were collected in the summer of 2011 as fire briquettes from the drying yards of two neighboring olive mills. After several weeks of ex-

Table 1. Description of the olive pomace samples used in the present study.

Sample Identity	Sample Mass/Kg	Olive Harvest Season	Physical State	Sample Source
KT120	2.7	2009	3 Briquettes	Al-kfarat region
KT121	2.9	2010	3 Briquettes	[a]Old Sa'doon mill drying yard (pile top)
KT122	6.1	2010	5 Briquettes	[b]New Sa'doon mill drying yard
KT123	5.3	2010	5 Briquettes	Old Sa'doon mill drying yard (pile bottom)
RR1	2.5	2011	Wet loose form	Al-Korah region (olive mill line)
RR2	4.0	2011	Wet loose form	Irbid (olive mill line)

[a]Nuiemeh intersection, Irbid-Amman highway. [b]3 km south of Nuiemeh intersection.

posure to the ambient of the laboratory, samples KT121 and KT123 attained a moisture level of about 2% while that of sample KT122 was 11%. Relatively short laboratory air-drying of samples RR1 and RR2 had resulted in a moisture level of about 43% and 55%, respectively. However, prolonged laboratory air-drying brought the moisture level of these two samples down to about 10%. As with regard to the description of an olive pomace fire briquette, it is a machine-fabricated solid object, cylindrical in shape with base diameter of about 30 cm, a height of about 20 cm, and a mass of about one Kg. Before starting the measurements reported in the present study, about 500 g of each sample were ground to a fine powder by using an electric grinder driven by a motor with a power of 600 W. The fine powder had passed a 750 μm sieve. All the measurements reported in this work were carried out on such powder.

The moisture content of the powder samples was determined by the use of a conventional drying oven at 110°C for about 15 hours. At least duplicate runs were carried out for each sample with a mass of about one to two grams.

The determination of the ash content, which represents the mineral component of the olive pomace, was obtained by igniting about one gram of the test sample in a muffle furnace at 600°C for five hours. At least triplicate runs were performed for each sample. Porcelain crucibles were used for sample ignition.

The exhausted (de-oiled) olive pomace samples were prepared from their parent crude olive pomace via Soxhlet hexane-extraction process. The letter "h" is added as a suffix to the symbol of the crude pomace to denote the residue of the hexane-extracted sample. For example KT120h is both the residue left after subjecting sample KT120 to a hexane-extraction treatment and an exhausted olive pomace sample. The procedure followed in the preparation of the de-oiled olive pomace was as follows. The required mass of a dry crude olive pomace was placed in the cellulose thimble. The thimble was then placed inside the Soxhlet extractor. About 150 ml of n-hexane (purity 95%) were placed in the 250-ml round-bottom distillation flask. After that the Soxhlet extractor was inserted between the condenser and the distillation flask. A heating mantle was used for distillation of hexane. Other relevant precautions included leaving enough space above the sample to avoid spillage of the fine fraction of the sample when the solvent runs down from the soxhlet chamber to the distillation flask, adjustment of the siphoning rate of hexane to achieve optimal contact time between solvent and sample, and adjustment of the settings of the heating mantle and the water tap to minimize loss of solvent. The run time was about 12 hours after which the color of the extractives solution in the distillation flask was noted and the thimble was taken out of the Soxhlet extractor for drying at 80°C. When the thimble and its contents attained constant mass, the % of extracted fatty material (extractives) was calculated. The density of the extractives solution was also measured.

The heat of combustion of both the de-oiled and its parent sample was measured by using an adiabatic oxygen bomb calorimeter (IKA Calorimeter System C 2000 Basic), stationed at the laboratories of the Department of Industrial Chemistry, the Royal Scientific Society (RSS). The heat capacity (Energy Equivalent) of the calorimeter is 8985 J/K. About one gram of the test sample was used for calorimetric measurements. When the combustion of the sample was completed, the oxygen bomb was taken out for an acid wash. The acid wash step entailed thorough washing of the internal surface of the oxygen bomb by distilled water in order to collect the hydrated acids such as HNO_3 and H_2SO_4 that come from the oxidation of nitrogen and sulfur contents of the sample. The acid wash solution was titrated by using 0.0846 M NaOH and phenolphthalein as indicator. As will be explained later, the consumed moles of NaOH and the correction factor for the acid contribution are used with other corrections to adjust the gross calorific value (GCV). The practical value of the heat of combustion is known as net calorific value (NCV) which represents the actual heating value of a fuel when combusted in boilers and industrial applications under normal conditions of fuel combustion. The value of the NCV is usually calculated from the value of the GCV. However, both quantities are prime indices for fuel characterization.

The measurements of elemental analysis were carried out at the laboratories of the Chemistry Department, Jordan University (JU). The elemental analyzer that we used is Euro Ea 3000 elemental analyzer. Few milligrams of a dry sample were used for finding the mass percentages of carbon, hydrogen, and nitrogen. Duplicate runs were carried out for each sample. Analysis for sulfur was not possible at the time we used the elemental analyzer.

The thermogravimeter (Netzsch TGA 409 PC Luxx) was used for conducting the pyrolysis experiments (TG). This instrument is stationed at the laboratories of the Chemistry Department, Jordan University. Aluminum crucibles with pierced lids were used. A stream of flowing nitrogen at a flow rate of 100 ml/min was used to provide an inert atmosphere during the thermal degradation of the test sample. The heating rate adopted for the TG runs was chosen to be 20°C/min. Because of the softening of aluminum at temperatures close to 600°C, the upper temperature of a TG run was limited to near 600°C. The mass of a test sample was in the range 9 to 25 mg for the TG measurements. All samples were dried at 110°C before conducting the measurements. Microsoft Excel was used for generating the thermograms of a TG run.

3. Results and Discussion

3.1. Results of the Hexane-Extraction Experiments

The data shown in **Table 2** for sample RR1 represent a series of triplicates of typical hexane-extraction experiments. Triplicates of the other samples were treated in a manner similar to those of sample RR1. The data given in **Table 3** are based on averages of triplicates of extraction experiments for all samples. It is worth mentioning the relationship between the siphoning rat of the solvent and the duration of an extraction experiment. By comparing the siphoning rates and the durations of runs 1 and 3 with their corresponding counterparts of run 2 in **Table 2**, we conclude that, for nearly the same % extractives, short duration runs require slow solvent siphoning rates while fast siphoning rates require longer durations. As with regard to the color of the extractives solution, the yellowish color indicates that olive oil has been extracted by hexane from the crude olive pomace while the brown color of the extractives solution of sample KT120w is understandable in view of the fact that the color of waste water emerging from an olive mill is usually brown. However, the brown color of the extractives solution of sample KT123 presents an apparent contradiction to this line of reasoning. Since sample KT123 was collected from the bottom of an olive pomace pile, it did not suffer leaching by rain and might retained plenty of moisture saturated with brown-coloring material that can mask the yellowish color of any extracted olive oil. As can be seen from **Table 3** the percentage of the extracted matter ranges between 6.0 and 21.3. Such range was also observed previously for hexane-extracted pomace as well as for water-extracted olive pomace [3]. Several factors can affect the amount of fat material that can be extracted by the non-polar solvent. Examples of these factors are the type of olive cultivar, the degree of ripeness of the olive fruits, the working conditions of the olive mill, the weather conditions prevailing in the storage yards of olive mills, particle size of the test sample, and the possible degradation or fermentation by microorganisms. Most of these variables are not easy to control in a public olive mill. Despite all these considerations, we notice from **Table 3** that the history of a sample might play an important role in determining the amount of its extractable residual oil. For example sample KT120 is the oldest sample with as-received moisture content of about 1% has the lowest percentage of extractives matter, while the fresh samples RR1 and RR2 have higher percentages of extractable matter. Old samples seem to lose significant amounts of their fat content due to physical and chemical processes. In accordance with this line of reasoning are the differences in the extractives percentages of samples KT121 and KT123. These two samples were collected from the drying yard of the same olive mill. KT121 is a pomace sample taken from the top of the olive pomace pile in the storage yard of the mill and seem to have lost a significant amount of volatiles due to its exposure to weather conditions for about five months; on the other hand sample KT123 was taken from the bottom of the pomace pile; therefore KT123 is protected from weather conditions that prevail from the end of the olive harvest season in December up to the beginning of the sun-drying period in May. Consequently, the amount of extractable matter of sample KT123 is expected to be higher than that of sample KT121. The extraction results reported in **Table 3** can be compared with some literature values obtained by exhaustive Soxhlet extraction. The relevant literature values are 5.82% in hexane, 5.22% in ethanol [19], 12.21% in pure hexane, and 9% in acidified hexane [20]. The water-extracted matter is certainly not a fatty material; instead the extractives are very likely to be soil, debris, minerals, and other water soluble degraded polysaccharides. The % RD (% relative deviation) that appears in the last row of **Table 3** is defined by Equation (1). Its value can be considered as a precision indicator.

$$\%\ RD = \frac{1}{n}\sum_{1}^{n}\left|\frac{M-X}{M}\right|\times 100 \qquad (1)$$

where n is the number of replicates, M is the mean of the replicates, and X is the value of a replicate.

3.2. Determination of the Ash Content of the De-Oiled Olive Pomace

The ash content of the de-oiled olive pomace samples is given in **Table 4** on dry basis (for details see **Table S1** in the supplementary material section).There are variations in the percentages of ash content as reported in this table where samples KT120h, KT122h, and KT120w have higher ash content relative to the other samples. The history of each sample seems to be an important factor in determining its ash content. To the best of our knowledge, previous studies on the ash of de-oiled Jordanian olive pomace have not been reported.

3.3. Determination of the Ash Content of the Crude Olive Pomace

The ash content of the crude olive pomace samples is given in **Table 5** on dry basis (for details see **Table S2** in the supplementary material section).

With the exception of samples KT120 and KT122 that show the highest ash content, the ash content of the other samples is in reasonable agreement with the results of previous studies on the crude olive pomace of Jordan [3,4].

A comparison is made between the ash content of the de-oiled pomace and the ash content of the crude pomace

Table 2. Results of typical hexane-extraction experiments for sample RR1.

Test parameters	Run 1	Run 2	Run 3
Hexane volume/ml	150	150	150
Mass of empty thimble/g	1.9306	4.5735	2.0079
Initial mass of thimble + sample/g	4.7231	8.0559	4.405
Sample initial mass/g	2.7925	3.4824	2.3971
Solvent siphoning rate/cycles per hour	8	3	7
Duration of run/h	7.5	3.2	6.7
Solution colour at end of run	Yellowish	Yellowish	Yellowish
Final mass of dry thimble + residue/g	4.2801	7.4463	4.011
Mass of extractives/g	0.4430	0.6096	0.3940
% Extractives	15.86	17.51	16.44
Mass of extractives solution/g	23.3462	52.6634	45.9405
Final volume of extractives solution/ml	35.95	81.45	71.05
Density of extractives solution/(g/ml)	0.6496	0.6467	0.6467
Concentration of extractives solution/(mg/ml)	12.3	7.5	5.6

Table 3. Percentages of extractives matter of crude olive pomace samples extracted by hexane or water.

Sample identity	KT120	KT121	KT122	KT123	RR1	RR2	[a]KT120w
Average sample mass/g	5.14	5.04	4.03	5.46	2.89	2.82	2.28
Run average duration/h	12	10	10	12	5.8	12	24
Average siphoning rate/cycles per h	7	7	7	6	6	5	1
Color of extractives solution	yellowish	yellowish	yellow	brown	yellowish	yellow	brownish
% Extractives	6.02	7.67	8.02	10.45	16.60	21.30	14.20
% RD	0.50	1.88	6.69	0.48	3.63	2.50	12.11

[a]Sample KT120 was extracted by distilled water in addition to extraction by hexane.

Table 4. The percentages of the ash content of the hexane-extracted and water-extracted olive pomace samples.

Sample	No. of Replicates	Average % Ash	% RD
KT120h	4	7.84	8.27
KT121h	3	2.10	5.44
KT122h	5	7.49	7.07
KT123h	3	1.62	11.29
RR1h	3	1.76	12.19
RR2h	3	1.71	10.25
KT120w	4	6.26	4.42

Table 5. The percentages of ash content of the crude olive pomace samples.

Sample	No. of Replicates	Average % Ash	% RD
KT120	3	10.06	1.74
KT121	3	2.51	1.94
KT122	3	10.56	0.94
KT123	3	2.41	2.81
RR1	3	2.57	1.90
RR2	3	2.89	3.49

and the results are given in **Table 6**. According to this table the ash content of the de-oiled pomace is less than that of the crude pomace for all samples. We have no clear cut explanation for such difference in ash content. All samples extracted and non-extracted gave a white ash which might rule out the possibility of having unburned organic matter. Nevertheless, one can visualize a porous structure resulting from the leaching of the fatty material

Table 6. Comparison of the ash content of the de-oiled pomace with the ash content of the crude pomace.

Sample pair	KT120h/KT120	KT121h/KT121	KT122h/KT122	KT123h/KT123	RR1h/RR1	RR2h/RR2	KT120w/KT120
Ash Ratio	0.78	0.84	0.71	0.76	0.68	0.59	0.62

that could allow extensive thermal decomposition of the mineral matter of the de-oiled pomace. Such porous structure does not exist in the case of the crude pomace. In the case of the sample extracted with water, the removal of the non-fat material might also lead to a porous structure that enhances the decomposition of the water-insoluble inorganic matter. If high ash formation is considered as disadvantage, our results show that the exhausted olive pomace is superior to crude olive pomace when used as a solid fuel.

3.4. Ultimate Analysis of Crude Olive Pomace

The ultimate analysis of the crude olive pomace is given in **Table 7** on dry basis (for details see **Tables S2** and **S3** in the supplementary material section). % Oxygen was calculated by difference on dry basis. We were unable to determine the sulfur content of the samples due to the unavailability of such analysis by the instrument we used. However, the sulfur content of the briquettes of a Jordanian crude olive pomace sample was reported to be about 0.1% [4]. According to the data given in **Table 7**, samples KT120 and KT122 have their carbon and oxygen percentages lower than the averages given in the last row of **Table 7** while their ash percentages are higher than the overall average. It is interesting to note the inverse relationship between the percentage of carbon and the percentage of ash for these two samples; at the moment we have no explanation for this observation. In terms of specifications, the values given in the last row of **Table 7** can be reliably considered for any future attempts concerned with setting specifications for crude olive pomace in Jordan.

3.5. Ultimate Analysis of Exhausted Olive Pomace

The ultimate analysis of the exhausted olive pomace is given in **Table 8** on dry basis (for details see **Tables S1** and **S4** in the supplementary material section). By comparing the ultimate analyses given in **Tables 7** and **8**, we find that the percentages of carbon and hydrogen for the de-oiled pomace are slightly lower than those of the crude olive pomace. This is an expected result because extraction by hexane leads to loss of fat material which is carbon and hydrogen containing matter. The difference in oxygen content reflects the difference in ash content of the two types of the olive pomace. As with regard to the nitrogen content the two types of olive pomace have no difference within the limits of the precision in the nitrogen determination. In addition to these observations, the inverse relationship between the ash and carbon percentages noted earlier for samples KT120 and KT122 is still maintained after being extracted by hexane (samples KT120h and KT122h in **Table 8**).

Table 7. The ultimate analysis of crude olive pomace.

Sample	% Nitrogen	% Carbon	% Hydrogen	% Ash	% Oxygen
KT120	1.21	48.65	6.26	10.06	33.82
KT121	1.23	52.02	6.50	2.51	37.74
KT122	0.97	48.42	5.96	10.56	34.09
KT123	1.38	53.10	5.93	2.41	37.18
RR1	1.01	52.44	6.74	2.57	37.24
RR2	1.13	53.55	7.11	2.89	35.32
Average %	1.16	51.36	6.42	5.17	35.90

Table 8. The ultimate analysis of the exhausted olive pomace.

Sample	% Nitrogen	% Carbon	% Hydrogen	% Ash	% Oxygen
KT120h	1.39	46.56	5.41	7.84	38.80
KT121h	1.24	49.69	5.93	2.10	41.04
KT122h	0.91	48.32	6.02	7.49	37.26
KT123h	1.59	50.91	5.85	1.62	40.03
RR1h	1.41	49.81	6.01	1.76	41.01
RR2h	1.82	49.50	6.00	1.71	40.97
Average %	1.39	49.13	5.87	3.75	39.85

3.6. Results of the Heat of Combustion

The results of heat of combustion are given in **Table 9** for both the crude and the exhausted olive pomace samples on dry basis. The oxygen bomb calorimeter used in the present study has a precision of 0.01% when used for finding the GCV of a variety of samples including olive pomace, oak wood, olive wood, pine wood, olive oil, kerosene, and diesel [3]. Based on this high precision, only one measurement was taken for each sample in the present study. In addition to the GCV, the experimental and expected values of the acid wash, expressed as volumes of NaOH, are also included in **Table 9**. The expected volume of NaOH reported in the last column of **Table 9** was calculated on the basis that the % of S is 0.1 [4] and the % of N is the experimentally-determined value for a test sample.

Under the experimental conditions of the oxygen

Table 9. Heat of combustion data for crude and exhausted olive pomace samples.

Sample	Sample Mass/g	GCV (J/g)	(V_{NaOH}/ml), (Acid Wash)	(V_{NaOH}/ml), (Expected)
KT120	0.7183	20,121	7.7	7.9
KT121	0.8012	21,765	3.9	8.9
KT122	1.1224	17,086	4.8	10.0
KT123	0.9171	22,023	4.4	11.4
RR1	0.7921	22,725	4.1	7.4
RR2	0.8215	23,418	4.5	8.4
KT120h	0.6363	19,286	3.6	7.9
KT121h	0.7950	20,413	3.0	8.9
KT122h	0.8435	19,370	4.2	7.1
KT123h	0.8041	20,785	3.9	11.4
RR1h	0.9732	20,350	2.3	12.3
RR2h	0.7539	20,210	3.3	12.2
KT120w	0.4648	20,570	3.8	5.3

bomb calorimetry, the GCV values reported in **Table 9** cannot be credited to the sample because heat is generated by the condensation of water coming from sample moisture and water that has been chemically generated due to combustion of the hydrogen content of the sample. This contribution can be accounted for if the percentages of hydrogen and moisture content of a sample are known. In addition to the heat coming from condensation of water, the heat of hydration of acids that originate from nitrogen, sulfur, and halogen contents of the sample should also be subtracted from the GCV. Another conceivable source of heat might come from the interaction of the condensed water with the mineral residue left after the burning of the sample. In the present study we only dealt with corrections related to sulfur, nitrogen, and hydrogen of a sample on dry basis. The acid wash reported in **Table 9** as volume of NaOH represents the amount of NaOH needed to neutralize the acids formed during the combustion of the sample. The entries in the last column of **Table 9** are the expected values of the volume of NaOH required for neutralizing HNO_3 and H_2SO_4 based on the percentages of N and S. For the case of acid correction, the current information regarding the thermochemical corrections for oxygen bomb calorimetry [21, 22] only heat resulting from combination of water vapor with H_2SO_4 and HNO_3 can be accounted for. Under the bomb conditions, the formation of HNO_3 and H_2SO_4 generates heat energy of 59.7 and 297.2 kJ/mol for HNO_3 and H_2SO_4, respectively [21,22]. With an atomic mass of S equal to 32.065 g/mol, the sulfuric acid correction is 92.7 J for each percentage point of sulfur per gram of sample and with atomic mass of N taken as 14.0067 g/mol, the nitric acid correction is 42.6 J for each percentage point of nitrogen per gram of sample. The heat generated due to condensation of sample moisture (W) is given by 2.442 (%W/100) kJ/g of condensed water per gram of wet sample and the corresponding value coming from condensation of water vapor due to oxidation of sample hydrogen is given as 2.442 (%H/100) (9.01) (1-%W/100) per gram of wet sample. The factor 2.442 is the heat of condensation of water vapor in units of kJ/g of condensed water based on a value of 43.990 kJ per mol of condensed water at 25˚C [23].

Based on this information we calculated the values of NCV on dry basis in units of Joule per gram of dry sample according to the following formula.

$$NCV = GCV - \left\{ (2442)\left(\frac{\%H}{100}\right)(9.01) + (92.7)(\%S) + (42.6)(\%N) \right\} \quad (2)$$

By considering $\%S = 0.1$ [4] and a value of 1.3% N representing the average of % N reported in **Tables 7** and **8**, we found that the HNO_3 correction is about 6 times larger than the H_2SO_4 correction. According to Equation (2) the correction due to the presence of 0.1% S is 9.27 J per gram of sample. We have included this value in generating the data given in **Table 10**. In this table, H_1/(J/g) is an acid correction based on experimental acid wash under the assumption that only HNO_3 is formed during sample combustion, H_2/(J/g) is an acid correction based on calculated acid wash under the assumption that both HNO_3 and H_2SO_4 are formed during sample combustion, and H_3/(kJ/g) is a correction for condensation of the vapor of the chemically-generated water based on % hydrogen on dry basis.

The values of NCV reported in the sixth column of **Table 10** were obtained by applying Equation (2). Each value of NCV represents the heat of combustion of a fuel under normal conditions where the sulfur of a sample emerges as SO_2 gas, nitrogen emerges as elemental nitrogen, and water due to sample moisture and combustion of sample hydrogen leaves the combustion facility as vapor. The ratios given in the last column of **Table 10** indicate that the NCV is nearly 7% lower than the GCV. In order to check the reliability of using the experimentally obtained volumes of NaOH in calculating the acid correction, we have calculated the expected volume of NaOH (in ml) required for neutralizing HNO_3 and H_2SO_4 in the acid wash according to Equation (3) and the stoichiometric relation: moles of NaOH at the end point of titration = moles of HNO_3 + 2 (moles of H_2SO_4).

$$V_{NaOH} = (10)\left(\frac{m}{M}\right)\left\{\left(\frac{\%N}{14.0067}\right) + (2)\left(\frac{\%S}{32.065}\right)\right\} \quad (3)$$

where M and m stand for the molarity of NaOH and sample mass in grams, respectively. The calculated volumes are given in the last column of **Table 9**. With the

Table 10. Data for the calculation of the net calorific value (NCV) of crude and exhausted olive pomace samples on dry basis.

Sample	GCV (J/g)	H_1/(J/g)	H_2/(J/g)	H_3/(kJ/g)	NCV (J/g)	NCV/GCV
KT120	20121	54.1	61.0	1.38	18683	0.93
KT121	21765	24.6	61.7	1.43	20274	0.93
KT122	17086	21.6	50.4	1.31	15724	0.92
KT123	22023	24.2	68.2	1.32	20640	0.94
RR1	22725	26.1	52.5	1.48	21188	0.93
RR2	23418	27.7	57.4	1.56	21797	0.93
KT120h	19286	28.6	68.5	1.19	18027	0.93
KT121h	20413	19.1	62.3	1.30	19046	0.93
KT122h	19370	25.1	48.0	1.32	17998	0.93
KT123h	20785	24.5	77.0	1.29	19420	0.93
RR1h	20350	11.9	69.5	1.32	18959	0.93
RR2h	20210	22.1	87.0	1.32	18804	0.93
KT120w	20570	41.3	63.4	1.32	19191	0.93

exception of samples KT120 and KT120w that have their experimental acid wash values comparable with the calculated acid wash values, all other samples have their acid wash volumes incorrectly reported by the calorimeter operator. Because of this observation we have relied on the data of elemental analysis and the thermochemical corrections to account for heat generated by formation of HNO_3 and H_2SO_4 as given in the fourth column of **Table 10**. The ratios GCV (exhausted)/GCV (crude) and NCV (exhausted)/NCV (crude), as calculated from **Table 10**, have their averages equal to 0.96. This means that the calorific value of the exhausted pomace is about 4% lower than that of the crude pomace. This is a consequence of the loss of the fat material by hexane extraction. On the other hand extraction by water (sample KT120w in **Table 10**) seems to increase the values of the GCV and NCV of the parent crude pomace (sample KT120 in **Table 10**) by a factor of about 2.5%, presumably due to a decrease in the dilution effect of the inorganic component of a pomace sample. For comparison purposes, the values of the GCV for the crude pomace with an average of 21.19 kJ/g are in fair agreement with literature values of 22.1 [3], 23 [4] and 22.24 kJ/g [23]. The values of NCV for the crude olive pomace with an average of 19.72 kJ/g can be compared with literature values of 20.2 [6] and 20.61 kJ/g [25]. As with regard to the data of the exhausted olive pomace reported in **Table 10** we have an average value of 20.07 kJ/g for GCV which can be compared with the literature values 19.7 [3] and 19.60 kJ/g [26]. The average value of NCV for exhausted olive pomace is 18.71 kJ/g which is in fair agreement with the literature values of 17.7 [3] and 17.9 kJ/g [26]. All of these cited values are on dry basis.

To conclude this section, we have considered some correlation formulas that have been reported in the literature for estimating the value of the GCV from the elemental analysis data of a fuel. We have used the correlation formula suggested by Meraz *et al.* [27] as given in Equation (4).

$$\begin{aligned} GCV = \left(1-\frac{\%H_2O}{100}\right)\{&-(0.3708)(\%C) \\ &-(1.1124)(\%H) \\ &+(0.1391)(\%O) \quad \text{kJ/g} \\ &-(0.3178)(\%N) \\ &-(0.1394)(\%S)\} \end{aligned} \quad (4)$$

The statistical parameters associated with the predictions of Equation (4) are the mean absolute error (MAE) and the mean bias error (MBE) defined as follows.

$$MAE = \frac{1}{n}\sum_{1}^{n}\left|\frac{GCV_c - GCV}{GCV}\right| \times 100\% \quad (5)$$

$$MBE = \frac{1}{n}\sum_{1}^{n}\frac{GCV_c - GCV}{GCV} \times 100\% \quad (6)$$

where n is the number of data points which is 13 in our case and GCV_c is the predicted value of GCV. If the elemental analysis of the sample is given on dry basis, the first factor on the right hand side of Equation (4) is unity. When the algebraic signs are kept as such in this equation, the resulting sign of GCV will be negative in accord with the thermodynamic convention. But we have reversed these algebraic signs in order to get a positive sign for GCV which is in accord with the convention of reporting the calorific quantities of fuels as positive values. Based on Equation (4), **Table 11** shows the results of our calculations of the calculated gross calorific values ($GCV_{calc.}$) and the ratio of the calculated value relative to

Table 11. Comparison between experimental and calculated values of GCV.

Sample	GCV_{exp} (kJ/g)	GCV_{calc} (kJ/g)	GCV_{calc}/GCV_{exp}
KT120	20.121	20.681	1.03
KT121	21.765	21.658	1.00
KT122	17.086	20.149	1.18
KT123	22.023	21.611	0.98
RR1	22.725	22.091	0.97
RR2	23.418	23.207	0.99
KT120h	19.286	18.332	0.95
KT121h	20.413	19.707	0.97
KT122h	19.370	19.715	1.02
KT123h	20.785	20.323	0.98
RR1h	20.350	19.897	0.98
RR2h	20.210	19.905	0.98
KT120w	20.570	19.727	0.96

the experimental value ($GCV_{calc.}/GCV_{exp}$). The data used for these calculations were taken from the ultimate analyses given in **Tables 7** and **8** with the % of S being ignored due to its insignificance in the calculations. As with regard to the values of the statistical parameters, we found the values 3.62% and −0.15% for MAE and MBE, respectively. This means that our results for the elemental analysis and the *GCV* correlate very well with the predictions of Equation (4). We have also used a correlation formula suggested by C.Y. Yin [28] that predicts the value of *GCV* (kJ/g) from ultimate analysis. This correlation formula is given as follows.

$$GCV = 0.2949(\%C) + 0.8250(\%H) \quad (7)$$

Application of Equation (7) gave error levels of 5.41% and −3.51% for MAE and MBE, respectively. Based on the values of MAE and MBE, Equation (4) provides good predictions for our experimental values of *GCV* when compared with predictions of Equation (7).

3.7. Thermogravimetric Measurements

We have organized the pyrolysis thermograms of the investigated samples into three groups. Group (I) includes thermograms of samples whose source samples were compacted cylindrical briquettes, sun-dried in the storage yards of an olive mill (samples KT121, KT121h, KT122, KT122h, KT123, and KT123h); group (II) includes thermograms of samples whose source samples were obtained from the olive mill operation line and were air-dried in the laboratory in their loose form (samples RR1, RR1h, RR2, and RR2h); group (III) includes thermograms of samples KT120, KT120h, and KT120w whose source sample is KT120 which has history similar to that of group (I). **Figures 1-3** show the thermograms of groups I, II, and III, respectively. Each TG curve

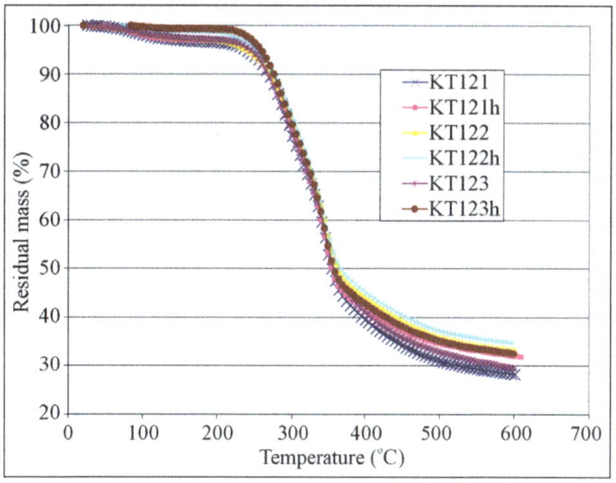

Figure 1. Pyrolysis thermograms of crude and hexane-extracted olive pomace samples given in group (I).

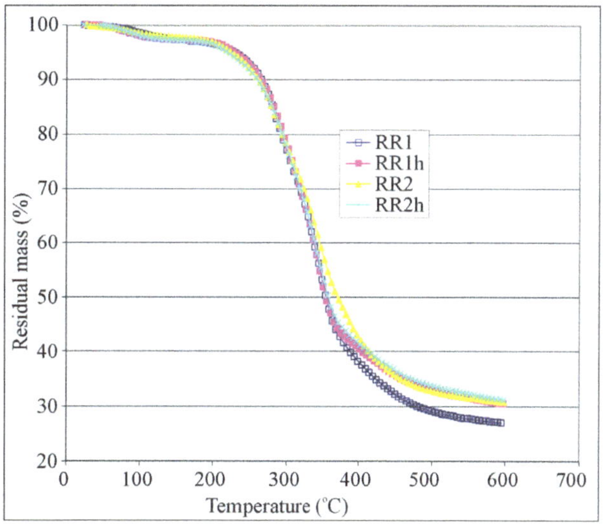

Figure 2. Pyrolysis thermograms of crude and hexane-extracted olive pomace samples given in group (II).

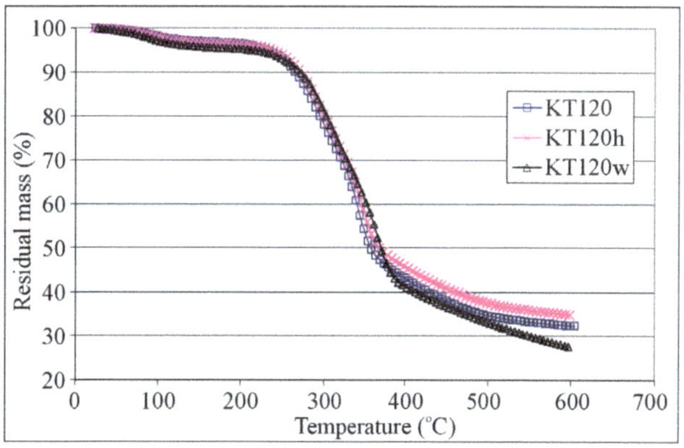

Figure 3. Pyrolysis thermograms of crude, hexane-extracted, and water-extracted olive pomace samples of group III.

shows large mass loss in the temperature range 200 - 400°C which is believed to be due to loss of high molecular mass volatiles and major degradation of the structural components of the olive pomace, namely: hemicellulose, cellulose, and lignin. The high temperature tailing part of the TG curve is thought to be a continuation of the slow degradation of lignin. Gaur and Reed [24] stated that hemicellulose degradation occurs between 200 and 350°C with maximum rate of degradation at 269°C while cellulose degradation starts above 320°C with maximum rate of degradation at 340°C. As with regard to lignin, its DSC (differential scanning calorimetry) curve showed flat endothermic peak at 425°C [24]. The 200 to 300°C temperature range is usually used for torrefaction of biomass in order to improve its heat value and to reduce its moisture content. In this torrefaction process, the hemicellulose of a biomass is degraded and most of its cellulose and lignin content is maintained [29]. Careful examination of the TG curves given in **Figures 1-3** had led us to locate four points on the TG curve. The first point (T_1) represents the temperature at which an abrupt mass loss takes place; the second point (T_2) is an inflection point at about 360°C; the third point occurs at temperature T_3 which represents a significant change in the slope of the TG curve (the rate of mass loss becomes slower after this temperature); the fourth point is symbolized by T_4 which represents the termination temperature of the TG run. The values of these temperatures and the percentages of the remaining mass (M_1 to M_4) of the test sample at these temperatures are given in **Table 12** for both the crude and the exhausted olive pomace. The value of M_4 in this table is the residual mass percentage of the sample at the run termination temperature. By reference to **Table 12**, the data pair T_1 and M_1 indicates that the mass loss is less than 10% when the sample is heated from room temperature up to about 250°C. This mass loss might be due to evolution of moisture and light organic volatiles trapped in the pores of the sample in addition to any surface moisture. The data pair T_2 and M_2 indicates that about 50% of the sample mass is lost when its temperature attains a value of about 360°C. Also a change in the magnitude of the rate of mass loss takes place at this temperature which could be due to the occurrence of a significant change in the ratio of the lignocellulosic structural components at that temperature. The data pair T_3 and M_3 indicates that about 66% of the sample mass is lost at a temperature of about 508°C. Also at temperature T_3 the rate of mass loss becomes slower; most likely due to the dominance of lignin in the remaining fraction of the sample which resists pyrolysis. The final data pair to be considered according to this line of reasoning is the T_4 and M_4 pair. T_4 is the temperature at which the TG run is terminated which amounts to about 600°C. At this temperature the percentage of the residual mass (M_4) is nearly 31%. It is worth mentioning that, on the average, the difference between the residual masses M_3 and M_4 is nearly 3% over a temperature range of about 90°C. This observation implies that the pyrolysis process is slowly approaching its end. Another conclusion that can be drawn from the M_3 and M_4 data given in the last two columns of **Table 12** is that, in general, the hexane-extracted samples have higher remaining mass percentage when compared with their un-extracted counter parts. This observation seems to hold true for the TG segment located between temperatures T_3 and T_4 as can be seen in **Figures 1** and **2**. We explain this observation on the basis that extraction by hexane frees the sample from its content of the easily pyrolyzed fat material; thereby concentrates the sample content with the pyrolysis-resistant lignin.

Table 12. The values of temperatures and the percentages of the remaining masses at four notable points on the TG pyrolysis thermograms of olive pomace samples.

Sample	T_1/°C	T_2/°C	T_3/°C	T_4/°C	M_1/%	M_2/%	M_3/%	M_4/%
KT120	255	360	465	605	92	50	37	32
KT120h	254	360	474	594	94	53	39	35
KT120w	252	382	507	597	93	44	32	27
KT121	251	346	510	601	94	56	31	28
KT121h	249	360	529	609	94	48	34	32
KT122	250	366	521	600	94	49	35	33
KT122h	253	363	528	598	95	51	36	35
KT123	250	360	520	600	94	49	32	29
KT123h	250	351	510	600	97	56	35	32
RR1	245	365	510	595	93	46	29	27
RR1h	247	362	512	597	93	47	32	30
RR2	213	363	523	598	96	53	32	31
RR2h	243	368	498	598	93	47	34	31

In order to explore the difference between extraction by a polar solvent and extraction by a nonpolar solvent, we subjected a crude olive sample for extraction by distilled water and extraction by hexane. Sample KT120 was used for this purpose. In this case three TG curves can be compared; one for the crude sample (KT120), one for the hexane-extracted sample (KT120h), and one for the water-extracted sample (KT120w). The graph given in **Figure 3** shows the TG thermograms of these samples. If we focus our attention on the segments of the TG curve between 350 and 600˚C, we find that the segment of sample KT120h lies above that of sample KT120. This observation is in line with our previous observation concerning the samples of groups I and II. On the other hand, we find the segment of sample KT120w lies below the other two segments. To explain this observation, we consider the following logical consequences of exposing a crude olive sample to water. When extracted by water, the crude olive pomace is freed from water soluble soil and other debris objects that originate from the storage yards of olive mills. In addition to this, dissolution of some inorganic matter and certain fractions of hemicellulose can take place. All these processes are believed to enhance the pyrolysis of the residue of the water-extracted pomace. By carrying TG runs at three different heating rates, García *et al.* [30] reported that the water-leached olive pomace and the un-leached pomace gave different thermal degradation rates in both inert and oxidizing atmospheres. Our results for KT120 and KT120w seem to be in agreement with this literature information.

3.8. Proximate Analysis of Olive Pomace

The determination of the proximate analysis of a fuel is necessary when a fuel is assessed for energy considerations. The quantities usually reported for the proximate analysis of a fuel are the percentages of moisture, volatile matter, fixed carbon, and ash. The traditional method for finding the percentages of volatile matter and fixed carbon by a non-automated method is the ASTM method [31]. In order to avoid the longer time of the ASTM method, the TGA technique is now used for getting the percentages of moisture, volatile matter, and fixed carbon [32]. An article devoted to determination of biomass proximate analysis by using thermogravimetry has appeared very recently [33].In the present study we were unable to carry out the usual steps of holding times and switching between inert and oxidizing atmospheres during a TG run due to a temperature limitation of not exceeding 600˚C. To do the best we can, we have used the TG pyrolysis curve of a test sample to estimate its % moisture (% *w*), % volatile matter (% VM), and % of fixed carbon (% FC) according to the following equations.

$$\%VM = \left(100 - M_3\right)\left(\frac{100}{100 - \%w}\right) \tag{8}$$

$$\%FC = \left(M_4 - \%ash\right)\left(\frac{100}{100 - \%w}\right) \tag{9}$$

The symbols M_3 and M_4 appearing in these two equations are residual masses with values given in **Table 12**. The values of % ash were taken from **Tables 4** and **5**. The calculated values of % *VM* and % *FC* are reported in **Table 13**. The values of % *w* were obtained from the TG curve. The value of % *w* for a given sample was found to be nearly half the percentage of the mass lost from room temperature to T_1 (which is $100 - M_1$). The average value of the ratio $(100 - M_1)/\%w$ is given by Equation (10).

$$\frac{1}{n}\sum_{1}^{13}\left(\frac{100 - M_1}{\%w}\right) = 1.7 \tag{10}$$

where n is the number of samples (n = 13)and M_1 is a residual mass with values given in **Table 12**.

Inorder to check the adequacy of applying Equations (8) and (9) for calculating the values of % *VM* and % *FC* given in **Table 13**, we searched the literature to find data on proximate analysis of lignocellulosic biomass and to find empirical equations that predict the value of GCV from a known proximate analysis. As with regard to reported proximate analysis, Al-Widyan *et al.* [4] have reported the values % *VM* = 86.51, % *FC* = 10.7, and % *ash* = 2.76% on dry basis for briquettes of a Jordanian crude olive pomace sample. The values of % *VM* and % *FC* as reported in reference [4] are somewhat inconsistent with the values usually reported for a lignocellulosic biomass. The % *VM* is higher than usual and the % *FC* is lower than usually reported for lignocellulosic biomass as can be seen in references [25,34] and the literature data cited in references [28,35]. As examples of literature values we present two examples. For the first example we have %*VM* of 80.94, 79.10, 77.77; %*FC* of 18.50, 15.30, 17.92; %*ash* of 0.56, 5.60, and 4.31 on dry basis for pit, pulp, and residue of olive pomace, respectively [25]. For the second example we present the values % *VM* = 76.3, % *FC* = 19.2, %*ash* = 4.5 on dry basis for olive pomace samples [34]. In our case most of the values of % *FC* are substantially higher than the literature values while the values of %*VM* are somewhat close to literature values given in the two examples. Despite the fact that the sum of the percentages of VM, FC, and ash reported in **Table 13** is very close to the theoretical value of 100%, we tried to look for another avenue that could support our findings given in **Table 13**. For this purpose we have chosen two empirical correlation formulas used for predicting the gross calorific value of a fuel from its given proximate analysis. The first correlation formula we tried is a versatile correlation formula, reported by-

Table 13. The proximate analysis of crude, hexane-extracted, and water-extracted olive pomace samples on dry basis.

Sample	% VM	% FC	% Ash	∑%	FC/VM
KT120	65.6	22.8	10.1	98.5	0.35
KT120h	63.5	28.3	7.8	99.7	0.45
KT120w	71.6	21.8	6.3	99.7	0.30
KT121	71.9	26.6	2.5	100.9	0.37
KT121h	68.0	30.8	2.1	101.0	0.45
KT122	67.0	23.1	10.6	100.7	0.34
KT122h	66.0	28.4	7.5	101.8	0.43
KT123	70.1	27.4	2.4	99.9	0.39
KT123h	66.3	31.0	1.6	99.0	0.47
RR1	74.0	25.4	2.6	102	0.34
RR1h	70.8	29. 4	1.8	102	0.41
RR2	70.8	29.3	2.9	103	0.41
RR2h	68.8	30.5	1.7	101	0.44

Parikh *et al.* [35], which gave minimum error levels over 450 data points. This correlation formula is given by Equation (11) while its error measures are given by Equations (12) and (13).

$$HHV(=GCV) = (0.3536)(FC)+(0.1559)(VM) - (0.0078)(Ash) \text{ MJ/Kg} \quad (11)$$

$$MAE = \frac{1}{n}\sum_{1}^{n}\left|\frac{HHV_c - HHV}{HHV}\right| \times 100\% \quad (12)$$

$$MBE = \frac{1}{n}\sum_{1}^{n}\left(\frac{HHV_c - HHV}{HHV}\right) \times 100\% \quad (13)$$

where *MAE* and *MBE* are the abbreviations defined by Equations (5) and (6) in Section 3.6. The abbreviation HHV_c stands for the calculated higher heating value while the abbreviation *HHV* stands for the measured value. In the present study we used the abbreviation *GCV* instead of the *HHV*; but both denote the same quantity.

The values of the average absolute error and the average bias error with respect to the measured values of *HHV*, as reported by Parikh *et al.* [35], are 3.74% and 0.12%, respectively. We have applied Equation (11) to calculate a value for the *GCV* by inserting the values of % *VM* and % *FC* given in **Table 13** for 13 data points (n = 13 in Equations (12) and (13)). The measured *GCV* data given in the second column of **Table 11** were used in calculating the error parameters defined by Equations (12) and (13). The average absolute error and the average bias error were found to be 6.49% and −0.54%, respectively. This finding gives support to the adequacy of the TG procedure that generated the values of % *VM* and % of *FC*. The second empirical formula we used was reported by C.Y. Yin [28]. The formula is given by Equation (14).

$$HHV = 0.1905(VM)+0.2521(FC) \quad (14)$$

By inserting the values of % *VM* and % *FC* given in **Table 13**, we calculated 13 *HHV*. These calculated values together with the measured values of *GCV*, given in **Table 11**, were inserted in Equations (12) and (13) to find the values of *MAE* and *MBE* associated with the predictions of Equation (14). The values were found to be 5.57% and −2.98% for *MAE* and *MBE*, respectively. As a comment, the correlation formula given in Equation (14) has a lower prediction accuracy when compared with Equation (11). The overall conclusion to be drawn from the application of Equation (11) is that the proximate analysis given in **Table 13** has been justified.

The values reported in the last column of **Table 13** represent the ratio of % fixed carbon divided by % volatiles. It is evident from this table that the exhausted olive pomace samples have their ratios of % fixed carbon to % volatiles higher than those of their parent crude pomace samples. The final point concerning the data of **Table 13** has to do with proximate analysis for specification purposes. Based on the data given in **Table 13** the average proximate analysis of the crude olive pomace samples is 69.9% volatiles, 25.8% fixed carbon and 5.2% ash and the proximate analysis of the exhausted olive pomace (sample KT120w not included) is 67.2% volatiles, 29.7% fixed carbon, and 3.8% ash.

4. Acknowledgments

The authors acknowledge the financial support of Yarmouk University since all the paid services at RSS and UJ were part of the fund allocated for project No. 28/2010. The authors would like to thank Mrs. Leena Al-Otoom and Mrs. Nisreen Al-Khateeb (RSS) for their assistance in the GCV measurements, Miss Ruba Zaloom (UJ) for her assistance in the TG measurements and Miss Mona Tayyem (UJ) for her assistance in the elemental analysis measurements.

REFERENCES

[1] Department of General Statistics, "Agricultural Statistics," Jordan, 2012.

[2] S. M. Al-Shdeifat and H. A. Al-Bdour, Annual Report, National Center for Scientific Research and Agricultural Extension, Jordan, 2009.

[3] K. M. Tawarah, "Thermochemical Characterization of Jordanian Olive Pomace," Manuscript under Preparation for Publication.

[4] M. I. Al-Widyan, G. Tashtoush and A. I. Khdair, "Briquettes of Olive Cake as a Potential Source of Thermal Energy," *Journal of Solid Waste Technology and Management*, Vol. 28, No. 2, 2002, pp. 51-59.

[5] M. S. Haddadin, J. Haddadin, O. I. Arabiyat and B. Hattar, "Biological Conversion of Olive Pomace into Compost by Using Trichoderma Harzianum and Phanerochaete Chrysosporium," *Bioresource Technology*, Vol. 100, No. 20, 2009, pp. 4773-478.

[6] M. S. Haddadin and S. M. Abdulrahim, "Solid State Fermentation of Waste Pomace from Olive Processing," *Journal of Chemical Technology and Biotechnology*, Vol. 74, No. 7, 1999, pp. 613-618.

[7] N. Mameri, F. Aioueche, D. Belhocine, H. Grib, H. lounici and D. L. Piron, "Preparation of Activated Carbon from Olive Mill Solid Residue," *Journal of Chemical Technology and Biotechnology*, Vol. 75, No. 7, 2000, pp. 625-631.

[8] M. L. Cayuela, P. D. Millner, S. L. Meyer and A. Roig, "Potential of Olive Mill Waste and Compost as Bio-Based Pesticides against Weeds, Fungi, and Nematodes," *Science of the Total Environ*, Vol. 399, No. 1-3, 2008, pp. 8-11.

[9] K. R. Cliffe and S. Patumsawad, "Co-Combustion of Waste from Olive Oil Production with Coal in a Fluidized Bed," *Waste Management*, Vol. 21, No. 1, 2001, pp. 49-53.

[10] D. Fisher, "Types of Olive Oil for Soap Making Recipes," 2013. http://candleandsoap.about.com/od/soapmakingoils/a/Types-Of-Olive-Oil-For-Soap-Making-Recipes.htm

[11] F. Pagnanellia, S. Mainellia, F. Vegli and L. Toroa, "Heavy Metal Removal by Olive Pomace: Biosorbent Characterisation and Equilibrium Modelling," *Chemical Engineering Science*, Vol. 58, No. 20, 2003, pp. 4709-4717.

[12] F. Pagnanelli, L. Toro and F. Veglio, "Olive Mill Solid Residues as Heavy Metal Sorbent Material: A preliminary Study," *Waste Management*, Vol. 22, No. 8, 2002, pp. 901-907.

[13] P. S. Moral and M. V. Méndez, "Production of Pomace Olive Oil," *Grasas y Aceites*, Vol. 57, No. 1, 2006, pp. 47-55.

[14] M. Abu-Qudais, "Fluidized-Bed Combustion for Energy Production from Olive Cake," *Energy*, Vol. 21, No. 3, 1996, pp. 173-178.

[15] M. Abu-Qudais and G. Okasha, "Diesel Fuel and Olive-Cake Slurry: Atomization and Combustion Performance," *Applied Energy*, Vol. 54, No. 4, 1996, pp. 315-326.

[16] T. M. Alkhamis and M. M. Kablan, "Olive Cake as an Energy Source and Catalyst for Oil Shale Production of Energy and Its Impact on the Environment," *Energy Conversion & Management*, Vol. 40, No. 17, 1999, pp. 1863-1870.

[17] M. I. Al-Widyan, H. F. Al-Jalil, M. M. Abu-Zreig and N. H. Abu-Hamdeh, "Physical Durability and Stability of Olive Cake Briquettes," *Canadian Biosystems Engineering*, Vol. 44, 2002, pp. 341-345.

[18] Y. H. Khraisha, M. A. Hamdan and H. S. Qalalweh, "Direct Combustion of Oive Cake Using Fluidized Bed Combustor," *Energy Sources*, Vol. 21, No. 4, 1999, pp. 319-327.

[19] S. Meziane, H. Kadi and O. Lamrous, "Kinetic Study of Oil Extraction from Olive Foot Cake, " *Grasas y Aceites*, Vol. 57, No. 2, 2006, pp. 175-179.

[20] S. Kmieciak, S. Meziane, H. Kadi and R. l. Uloussaoui, "Oil Extraction from Olive Foot Cake with Acidic Hexane," *Grasas y Aceites*, Vol. 42, No. 1, 1991, pp. 46-50.

[21] "Parr 587 M-6400 Oxygen Bomb Calorimeter Operating Instruction Manual," Chapter 7, Parr Instrument Company, Moline, 2011.

[22] ASTM Standard D5865-11A, "Standard Test Method for Gross Calorific Value of Coal and Coke," Appendix X1, ASTM International, West Conshohocken, 2011.

[23] K. N. Marsh, "Recommended Reference Materials for the Realization of Physicochemical Properties," Blackwell, Oxford, 1987.

[24] S. Gaur and T. B. Reed, "Thermal Data for Natural and Synthetic Fuels," Marcel Dekker, New York, 1998, p. 259.

[25] T. Miranda, A. Esteban, S. Rojas, I. Montero and A. Ruiz, "Combustion Analysis of Different Olive Residues," *International Journal of Molecular Science*, Vol. 9, No. 4, 2008, pp. 512-525.

[26] "Olive Cake Pellets, Typical Specifications," Integra Fuels, C/Conde De Aranda 8-1D 28001 Madrid.

[27] L. Meraz, A. Dominguez, I. Kornhauser and F. Roja, "A Thermochemical Concept-Based Equation to Estimate Waste Combustion Enthalpy from Elemental Composition," *Fuel*, Vol. 82, No. 12, 2003, pp. 1499-1507.

[28] C. Y. Yin, "Prediction of Higher Heating Value of Biomass from Proximate and Ultimate Analyses," *Fuel*, Vol. 90, No. 3, 2011, pp. 1128-1132.

[29] W. Chun-Te and L. Far-Ching, "The Properties of Torrefied Biomass from Six Major Bamboos in Taiwan," *Proceedings of the 55th International Convention of Society of Wood Science and Technology*, Beijing, 27-31 August 2012, 8 Pages.

[30] P. García-Ibañez, M. M. Sánchez and A. A. Cabanillas, "Thermogravimetric Analysis of Olive-Oil Residue in Air Atmosphere," *Fuel Processing Technology*, Vol. 87, No. 2, 2006, pp. 103-107.

[31] ASTM Standard D3172, "Standard Practice for Proximate Analysis of Coal and Coke," ASTM International, West Conshohocken, 2007. www.astm.org

[32] M. C. Mayoral, M. T. Izquierdo, J. M. Andres and B. Rubio, "Different Approaches to Proximate Analysis by Thermogravimetry Analysis," *Thermochimica Acta*, Vol. 370, No. 1-2, 2001, pp. 91-97.

[33] R. Garcia, C. Pizarro, A. G. Lavin and J. L. Bueno, "Biomass Proximate Analysis Using Thermogravimetry," *Bioresources Technology*, Vol. 139, 2013, pp. 1-4.

[34] M. Varol, "Combustion and Co-Combustion of Olive Cake and Coalina Fluidized Bed," Thesis, Middle East Technical University, Turkey, 2006.

[35] J. Parikha, S. A. Channiwalab and G. K. Ghosal, "A Correlation for Calculating HHV from Proximate Analysis of Solid Fuels," *Fuel*, Vol. 84, No. 5, 2005, pp. 487-494.

Supplementary Material

Table S1. The percentages of the ash content of the hexane-extracted and water-extracted olive pomace samples on dry basis.

Sample	Trial	(Empty Crucible)/g	(Crucible + Sample)/g	Sample/g	Crucible + Sample (after ignition)/g	Ash mass/g	% Ash
KT120h	1	22.2733	22.7634	0.4901	22.3095	0.0362	7.386
	2	13.1772	13.6230	0.4458	13.2133	0.0361	8.098
	3	22.0819	22.5570	0.4751	22.1151	0.0332	6.988
	4	14.5650	15.5493	0.9843	14.6523	0.0873	8.869
Average							7.835
KT121h	1	11.9696	12.6064	0.6368	11.9826	0.0130	2.041
	2	12.8042	13.3038	0.4996	12.8141	0.0099	1.982
	3	13.3955	13.8673	0.4718	13.4062	0.0107	2.268
Average							2.097
KT122h	1	13.316	13.8162	0.5002	13.3537	0.0377	7.537
	2	11.5936	12.0463	0.4527	11.6253	0.0317	7.002
	3	13.2681	13.8619	0.5938	13.3076	0.0395	6.652
	4	11.9734	13.038	1.0646	12.0607	0.0873	8.200
	5	13.1113	14.2821	1.1708	13.2056	0.0943	8.054
Average							7.489
KT123h	1	13.1371	13.6459	0.5088	13.1447	0.0076	1.494
	2	13.0094	13.5047	0.4953	13.0167	0.0073	1.474
	3	14.5642	15.0812	0.5170	14.574	0.0098	1.896
Average							1.621
RR1h	1	13.1110	13.6240	0.5130	13.1204	0.0094	1.832
	2	10.7543	11.2030	0.4487	10.7633	0.0090	2.006
	3	12.2390	12.7470	0.5080	12.2463	0.0073	1.437
Average							1.758
RR2h	1	21.6266	22.1484	0.5218	21.6369	0.0103	1.974
	2	13.4555	13.9390	0.4835	13.4636	0.0081	1.675
	3	13.3365	13.8420	0.5055	13.344	0.0075	1.484
Average							1.711
KT120w	1	20.9727	21.5050	0.5323	21.0052	0.0325	6.106
	2	20.7623	21.3035	0.5412	20.7960	0.0337	6.227
	3	10.6980	11.1280	0.4300	10.7233	0.0253	5.884
	4	20.7632	21.2581	0.4949	20.7969	0.0337	6.809
Average							6.256

Table S2. The percentages of ash content of the crude olive pomace samples on dry basis.

Sample	Trial	(Empty Crucible)/g	(Crucible + Sample)/g	Sample/g	Crucible + sample (after ignition)/g	Ash mass/g	% Ash
KT120	1	21.6276	22.6734	1.0458	21.7321	0.1045	9.992
	2	20.9727	21.9038	0.9311	21.0645	0.0918	9.859
	3	22.2769	23.4619	1.1850	22.3992	0.1223	10.321
Average							10.057
KT121	1	10.6947	11.9936	1.2989	10.7282	0.0335	2.579
	2	13.3950	14.3417	0.9467	13.4183	0.0233	2.461
	3	13.0082	14.0131	1.0049	13.0331	0.0249	2.478
Average							2.506
KT122	1	11.5931	12.9021	1.3090	11.7305	0.1374	10.497
	2	13.4550	14.7932	1.3382	13.5948	0.1398	10.477
	3	22.0817	23.3579	1.2762	22.2180	0.1365	10.696
Average							10.557
KT123	1	13.1771	14.015	0.8379	13.1981	0.021	2.506
	2	13.1386	14.4532	1.3146	13.1690	0.0304	2.312
	3	13.3171	14.3074	0.9903	13.3411	0.024	2.424
Average							2.414
RR1	1	12.2364	13.2909	1.0545	12.2643	0.0279	2.646
	2	12.8295	13.8590	1.0295	12.8559	0.0264	2.564
	3	13.2670	14.6270	1.3600	13.3011	0.0341	2.507
Average							2.573
RR2	1	13.3336	14.6965	1.3629	13.3745	0.0409	3.001
	2	10.7528	11.7782	1.0254	10.7829	0.0301	2.935
	3	12.8038	13.9784	1.1746	12.8360	0.0322	2.741
Average							2.893

Table S3. Elemental analysis of crude olive pomace samples on dry basis. The values in parenthesis are % RD.

Sample	Trial	% Nitrogen	% Carbon	% Hydrogen
KT120	1	1.319	49.166	6.293
	2	1.107	48.131	6.223
Average %, (% RD)		1.213, (8.739)	48.649, (1.064)	6.258, (0.535)
KT121	1	1.296	52.174	6.492
	2	1.165	51.869	6.501
Average %, (% RD)		1.231, (5.323)	52.022, (0.293)	6.497, (0.069)
KT122	1	0.838	48.644	6.038
	2	1.094	48.195	5.886
Average %, (% RD)		0.966, (13.251)	48.420, (0.464)	5.962, (1.275)
KT123	1	1.335	53.033	6.451
	2	1.432	53.157	5.412
Average %, (% RD)		1.384, (3.506)	53.095, (0.117)	5.932, (8.758)
RR1	1	1.121	52.108	6.652
	2	0.906	52.768	6.837
Average %, (% RD)		1.014, (10.607)	52.438, (0.629)	6.745, (1.371)
RR2	1	1.226	53.875	7.166
	2	1.031	53.224	7.045
Average %, (% RD)		1.129, (8.640)	53.550, (0.608)	7.106, (0.851)

Table S4. Elemental analysis of hexane-extracted and water-extracted olive pomace samples on dry basis. The values in parenthesis are % RD.

Sample	Trial	% Nitrogen	% Carbon	% Hydrogen
KT120h	1	1.368	46.847	5.410
	2	1.410	46.280	5.416
Average %, (% RD)		1.389, (1.512)	46.564, (0.610)	5.413, (0.055)
KT120w	1	1.603	48.153	5.913
	2	0.939	48.740	6.044
Average %, (% RD)		1.271, (26.121)	48.447, (0.606)	5.979, (1.100)
KT121h	1	0.952	49.483	6.020
	2	1.536	49.903	5.835
Average %, (% RD)		1.244, (23.473)	49.693, (0.423)	5.928, (1.560)
KT122h	1	0.918	47.645	5.900
	2	0.900	48.992	6.134
Average %, (% RD)		0.909, (0.990)	48.319, (1.394)	6.017, (1.944)
KT123h	1	1.889	51.070	5.763
	2	1.290	50.740	5.943
Average %, (% RD)		1.590, (18.836)	50.905, (0.324)	5.853, (1.538)
RR1h	1	1.182	49.864	6.283
	2	1.464	50.747	6.148
	3	1.340	49.369	5.692
	4	1.665	49.276	5.899
Average %, (% RD)		1.413, (10.741)	49.814, (0.987)	6.006, (3.497)
RR2h	1	1.585	49.669	6.074
	2	2.061	49.334	5.915
Average %, (% RD)		1.823, (13.055)	49.502, (0.338)	5.995, (1.326)

Permissions

The contributors of this book come from diverse backgrounds, making this book a truly international effort. This book will bring forth new frontiers with its revolutionizing research information and detailed analysis of the nascent developments around the world.

We would like to thank all the contributing authors for lending their expertise to make the book truly unique. They have played a crucial role in the development of this book. Without their invaluable contributions this book wouldn't have been possible. They have made vital efforts to compile up to date information on the varied aspects of this subject to make this book a valuable addition to the collection of many professionals and students.

This book was conceptualized with the vision of imparting up-to-date information and advanced data in this field. To ensure the same, a matchless editorial board was set up. Every individual on the board went through rigorous rounds of assessment to prove their worth. After which they invested a large part of their time researching and compiling the most relevant data for our readers. Conferences and sessions were held from time to time between the editorial board and the contributing authors to present the data in the most comprehensible form. The editorial team has worked tirelessly to provide valuable and valid information to help people across the globe.

Every chapter published in this book has been scrutinized by our experts. Their significance has been extensively debated. The topics covered herein carry significant findings which will fuel the growth of the discipline. They may even be implemented as practical applications or may be referred to as a beginning point for another development. Chapters in this book were first published by Scientific Research Publishing Inc.; hereby published with permission under the Creative Commons Attribution License or equivalent.

The editorial board has been involved in producing this book since its inception. They have spent rigorous hours researching and exploring the diverse topics which have resulted in the successful publishing of this book. They have passed on their knowledge of decades through this book. To expedite this challenging task, the publisher supported the team at every step. A small team of assistant editors was also appointed to further simplify the editing procedure and attain best results for the readers.

Our editorial team has been hand-picked from every corner of the world. Their multi-ethnicity adds dynamic inputs to the discussions which result in innovative outcomes. These outcomes are then further discussed with the researchers and contributors who give their valuable feedback and opinion regarding the same. The feedback is then collaborated with the researches and they are edited in a comprehensive manner to aid the understanding of the subject.

Apart from the editorial board, the designing team has also invested a significant amount of their time in understanding the subject and creating the most relevant covers. They scrutinized every image to scout for the most suitable representation of the subject and create an appropriate cover for the book.

The publishing team has been involved in this book since its early stages. They were actively engaged in every process, be it collecting the data, connecting with the contributors or procuring relevant information. The team has been an ardent support to the editorial, designing and production team. Their endless efforts to recruit the best for this project, has resulted in the accomplishment of this book. They are a veteran in the field of academics and their pool of knowledge is as vast as their experience in printing. Their expertise and guidance has proved useful at every step. Their uncompromising quality standards have made this book an exceptional effort. Their encouragement from time to time has been an inspiration for everyone.

The publisher and the editorial board hope that this book will prove to be a valuable piece of knowledge for researchers, students, practitioners and scholars across the globe.

List of Contributors

Khodabakhsh Niknam and Abolhassan Piran
Department of Chemistry, Faculty of Sciences, Persian Gulf University, Bushehr, Iran

Katsuya Konno
Hitachi Automotive Systems Ltd., Automotive Systems R&D Laboratory, Kawasaki, Japan
Department of Life and Environmental Sciences, Faculty of Engineering, Chiba Institute of Technology, Chiba, Japan

Kaoru Onoe, Yasuyuki Takiguchi and Tatsuaki Yamaguchi
Department of Life and Environmental Sciences, Faculty of Engineering, Chiba Institute of Technology, Chiba, Japan

Yi Liu, Ryuichi Sugimoto and Katsuhiro Sumi
School of Environmental Science and Engineering, Kochi University of Technology, Kami, Japan

Katakam N. Gangadhar, Manneganti Vijay, Rachapudi Badari Narayana Prasad Bhetala Lakshmi and Anu Prabhavathi Devi
Centre for Lipid Research, CSIR-Indian Institute of Chemical Technology, Hyderabad, India

Hishame Abd-Elnabi, Afaf Mohamed Abdel Hameed, Reham R. Awed and Kamal Usef Sadek
Department of Chemistry, Faculty of Science, El-Minia University, El-Minia, Egypt

Ramadan Ahmed Mekheimer
Department of Chemistry, Faculty of Science, El-Minia University, El-Minia, Egypt
Department of Chemistry, Faculty of Science for Girls, King Abdulaziz University, Jeddah, KSA

Sangita Makone and Shreyas Mahurkar
Chemical Sciences Research Laboratory, School of Chemical Sciences, Swami Ramanand Teerth Marathwada University, Nanded, India

Marielle Fassier, Céline Ducroquetz and Olivier Dupont
Centre Technique de Matériaux Naturels de Construction, Paris, France

Céline Faugeron and Vincent Gloaguen
Laboratoire de Chimie des Substances Naturelles, Limoges, France

Genoveva Rosano-Ortega, Carlos Vega Lebrún, Juan M. López Oglesby and Beatriz Pérez Armendariz
Research Department, Universidad Popular Autónoma del Estado de Puebla (UPAEP), Puebla, México

Sonia Martínez Gallegos
Division of Graduate Studies and Research, Instituto Tecnológico de Toluca, México City, México

Daniel Cruz-González
Chemical Engineering Faculty, Benemérita Universidad Autónoma de Puebla, Puebla, México

Djamel Ghernaout
Chemical Engineering Department, Saad Dahlab University of Blida, Blida, Algeria

Pranab Jyoti Das, Jupitara Das and Munni Ghosh
Department of Chemistry, Gauhati University, Guwahati, India

Sabera Sultana
Indian Institute of Technology, Guwahati, India

Sneha Naik and Vidya Desai
Department of Chemistry, Dnyanprassarak Mandal's College of Arts, Science and Commerce, Mapusa, India

Do-Hun Lee, Jung-Tai Hahn and Dai-Il Jung
Department of Chemistry, Dong-A University, Busan, South Korea
Department of Beautycare, Youngdong University, Youngdong, South Korea

Min Ji, Yiyong Zhou, Qunxing Luo and Tianxi Cai
School of Chemistry, Dalian University of Technology, Dalian, China

Sang-Eon Park
Laboratory of Nano-Green Catalysis and Nano Center for Fine Chemicals Fusion Technology, Department of Chemistry, Inha University, Incheon, South Korea

Pushkin Chakraborty and Subhas C. Roy
Department of Organic Chemistry, Indian Association for the Cultivation of Science, Jadavpur Kolkata, India

Jun-Ichi Kadokawa
Graduate School of Science and Engineering, Kagoshima University, Kagoshima, Japan

Silvia C. Gutiérrez-Gutiérrez, Genoveva Rosano-Ortega, Carlos Vega-Lebrún and Juan Manuel López
Research Department, Universidad Popular Autónoma del Estado de Puebla (UPAEP), Puebla, México

Pedro Ávila-Pérez
Technological Research Divisi6n, Instituto Nacional de Investigaciones Nucleares, Ocoyoacac, México

Sonia M. Martínez Gallegos
Division of Graduate Studies and Research, Instituto Tecnol6gico de Toluca, Toluca, México

Daniel Cruz González
Chemical Engineering Faculty, Benemérita Universidad Aut6noma de Puebla, Puebla, México

Eligeti Rajanarendar, Kammari Thirupathaiah, Saini Rama Krishna and Baireddy Kishore
Department of Chemistry, Kakatiya University, Warangal, India

Hualiang An, Lijuan Kang, Wei Gao, Xinqiang Zhao and Yanji Wang
Hebei Provincial Key Lab of Green Chemical Technology and High Efficient Energy Saving, Hebei University of Technology, Tianjin, China

Ville Kuokkanen and Toivo Kuokkanen
Department of Chemistry, University of Oulu, Oulu, Finland

Jaakko Rämö
Thule Institute, University of Oulu, Oulu, Finland

Ulla Lassi
Department of Chemistry, University of Oulu, Oulu, Finland
Kokkola University Consortium Chydenius, Kokkola, Finland

Xiaoji Cao, Lingxiao Shen, Xuemin Ye, Feifei Zhang, Jiaoyu Chen and Weimin Mo
Research Center of Analysis and Measurement, Zhejiang University of Technology, Hangzhou, China

Akif Hamid Azizov, Reyhan Valy Aliyeva and Khayala Haydar Asgerova
The Y.H. Mamedaliyev Institute of Petrochemical Processes of Azerbaijan National Academy of Sciences, Baku, Azerbaijan

Habauka M. Kwaambwa
School of Health and Applied Sciences, Polytechnic of Namibia, Windhoek, Namibia

Fiona M. Nermark
Department of Chemistry, University of Botswana, Gaborone, Botswana

Khalid M. Tawarah and Rajaa A. Rababah
Department of Chemistry, Yarmouk University, Irbid, Jordan